Adaptive Phytoremediation Practices

Adaptive Phytoremediation Practices

Resilience to Climate Change

Vimal Chandra Pandey
Department of Environmental Science,
Babasaheb Bhimrao Ambedkar University,
Lucknow, India

Gordana Gajić
Department of Ecology,
Institute for Biological Research "Sinisa Stankovic",
National Institute of Republic of Serbia, University of
Belgrade, Belgrade, Serbia

Pallavi Sharma
School of Environment and Sustainable Development,
Central University of Gujarat, Gandhinagar, India

Madhumita Roy
Department of Microbiology, Bose Institute,
Kolkata, India

ELSEVIER

Elsevier
Radarweg 29, PO Box 211, 1000 AE Amsterdam, Netherlands
The Boulevard, Langford Lane, Kidlington, Oxford OX5 1GB, United Kingdom
50 Hampshire Street, 5th Floor, Cambridge, MA 02139, United States

Notices
Knowledge and best practice in this field are constantly changing. As new research and experience
broaden our understanding, changes in research methods, professional practices, or medical treatment
may become necessary.

Practitioners and researchers must always rely on their own experience and knowledge in evaluating
and using any information, methods, compounds, or experiments described herein. In using such
information or methods they should be mindful of their own safety and the safety of others, including
parties for whom they have a professional responsibility.

To the fullest extent of the law, neither the Publisher nor the authors, contributors, or editors, assume
any liability for any injury and/or damage to persons or property as a matter of products liability,
negligence or otherwise, or from any use or operation of any methods, products, instructions, or ideas
contained in the material herein.

Library of Congress Cataloging-in-Publication Data
A catalog record for this book is available from the Library of Congress

British Library Cataloguing-in-Publication Data
A catalogue record for this book is available from the British Library

ISBN: 978-0-12-823831-8

For information on all Elsevier publications
visit our website at https://www.elsevier.com/books-and-journals

Publisher: Candice G. Janco
Acquisitions Editor: Marisa LaFleur
Editorial Project Manager: Aleksandra Packowska
Production Project Manager: Joy Christel Neumarin
  Honest Thangiah
Cover Designer: Vicky Pearson Esser

Typeset by STRAIVE, India

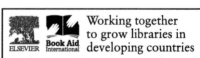

Working together
to grow libraries in
developing countries

www.elsevier.com • www.bookaid.org

Contents

About the authors

Vimal Chandra Pandey

Dr. Vimal Chandra Pandey featured in the world's top 2% scientists curated by Stanford University, United States, and has published papers in the *PLOS Biology* journal. Dr. Pandey is a leading researcher in the field of environmental engineering, particularly phytomanagement of polluted sites. His research focuses mainly on the remediation and management of degraded lands, including heavy metal-polluted lands and postindustrial lands polluted with fly ash, red mud, mine spoil, and others, to regain ecosystem services and support a bio-based economy with phytoproducts through affordable green technology (phytoremediation). His research interests also lie in exploring industrial crop-based phytoremediation to attain bioeconomy security and restoration, adaptive phytoremediation practices, phytoremediation-based biofortification, carbon sequestration in waste dumpsites, fostering bioremediation for utilizing polluted lands, and attaining UN Sustainable Development Goals. Recently, Dr. Pandey worked as a CSIR-Pool Scientist (Senior Research Associate) in the Department of Environmental Science at Babasaheb Bhimrao Ambedkar University, Lucknow, India. Dr. Pandey also worked as Consultant at the Council of Science and Technology, Uttar Pradesh; DST-Young Scientist in the Plant Ecology and Environmental Science Division at CSIR-National Botanical Research Institute, Lucknow; and DS Kothari Postdoctoral Fellow in the Department of Environmental Science at Babasaheb Bhimrao Ambedkar University, Lucknow. He is the recipient of a number of awards/honors/fellowships and is a member of the National Academy of Sciences, India. Dr. Pandey serves as a subject expert and panel member for the evaluation of research and professional activities in India and abroad for fostering environmental sustainability. He has published over 100 scientific articles/book chapters in peer-reviewed journals/books. Dr. Pandey is also the author and editor of seven books published by Elsevier with several more forthcoming. He is associate editor of *Land Degradation and Development* (Wiley); editor of *Restoration Ecology* (Wiley); associate editor of *Environment, Development and Sustainability* (Springer); associate editor of *Ecological Processes* (Springer Nature); advisory board member of *Ambio* (Springer); editorial board member of *Environmental Management* (Springer); and editorial board member of *Bulletin of Environmental Contamination and Toxicology* (Springer). He also works/worked as guest editor for *Energy, Ecology and Environment* (Springer); *Bulletin of Environmental Contamination and Toxicology* (Springer); *Sustainability* (MDPI); and *Land Degradation and Development* (Wiley). Email address: vimalcpandey@gmail.com, ORCID: https://orcid.org/0000-0003-2250-6726, Google Scholar: https://scholar.google.co.in/citations?user=B-5sDCoAAAAJ&hl.

Gordana Gajić
Dr. Gordana Gajić is senior research associate in the Department of Ecology, Institute for Biological Research "Siniša Stanković," University of Belgrade, Belgrade, Serbia. Dr. Gajić received her PhD from the Faculty of Biology, University of Belgrade, Serbia, on the topic "Ecophysiological Adaptation of Selected Herbaceous Plants Growing on the Fly Ash Deposits of Thermoelectric Plant 'Nikola Tesla-A' in Obrenovac." Her main research areas are phytoremediation, ecorestoration, revegetation, ecotoxicology, environmental pollution, fly ash and mine waste phytomanagement, urban pollution, monitoring and management of polluted sites, metal(loid) exposure, plant-soil system, plant ecophysiology/biochemistry, oxidative stress, antioxidants, stress tolerance, and plant adaptive response to stress. Dr. Gajić has been involved in six research projects. She has published over 30 research papers in reputed journals and 5 book chapters with Elsevier, Nova Science Publishers, and Studium Press. Dr. Gajić is a reviewer of several reputed national and international journals.

Pallavi Sharma
Dr. Pallavi Sharma is professor in the School of Environment and Sustainable Development, Central University of Gujarat, India. Dr. Sharma is currently engaged in deciphering the mechanisms of abiotic stress tolerance in plants. She obtained her PhD in biochemistry (2005) from the Department of Biochemistry, Banaras Hindu University, Varanasi, Uttar Pradesh. She worked as a postdoctoral research fellow in the National Institute of Plant Genome Research, New Delhi, India; Department of Physics, University of Arkansas, United States; and Department of Plant Sciences, College of Agriculture and Bioresources, University of Saskatchewan, Canada, where her research was focused on MAP kinase signaling cascade in rice plants and mechanisms of cold stress tolerance and fusarium head blight resistance in wheat plants. She has published 50 scientific articles/book chapters in peer-reviewed journals/books with more than 7600 citations.

Madhumita Roy
Dr. Madhumita Roy is DST-Women Scientist in the Department of Microbiology, Bose Institute, India. Dr. Roy did her PhD work from the Microbiology Department of Bose Institute, Kolkata, and was awarded her degree from Jadavpur University in 2011. Thereafter, she did her postdoc from CSIR-IICB, Kolkata. She also worked as assistant professor at Techno India University, West Bengal. Her early research was centered on bioremediation of various kinds of heavy metals, metalloids, and xenobiotics like polycyclic aromatic hydrocarbons and then shifted to plant-based bioremediation or phytoremediation. She has published 19 articles in peer-reviewed journals, many conference proceedings, and 5 book chapters.

Foreword by Dr. Jan Frouz

Environmental pollution and climate change are serious global threats that affect ecosystem functioning as well as the well-being of humans. Phytoremediation—the use of plants to remove pollutants from soil—may help to mitigate these threats. However, phytoremediation techniques are affected by ongoing global changes. Moreover, plants enter into complex interactions with not only soil microflora but also other organisms, which make their interactions even more complex owing to the ongoing global changes. Therefore, there is a pressing need for plant scientists to develop a holistic approach to reduce the ever-increasing number of pollutants in the scenario of changing climate. In this context, this book, *Adaptive Phytoremediation Practices: Resilience to Climate Change*, offers a promising way against pollutants in the changing climate as an ultimate hope for the sustainability of nature. The book also pays considerable attention to the multiple benefits of ecosystem services of phytoremediated polluted sites and also provides insights into how plants, polluted sites, and adaptive phytomanagement practices can be used to improve resilience to untoward incidents such as drought, salinity, heavy metal toxicity, which has become very necessary for effective phytoremediation in the changing climate.

Climate-resilient phytoremediation is gaining attention globally due to the mixing of climate change and ecosystem pollution. Changing climatic conditions also cause various biotic and abiotic stresses in plants and thereby negatively affect plant establishment, growth, and yield. Therefore, the integration of suitable climate-resilient plants and adaptive remedial practices along with proper agro-biotechnological interventions are of paramount importance to mitigate the rapidly growing pollution. Application of soil amendments, inoculation of microbes (fungi and plant growth-promoting rhizobacteria), selection of climate-resilient plants, and genetic engineering are techniques used to sustain phytoremediation in the changing climate. Currently, there is no book available in the market that can cover this novel topic, *Adaptive Phytoremediation Practices: Resilience to Climate Change*. This book can be considered a single source of phytoremediation knowledge on adaptive phytoremediation practices in a changing climate. It is the first book that covers this unique and novel topic and will enjoy a wide readership. Overall, this book is an excellent, up-to-date, and highly informative resource; it is a timely contribution that has a simple-to-understand and easy-to-read format. It contains rich literary text of great depth, clarity, and coverage from which scientific knowledge can grow and widen in this field.

I appreciate the efforts of the authors, Dr. Vimal Chandra Pandey, Dr. Gordana Gajić, Dr. Pallavi Sharma, and Dr. Madhumita Roy, in bringing out this valuable book with the help of the leading global publisher, Elsevier. The book comprises eight chapters covering various aspects of the integration of phytoremediation and climate change. I hope this book will be a remarkable asset for researchers, PhD scholars, plant scientists, policy makers, practitioners, entrepreneurs, and other stakeholders alike.

Jan Frouz
Environmental Centre, Charles University, Prague, Czech Republic

Foreword by Dr. ir. Filip M G Tack

Concerns about the pollution of our environment have recently been augmented by significant worries about the effects of human activities on global climate change. Since the early 1970s, awareness rose that uncontrolled release of contaminants into the environment is threatening the ecosystem and human health. Only very recently, and despite early warnings, it has been universally accepted that massive emissions of carbon dioxide and methane during the industrial development and agricultural intensification of modern society are one of the main factors causing a rapid unfavorable worldwide change in climate as a result of global warming. The transition will be hard. It will need to be fast, or humanity will fail.

Meticulously preserving our 20% wilderness that remains and restoring additional wildland will be crucial to restore the resilience and stability of the earth's climate system. Harm to the environment was also done at many other levels and scales. Plants will play a central role in mitigating or undoing the harm. The need for science to develop more integrated approaches to mitigate environmental damage and ultimately climate change has never been so urgent. In this context, the present book, *Adaptive Phytoremediation Practices: Resilience to Climate Change*, highlights the potential role of plants in environmental technology to counter adverse environmental effects that resulted from unsustainable human development. It highlights the benefits of phytoremediated polluted sites, which can provide multiple ecosystem services. The book provides insight into how adaptive phytomanagement practices can lead to remediation of degraded and contaminated sites through the establishment of plant systems that are resilient against untoward incidents such as drought, salinity, and heavy metal toxicity.

Climate-resilient phytoremediation is gaining attention worldwide due to the links between the problems of climate change and of ecosystem pollution. Changing climatic conditions cause various biotic and abiotic stresses in plants, negatively affecting the establishment of plants and their health, strength, and growth performance. Selection of suitable climate-resilient plants must be combined with adequate remedial practices including adapted agronomic and (bio)technological interventions. Application of soil amendments, inoculation of microbes, selection of climate-resilient plants, and genetic engineering are tools within sustainable phytomanagement strategies in a changing climate. No earlier publications made such a strong connection between issues of contaminant management and threats to humanity related to climate change. The book *Adaptive Phytoremediation Practices: Resilience to Climate Change* is an up-to-date and timely contribution, presented in a way that is accessible to a wide multidisciplinary audience.

I am happy to recognize the efforts of the authors, Dr. Vimal Chandra Pandey, Dr. Gordana Gajić, Dr. Pallavi Sharma, and Dr. Madhumita Roy, in creating this valuable work with the help of the leading global publisher, Elsevier. The book comprises eight chapters covering various aspects of phytoremediation in connection with climate change. I believe this book will be a valuable asset for students, researchers, policy makers, practitioners, entrepreneurs, and other stakeholders alike.

Filip M.G. Tack

Department of Green Chemistry and Technology, Ghent University, Ghent, Belgium

Preface

Adaptive Phytoremediation Practices: Resilience to Climate Change presents phytoremediation practices in the scenario of changing climate and plant responses to stress. The book provides a brief overview of different aspects of adaptive phytoremediation practice and responses of plants to climatic stressors and pollution. It presents the current state of research on structural and functional characteristics of resilient plants and advancements in the knowledge on lipidomics and designer plants that can be applied in agriculture and phytoremediation using gene editing tools such as the CRISPR/Cas 9 system. The book covers all aspects of soil and phytomanagement and offers prospects for the development of climate-resilient economic crops. It aims to provide knowledge on the bioeconomy of contaminated biomass, ecosystem services, and policy frameworks. Appropriate green remediation strategies are essential for climate, energy, agricultural, and food policy domains linking the environment, people, markets, assets, and finance to conserve natural resources and improve human health and well-being worldwide. This book will be ideal for students, researchers, environmentalists, ecological engineers, regulatory agencies, policy makers, and other stakeholders.

Chapter 1 provides general information about the impacts of climate change on plant growth and metal uptake together with plant-microbe interaction to help readers better understand the impacts of elevated CO_2, temperature, and drought on plants. Chapter 2 covers plant responses to abiotic and biotic stresses, such as high temperatures, drought, salinity, cold, flood, heavy metals, and pests, highlighting plant tolerance. Chapter 3 elaborates the functional and structural responses of resilient plants to climate change and pollution with a focus on plant physiology, biochemistry, anatomy, and morphology. Chapter 4 provides a detailed overview of soil management using arbuscular mycorrhizae against drought and salinity stress and rhizobacteria and industrial and organic wastes in the management of degraded soils, with special emphasis on biofuel, fiber, aromatic essential oil, and fortified crop production and their role in phytoremediation. Chapter 5 focuses on phytoremediation strategies and ecosystem services for polluted sites through provisioning, regulating, and supporting services. A detailed review of biotechnological strategies for the generation of designer plants for climate-resilient phytoremediation involving the application of CRISPER for cold, salt, heat, and drought plant tolerance, and improvement of their yields, quality, and nutrition, is presented in Chapter 6. Making biomass from phytoremediation; detailed information on biochemical and thermochemical processes; the recovery of biodiesel, biogas, bioethanol, and dye from contaminated biomass; the conversion of biomass to bioelectricity; and making bioenergy from farm waste are covered in Chapter 7. Finally, Chapter 8 describes the status of remediation and policy implications, climate smart agriculture, and phytoremediation all of which aim to create a greener and sustainable economy for the protection of the environment and human health around the globe.

Vimal Chandra Pandey
Gordana Gajić
Pallavi Sharma
Madhumita Roy

Acknowledgments

We sincerely thank Peter J. Llewellyn and Marisa LaFleur (acquisitions editor), Aleksandra Packowska (editorial project manager), and Swapna Praveen (copyrights coordinator) and Joy Christel Neumarin Honest Thangiah (production project manager) from Elsevier for their excellent support, guidance, and coordination during the production of this fascinating project. We would like to thank all the reviewers for their time and expertise in reviewing the chapters of this book. Vimal Chandra Pandey is grateful to the Council of Scientific and Industrial Research, Government of India, New Delhi, for the support under Scientist's Pool Scheme (Pool No. 13 (8931-A)/2017). Gordana Gajić is thankful to the Ministry of Education, Science and Technological Development of the Republic of Serbia (No. 451-03-9/2021-14/200007) for their support. The authors are thankful to Dr. Jan Frouz, Professor and Director, Environmental Centre, Charles University, Prague, Czech Republic, and Dr. ir. Filip M.G. Tack, Professor in Biogeochemistry of Trace Elements, Department of Green Chemistry and Technology, Ghent University, Belgium, for writing the foreword at short notice. Finally, we thank our respective families for their unending support, interest, and encouragement, and apologize for the many missed dinners!

Phytoremediation in a changing climate

![1]

Chapter outline

1   Introduction

Various human activities, including industrial, mining, and agricultural, result in contamination of air, soil, and water with various pollutants (D'Surney and Smith, 2005; Mishra et al., 2020; Majhi et al., 2021; Pandey and Bauddh, 2019; Pandey, 2020a,b). These contaminants show an adverse impact on living organisms (Alloway, 2013; Nsanganwimana et al., 2014; Pandey, 2020b). Therefore, sustainable and effective management of polluted lands is indispensable (Pandey and Bauddh, 2019; Pandey, 2020a). Traditionally, chemical and physical technologies, including chemical desorption, soil washing, incineration, and solidification, have been utilized for the remediation of contaminated sites (González-Moscoso et al., 2019). Although suitable for point source pollution, these technologies are unfeasible for handling large areas. Also, these procedures are costly and noneco-friendly.

Phytoremediation is a green, eco-friendly, and cost-effective alternative for soil remediation in large-polluted areas with multiple benefits (Pilon-Smits, 2005; Yan et al., 2020; Pandey and Bajpai, 2019; Pandey and Souza-Alonso, 2019; Pathak et al., 2020; Mishra et al., 2020; Praveen and Pandey, 2020) and has received great attention. The approach depends on plant's capability to neutralize or mitigate the harmful effect of contaminants (Salt et al., 1995; Pandey et al., 2015). Phytoremediation includes many different plant processes, including (i) phytovolatilization, which involves uptake and transfer of the harmful substance to the air; (ii) phytostabilization, which reduces the mobilization of the contaminants and leaching; (iii) phytoextraction, which involves the taking up and accumulation of contaminants in the tissues. Microbe-assisted phytoremediation (MAP) is considered a very effective and useful approach to remediate soil contaminated with metals. The cohesive interaction between microbes and plants may trigger plant growth, provide enhanced resistance to abiotic stresses and disease,

Adaptive Phytoremediation Practices. https://doi.org/10.1016/B978-0-12-823831-8.00004-9

and promote adaptation to contaminated environments, and thus can play a vital part in phytoremediation (Ashraf et al., 2017).

Climate change is of utmost environmental concern which the whole world is facing due to its adverse impacts on all living organisms, including plant. Volcanic eruptions, Sun's intensity, and alterations in concentration of greenhouse gas (GHG) contribute to earth's climate change (IPCC, 2013). GHG emissions due to anthropogenic activities are the principal reason of the fast-changing climate. Natural sinks of carbon such as planet's forest (photosynthesis) and oceans are not able to keep up with the rising emissions of GHG so there is build-up of GHG, which is causing disturbingly rapid warming worldwide (Denchak, 2017). Major threats associated with climate change include drought, flooding, high-temperature land salinization etc. It has potential to make environmental stresses more intense and accelerate the processes to make them set in regions where they were not present earlier (El Massah and Omran, 2015).

Climatic change and associated stresses have potential to affect growth, development, and metal uptake by plants (Fig. 1). It can alter the allocation of photosynthate in rhizosphere resulting in changed microbial composition, and thus influence the uptake of metals by plants and hence efficiency of phytoremediation (Compant et al., 2010;

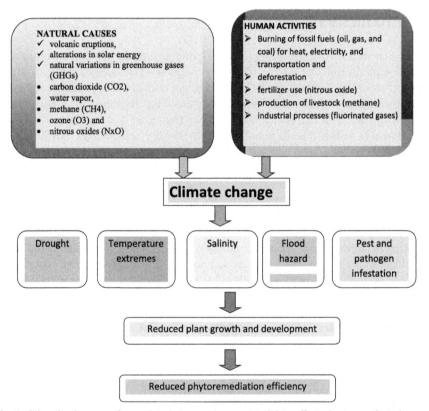

Fig. 1 Climatic change and associated stresses have potential to affect plant growth and development and hence phytoremediation efficiency of plants.

Guenet et al., 2012; Dutta and Dutta, 2016). This chapter will give an overview of the global climate change, associated environmental constraints, and their impact on plant growth, metal uptake, plant-microbe-metal interaction, and phytoremediation.

2 Global climate change

Climate changes before industrial revolution during the 1700s may be attributed to natural causes, including volcanic eruptions, alterations in solar energy, and natural variations in GHG such as carbon dioxide (CO_2), water vapor, methane (CH_4), ozone (O_3), and nitrous oxides (N_xO) (IPCC, 2013). However, natural causes singly cannot explain current climate changes. Warming from the mid of 20th century has been attributed mainly to human activities. Fossil fuel (coal, oil, and gas) burning for heat, electricity, and transportation and deforestation are leading sources of anthropogenic emissions of GHG. Other anthropogenic activities like fertilizer usage (nitrous oxide), production of livestock (CH_4), and some industrial processes (fluorinated gases) also lead to GHG production. GHG absorb thermal infrared (IR) radiation, which is discharged by the surface of Earth and atmosphere and, in turn, emits IR radiation, with a substantial part of this energy used in warming the surface of Earth and lower atmosphere (Ledley et al., 1999). Increased GHG emissions can provoke extreme climate changes such as heat, floods, and droughts. Some GHGs such as CH_4 can persist for up to decades, whereas CO_2 stays for centuries to millennia (Zickfeld et al., 2017). The concentration of CO_2, which is the earth's main climate change contributor, has enhanced by $> 40\%$ due to various human activities since preindustrial times. The atmospheric global CO_2 growth rate was around 0.6 ± 0.1 ppm year^{-1} during the 1960s. However, between 2009 and 2018, the rate of growth was 2.3 ppm year^{-1}. The CO_2 was around 409.8 ± 0.1 ppm (global average) in year 2019, an increase of 2.5 ± 0.1 ppm compared to 2018. By this century end, atmospheric CO_2 is expected to be around 700 ppm (Lindsey, 2020).

Rising CO_2 can affect plant metabolism both directly and indirectly. CO_2 has an important role in photosynthesis so it can directly affect photosynthesis. Indirectly, it can affect plant performance via its impact on temperature and drought (Dusenge et al., 2019). Increased CO_2 concentration amplifies Earth's greenhouse effect. Strong correlation between CO_2 concentration and air temperature has been noted throughout glacial cycles in past numerous hundred thousand of years. When the CO_2 concentration increases, temperature also goes up. Also, anthropogenic activity-induced climate change has also led to incidence and intensity of temperature extremes (Meehl et al., 2007; Seong et al., 2021). IPCC estimated that by the year 2100, an increase in temperature will be between 1.8°C and 4.0°C (IPCC, 2014). Heat waves or extreme temperature events are projected to become more intense, more frequent, and last longer than what is being currently observed (Meehl et al., 2007). Daily minimum temperatures are suggested to increase more quickly compared to daily maximum temperatures causing increased daily mean temperatures and a higher risk of extreme events, which can cause damaging effects on plant (Meehl et al., 2007). As temperature is among the key environmental factor controlling the plant growth and development

(Hatfield and Prueger, 2015), the responses of plant to temperature variations have become a major concern nowadays.

Inconsistency in temperature and occurrence of precipitation may lead to a rise in water scarcity. Therefore, due to global warming and change in climatic conditions, changes in distribution, the frequency, severity, and duration of drought are expected to increase in the future. For 21st century, a steady rise in aridity has been projected by Global climate models (Scheff and Frierson, 2015; Zarch et al., 2017; Greve et al., 2019). According to these projections, drylands will possibly expand due to surges in evaporative demand and a hydrological cycle having extended and more severe dry periods at the worldwide scale. Also, an enhancement in soil salt concentration and a reduction in water quality may lead to soil salinization, which may cause land degradation (Tomaz et al., 2020). River flood hazard may also increase owing to climate alteration (Prudhomme et al., 2003; Hirabayashi et al., 2013; Avand et al., 2021). Although warming may not trigger floods, it influences snowmelt and rainfall that can contribute to flood. Climate change can have positive as well as negative effects on pathogens and pests. It can lead to alterations in their population dynamics, abundance and diversity, biotypes, geographical distribution, interactions with plant, natural enemies, and extinction. Increased pests and pathogens infestation due to climate change are projected to cause reduced crop yields (Dhankher and Foyer, 2018).

3 Impacts of climate change on growth, uptake of metals, and phytoremediation potential of plants

Phytoremediation potential (metal extraction and stabilization of contaminants) of plants depends on different factors like bioavailability of metal in soils, plant biomass, metal accumulation capability, microbiome, and climatic factors (Glick, 2010; Grčman et al., 2001; Miransari, 2011). Any abiotic parameter that influences these factors should also affect phytoremediation efficacy. Climate change and stresses associated with it such as elevated CO_2 (EC), temperature, and drought can alter the plant growth and metal uptake and accumulation potential of plants. Different effects of EC, drought or warming, or a combination of climatic stress on metal accumulation and growth of plants differ greatly across plants used and chemical, biological, and physical characteristics of environment. These properties include pH of the soil, type of heavy metals and their bioavailable concentrations, diversity of microorganisms, and effects of climatic factors interaction (Rajkumar et al., 2013). Consequently, it is important to understand the interaction between climate change, growth of plant, and metal uptake and accumulation. Understanding the climate change effect on plant growth and metal accumulation can be beneficial when developing phytoremediation strategy for particular contaminated sites (Brunham and Bendell, 2011).

3.1 Impacts of elevated concentration of CO₂

Elevated concentration of CO_2 or EC alters the growth, metal uptake, and phytoremediation efficiency of plants (Table 1). It is widely recognized that EC can stimulate

Table 1 Effects of elevated CO_2 on the heavy metal phytoremediation potential of plants.

Plant	Elevated CO_2 treatment	Metal treatment	Effect	Reference
Populus × euramericana (Dode) cv. "Nanlin-95" (NL95) one willow genotype (*Salix jiangsuensis* CL. "172" (J172))	$800 + 50$ μL L^{-1}	0.06, 5.2, and 25.3 mg Cd kg^{-1}	Increased Cd phytoremediation efficiency especially at high Cd concentration due to increased biomass	Guo et al. (2015)
Marvdasht cv. for wheat and Sepideh cv. for sorghum	900 ± 50 μL L^{-1}	0, 10, 20, and 40 mg Cd kg^{-1}	EC reduced the concentration of Cd in the roots and shoots of Cd-tolerant wheat, while it enhanced for Cd-tolerant sorghum plants	Khanboluki et al. (2018)
Noccaea caerulescens	550 ± 50 ppm	Cd 0.93 ± 0.11 mg kg^{-1} Pb 91.2 ± 5.6 mg kg^{-1} Cu 72.6 ± 6.2 mg kg^{-1} Zn 188.2 ± 11.8 mg kg^{-1}	Accumulation of metals and biomass	Luo et al. (2019)
Sorghum vulgare × Sorghum vulgare var. *sudanense* hybrid and *Trifolium pratense*	860 μL L^{-1}	0, 300, 1500, and 3000 mg Cs kg^{-1}	Increased biomass and Cs accumulation	Wu et al. (2009)
Brassica juncea *Helianthus annuus*	800 and 1200 μL L^{-1}	100 and 200 mg Cu kg^{-1}	At EC (1200 μL L^{-1}), both Indian mustard and sunflower showed increased biomass and Cu extraction capability	Tang et al. (2003)
Lemna minor	700 ppm	0, 1.5, 2.5, and 5 mg Cd L^{-1}	Cd concentration enhanced in plant	Pietrini et al. (2016)
Robinia pseudoacacia	700 ± 23 μmol mol^{-1}	0, 1, and 5 mg kg dry $soil^{-1}$	Three years of elevated CO_2 decreased Cd uptake into leaves	Jia et al. (2017)
Robinia pseudoacacia	700 ± 23 μmol mol^{-1}	500.0 mg Pb kg dry weight $soil^{-1}$	3 years of elevated CO2 increased Pb concentration	Jia et al. (2018)
Lolium perenne and *Lolium multiflorum*	1000 ± 80 μL L^{-1}	4, and 16 mg Cd L^{-1}	Increased growth and Cd uptake	Jia et al. (2011)
Arabidopsis thaliana	800 μmol mol^{-1}	50, and 110 mM As	Aboveground biomass and As concentration decreased	Fernandez et al. (2018)

photosynthesis and plant biomass with the degree of stimulation being species or cultivar dependent (Ziska et al., 2012). Faster growth and greater biomass, in turn, could enhance phytoextraction and toxin uptake in aerial plant tissue. EC can augment plant biomass by 50% via CO_2 fixation (Dier et al., 2018). It can enhance plant height, dry weight, and the leaf area by stimulating photosynthesis (Khanboluki et al., 2018). As CO_2 increases, the stomatal pores don't open as wide, leading to decreased conductance of stomata (gs), lower transpiration, and increased intrinsic water-use efficiency (WUEi) (Jarvis et al., 1999; Aranjuelo et al., 2006; Erice et al., 2006; Long et al., 2006). Högy et al. (2010) reported that EC (409 ppm versus 537) enhanced biomass and concentration of metabolites of low molecular mass in several weed species. In comparison with herbaceous species, heavy metal accumulator plants, poplars, and willows were found to be more EC responsive (Ainsworth and Long, 2005), as evident by their biomass, height, and carbon assimilation, which was suggested to be the result of stimulatory effect of EC on the photosynthetic rate and expansion of leaf (Curtis and Wang, 1998; Nowak et al., 2004). However, long-term continuous EC exposure may lead to a reduction (downregulation) of photosynthetic capacity. Often this decrease is accompanied by decreased photosynthetic enzymes and pigments such as ribulose-1,5-bisphosphate carboxylase/oxygenase (Rubisco) (Faria et al., 1996; Centritto and Jarvis, 1999). Goufo et al. (2014) observed that elevated CO_2 increased the accumulation of carbon in rice throughout its whole life cycle.

ROS have been related to the inhibition of growth, utilization of nutrients, and photorespiration of plants. Since heavy metals increase ROS generation, it is hypothesized that CO_2 can reduce damaging effects of metals on plants. EC levels led to increased plant growth, uptake of Cd, rate of assimilation CO_2, efficiency of intrinsic water use and activities of antioxidative enzyme, decreased gs, transpiration rate and malondialdehyde (MDA), and no change in Cd concentrations in poplar genotype (NL95) and willow genotype (J172). It was proposed that EC promoted the growth of plant by boosting photosynthesis and increased phytoremediation efficiency, mainly at high Cd content (Guo et al., 2015). Cd accumulation due to EC differed in Cd-tolerant wheat and sorghum. EC reduced Cd contents in roots and shoots of Cd-tolerant wheat, while it enhanced for Cd-tolerant sorghum plants (Khanboluki et al., 2018).

Accumulation of metal and biomass in *Noccaea caerulescens* increased under EC (550±50 ppm) compared to controls (400 ppm CO_2), suggesting that the EC could increase phytoremediation efficiency. Concentrations of water-soluble and exchangeable Cu and Pb decreased in soil under EC conditions, which decreased the risks of leaching of these metals. MDA concentrations in *N. caerulescens* reduced to varying degrees with the EC. It was hence suggested that EC can decrease oxidative damage resulting from metals in this species (Luo et al., 2019). EC (860 $\mu L\ L^{-1}$) caused 32%–111% increased aerial biomass of the *Sorghum* and 8%–11% in *Trifolium* species in comparison with ambient CO_2. It also caused up to 73% increased Cs accumulation in *Sorghum* and 43% in *Trifolium* species (Wu et al., 2009). At EC (1200 $\mu L\ L^{-1}$), both Indian mustard and sunflower showed increased leaf area, biomass, and Cu extraction capability compared to when grown under ambient CO_2 (Tang et al., 2003). Increased photosynthetic efficiency, Cd accumulation, antioxidant capacity, growth, and Cd phytoremediation capability of *Lemna minor* with increased Cd treatments, time of exposure, and EC (350 and 700 ppm) were observed (Pietrini et al., 2016). Cd toxicity alleviation at low Cd treatments at EC in *L. minor* may be credited to enhanced photosynthesis and increased antioxidant capacity. Long-term EC increased the phytoextraction of Pb by *Robinia pseudoacacia* from contaminated soils (500 mg kg^{-1} soil) (Jia et al., 2018). Increased accumulation of Pb in leaves and removal of Pb from soils were observed. Width and height of seedlings increased due to EC in comparison with ambient CO_2. EC led to increased SOD and CAT activities and higher concentrations of cystine, glutathione, phytochelatin, phenolics, proline, and flavonoids in plants under Pb treatment. EC (800 μL L^{-1}) stimulated the growth of *Sedum alfredii* nonhyperaccumulating and hyperaccumulating ecotype but an increase was much greater in hyperaccumulating ecotype. EC was proposed to be an approach to increase phytoremediation efficiency of *S. alfredii* hyperaccumulating ecotype (Li et al., 2013).

EC increased *Arabidopsis thaliana* biomass (aerial) significantly. However, the level of stimulation of biomass was dependent on ecotype. At As concentration of 110 μM, the relative effect of EC was to reduce both As concentration and As uptake per plant; however, genetic variation was also obvious among *A. thaliana* for As phytoextraction at existing and expected CO_2 levels (Fernandez et al., 2018). EC (360 and 1000 μL L^{-1}) led to more increase in shoot biomass than in root biomass of *Lolium perenne* and *Lolium multiflorum* exposed to 4 and 16 mg Cd L^{-1} but reduced Cd concentrations in all plant tissues (Jia et al., 2011).

3.2 Impacts of elevated temperature

Temperature influences the growth and development rate of plants. It can also influence metal uptake by plants (Table 2). Each plant species has a particular temperature range characterized by an optimum, maximum, and minimum. Responses of plants to temperature increase depend on the plant development stage. For most species of plants, vegetative development generally has a greater optimum temperature compared to reproductive development. Temperature may affect water chemistry and growth of plants (Fritioff et al., 2005). It alters plant transpiration rate, metabolism, and growth and thus influences uptake and elimination of contaminants (Yu et al., 2005). Changes in temperature show the large impact on the uptake and accumulation of metals by plants. Therefore, it is imperative to understand the relation between temperature change, plant growth, and uptake and accumulation of metal when planning for phytoremediation in different bioclimatic zones (Brunham and Bendell, 2011). Fastest growth happens around 15 and 30°C in most plants (Larcher, 1995). Application of higher temperatures in the field can minimize the time duration required to attain clean-up aims. Baghour et al. (2001) observed higher Cr uptake by Spunta *variety of Solanum tuberosum* at high temperature compared to low temperature. Metal transfer from soil to plant was more at higher temperature in *Potamogeton natans, Elodea canadensis* (Fritioff et al., 2005), willow (Yu et al., 2010), and safflower (Pourghasemian et al., 2013) for various metals (Cd, Cu, Cr, Zn, Pb). This was suggested to be due to enhanced plant growth and uptake of metals by roots under warmer temperatures. Temperature increase (26 and 30°C) enhanced Cr toxicity by reducing photosynthesis and growth and increasing activities of antioxidative

Table 2 Effects of elevated temperature and water deficit on the heavy metal phytoremediation potential of plants.

Plant species	Elevated temperature/water deficit	Metal	Effect	Reference
Atriplex halimus	Elevated temperature 26 and 30°C	1 mM Cr	Enhanced Cr uptake and accumulation	Mesnoua et al. (2018)
Oryza sativa	Elevated temperature 40°C for 2 h	50 μM 100 and 500 μM Cd	Increased Cd uptake and accumulation due to heat treatment in vacuoles, parenchyma, and vascular cylinder of root tissue	Shah et al. (2013)
Salix matsudana Koidz × *alba* L.	Elevated temperature 11, 14, 17, 20, 22, 24, 26, 28, 30, and 32°C	1.51 (± 0.01) and 1.53 (± 0.01) mg Cr L^{-1} for the treatment with Cr(VI) and Cr(III). respectively	The removal rates of Cr(III) and Cr(VI) by plants showed a linear relation with temperature increase	Yu et al. (2010)
Solanum tuberosum var. Spunta	16, 20, 23, 27, and 30°C	Ag: 1 μg L^{-1} As: 2 μg L^{-1} Sb: 1 μg L^{-1} Cr: 2 μg L^{-1}	Root zone temperature of 23–30°C most favorable for uptake and accumulation of Ag, As, Sb, and Cr	Baghour et al. (2001)
Carthamus spp.	11–32°C	0, 0.5, 1, 5, 10, 20, 50, 100, and 500 μM Cd	Enhanced relative uptake and concentration of Cd with increase in temperature	Pourghasemian et al. (2013)
Festuca arundinacea	Water deficit Soil moisture registered below 50% of control	0.64 + 0.05 μg Se g^{-1}	Increased Se concentration	Tennant et al. (2000)
Lolium multiflorum	Soil was supplied with a fraction of water evapotranspired in well-watered pots.	Rural sewage sludge Cu 204.5 mg kg^{-1} Zn 730.5 mg kg^{-1} Ni < 25.0 mg kg^{-1} Mn 225.8 mg kg^{-1}	Reduced concentration coefficient of Cu, Zn, Ni, and Mn in roots under drought conditions	Pascual et al. (2004)

Species	Treatment	Metal concentration	Findings	Reference
Triticum aestivum	70% of water-holding capacity	7.67 mg Cd kg^{-1}	Drought treatment further reduced the biomass of plant and resulted in oxidative stress. In wheat tissue, drought increased Cd content	Khan et al. (2019)
Beta vulgaris	T1: applied 300 L/block each time, T2: applied 200 L/block each time, T3: applied 100 L/block each time; T1 was control	1.87 mg Cd kg^{-1}	Shoot biomass and efficiency of Cd remediation potential also increased	Tang et al. (2019b)
Brassica oxyrrhina	200 mg L^{-1} PEG	Cu, Zn ranging from 300 to 1500 mg L^{-1}	Metal concentration decreased. Plant growth did not differ, greatly suggesting that drought stress did not make plants extra vulnerable to metal stress	Ma et al. (2016b)
Zea mays cv. single cross 704	Irrigated with different Cd concentration at 1-, 3-, and 7-day intervals	10 and 20 mg Cd L^{-1}	Cd accumulation was more	Azizian et al. (2013)

enzymes in *Atriplex halimus*. It also increased nutrient uptake and translocation but decreased Cr translocation. At 1 mM Cr treatment, Cr concentration in roots at 26°C was 7.2 g kg^{-1} and at 30°C was 9.1 g kg^{-1}, respectively, whereas Cr concentration in the shoot was 0.45 at 26°C and 0.44 g kg^{-1} at 30°C (Mesnoua et al., 2018).

Shah et al. (2013) reported enhanced Cd uptake and accumulation in vacuoles, parenchyma, and vascular cylinder and stimulated antioxidant system in rice cv. Bh-1 roots subjected to a combination of heat stress and suggested that heat stress may mitigate mild Cd toxicity directly by an increased formation of Cd^{2+}-MT providing Cd^{2+} tolerance. The removal rates of Cr(VI) and Cr(III) by plants showed a linear relation with a temperature rise from 11°C to 32°C. The rate of Cr(III) and Cr(VI) removal was the highest at 32°C. Cr(VI) and Cr(III) translocation was observed from the root to the lower stem. Cr(VI) removal using hybrid willow was more vulnerable to alterations in temperature compared to Cr(III) (Yu et al., 2010).

Efficient transfer of metals from soil to plants is good for phytoremediation purpose. Therefore, improved understanding of how the temperature affects the metal bioavailability in cultivated soils is important. Dissolved organic matter (DOM) is the main metal complexing substance in soil, whereas soil organic matter (SOM) is the prominent metal-carrying phase in soil. Content and properties of the organic matter determine the metal complexation and mobility in soil. The temperature increase is suggested to stimulate the metal mobilization via influencing the SOM degradation and altering the metal complexation in water present in pores by changing the nature and quality of DOM (Cornu et al., 2011). In lettuce crop, a consistent decrease in bioavailability of Zn and Cd was reported from 10°C to 30°C, and this was credited to a strong metal complexation in the pore water because of a rise in soil temperature at higher temperatures. However, Zn and Cd transfer from soil to plant was stimulated at higher soil temperatures, suggesting that this process was affected most at higher temperature (Cornu et al., 2016).

In many cases, plant growing under high temperature condition showed the promotion of plant growth, increased photosynthetic rate, and accumulation of greater amount of heavy metals. Temperature-led indirect effects, such as SOM decomposition, on metal mobilization and/or uptake have been suggested the reason for higher metal accumulation (Sardans et al., 2008; Van Gestel, 2008; Conant et al., 2011).

3.3 Impacts of drought

Drought can be critical in areas with low availability of water or random alteration in weather during plant growth period. The effects of drought are likely to increase with climate change and growing water scarcity. It is a major environmental stress factor that affects the growth and development of plants. Understanding of relation of drought, plant growth, and metal uptake is necessary for phytoremediation under climate change. Variable effects of drought on phytoremediation potential of plants were reported in metal-polluted soils. In tall fescue, low soil moisture increased the accumulation of Se and reduced the production of biomass. Overall, drought reduces the heavy metal phytoremediation potential of plants (Tennant and Wu, 2000). High drought stress may harm many functions of plants but the main consequence is a decrease of carbon fixation, which causes a decreased plant growth depending on duration of the stress, the stage, and the existence of other stresses. However, Angle et al. (2003) observed higher biomass and uptake of metal such as Zn or Ni by plants at higher soil moisture leading to greater concentration of metals extracted from soil in *Alyssum murale*, *Berkheya coddii*, and *Thlaspi caerulescens*. *Beta vulgaris* plants grown in Cd-polluted soil [controlled deficit irrigation (300 L: T1, 200 L: T2, 100 L: T3)/block at each irrigation] throughout the organogenesis stage. The shoot biomass increased by 15.8%, and efficiency of Cd phytoremediation potential under T2 treatment (5.42 g ha^{-1}) was 39.7% higher compared to T1 and 61.8% higher than T3 (Tang et al., 2019b).

Water is needed for many biochemical and physiological processes, including germination, division and elongation of cells, various metabolic activities including organic compounds synthesis, respiration, and photosynthesis (Lange et al., 2012). Drought leads to lower height, decreased leaf number and size, and less production of fruit. It also causes stomatal closure and increased resistance to CO_2 diffusion in leaves (Alizadeh et al., 2015). Study on the structure of xylem and hydraulic conductivity revealed that metal stress exacerbates water stress in an additive manner and thus makes the *Acer rubrum* plants more sensitive to drought (de Silva et al., 2012). However, the effect of metal stress and combination of drought and metal toxicity on the plant growth did not differ greatly, suggesting that drought stress did not make plants extra vulnerable to metal stress (Ma et al., 2016b). Photosynthetic performance such as gs, chlorophyll fluorescence, and chlorophyll concentration declined. Water status of plant was the last affected parameter by the drought treatment (Ings et al., 2013). Water deficit also reduced the transfer and concentration coefficient of different metals, including Cu, Zn, Ni, and Mn in *L. multiflorum* Lam. roots in the soil in which rural sewage sludge was amended in high amount (Pascual et al., 2004). Drought reduced the transfer and concentration coefficient of some heavy metals in the root of sewage sludge (SS), mineral fertilizer (M)-treated plants, indicating that heavy metals in sewage sludge were preserved in chemical forms with low availability (Pascual et al., 2004).

Cd-treated wheat plants grown for 125 days after seed sowing had lesser biomass and greater oxidative stress and drought treatment further reduced the biomass of plant and resulted in oxidative stress. In wheat tissues, drought enhanced the Cd content (Khan et al., 2019). In Indian mustard plant, drought stress at 90 day after sowing led to a decreased Cd uptake (Bauddh and Singh, 2012). When corn plants were irrigated with water containing 0, 5, 10, and 20 mg Cd L^{-1} at 1-, 3-, and 7-day irrigation intervals, toxicity symptoms of Cd appeared clearly on the plant. Stem dry weight, transpiration, and height of plant were most sensitive to Cd levels. Cd accumulation was 24%, 56%, and 27% more at 1-, 3-, and 7-day irrigation interval in leaves compared to stems, indicating that corn can be useful for Cd phytoremediation under optimum water conditions and mild pollution of the soil (Azizian et al., 2013). As the success of phytoremediation technique relies on the plants' capability of heavy metal accumulation and sufficient biomass yield, these studies revealed that drought might not exert significant positive effects on the phytoremediation of heavy metals.

4 Plant-microbe interaction and their effect on metal uptake and phytoremediation in a changing climate

Plant-associated microorganisms are an important factor influencing the metal uptake and growth response of plants to climate change (Compant et al., 2010) (Fig. 2). Exudates of plant roots are reported to increase the mobility of nutrients and heavy metals by inducing microbial activity indirectly and thus enhance the efficiency of phytoremediation (Ström et al., 2002; Pérez-Esteban et al., 2013; Sessitsch et al., 2013). Plant rhizosphere-associated microbial communities play a beneficial role in the growth of plants in metal-polluted soils (Xiong et al., 2010; Rajkumar et al., 2012). Many of these microbes boost the plant growth and uptake of nutrient by making nutrient available to plant through the solubilization of mineral nutrients like P, N, Fe, K, etc., production and excretion of plant growth-promoting compounds such as phytohormones, and release of specific enzymes (e.g., 1-aminocyclopropane-1-carboxylate deaminase). They also reduce metal led toxicity indirectly by stimulating defense mechanisms of plants. Microbes alter the bioavailability of heavy metals in soil by several mechanisms, including alteration of soil pH, metal translocation, chelation, complexation, acidification, precipitation, and redox reactions (Rajkumar et al., 2012; Guo et al., 2014; Ma et al., 2016a), and thus help in their uptake, sequestration, and detoxification from polluted sites. Different microbes adopt different mechanisms to activate either immobilization or mobilization and hence contribute directly to phytostabilization or phytoextraction (Ma et al., 2011). Some plant growth-promoting

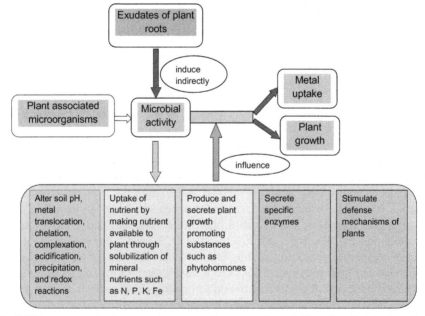

Fig. 2 Plant-associated microorganisms are an important factor influencing the metal uptake and growth response of plants to climate change.

bacteria (PGPB) produce siderophores, phytohormones, and biochelators; fix nitrogen; and solubilize mineral nutrients such as phosphate or potassium, which are beneficial for MAP (Ma et al., 2011; Rajkumar et al., 2012; Ahemad and Kibret, 2014). The impacts of adverse environmental conditions accompanying climate change, such like EC, drought, and warming, on beneficial plant-microorganism-metal interactions are increasingly being explored. Several studies have revealed the impact of climate change on plant-microbe interactions and how this affects phytoremediation.

4.1 Impact of elevated CO_2

EC in general enhances the biomass production and metal accumulation in plants under climate changes. It generally helps plants to support greater microbial population and/or protects diverse microorganisms against heavy metal stress (Rajkumar et al., 2013). Generally, EC increases the growth and productivity of plant, resulting in an enhancement in input of carbon into soil, alteration in soil microenvironments including enhanced soil moisture, and promotion of activity and growth of microbes (Brett Runion et al., 2009; Lipson et al., 2005; Nelson et al., 2004).

In *R. pseudoacacia* seedlings, heavy metals led to decreased population and biomass of microbes in rhizosphere soils and altered microbial community too. Enhanced microbial populations, biomass, and changes in microbial community in *R. pseudoacacia* seedlings grown under EC and the presence of Pb, Cd, or Cd + Pb were reported. EC also enhanced the ratio of removal of Pb and Cd in rhizospheric soils. It can be concluded that EC can positively affect soil fertility, rhizospheric microenvironment, and heavy metal removal from soil (Huang et al., 2017).

In metal-rich soils also, EC leads to alteration in diversity and activity of microbes associated with plant roots. Generally, EC has a positive impact on abundance of ectomycorrhizal and arbuscular fungi but the effects on PGPB and endophytic fungi are variable. Root of *Trifolium* and *Sorghum* under EC in Cs-enriched soils consisted of large fungal and bacterial population but not actinomycetes compared to EC and Cs-contaminated soil. Microbial associations enhanced Cs accumulation by the plant without reducing growth (Wu et al., 2009). Song et al. (2012) observed increased actinomycetes, bacterial, and fungal populations and enhanced microbial N and C in the rhizospheric soils of *Amaranthus cruentus* and *Phytolacca americana*. EC increased biomass and triggered more Cs accumulation by plants. EC and inoculation of endophyte led to the stimulation of plant growth (enhancement in morphological properties of root and exudation, altered uptake, and distribution of Cd of both *S. alfredii* ecotypes) (Tang et al., 2019a). In rice rhizosphere, EC led to increased microbial respiration leading to reductive Fe(III) (hydr)oxide dissolution and oxidation of arsenite (Oremland and Stolz, 2003), as suggested by increases in Fe and arsenate concentrations (Muehe et al., 2019). Application of *Burkholderia* sp. D54 and EC enhanced the production of biomass and the accumulation of Cs in *A. cruentus* and *P. americana* than in control (Tang et al., 2011). Similarly, enhanced growth and phytoextraction efficiency were observed in rye grass grown in the presence of multiple heavy metal/metalloid (Zn, As, Pb, and Cd)-polluted soil (Guo et al., 2014). Perhaps EC reduces the negative effects exerted by heavy metal on rhizosphere-associated microbial population and thus leads to the beneficial influence on plants and metal uptake under altered environmental conditions (Song et al., 2012; Tang et al., 2011; Wu et al., 2009). Further work is needed to elucidate how climate change interacts with microbes and heavy metal uptake in plants in natural environment.

4.2 Impact of drought

Climate change can lead to reduced rainfall and warmer temperatures, which can result in drought. Water deficit can affect the growth of plants and microbes and metal uptake. Under drought, plants display various physiological responses in order to defend itself from the detrimental effects. Alteration in constituent of root exudate, the main way by which plants employee microbes, is one such mechanism. Drought condition alters the photosynthates distribution in rhizosphere, leading to an alteration in microbial associations and their impact on the growth of plant (Compant et al., 2010; Guenet et al., 2012). Drought can impact the microbial community in a direct or indirect way. It can affect microbial community by supporting the growth of desiccation-tolerant microbes or altering soil chemistry and diffusion rates (Naylor and Coleman-Derr, 2018). Influence of drought on microbes depends on the type and strain (Guenet et al., 2012). Under drought, many metabolites and monoderm bacteria are enriched in sorghum root. Monoderms present in the drought-treated rhizosphere exhibited an enhanced activity of transporters related to these metabolites (Xu et al., 2018). Improved plant drought tolerance of *Trifolium repens* due to inoculation of *Pseudomonas putida* and *Bacillus megaterium* under drought conditions was associated with high indole-3-acetic acid (IAA) production by these strains, which can stimulate the growth of root and thus the water uptake capability (Marulanda et al., 2009). Drought stress caused an increased structure and activity of fungal community, but the bacterial community structure and activity were decreased. Perhaps the fungi are capable of exploring the dry environments via hyphal spread (de Boer et al., 2005). Fungal communities can change from one predominant member to other because of small change in the availability of soil moisture ($<$ 30% decrease in water holding capacity), whereas bacterial communities stay constant, indicating higher fungal plasticity than bacterial plasticity (Kaisermann et al., 2015). Mycorrhizal colonization may lead to enhanced xylem pressure by enhancing root biomass, which may be considered a mechanism to increase water uptake under drought conditions (Augé et al., 2001). Some beneficial microbes can reduce oxidative damage due to drought induced-ROS formation by increasing the activities of antioxidant enzymes. *Glomus versiforme* inoculation enhanced SOD, GPX, and CAT activities and drought tolerance in water-stressed *Poncirus trifoliata* (Wu and Zou, 2009).

Interaction of plant, heavy metal uptake, and microbes under drought conditions has been studied widely. It was reported that under water stress, drought-tolerant *Streptomyces pactum* Act12, an actinomycete strain, increased plant height, root length, and shoot biomass, Cd concentrations, Cd translocation from the root to the shoot, and bioconcentration factor in amaranth (*Amaranthus hypochondriacus*) (Cao et al., 2016). Physiochemical analysis showed that Act12 increased plant tolerance to Cd by increasing glutathione concentration, enhancing activities of CAT and SOD, and decreasing content of MDA in the leaves under water-deficit condition. Fernández et al. (2012) observed that in mine soil, rhizosphere-associated fungal communities of *Crithmum maritimum* and *Tetraclinis articulata* resisted water stress, whereas bacterial populations were drought susceptible. Ma et al. (2017) demonstrated that ASS1, a serpentine endophytic bacterium, can protect and thus assist plants to flourish in semiarid regions and increase the process of metal phytoremediation.

4.3 Impact of high temperature

Alteration in composition of plant-associated microbial community depends on length of the increased temperature treatment, experimental approach, availability of substrate, ecosystem, and climatic factors (Frey et al., 2008; Compant et al., 2010; Deslippe et al., 2011). Several investigations have been focused on how a rise in temperature can affect microbial diversity, activity, and their interaction with plants (Frey et al., 2008; Compant et al., 2010; Deslippe et al., 2011). Soil warming by 5°C for 12 years caused a significant decrease in biomass of microbes and consumption of carbohydrates, amino acids, and carboxylic acids (Frey et al., 2008). Compant et al. (2010) observed significant effects of augmented temperature on plant-associated microorganisms and the effects varied significantly with the system and the range of temperature studied. Warming led to a substantial enhancement in large biomass fungi having proteolytic capability, specifically *Cortinarius* spp., and decrease in fungi having affinity for labile N, particularly *Russula* spp. It was suggested that warming greatly modifies cycling of nutrients in tundra and could promote the expansion of *Betula nana* by promoting large mycorrhizal networks (Deslippe et al., 2011).

Even warming by approximately 1°C directly stimulates the microbial activity (Tyagi et al., 2014). Hasegawa et al. (2005) reported an increase in virulence of EC153 strain of *Erwinia carotovora* at 34.5°C than at 28°C and suggested it to be due to higher production of enzymes involved in cell wall degradation, signals associated with quorum-sensing and extracellular proteins. Elevated temperature positively influenced plant-microbe interaction in some cases (growth of hypha and colonization of plant by arbuscular mycorrhizal fungi (AMF), perhaps because of quicker distribution of carbon to the rhizosphere of plant where AMF lives (Compant et al., 2010). Heat stress also affected the microbe's survival directly (Fahimipour et al., 2018).

Heavy metal contamination decreased the bacterial growth, whereas the application of combined stress of heavy metals and heat showed no considerable effect on growth of bacteria and physiological profile of its community (Kools et al., 2008). Four years of elevated temperature (+ 1.99°C) exposure led to greater influence on the community profile of ammonia-oxidizing bacteria, ammonia-oxidizing archaea, bacteria, and fungi present in the rhizospheric region of the seedlings of *R. pseudoacacia* growing under Cd treatment (Jia et al., 2019). Increased temperature resulted in an enhanced quantity of genera like *Stenotrophomonas*, *Methylobacterium*, and *Archangium* but reduced quantity of genera like *Ramlibacter*, *Nitrosospira*, and *Microascus*, when present in combination with Cd. Presence of high concentrations of soil heavy metals and elevated air temperature can influence the microbial community present in rhizospheric region of soils by changing the patterns of supply of photosynthates to roots.

5   Conclusions and future prospects

Direct or indirect impacts of climatic stress are expected to have large influence on phytoremediation process since the plants and microbes greatly modify their responses to heavy metals and climatic stress. Also, climate changes influence the mobility of heavy metal in soils. Certain climate alterations such as EC are known to have a positive effect on the plant growth and development, structure and activity of rhizospheric microbial community, and heavy metal phytoremediation. EC in general have a positive influence on root exudation and/or its consequence on rhizosphere microbial diversity and activity, which facilitates heavy metal phytoremediation (depending on both the plants and heavy metals) by increasing nutrient cycling and availability, metal mobilization/immobilization, and plant growth, and/or by improving soil quality. However, the impact of temperature rise, drought, and combined climatic stressors on plant growth, metal uptake, and plant-microbe-heavy metal interaction varies substantially across plant used and chemical, physical, and biological characteristics (soil pH, heavy metal type, and its bioavailable concentrations, microbial diversity) of the environment. Thus, the effects of combination of climatic stresses on various plant species under metal-contaminated field soils need to be understood to plan future phytoremediation strategies.

References

Ahemad, M., Kibret, M., 2014. Mechanisms and applications of plant growth promoting rhizobacteria: current perspective. J. King Saud Univ. Sci. 26, 1–20.

Ainsworth, E.A., Long, S.P., 2005. What have we learned from 15 years of free-air CO2 enrichment (FACE)? A meta-analytic review of the responses of photosynthesis, canopy properties and plant production to rising CO2. New Phytol. 165, 351–372.

Alizadeh, V., Shokri, V., Soltani, A., Yousefi, M.A., 2015. Effects of climate change and drought-stress on plant physiology. Int. J. Adv. Biol. Biomed. Res. 3, 38–42.

Alloway, B.J., 2013. Sources of heavy metals and metalloids in soils. In: Alloway, B.J. (Ed.), Heavy Metals in Soils. Springer, Dordrecht, pp. 11–50.

Angle, J.S., Baker, A.J., Whiting, S.N., Chaney, R.L., 2003. Soil moisture effects on uptake of metals by Thlaspi, Alyssum, and Berkheya. Plant Soil 256, 325–332.

Aranjuelo, I., Irigoyen, J.J., Perez, P., Martinez-Carrasco, R., Sanchez-Diaz, M., 2006. Response of nodulated alfalfa to water supply, temperature and elevated CO_2: productivity and water relations. Environ. Exp. Bot. 55, 130–141.

Ashraf, M.A., Hussain, I., Rasheed, R., Iqbal, M., Riaz, M., Arif, M.S., 2017. Advances in microbe-assisted reclamation of heavy metal contaminated soils over the last decade: a review. J. Environ. Manag. 198, 132–143.

Augé, R.M., Stodola, A.J., Tims, J.E., Saxton, A.M., 2001. Moisture retention properties of a mycorrhizal soil. Plant Soil 230, 87–97.

Avand, M., Moradi, H.R., Ramazanzadeh Lasboyee, M., 2021. Spatial prediction of future flood risk: an approach to the effects of climate change. Geosciences 11, 25. https://doi.org/10.3390/geosciences11010025.

Azizian, A., Amin, S., Maftoun, M., Emam, Y., Noshadi, M., 2013. Response of corn to cadmium and drought stress and its potential use for phytoremediation. J. Agric. Sci. Technol. 15, 357–368.

Baghour, M., Moreno, D.A., Víllora, G., Hernández, J., Castilla, N., Romero, L., 2001. Phytoextraction of Cd and Pb and physiological effects in potato plants (*Solanum tuberosum* var. Spunta): importance of root temperature. J. Agric. Food Chem. 49, 5356–5363.

Bauddh, K., Singh, R.P., 2012. Growth, tolerance efficiency and phytoremediation potential of *Ricinus communis* (L.) and *Brassica juncea* (L.) in salinity and drought affected cadmium contaminated soil. Ecotoxicol. Environ. Saf. 85, 13–22.

Brett Runion, G., Allen Torbert, H., Prior, S.A., Rogers, H.H., 2009. Effects of elevated atmospheric carbon dioxide on soil carbon in terrestrial ecosystems of the southeastern United States. In: Soil Carbon Sequestration and the Greenhouse Effect. vol. 57. American Society of Agronomy, Madison, WI, pp. 233–262.

Brunham, W., Bendell, L.I., 2011. The effect of temperature on the accumulation of cadmium, copper, zinc, and lead by *Scirpus acutus* and *Typha latifolia*: a comparative analysis. Water Air Soil Pollut. 219, 417–428.

Cao, S., Wang, W., Wang, F., Zhang, J., Wang, Z., Yang, S., Xue, Q., 2016. Drought-tolerant *Streptomyces pactum* Act12 assist phytoremediation of cadmium-contaminated soil by *Amaranthus hypochondriacus*: great potential application in arid/semi-arid areas. Environ. Sci. Pollut. Res. 23, 14898–14907.

Centritto, M., Jarvis, P.G., 1999. Long-term effects of elevated carbon dioxide concentration and provenance on four clones of Sitka spruce (*Picea sitchensis*). II. Photosynthetic capacity and nitrogen use efficiency. Tree Physiol. 19, 807–814.

Compant, S., Van Der Heijden, M.G., Sessitsch, A., 2010. Climate change effects on beneficial plant–microorganism interactions. FEMS Microbiol. Ecol. 73, 197–214.

Conant, R.T., Ryan, M.G., Agren, G.I., Birge, H.E., Davidson, E.A., Eliasson, P.E., et al., 2011. Temperature and soil organic matter decomposition rates–synthesis of current knowledge and a way forward. Glob. Chang. Biol. 17, 3392–3404.

Cornu, J.Y., Schneider, A., Jezequel, K., Denaix, L., 2011. Modelling the complexation of Cd in soil solution at different temperatures using the UV-absorbance of dissolved organic matter. Geoderma 162, 65–70.

Cornu, J.Y., Denaix, L., Lacoste, J., Sappin-Didier, V., Nguyen, C., Schneider, A., 2016. Impact of temperature on the dynamics of organic matter and on the soil-to-plant transfer of Cd, Zn and Pb in a contaminated agricultural soil. Environ. Sci. Pollut. Res. 23, 2997–3007.

Curtis, P.S., Wang, X., 1998. A meta-analysis of elevated CO_2 effects on woody plant mass, form, and physiology. Oecologia 113, 299–313.

D'Surney, S.J., Smith, M.D., 2005. Chemicals of Environmental Concern. pp. 526–530, https://doi.org/10.1016/B0-12-369400-0/00206-4.

de Boer, W., Folman, L.B., Summerbell, R.C., et al., 2005. Living in a fungal world: impact of fungi on soil bacterial niche development. FEMS Microbiol. Rev. 29, 795–811.

de Silva, N.D.G., Cholewa, E., Ryser, P., 2012. Effects of combined drought and heavy metal stresses on xylem structure and hydraulic conductivity in red maple (*Acer rubrum* L.). J. Exp. Bot. 63, 5957–5966.

Denchak, M., 2017. Global Climate Change: What You Need to Know. NRDC. Available at: https://www.nrdc.org/stories/global-climate-change-what-you-need-know. (Accessed 26 January 2021).

Deslippe, J.R., Hartmann, M., Mohn, W.W., Simard, S.W., 2011. Long-term experimental manipulation of climate alters the ectomycorrhizal community of *Betula nana* in Arctic tundra. Glob. Chang. Biol. 17, 1625–1636.

Dhankher, O.P., Foyer, C.H., 2018. Climate resilient crops for improving global food security and safety. Plant Cell Environ. 41, 877–884.

Dier, M., Meinen, R., Erbs, M., Kollhorst, L., Baillie, C.K., Kaufholdt, D., Kücke, M., Weigel, H.J., Zörb, C., Hänsch, R., Manderscheid, R., 2018. Effects of free air carbon dioxide enrichment (FACE) on nitrogen assimilation and growth of winter wheat under nitrate and ammonium fertilization. Glob. Chang. Biol. https://doi.org/10.1111/gcb.13819 (Assessed 10 March 2021).

Dusenge, M.E., Duarte, A.G., Way, D.A., 2019. Plant carbon metabolism and climate change: elevated CO2 and temperature impacts on photosynthesis, photorespiration and respiration. New Phytol. 221, 32–49.

Dutta, H., Dutta, A., 2016. The microbial aspect of climate change. Energy Ecol. Environ. 1, 209–232.

El Massah, S., Omran, G., 2015. Would climate change affect the imports of cereals? The case of Egypt. In: Leal Filho, W. (Ed.), Handbook of Climate Change Adaptation. Springer, Heidelberg, https://doi.org/10.1007/978-3-642-38670-1_61.

Erice, G., Irigoyen, J.J., Pérez, P., Martínez-Carrasco, R., Sánchez-Díaz, M., 2006. Effect of elevated CO2, temperature and drought on photosynthesis of nodulated alfalfa during a cutting regrowth cycle. Physiol. Plant. 126, 458–468.

Fahimipour, A.K., Hartmann, E.M., Siemens, A., Kline, J., Levin, D.A., Wilson, H., Betancourt-Román, C.M., Brown, G.Z., Fretz, M., Northcutt, D., Siemens, K.N., 2018. Daylight exposure modulates bacterial communities associated with household dust. Microbiome 6, 1–13.

Faria, T., Wilkins, D., Besford, R.T., Vaz, M., Pereira, J.S., Chaves, M.M., 1996. Growth at elevated CO2 leads to down-regulation of photosynthesis and altered response to high temperature in *Quercus suber* L. seedlings. J. Exp. Bot. 47, 1755–1761.

Fernández, D.A., Roldán, A., Azcón, R., Caravaca, F., Bååth, E., 2012. Effects of water stress, organic amendment and mycorrhizal inoculation on soil microbial community structure and activity during the establishment of two heavy metal-tolerant native plant species. Microb. Ecol. 63, 794–803.

Fernandez, V., Barnaby, J.Y., Tomecek, M., Codling, E.E., Ziska, L.H., 2018. Elevated CO_2 may reduce arsenic accumulation in diverse ecotypes of *Arabidopsis thaliana*. J. Plant Nutr. 41, 645–653.

Frey, S.D., Drijber, R., Smith, H., Melillo, J., 2008. Microbial biomass, functional capacity, and community structure after 12 years of soil warming. Soil Biol. Biochem. 40, 2904–2907.

Fritioff, Å., Kautsky, L., Greger, M., 2005. Influence of temperature and salinity on heavy metal uptake by submersed plants. Environ. Pollut. 133 (2), 265–274.

Glick, B.R., 2010. Using soil bacteria to facilitate phytoremediation. Biotechnol. Adv. 28, 367–374.

González-Moscoso, M., Rivera-Cruz, M.D.C., Trujillo-Narcía, A., 2019. Decontamination of soil containing oil by natural attenuation, phytoremediation and chemical desorption. Int. J. Phytoremediation 21, 768–776.

Goufo, P., Pereira, J., Moutinho-Pereira, J., Correia, C.M., Figueiredo, N., Carranca, C., Rosa, E.A., Trindade, H., 2014. Rice (*Oryza sativa* L.) phenolic compounds under elevated carbon dioxide (CO_2) concentration. Env. Exp. Bot. 99, 28–37.

Grčman, H., Velikonja-Bolta, Š., Vodnik, D., Kos, B., Leštan, D., 2001. EDTA enhanced heavy metal phytoextraction: metal accumulation, leaching and toxicity. Plant Soil 235, 105–114.

Greve, P., Roderick, M.L., Ukkola, A.M., Wada, Y., 2019. The aridity index under global warming. Environ. Res. Lett. 14, 124006.

Guenet, B., Lenhart, K., Leloup, J., Giusti-Miller, S., Pouteau, V., Mora, P., Nunan, N., Abbadie, L., 2012. The impact of long-term CO2 enrichment and moisture levels on soil microbial community structure and enzyme activities. Geoderma 170, 331–336.

Guo, J., Feng, R., Ding, Y., Wang, R., 2014. Applying carbon dioxide, plant growth-promoting rhizobacterium and EDTA can enhance the phytoremediation efficiency of ryegrass in a soil polluted with zinc, arsenic, cadmium and lead. J. Environ. Manag. 141, 1–8.

Guo, B., Dai, S., Wang, R., Guo, J., Ding, Y., Xu, Y., 2015. Combined effects of elevated CO_2 and Cd-contaminated soil on the growth, gas exchange, antioxidant defense, and Cd accumulation of poplars and willows. Environ. Exp. Bot. 115, 1–10.

Hasegawa, H., Chatterjee, A., Cui, Y., Chatterjee, A.K., 2005. Elevated temperature enhances virulence of *Erwinia carotovora* subsp. *carotovora* strain EC153 to plants and stimulates production of the quorum sensing signal, N-acyl homoserine lactone, and extracellular proteins. Appl. Environ. Microbiol. 71, 4655–4663.

Hatfield, J.L., Prueger, J.H., 2015. Temperature extremes: effect on plant growth and development. Weather Clim. Extrem. 10, 4–10.

Hirabayashi, Y., Mahendran, R., Koirala, S., Konoshima, L., Yamazaki, D., Watanabe, S., Kim, H., Kanae, S., 2013. Global flood risk under climate change. Nat. Clim. Chang. 3, 816–821.

Högy, P., Keck, M., Niehaus, K., Franzaring, J., Fangmeier, A., 2010. Effects of atmospheric CO_2 enrichment on biomass, yield and low molecular weight metabolites in wheat grain. J. Cereal Sci. 52, 215–220.

Huang, S., Jia, X., Zhao, Y., Bai, B., Chang, Y., 2017. Elevated CO2 benefits the soil microenvironment in the rhizosphere of *Robinia pseudoacacia* L. seedlings in Cd-and Pb-contaminated soils. Chemosphere 168, 606–616.

Ings, J., Mur, L.A., Robson, P.R., Bosch, M., 2013. Physiological and growth responses to water deficit in the bioenergy crop *Miscanthus* x *giganteus*. Front. Plant Sci. 4, 468. https://doi.org/10.3389/fpls.2013.00468.

IPCC, 2013. Climate change 2013: the physical science basis. In: Stocker, T.F., Qin, D., Plattner, G.-K., Tignor, M., Allen, S.K., Boschung, J., Midgley, P.M. (Eds.), Contribution of Working Group I to the Fifth Assessment Report of the Intergovernmental Panel on Climate Change. Cambridge University Press, Cambridge and New York, NY.

IPCC, 2014. Climate change 2014: synthesis report. In: Core Writing Team, Pachauri, R.K., Meyer, L.A. (Eds.), Contribution of Working Groups I, II and III to the Fifth Assessment Report of the Intergovernmental Panel on Climate Change. IPCC, Geneva. 151 pp.

Jarvis, A.J., Mansfield, T.A., Davies, W.J., 1999. Stomatal behaviour, photosynthesis and transpiration under rising CO_2. Plant Cell Environ. 22, 639–648.

Jia, Y., Tang, S.R., Ju, X.H., Shu, L.N., Tu, S.X., Feng, R.W., Giusti, L., 2011. Effects of elevated CO2 levels on root morphological traits and Cd uptakes of two Lolium species under Cd stress. J Zhejiang Univ Sci B 12, 313–325.

Jia, X., Zhao, Y.H., Liu, T., He, Y.H., 2017. Leaf defense system of *Robinia pseudoacacia* L. seedlings exposed to 3 years of elevated atmospheric CO2 and Cd-contaminated soils. Sci. Total Environ. 605, 48–57.

Jia, X., Zhang, C., Zhao, Y., Liu, T., He, Y., 2018. Three years of exposure to lead and elevated CO2 affects lead accumulation and leaf defenses in *Robinia pseudoacacia* L. seedlings. J. Hazard. Mater. 349, 215–223.

Jia, X., Li, X.D., Zhao, Y.H., Wang, L., Zhang, C.Y., 2019. Soil microbial community structure in the rhizosphere of *Robinia pseudoacacia* L. seedlings exposed to elevated air temperature and cadmium-contaminated soils for 4 years. Sci. Total Environ. 650, 2355–2363.

Kaisermann, A., Maron, P.A., Beaumelle, L., Lata, J.C., 2015. Fungal communities are more sensitive indicators to non-extreme soil moisture variations than bacterial communities. Appl. Soil Ecol. 86, 158–164.

Khan, Z.S., Rizwan, M., Hafeez, M., et al., 2019. The accumulation of cadmium in wheat (*Triticum aestivum*) as influenced by zinc oxide nanoparticles and soil moisture conditions. Environ. Sci. Pollut. Res. 26, 19859–19870. https://doi.org/10.1007/s11356-019-05333-5.

Khanboluki, G., Hosseini, H.M., Holford, P., Moteshare zadeh, B., Milham, P.J., 2018. Effect of elevated atmospheric CO_2 concentration on growth and physiology of wheat and sorghum under cadmium stress. Commun. Soil Sci. Plant Anal. 49, 2867–2882.

Kools, S.A., Berg, M.P., Boivin, M.E.Y., Kuenen, F.J., van der Wurff, A.W., van Gestel, C.A., van Straalen, N.M., 2008. Stress responses investigated; application of zinc and heat to terrestrial model ecosystems from heavy metal polluted grassland. Sci. Total Environ. 406, 462–468.

Lange, O.L., Kappen, L., Schulze, E.D. (Eds.), 2012. Water and Plant Life: Problems and Modern Approaches. vol. 19. Springer Science & Business Media.

Larcher, W., 1995. Photosynthesis as a tool for indicating temperature stress events. In: Ecophysiology of Photosynthesis. Springer, Berlin, Heidelberg, pp. 261–277.

Ledley, T.S., Sundquist, E.T., Schwartz, S.E., Hall, D.K., Fellows, J.D., Killeen, T.L., 1999. Climate change and greenhouse gases. Eos Trans. Am. Geophys. Union 80, 453–458.

Li, T., Tao, Q., Liang, C., Shohag, M.J.I., Yang, X., Sparks, D.L., 2013. Complexation with dissolved organic matter and mobility control of heavy metals in the rhizosphere of hyperaccumulator *Sedum alfredii*. Environ. Pollut. 182, 248–255.

Lindsey, R., 2020. Climate Change: Atmospheric Carbon Dioxide. https://www.climate.gov/news-features/understanding-climate/climate-change-atmospheric-carbon-dioxide. (Assessed 19 December 2021).

Lipson, D.A., Wilson, R.F., Oechel, W.C., 2005. Effects of elevated atmospheric CO2 on soil microbial biomass, activity, and diversity in a chaparral ecosystem. Appl. Environ. Microbiol. 71, 8573–8580.

Long, S.P., Ainsworth, E.A., Leakey, A.D., Nösberger, J., Ort, D.R., 2006. Food for thought: lower-than-expected crop yield stimulation with rising CO_2 concentrations. Science 312, 1918–1921. https://doi.org/10.1016/j.plantsci.2020.110432.

Luo, J., Yang, G., Igalavithana, A.D., He, W., Gao, B., Tsang, D.C., Ok, Y.S., 2019. Effects of elevated CO2 on the phytoremediation efficiency of *Noccaea caerulescens*. Environ. Pollut. 255, 113169. https://doi.org/10.1016/j.envpol.2019.113169.

Ma, Y., Prasad, M.N.V., Rajkumar, M., Freitas, H., 2011. Plant growth promoting rhizobacteria and endophytes accelerate phytoremediation of metalliferous soils. Biotechnol. Adv. 29, 248–258.

Ma, Y., Oliveira, R.S., Freitas, H., Zhang, C., 2016a. Biochemical and molecular mechanisms of plant-microbe-metal interactions: relevance for phytoremediation. Front. Plant Sci. 7, 918. https://doi.org/10.3389/fpls.2016.00918.

Ma, Y., Rajkumar, M., Zhang, C., Freitas, H., 2016b. Inoculation of *Brassica oxyrrhina* with plant growth promoting bacteria for the improvement of heavy metal phytoremediation under drought conditions. J. Hazard. Mater. 320, 36–44.

Ma, Y., Rajkumar, M., Moreno, A., Zhang, C., Freitas, H., 2017. Serpentine endophytic bacterium *Pseudomonas azotoformans* ASS1 accelerates phytoremediation of soil metals under drought stress. Chemosphere 185, 75–85.

Majhi, P.K., Kothari, R., Arora, N.K., Pandey, V.C., Tyagi, V.V., 2021. Impact of pH on pollutional parameters of textile industry wastewater with use of *Chlorella pyrenoidosa* at lab-scale: a green approach. Bull. Environ. Contam. Toxicol. https://doi.org/10.1007/s00128-021-03208-5.

Marulanda, A., Barea, J.M., Azcón, R., 2009. Stimulation of plant growth and drought tolerance by native microorganisms (AM fungi and bacteria) from dry environments: mechanisms related to bacterial effectiveness. J. Plant Growth Regul. 28, 115–124.

Meehl, G.A., Arblaster, J.M., Tebaldi, C., 2007. Contributions of natural and anthropogenic forcing to changes in temperature extremes over the United States. Geophys. Res. Lett. 34. https://doi.org/10.1029/2007GL030948.

Mesnoua, M., Mateos-Naranjo, E., Pérez-Romero, J.A., Barcia-Piedras, J.M., Lotmani, B., Redondo-Gómez, S., 2018. Combined effect of Cr-toxicity and temperature rise on physiological and biochemical responses of *Atriplex halimus* L. Plant Physiol. Biochem. 132, 675–682.

Miransari, M., 2011. Hyperaccumulators, arbuscular mycorrhizal fungi and stress of heavy metals. Biotechnol. Adv. 29, 645–653.

Mishra, T., Pandey, V.C., Praveen, A., Singh, N.B., Singh, N., Singh, D.P., 2020. Phytoremediation ability of naturally growing plant species on the electroplating wastewater-contaminated site. Environ. Geochem. Health 42, 4101–4111. https://doi.org/10.1007/s10653-020-00529-y.

Muehe, E.M., Wang, T., Kerl, C.F., Planer-Friedrich, B., Fendorf, S., 2019. Rice production threatened by coupled stresses of climate and soil arsenic. Nat. Commun. 10, 1–10.

Naylor, D., Coleman-Derr, D., 2018. Drought stress and root-associated bacterial communities. Front. Plant Sci. 8, 2223. https://doi.org/10.3389/fpls.2017.02223.

Nelson, J.A., Morgan, J.A., LeCain, D.R., et al., 2004. Elevated CO2 increases soil moisture and enhances plant water relations in a long-term field study in semi-arid shortgrass steppe of Colorado. Plant Soil 259, 169–179.

Nowak, R.S., Ellsworth, D.S., Smith, S.D., 2004. Functional responses of plants to elevated atmospheric CO2–do photosynthetic and productivity data from FACE experiments support early predictions? New Phytol. 162, 253–280.

Nsanganwimana, F., Pourrut, B., Mench, M., Douay, F., 2014. Suitability of Miscanthus species for managing inorganic and organic contaminated land and restoring ecosystem services. A review. J. Environ. Manag. 143, 123–134.

Oremland, R.S., Stolz, J.F., 2003. The ecology of arsenic. Science 300, 939–944.

Pandey, V.C., 2020a. Phytomanagement of Fly Ash. Elsevier, Amsterdam, p. 334, https://doi.org/10.1016/C2018-0-01318-3.

Pandey, V.C., 2020b. Fly ash properties, multiple uses, threats, and management: an introduction. In: Phytomanagement of Fly Ash. Elsevier, Amsterdam, pp. 1–34, https://doi.org/10.1016/B978-0-12-818544-5.00001-8.

Pandey, V.V., Bajpai, O., 2019. Phytoremediation: from theory toward practice. In: Pandey, V.C., Bauddh, K. (Eds.), Phytomanagement of Polluted Sites: Market Opportunities in Sustainable Phytoremediation. Elsevier, Amsterdam, pp. 1–49, https://doi.org/10.1016/B978-0-12-813912-7.00001-6.

Pandey, V.C., Bauddh, K., 2019. Phytomanagement of Polluted Sites: Market Opportunities in Sustainable Phytoremediation. Elsevier, Amsterdam, p. 602.

Pandey, V.C., Souza-Alonso, P., 2019. Market opportunities in sustainable phytoremediation. In: Pandey, V.C., Bauddh, K. (Eds.), Phytomanagement of Polluted Sites: Market Opportunities in Sustainable Phytoremediation. Elsevier, Amsterdam, pp. 51–82, https://doi.org/10.1016/B978-0-12-813912-7.00002-8.

Pandey, V.C., Pandey, D.N., Singh, N., 2015. Sustainable phytoremediation based on naturally colonizing and economically valuable plants. J. Clean. Prod. 86, 37–39.

Pascual, I., Antolín, M.C., García, C., Polo, A., Sánchez-Díaz, M., 2004. Plant availability of heavy metals in a soil amended with a high dose of sewage sludge under drought conditions. Biol. Fertil. Soils 40, 291–299.

Pathak, S., Agarwal, A.V., Pandey, V.C., 2020. Phytoremediation—a holistic approach for remediation of heavy metals and metalloids. In: Pandey, V.C., Singh, V. (Eds.), Bioremediation of Pollutants: From Genetic Engineering to Genome Engineering. Elsevier, Amsterdam, pp. 3–16, https://doi.org/10.1016/B978-0-12-819025-8.00001-6.

Pérez-Esteban, J., Escolástico, C., Moliner, A., Masaguer, A., 2013. Chemical speciation and mobilization of copper and zinc in naturally contaminated mine soils with citric and tartaric acids. Chemosphere 90, 276–283.

Pietrini, F., Bianconi, D., Massacci, A., Iannelli, M.A., 2016. Combined effects of elevated CO2 and Cd-contaminated water on growth, photosynthetic response, Cd accumulation and thiolic components status in *Lemna minor* L. J. Hazard. Mater. 309, 77–86.

Pilon-Smits, E., 2005. Phytoremediation. Annu. Rev. Plant Biol. 56, 15–39.

Pourghasemian, N., Ehsanzadeh, P., Greger, M., 2013. Genotypic variation in safflower (Carthamus spp.) cadmium accumulation and tolerance affected by temperature and cadmium levels. Environ. Exp. Bot. 87, 218–226.

Praveen, A., Pandey, V.C., 2020. Pteridophytes in phytoremediation. Environ. Geochem. Health 42, 2399–2411. https://doi.org/10.1007/s10653-019-00425-0.

Prudhomme, C., Jakob, D., Svensson, C., 2003. Uncertainty and climate change impact on the flood regime of small UK catchments. J. Hydrol. 277, 1–23.

Rajkumar, M., Sandhya, S., Prasad, M.N.V., Freitas, H., 2012. Perspectives of plant-associated microbes in heavy metal phytoremediation. Biotechnol. Adv. 30, 1562–1574.

Rajkumar, M., Prasad, M.N.V., Swaminathan, S., Freitas, H., 2013. Climate change driven plant–metal–microbe interactions. Environ. Int. 53, 74–86.

Salt, D.E., Blaylock, M., Kumar, N.P., Dushenkov, V., Ensley, B.D., Chet, I., Raskin, I., 1995. Phytoremediation: a novel strategy for the removal of toxic metals from the environment using plants. Biotechnology 13, 468–474.

Sardans, J., Peñuelas, J., Estiarte, M., 2008. Changes in soil enzymes related to C and N cycle and in soil C and N content under prolonged warming and drought in a Mediterranean shrubland. Appl. Soil Ecol. 39, 223–235.

Scheff, J., Frierson, D.M., 2015. Terrestrial aridity and its response to greenhouse warming across CMIP5 climate models. J. Clim. 28, 5583–5600.

Seong, M.G., Min, S.K., Kim, Y.H., Zhang, X., Sun, Y., 2021. Anthropogenic greenhouse gas and aerosol contributions to extreme temperature changes during 1951–2015. J. Clim. 34, 857–870.

Sessitsch, A., Kuffner, M., Kidd, P., Vangronsveld, J., Wenzel, W.W., Fallmann, K., Puschenreiter, M., 2013. The role of plant-associated bacteria in the mobilization and phytoextraction of trace elements in contaminated soils. Soil Biol. Biochem. 60, 182–194.

Shah, K., Singh, P., Nahakpam, S., 2013. Effect of cadmium uptake and heat stress on root ultrastructure, membrane damage and antioxidative response in rice seedlings. J. Plant Biochem. Biotechnol. 22, 103–112.

Song, N., Zhang, X., Wang, F., Zhang, C., Tang, S., 2012. Elevated CO2 increases Cs uptake and alters microbial communities and biomass in the rhizosphere of *Phytolacca americana* Linn (pokeweed) and *Amaranthus cruentus* L. (purple amaranth) grown on soils spiked with various levels of Cs. J. Environ. Radioact. 112, 29–37.

Ström, L., Owen, A.G., Godbold, D.L., Jones, D.L., 2002. Organic acid mediated P mobilization in the rhizosphere and uptake by maize roots. Soil Biol. Biochem. 34, 703–710.

Tang, S., Xi, L., Zheng, J., Li, H., 2003. Response to elevated CO_2 of Indian mustard and sunflower growing on copper contaminated soil. Bull. Environ. Contam. Toxicol. 71, 988–997.

Tang, S., Liao, S., Guo, J., Song, Z., Wang, R., Zhou, X., 2011. Growth and cesium uptake responses of *Phytolacca americana* Linn. and *Amaranthus cruentus* L. grown on cesium contaminated soil to elevated CO2 or inoculation with a plant growth promoting rhizobacterium Burkholderia sp. D54, or in combination. J. Hazard. Mater. 198, 188–197.

Tang, L., Hamid, Y., Sahito, Z.A., Gurajala, H.K., He, Z., Yang, X., 2019a. Effects of CO_2 application coupled with endophyte inoculation on rhizosphere characteristics and cadmium uptake by *Sedum alfredii* Hance in response to cadmium stress. J. Environ. Manag. 239, 287–298.

Tang, X., Song, Y., He, X., Yi, L., 2019b. Enhancing phytoremediation efficiency using regulated deficit irrigation. Pol. J. Environ. Stud. 28, 2399–2405.

Tennant, T., Wu, L., 2000. Effects of water stress on selenium accumulation in tall fescue (*Festuca arundinacea* schreb) from a selenium-contaminated soil. Arch. Environ. Contam. Toxicol. 38, 32–39.

Tomaz, A., Palma, P., Alvarenga, P., Gonçalves, M.C., 2020. Soil salinity risk in a climate change scenario and its effect on crop yield. In: Climate Change and Soil Interactions. Elsevier, pp. 351–396.

Tyagi, S., Singh, R., Javeria, S., 2014. Effect of climate change on plant-microbe interaction: an overview. Eur. J. Mol. Biotechnol. 5, 149–156.

Van Gestel, C.A., 2008. Physico-chemical and biological parameters determine metal bioavailability in soils. Sci. Total Environ. 406, 385–395.

Wu, Q.S., Zou, Y.N., 2009. Mycorrhiza has a direct effect on reactive oxygen metabolism of drought-stressed citrus. Plant Soil Environ. 55, 436–442.

Wu, H., Tang, S., Zhang, X., Guo, J., Song, Z., Tian, S., Smith, D.L., 2009. Using elevated CO2 to increase the biomass of a *Sorghum vulgare* × *Sorghum vulgare* var. *sudanense* hybrid and *Trifolium pratense* L. and to trigger hyperaccumulation of cesium. J. Hazard. Mater. 170, 861–870.

Xiong, J., Wu, L., Tu, S., Van Nostrand, J.D., He, Z., Zhou, J., Wang, G., 2010. Microbial communities and functional genes associated with soil arsenic contamination and the rhizosphere of the arsenic-hyperaccumulating plant *Pteris vittata* L. Appl. Environ. Microbiol. 76, 7277–7284.

Xu, L., Naylor, D., Dong, Z., Simmons, T., Pierroz, G., Hixson, K.K., Kim, Y.M., Zink, E.M., Engbrecht, K.M., Wang, Y., Gao, C., 2018. Drought delays development of the sorghum root microbiome and enriches for monoderm bacteria. Proc. Natl. Acad. Sci. 115, E4284–E4293.

Yan, A., Wang, Y., Tan, S.N., Yusof, M.L.M., Ghosh, S., Chen, Z., 2020. Phytoremediation: a promising approach for revegetation of heavy metal-polluted land. Front. Plant Sci. 11. https://doi.org/10.3389/fpls.2020.00359.

Yu, X.Z., Trapp, S., Zhou, P.H., Hu, H., 2005. The effect of temperature on the rates of cyanide metabolism of two woody plants. Chemosphere 59, 1099–1104.

Yu, X.Z., Peng, X.Y., Xing, L.Q., 2010. Effect of temperature on phytoextraction of hexavalent and trivalent chromium by hybrid willows. Ecotoxicology 19, 61–68.

Zarch, M.A.A., Sivakumar, B., Malekinezhad, H., Sharma, A., 2017. Future aridity under conditions of global climate change. J. Hydrol. 554, 451–469.

Zickfeld, K., Solomon, S., Gilford, D.M., 2017. Centuries of thermal sea-level rise due to anthropogenic emissions of short-lived greenhouse gases. Proc. Natl. Acad. Sci. 114, 657–662.

Ziska, L.H., Bunce, J.A., Shimono, H., Gealy, D.R., Baker, J.T., Newton, P.C., Reynolds, M.P., Jagadish, K.S., Zhu, C., Howden, M., Wilson, L.T., 2012. Food security and climate change: on the potential to adapt global crop production by active selection to rising atmospheric carbon dioxide. Proc. R. Soc. B Biol. Sci. 279, 4097–4105.

Plant responses toward climatic stressors individually and in combination with soil heavy metals

2

Chapter outline

1 Introduction

Climate change is known to induce various environmental stresses such as drought, heat, low temperature, salinity, flood, pest, and pathogen infestation, etc. These stresses and their complex interactions can directly or indirectly influence plant growth and development (Alnsour and Ludwig-Müller, 2015; Dusenge et al., 2019). During evolution, plants have evolved general and specific mechanisms toward various stresses leading to adaptation and survival under various stressful conditions (Raza et al., 2019). Different species have shown different acclimation capacities, partly dependent on the climate to which they are adapted (Somero, 2010).

Adaptation of plants to abiotic and biotic stresses involves the activation of molecular network cascades associated with perception of stress, signal transduction, and the expression of general and specific stress-linked genes and metabolites (Huang et al., 2012). A key step in plant defense is the well-timed stress perception to respond rapidly and efficiently. After recognizing stress, the plants' basal defense mechanisms activate complex defense signaling cascades differing among stresses (Rejeb et al., 2014). Activation of specific kinases, nitric oxide (NO), reactive oxygen species (ROS), and ion channels; accumulation of various phytohormones such as salicylic acid (SA), abscisic acid (ABA), gibberellic acid (GA), cytokinin (CK), auxin (AUX), ethylene (ET), and jasmonic acid (JA); and genetic machinery reprogramming lead to defense responses and enhanced plant tolerance to reduce the harmful effect of stresses (Sharma and Dubey, 2006, 2008, 2011, 2019; Kumar et al., 2008; Spoel and Dong, 2008;

Adaptive Phytoremediation Practices. https://doi.org/10.1016/B978-0-12-823831-8.00003-7

Sharma et al., 2009; Fraire-Velázquez et al., 2011; Rejeb et al., 2014; Huang et al., 2019; Jha et al., 2017; Jha and Sharma, 2019; Zhao et al., 2021a).

Many downstream responses, such as accumulation of osmolytes [glycine betaine (GB), proline, sugars] and heat shock proteins (HSPs), and activation of antioxidant defense system are common for different stresses (Lamers et al., 2020). Enhanced generation of ROS is an important indicator of environmental stress in plants. Excessive accumulation of ROS can cause cellular damage. Antioxidant molecules like ascorbate (AsA) and glutathione (GSH) and enzymes, such as superoxide dismutase (SOD), catalase (CAT), peroxidase (POD), glutathione reductase (GR), and glutathione-s-transferase (GST) present in plants, keep ROS concentrations restricted and hence contribute in the maintenance of cellular homeostasis under environmental stress (Sharma and Dubey, 2007; Sharma et al., 2010, 2012, 2019). In addition to general stress responses, specific stress results in the activation of specific events (Haak et al., 2017).

Combination of climatic stressors and soil heavy metal stress is bound to lead to a much more complex scenario. Multistress-tolerant plant will be beneficial for phytoremediation under climate change. For developing resilient crop for phytoremediation in changing climate, it is important to understand plant responses to abiotic or biotic stresses. Combination of these stresses can have a negative and additive effect on plants. It has been observed that the plants facing one stress and showing good defending capability can resist other stresses well (cross-tolerance) (Bowler and Fluhr, 2000). Plant's response to multiple stresses is specific, and prediction is not possible on the basis of response of plants to individual stresses (Atkinson and Urwin, 2012). Therefore, studying plant responses to climatic stressors individually and in combination with soil heavy metals is important to facilitate the development of crops with enhanced phytoremediation efficiency under changing climate (Colmenero-Flores and Rosales, 2014). For this, unraveling stress tolerance-associated gene resources and metabolites from plants is important. Detailed understanding of plant responses to stresses individually and in combination with soil heavy metals will help in developing plant varieties with high phytoremediation efficiency (Fig. 1).

2 Coping against abiotic stress condition

2.1 High-temperature tolerance

Global warming has enhanced the occurrence of extreme high-temperature events. It is projected to negatively affect the growth of plants (Bita and Gerats, 2013). High temperature exerts various effects on plants in terms of physiology, biochemistry, and gene regulation pathways, which causes the altered growth and development of plants. Plants' survival under heat stress depends on their ability to perceive stimulus, generate and transmit signals, and initiate suitable physiological and biochemical alterations. Plants possess various strategies to persist under high temperatures. These mechanisms are categorized into long-duration morphological and phenological evolutionary adaptations like altering orientation of leaf, transpirational cooling, or changes in the composition of membrane lipid, or short-term stress avoidance and acclimation

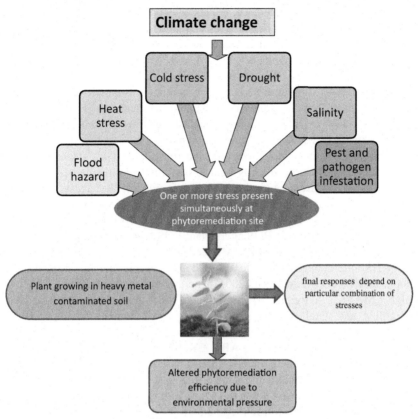

Fig. 1 Various abiotic stresses and biotic stresses usually arise together in the field; it is important to develop climate-resilient crops to reduce the pressure of environmental changes and increase the phytoremediation efficiency by identifying tolerance mechanisms in plants.

mechanisms (Hasanuzzaman et al., 2013). General mechanisms of heat stress tolerance are the production of stress proteins, osmoprotectants, ion transporters, antioxidants, and various factors associated with signaling cascades and transcriptional control that are involved in counteracting effects of stresses (Rodríguez et al., 2005). Maintenance of higher photosynthetic rates, leaf gas exchange, increased membrane thermostability, and heat avoidance have been correlated with high-temperature tolerance in plants (Scafaro et al., 2010; Bhattarai et al., 2021). In response to heat stress, heat shock TFs (HSFs) activate a group of genes that encode HSPs. Heat stress also induces the endoplasmic reticulum (ER)-localized unfolded protein response (UPR), which activates TFs that upregulate a different family of stress-responsive genes. It activates hormonal responses and alternative splicing of RNA which participate in thermotolerance (Li and Howell, 2021). Saidi et al. (2011) suggested a model in which plasma membrane and different secondary messengers, such as Ca^{2+} ions, NO, and H_2O_2, act as primary sensory role and calmodulins, MAPKs, and Hsp90 act as downstream components for activating HSFs. The co-regulatory investigation demonstrated novel genes as well as pathways [ferredoxin-PODs, NADP reductase, MAPKKK, and HSF-Hsp network] that are associated with heat tolerance. Enhanced expression of HSP and HSF family members in the thermotolerant rice hybrid compared to its parents suggests their role in maintenance of higher antioxidative defense systems and photosynthesis in hybrid, indicating that the intricate regulatory network involving HSF and HSP contributes to heat tolerance in hybrid rice (Wang et al., 2020).

High temperatures can cause a change in membrane fluidity, which can lead to the activation of Ca^{2+} channels present on the plasma membrane (Finka et al., 2012) triggering Ca^{2+} influx. Ca^{2+} stimulates apoplastic ROS overproduction by activating NADPH oxidase. The Ca^{2+}/calmodulin signaling pathway facilitates heat stress response and leads to thermotolerance (Wu and Jinn, 2012). Ca^{2+}-dependent protein kinases (CDPKs) and CIPKs directly bind to Ca^{2+}, calcium-regulated proteins, and regulate NADPH oxidase activity through phosphorylation (Kobayashi et al., 2007; Drerup et al., 2013). In Arabidopsis, NADPH oxidase functional loss caused an impairment in heat stress response (Larkindale et al., 2005; Miller et al., 2009). As a signaling molecule, apoplastic ROS alters plant ROS removal capacity by regulating the transcription of antioxidant genes (Mittler et al., 2004). Heat stress responses mediated by HSF are believed to be closely related to sensing and signal amplification by apoplastic ROS (Davletova et al., 2005). Thermotolerant wheat *cv*. HD3086, which had higher MAPK activity and expression under heat stress, demonstrated enhanced H_2O_2, proline, and starch accumulation and integrity of granules in comparison with thermosusceptible *cv*. BT-Schomburgk (Kumar et al., 2021).

Several plant hormones, including ABA, CK, SA, JA, AUX, and ET, play an essential role in providing thermotolerance to plants (Kotak et al., 2007; Lubovská et al., 2014). ABA induction is the main factor of thermotolerance (Maestri et al., 2002). ABA induces ZmCDPK7, which functions both upstream and downstream of RBOH and is involved in tolerance to high temperature in maize by facilitating sHSP17.4 phosphorylation, which might be essential for its chaperone function (Zhao et al., 2021c). Pan et al. (2006) observed that the inhibition of SA synthesis under heat stress led to not only reduction in endogenous SA concentration but also in heat tolerance level. HSPs and ET signaling-related genes are involved in the complex signal transduction network for heat tolerance (Wu and Yang, 2019). ET has a key role in plant heat stress responses and participates in thermotolerance (Poór et al., 2021). Exogenous treatment of ET provides thermotolerance in plants by triggering several stress-related proteins involved in maintenance of plant cells' functional integrity and stability (Jegadeesan et al., 2018). ET signaling enhanced thermotolerance through oxidative stress reduction and chlorophyll content maintenance (Wu and Yang, 2019). ET that is rapidly produced during HS is regulated by both SA and JA, suggesting that JA acts with SA, providing basal thermotolerance (Clarke et al., 2009). CK reduced oxidative stress leading to increased thermotolerance in plants. Heat tolerance by CK in creeping bentgrass requires complex mechanisms that operate at the transcriptional, posttranscriptional, and posttranslational levels (Xu et al., 2010). In Arabidopsis and tomato, brassinosteroids provide high-temperature tolerance by enhancing major HSPs biosynthesis (Ogweno et al., 2008; Bajguz and Hayat, 2009). Similarly, BRASSINAZOLE RESISTANT 1 (BZR1) regulates BR response and participates in heat stress tolerance. BZR1 regulation of heat stress responses in tomato is via ROS signaling, which is RBOH1 dependent (Yin et al., 2018). In *Pinus radiata*, SA and ABA play a key role in heat-induced thermotolerance in the first phase of heat stress response. It may be due to urgent need of plants for the regulation of stomatal closure and counteraction of enhancement in oxidative membrane damage observed in shorter-duration exposures. However, during longer-duration exposures and recovery, total sugars, proline, IAA, and CKs appeared to be more important (Escandón et al., 2016).

One of the key components of heat tolerance is the maintenance of membrane integrity. Heat stress leads to oxidative burst that enhances the lipid peroxidation of membrane and hence leads to a reduction in membrane thermostability and an enhancement in electrolyte leakage (Hasanuzzaman et al., 2013; Khan et al., 2013). NO is a free radicle and thus reacts with superoxide or lipid hydroperoxyl radicals and protects membranes (Bloodsworth et al., 2000). NO functions as signaling molecule and acts upstream of AtCaM3 and thus plays a role in thermotolerance by increasing HSF DNA-binding activity and HSP accumulation (Xuan et al., 2010). NO also functions as a promotor of proline biosynthesis leading to thermotolerance (Alamri et al., 2019).

Osmolytes accumulation in plants under high temperature is a key adaptive mechanism (Sakamoto and Murata, 2000). One of the immediate responses of heat is a decrease in water potential of leaf, which can be counterbalanced by early synthesis of GB, proline, and soluble sugars (Wahid, 2007). The tight correlation between ψ_p increase and proline, GB, and sugar accumulation indicated their probable association with osmotic adjustment in heat-stressed plants (Wahid and Close, 2007). Enhanced concentration of GB, proline, and soluble sugars is important for regulating osmotic potential and protecting cellular structures from high temperatures by maintaining cell water balance, stabilizing membrane, and buffering redox potential of the cells (Farooq et al., 2008). GB accumulation is common phenomenon in plants under high temperature (Wahid and Close, 2007; Wang et al., 2010). Under high temperature, GB production in chloroplasts sustains the Rubisco activation by confiscating Rubisco activase nearby the thylakoids and thus avoiding inactivation due to high temperature (Sakamoto and Murata, 2002). GB overaccumulation increased tolerance of the wheat photosynthetic apparatus to high temperature (Wang et al., 2010). It can alleviate the harmful impacts of moderate heat stress on photosystem II repairing during photoinhibition (Allakhverdiev et al., 2007).

At anthesis stage, depression in the production of osmoprotectants, antioxidants, and chlorophyll concentrations and overproduction of malondialdehyde were observed. Exogenous proline application enhanced the concentration of leaf proline and GB; increased activities of SOD, POD, CAT, and total soluble proteins; and decreased lipid peroxidation causing enhancement in contents of chlorophyll and per plant yield (Hanif et al., 2021). Activities of GST and CAT increased in both heat-tolerant (Mv Magma) and heat-sensitive (Plainsman V) wheat varieties exposed to heat stress. APX activity increased only in Plainsman V. Wheat varieties tolerant toward heat stress seemed to be related to the antioxidant level (Balla et al., 2009). Greater increase in enzyme activity in Plainsman V. compared to Mv Magma was correlated with higher heat sensitivity (Balla et al., 2009). SA application altered the activities of antioxidant enzyme and antioxidant concentration and conferred thermotolerance in the leaves of grape plant (Wang and Li, 2006). Among the three mulberry cultivars studied, cv. BC2-59 had comparatively efficient antioxidant system, which was correlated with the reduced oxidative damage in the leaves due to heat stress. Heat-tolerant rice genotype (IET 22218) showed higher accumulation of polyamines, antioxidant enzyme activity, and lower H_2O_2 accumulation, and membrane damage under heat stress. It was concluded that the enhanced content of polyamines and activity of antioxidant enzymes in IET 22218 under heat stress may lead to lower oxidative stress and sustained higher viability of pollen and fertility of spikelets (Karwa et al., 2020).

Modulation of starch and sucrose metabolism by SA and availability of carbohydrate during heat stress stimulates thermotolerance in plants (Liu and Huang, 2000; Kaur et al., 2019). Some sugars act as signaling molecules and antioxidants (Sugio et al., 2009; Lang-Mladek et al., 2010). Secondary metabolites, including flavonoids, phenolics, anthocyanins, and plant steroids, are important for plant responses toward heat stress and are associated with tolerance to heat (Wahid, 2007). It was predicted that the accumulation of soluble phenolics enhanced the activity of phenylalanine ammonia lyase and decreased the activity of PODs and polyphenol oxidase under high temperature, which may be related to the heat acclimation in tomato plants (Rivero et al., 2001). Transient increase in concentration of phenolic acid during short-duration heat shock indicates its role as a signaling molecule, whereas a dramatic increase under long duration indicates its role as a protectant (Wang et al., 2019a). Wahid (2007) suggested that changes in spreading of alteration in metabolite levels over space and time were crucial in enhancing net assimilation and tolerance to heat in sugarcane. Accumulation of anthocyanin under high temperature decreased the osmotic potential of leaf, causing enhanced uptake of water, but inhibited the transpirational loss (Wahid, 2007). Zeaxanthin participates in heat stress tolerance by reducing the membrane oxidative damage (Meiri et al., 2010).

Late embryogenesis abundant (LEA) proteins play vital roles in providing protection from heat stress. LEA proteins can lead to the prevention of aggregation and protection of enzymes and proteins. Prolonged retention of dehydrins (DHNs) in the leaves of CP-4333, heat-tolerant than HSF-240, heat-sensitive clone of *Saccharum officinarum*, appeared to be an adaptive benefit to tolerate heat stress (Galani et al., 2013). Repairing of heat-damaged components gives plant's capability to counteract heat stress. *HSFs* induce the *HSPs* expression rapidly, and both *HSFs* and *HSPs* have a role in thermotolerance. However, single *HSF* or *HSP* gene overexpression has very little effect on thermotolerance, indicating that *HSFs* and *HSPs* synergistically act to provide HS resistance (Zhao et al., 2021b. These HSPs are conserved and act as chaperons and have a multipurpose function during several stresses, including heat stress. They participate in maintaining the integrity of cell membrane, scavenging of ROS and osmolyte, and antioxidant production (Khan and Shahwar, 2020). ClpB-C confers thermotolerance in plants (Lee et al., 2006), and it is essential for plants' survival under heat stress. On the contrary, other ClpB proteins were not as much of significance (Mishra and Grover, 2019). ClpB proteins participate in basal and induced heat stress tolerance (Panzade et al., 2020). Synthesis of ubiquitin and conjugated ubiquitin in the first 30 min of heat treatment in soybean and mesquite was suggested as key heat tolerance mechanism (Huang and Xu, 2008). SlCHIP ubiquitin E3 ligase plays a key role in responses to heat stress in tomato probably by aiming the degradation of proteins misfolded during heat stress (Zhang et al., 2021).

2.2 Drought tolerance

The climate change is going to change the intensity and frequency of droughts (Strzepek et al., 2010; Cook et al., 2018; Mukherjee et al., 2018). Plant growth is significantly affected by water deficit. Some plants respond, adapt, and survive under drought condition by inducing several morphological, physiological, and biochemical responses (Nguyen et al., 1997; Kavar et al., 2007; Sharma and Dubey, 2005; Li et al., 2018; Liu et al., 2020a). Plants resist drought by escape, avoidance, and tolerance mechanisms. Drought escape is done via shortened life cycle, which allows plants to do reproduction before drought ensues (Araus et al., 2002). Drought avoidance reduces the loss of water from plants by controlling stomatal transpiration and maintaining uptake of water via deep and large root systems (Turner et al., 2001; Kavar et al., 2007). An ability to maintain growth, flower, and have economic yield under drought stress is known as drought tolerance.

Drought-tolerant plant shows changes at molecule, cell, tissue, physiology, and entire plant level. Single or numerous alterations such as a decrease in water loss through enhanced stomatal resistance, increased water uptake by the development of extensive root systems, and osmolyte accumulation regulate plant capacity to develop and stay alive under restricted water conditions. Roots, which are the sole source to get water through soil, are the important plant organ for drought adaptation (Kavar et al., 2007). Extensive root system with long length, high biomass, and density maintains favorable plant water status (Nguyen et al., 1997). Deep, thick, and extensive root system which can access water present in deep soil are considered to provide drought resistance to upland rice (Kavar et al., 2007). Decrease in area and number of leaf because of drought is associated with cutting down of water budget. Amount of cuticular wax has been associated with drought tolerance (Zhou et al., 2013; Guo et al., 2016). Heavy water loss results in mechanical disruption in some plants. In resurrection plants, tissue shrinks correspondingly to the water loss. Therefore, anatomical features such as curling of leaf and vessels that are flexible are considered important for drought tolerance (Abrams, 1990; Mansoor et al., 2019). General reactions to drought include stress detection, signaling, and initiation of acclimation and defense pathways. Drought tolerance involves Ca^{2+}, ROS, different kinases, hormones, and cross-talk between many TFs (Wrzaczek and Hirt, 2001).

Osmotic adjustment, ABA, and DHNs help in supporting high-water potential and thus bestowing drought tolerance to plants. ROS scavenging enzymes scavenge excess ROS produced during restricted water supply and hence confer drought tolerance. The miRNAs, which are 20–24-nt regulatory RNAs, have been involved in posttranslational regulation under drought (Bakhshi et al., 2016). Drought enhances cytoplasmic Ca^{2+} content in cells. Enhanced Ca^{2+} increased drought tolerance by enhancing γ-glutamyl kinase and reducing the activity of proline oxidase. Numerous potential calcium sensors, including SOS3 and CDPKs, may further transduce stress-induced calcium signals. ABA, SA, JA, AUX, CK, GA, ET, and BR hormones play a key part in tolerance of drought. Under drought, the concentration of CK, AUX, and GA usually decreases, whereas that of ET and ABA increases (Nilsen and Orcutte, 1996). ABA increases drought tolerance via regulating stomata and development of roots and initiating ABA-dependent pathway (Ullah et al., 2018). CK and ABA play contrasting roles under drought. ABA increase and CK decline cause closure of stomata and restricted water loss via transpirational regulation under drought (Morgan et al., 1990). ABA has been linked to an increased ratio of root-to-shoot dry weight, reduction in leaf area development, and development of extensive root systems. AUX breaks CK-induced apical dominance, thus inducing new root formation. Usually, endogenous auxins production is lowered when concentrations of ABA and ET increase. YUCCA participates in auxin biosynthesis. *AtYUC6* overexpression led to increased drought tolerance in potato via the regulation of ROS homeostasis (Park et al., 2013). Exogenous IAA application was shown to enhance stomatal conductance and net photosynthesis in cotton (Kumar et al., 2001). Improvement in white clover drought tolerance due to exogenous IAA was associated with endogenous concentration of plant hormone and alteration of genes participating in leaf senescence and responses to drought (Zhang et al., 2020).

CPK6 ABFs/AREBs phosphorylation facilitates ABA signaling and hence drought tolerance (Zhang et al., 2020). OsbZIP42 (bZIP TFs) participates in positive regulation of ABA signaling and provides ABA-dependent tolerance to drought in rice plant (Joo et al., 2019). Under drought, ET helps plant to optimize growth and provides tolerance (Pierik et al., 2007). ET controls leaf performance and also participates in drought-induced senescence (Young et al., 2004). TFs including WRKYs are also known to play a role in drought tolerance. In soybean, ABA receptor and gene SnRK2 kinase are activated by GmWRKY54. Other WRKY TFs usually regulate downstream genes participating in ABA pathway (Wei et al., 2019). Li et al. (2018) suggested that ethylene-responsive factor (OsERF71) provides drought tolerance by enhancing the expression of genes contributing in ABA signaling and biosynthesis of proline. Under drought stress, extensive root system is imperative for tolerance.

Osmoadjustment and osmoprotection are important bases for drought tolerance. The concentration of osmolytes has been shown to rise under drought and correlated well with drought tolerance. Osmolytes such as proline, GB, sucrose, and trehalose have been associated with the prevention of protein denaturation, detoxification of ROS, and stabilization of membranes. Their accumulation results in lower osmotic potential and helps in turgor. Rice possess a great ability to increase polyamine biosynthesis in leaves, particularly spermidine (SPD) and spermine (SPM) in free form and putrescine (PUT) in insoluble-conjugated form, as early response to drought, and this is viewed as a significant physiological trait associated with drought tolerance (Yang et al., 2007). Under drought condition, excessive ROS production in organelles, including chloroplasts, mitochondria, and peroxisomes, is an important factor responsible for impaired plant growth and productivity. The ROS target the peroxidation of cell membrane lipids and degradation of proteins, enzyme, and nucleic acids.

Raised antioxidant status is key strategy to tolerate drought in plants. Tolerant citrus rootstocks showed low H_2O_2 and malondialdehyde (MDA) content and high antioxidant enzyme activities (CAT, SOD, and POD) to deal with ROS generated during drought (Hussain et al., 2018). The high concentration of amino acids, like glycine, lysine, and histidine, can scavenge ROS by oxidative modification. Drought-induced accumulation of osmolytes and enhanced nonenzymatic and enzymatic defense systems reduced the harmful effect in maize hybrid Dong Dan 80 (Anjum et al., 2017). In *Cerasus humilis* plants, genotypic differences in drought-resistant (HR) and susceptible (ND4) have been attributed to reduced membrane injury because of lesser accumulation of ROS and MDA and ability to enhance antioxidant defense (increased SOD, AsA-GSH cycle, AsA, cytosolic APX, DHAR) under drought conditions in HR (Ren et al., 2016).

Expression of aquaporins and stress proteins are key mechanisms of drought tolerance. Increasing evidences in many crop plants point that AQPs are involved in fine control of water movement and thus play a significant role in providing tolerance to drought (Zargar et al., 2017). Stress proteins are mostly water soluble and are involved in cellular structure hydration and hence drought tolerance (Wahid and Close, 2007). DREB genes play a key role in drought tolerance. They belong to the erythroblastosis virus repressor factor gene family of TFs. Subclass DREB1 gene is inducible by cold, whereas DREB2 is inducible by dehydration (Choi et al., 2002). These genes are involved in abiotic stress signaling pathway. In plants, overexpression of DREB genes has led to increased drought tolerance (Kasuga et al., 1999; Yamaguchi-Shinozaki and Shinozaki, 2006). In transgenic *Festuca arundinacea*, increased expression of DREB1A gene led to an increased concentration of proline and drought tolerance (Zhao et al., 2007).

Drought stress leads to alteration in the expression levels of LEA/DHN-type genes and molecular chaperones that provide protection to the cellular proteins from denaturation. HSPs are chaperons and play a key role under drought tolerance. They are involved in ATP-dependent stabilization (protein unfolding or assembly) that prevent protein denaturation during stress (Gorantla et al., 2006). For drought tolerance in plants, LEA and membrane-stabilizing proteins are also important. These proteins sequester ions concentrated due to dehydration and also enhance water binding capacity (Gorantla et al., 2006). LEA proteins accumulate under drought stress and play crucial roles in drought tolerance in plants. DHNs bind membranes and proteins nonspecifically and have the ability to protect the structure and function of the membrane or protein from damage due to environmental stresses. They have lysine-rich domain, which is highly conserved and associated with hydrophobic interactions, resulting in macromolecule stabilization. DHNs have the capability to bind DNA and may repair or provide protection to DNA against environmental stresses (Liu et al., 2017). In wheat, TF TabHLH49 positively regulated the expression of DHN (WZY2) gene and increased drought tolerance (Liu et al., 2020a).

2.3 Salinity tolerance

Climate change creates a high risk of soil salinization (Dasgupta et al., 2015; Corwin, 2021). Various mechanisms exist in plants for tolerating high concentrations of salt. Alterations in several morphological, anatomical, biochemical, molecular, and physiological mechanisms (Poljakoff-Mayber, 1975; Hameed et al., 2009) lead to salt tolerance. Plant salinity tolerance may include limited salt uptake, tissue tolerance, salt accumulation in vacuoles, ion discrimination, and production of protective osmolytes, hormones, enzymes, antioxidants, etc. (Gorham and Wyn Jones, 1990; Munns and Tester, 2008). Morphological characteristics of the plant roots can restrict the entry of salts in plants in large quantities. Thick epidermis and sclerenchyma, developed bulliform cells, enhanced trichomes density, and increased capacity to retain moisture by increasing the size of the cell and volume of vacuoles are some of the morphological and anatomical adaptations of the plants in response to salinity (Akcin et al., 2015). Excretory structures including salt glands and vesicular hairs are key structural adaptation, which is vital for salt tolerance (Hameed et al., 2010).

Three important factors that influence plant growth under salinity stress are ion toxicity, water stress, and uptake and translocation of nutrients (Flowers et al., 1986; Flowers and Colmer, 2008). Various alterations at cell, organelle, and molecular level and whole plant level lead to salt tolerance in plants. Physiological and metabolic characteristics of plants can counter salts if they enter roots (Winicov, 1998). Salts at high concentrations disrupt the cells' ionic homeostasis. Maintenance of Na^+/K^+ homeostasis is a crucial mechanism for the survival of plants exposed to salinity stress. Avoidance via passive ion exclusion or ion dilution to keep salts away from metabolically active tissues (Allen et al., 1994; Munns and Tester, 2008) and (ii) vacuolar compartmentalization of salts (Munns, 2002) are mechanisms critical for reducing the harmful impacts of high concentrations of toxic ions. These mechanisms may cause variations in salt tolerance within species or genotypes. SOS signaling pathway plays a key part in homeostasis of ion and tolerance to salts (Yokoi et al., 2002; Zhu, 2003). It is situated downstream of cytosolic Ca^{2+} increase due to salt induction in roots and consists of three main proteins, SOS1, SOS2, and SOS3 (Mahajan et al., 2008; Ji et al., 2013). Ca^{2+} controls intracellular homeostasis of Na^+ along with SOS proteins. SOS3, which is a member of calcineurin B-like protein (CBL) family, acts as a sensor of Ca^{2+} and its interaction with CBL-interacting protein kinases (CIPKs) SOS2/CIPK24 freeing its self-inhibition and cause its activation (Halfter et al., 2000; Liu et al., 2000). Complex of SOS2 and SOS3 phosphorylates SOS1 on the plasma membrane. The SOS1 once phosphorylated causes an enhanced Na^+ efflux, reducing Na^+ toxicity (Quintero et al., 2011). Contrary to CBL4 which is particularly involved in stress tolerance, CBL10 is a multifunctional Ca^{2+} sensor. CBL10 plays a role in salt tolerance by controlling fluxes of Na^+ in vacuoles (Egea et al., 2018). In poplar, tonoplast localized two of the CBL10 homologs function under salinity by interacting with SOS2 in shoot (Tang et al., 2014). Interaction of CBL10 and CIPK6 with RBOHB localized at the plasma membrane regulates ROS production during effector-elicited immunity (de la Torre et al., 2013). Feng et al. (2018) investigated and suggested FERONIA receptor kinase association with Arabidopsis salinity tolerance. They showed that during the growth of plant exposed to salinity, weakening of cell wall activates FERONIA-mediated Ca^{2+} signaling to stop root from bursting. The FERONIA extracellular domain leads to physical contact with pectin, exhibiting a probable sensing mechanism.

Several genes including HKT (histidine kinase transporter) and NHX, which encode K^+ channels and transporters, are involved in homeostasis in many plant species.

Expressions of these genes involved in K^+ uptake increase under salt stress in the halophyte *Mesembryanthemum crystallinum* (Tran et al., 2020). HKT family transporters located on the plasma membrane regulate the transport of Na^+ and K^+ and play a key role in salt tolerance. Class 1 HKT transporters have been shown to prevent the excess accumulation of Na^+ in leaves in Arabidopsis. In rice, HKT transporter protects photosynthetic tissues by removing excess Na^+ from the xylem (Hauser and Horie, 2010). NHX proteins (intracellular) are Na^+, K^+/H^+ antiporters. They are involved in homeostasis of K^+, regulation of endosomal pH, and tolerance to salt. NHX isoforms NHX1 and NHX2, which are located on tonoplast, are considered necessary for active K^+ uptake for the regulation of turgor pressure and proper function of stomata.

Compatible solutes (osmolytes), which are low molecular weight compounds, are accumulated under salt stress. They are uncharged, polar, and water-soluble compounds that do not affect biochemical reactions even at very high concentrations (Singh et al., 2020). Concentrations of amino acids, including arginine, cysteine, and methionine, decreased, whereas the concentration of proline increased under salinity. Proline accumulation provides salinity tolerance and works as nitrogen reserve under stress recovery. It acts as O_2 quencher, thus showing an antioxidant capability (Matysik et al., 2002). It also increases the activity of antioxidative enzymes (Hoque et al., 2008). GB spraying on leaves led to the stabilization of pigment and enhanced photosynthetic rate under salinity stress (Cha-Um and Kirdmanee, 2010; Ahmad et al., 2013). Polyols, including sorbitol, mannitol, myo-inositol, or its methylated derivatives, are linked to salinity tolerance (Bohnert et al., 1995). They protect or stabilize membrane structures or enzymes that are dehydration-sensitive or are damaged due to enzymes. Carbohydrates, including sugars such as glucose, trehalose, fructans, and fructose, and starch, are found to be accumulated under salinity (Dubey and Singh, 1999; Pattanagul and Thitisaksakul, 2008). The key roles of these carbohydrates in salinity stress mitigation include carbon storage, osmoprotection, and ROS scavenging.

Plants exposed to environmental stresses, including salinity, generate ROS in large quantity, and these ROS are efficiently eliminated by antioxidant enzymes. Many studies have found differences in levels of expression or activity of antioxidant enzymes; these differences are sometimes associated with the more tolerant genotype and sometimes with the more sensitive genotype. In the sensitive variety (VA3) of *Amaranthus tricolor*, higher accumulation of ROS was observed than in the tolerant variety (VA14) due to salinity treatment (Sarker and Oba, 2020). Content of total carotenoids, proline, AsA, total flavonoid (TFC), total phenolic (TPC), and total antioxidant capacity (TAC) increased in both the varieties. In comparison with the tolerant variety, the rise in proline concentration was much more in the sensitive variety. Nonenzymatic antioxidants such as AsA, carotenoids, TPC, TFC, and TAC, and antioxidative enzymes APX and SOD appeared to play an important role in H_2O_2 detoxification in *A. tricolor* when H_2O_2 load was less. These were accompanied by guaiacol peroxidase (GPX) and CAT activities when H_2O_2 load was higher (in the salt-sensitive variety). Genotypic differences observed in the activity of antioxidant enzymes among different genotypes might be because of degree of stomatal closure and other parameters that change CO_2 fixation rate and photoinhibition rate (Munns and Tester, 2008). Pea plants treated with 0.15 M NaCl revealed an enhanced activity of S-nitrosylated APX and APX, and content of H_2O_2, S-nitrosothiol, and NO that can explain the increase in APX activity (Begara-Morales et al., 2014). Under salt stress, H_2O_2 and NO are proposed to serve as intermediary molecules in increasing salt resistance of *Populus euphratica* calluses by enhancing the K/Na ratio, due to the increased PM H^+-ATPase activity (Zhang et al., 2007).

Polyamines, which are aliphatic low molecular weight nitrogenous bases having two or more amino functional groups, play an important role under environmental stress (Chen et al., 2019). Their concentration is regulated by enzymes diamine oxidase and polyamine oxidase, which play a key role in providing tolerance to salinity (Takahashi and Kakehi, 2010; Cona et al., 2006). Biosynthesis and catabolism of polyamine is regulated by ABA (Shevyakova et al., 2013). Polyamines are associated with the preservation of integrity of membrane, control of expression of genes, biosynthesis of osmolytes, less generation of ROS, and control of Na^+ and Cl^- accumulation in various organs (Tisi et al., 2008; Navakoudis et al., 2003). Increased concentration of polyamines has been correlated with salinity stress tolerance (Begara-Morales et al., 2014). Under salinity stress, an enhanced expression of enzymes PUT synthesis and spermine synthase, which are involved in SPM synthesis, was observed, whereas mutant in polyamine biosynthetic genes shows salt sensitivity (Yamaguchi et al., 2006). Enhanced salt tolerance is reported in plants overproducing PUT, SPM, and SPD in tobacco, rice, and Arabidopsis (Roy and Wu, 2002).

Salinity stress causes water deficit and osmotic stress in plants. Enhanced ABA production is a key cellular signal for salt stress in roots and shoots. Accumulation of ABA (endogenous) is generally higher in the stress-tolerant varieties compared to their sensitive counterparts. ABA is a common intermediary in stress-induced signal transduction, involved in the regulation of ion channels or alterations in the expression of several stress-inducible genes that cause major alterations in proteome expressions. The positive correlation between the accumulation of ABA and tolerance to salinity has been partly credited to an increase in the concentrations of K^+, Ca^{2+}, some hydrophilic proteins like LEA and osmolytes like sugars and proline and, in vacuoles of roots, which counteract with the uptake of Na^+ and Cl^- and the modulation of expression of several genes linked to salt and water deficit response (Gupta and Huang, 2014). LEA proteins bind and replace H_2O_2, sequester ions, maintain protein and membrane structure, act as molecular chaperones, stabilize the membrane, and mediate the transportation of specific molecules to the nucleus. Upstream sequences of ABA inducible genes showed ABA-responsive element (ABRE). Fukuda and Tanaka (2006) demonstrated the effects of ABA on the expression of two genes, *HVP1* and *HVP10*, vacuolar H^+-inorganic pyrophosphatase and HvVHA-A, a catalytic subunit A of vacuolar H^+-ATPase in barley under salinity stress.

In wheat, MAPK4-like, tonoplast intrinsic protein-1 (TIP 1), and germin-like protein-1 (GLP 1) expression increased due to ABA treatment under salinity stress (Keskin et al., 2010). It was reported that SA induces tolerance to salinity in Arabidopsis via the restoration of membrane potential and a reduction of K^+ due to salinity through guard cell outward-rectifying K(+) (GORK) channels (Jayakannan et al., 2015). Arabidopsis seedling pretreated with SA showed upregulation of H^+-ATPase activity, thereby improving K^+ retention during salt stress; SA pretreatment did not prevent Na^+ accumulation in the roots but in the shoots, it somehow reduced Na^+ accumulation (Jayakannan et al., 2015). SA treatment alleviated damaging impacts of salinity and improved plant performance in two cultivars of Ethiopian mustard. Antioxidative enzyme activity increased in a concentration-dependent mode. Increase in the antioxidant enzyme activity in normal or salt-treated plants due to SA treatment suggests their key part in controlling redox balance and defending the plants from oxidative damage (Husen et al., 2018). SA improved salinity-induced injury to foliar function, growth of plant, and antioxidant defense system in *Brassica carinata*. BR application improved the activities of antioxidative enzymes like SOD, APX, GPX, and POD and the buildup of nonenzymatic antioxidant compounds such as ascorbate, reduced glutathione, and tocopherol. EBR reduced the damaging effects of salinity stress in young plants of *Eucalyptus urophylla*, refining homeostasis associated with K^+/Na^+ ratio and the nutrient concentration in the tissues. EBR is beneficial for growth and gas exchange and enhances the stomatal density, spongy parenchyma, and palisade parenchyma, suggesting that the EBR exogenous application provides salinity tolerance.

SNF-1 group of serine/threonine kinase activates/phosphorylates OSBZ8 in SPD presence during salt stress. Expression of OSBZ8 was higher in cultivars, which were salt-tolerant than salt-sensitive cultivar (Mukherjee et al., 2006). DREB1/CBF, DREB2, and AREB/ABF play a key role in salinity tolerance (Nakashima et al., 2000, 2009; Narusaka et al., 2003). TFs such as ZFP179 and OsNAC5 were upregulated under salinity and controlled biosynthesis and accumulation of LEA proteins, proline, and sugar, which play a vital role in stress tolerance (Sun et al., 2010; Takasaki et al., 2010). AtWRKY8 upregulation was observed in Arabidopsis under salinity. It binds directly with the RD29A promoter and hence is considered as one of the target genes of AtWRKY8 (Hu et al., 2013b). SALT-RESPONSIVE ethylene-responsive factor (SERF1) seems to be important for salt tolerance. It binds with the promoters of DREB2A, MAPK5, MAP3K6, and ZFP179. SERF1-dependent genes were H_2O_2-responsive. SERF1-deficient plants were more salt-sensitive than wild type, and constitutive SERF1 overexpression improved salt tolerance (Schmidt et al., 2013).

Splicing is considered imperative for plant growth, development, and responses of plants to environmental stresses (Laloum et al., 2018). NAC TF overexpression in both wheat and rice provided salt tolerance, thus indicating their involvement in stress tolerance (Jiang et al., 2019). PRMT5/SKB1 belonging to type II protein arginine methyltransferase impacted splicing in Arabidopsis and also affected the development and response to salinity by alteration in status of methylation H4R3sme2 (for symmetric dimethylation of histone H4 arginine 3) and LSm4 (Hernando et al., 2015). Cui et al. (2012) identified an endoplasmic reticulum-related protein degradation constituent (ubiquitin conjugase 32) in Arabidopsis that participates in BR-mediated tolerance to salinity.

Endogenous siRNAs and miRNAs play a vital role in response to salinity stress. Differential expression of various miRNAs is associated with a difference in salinity tolerance (Kulcheski et al., 2011; Li et al., 2011; Sun et al., 2016b). Gma-miR172a targets SSAC1 (AP2/EREBP-type TF) leading to indirect regulation of THI1 encoding a positive regulator of salinity tolerance (Pan et al., 2016). Overexpression of miR172c, which is also a member of miR172 family, resulted in an enhanced survival of salt stress-exposed Arabidopsis plants (Li et al., 2016).

2.4 Cold tolerance

Climate warming-led reduced snow cover might enhance exposure of living organisms to extreme cold (Bale and Hayward, 2010; Kearney, 2020). Tolerance induced by low-temperature exposure is known as cold acclimation and/or chilling tolerance. Enhanced tolerance to the physiochemical and physical impacts of freezing stress is cold acclimation (Guy, 1990; Thomashow, 1999), whereas plant's ability to tolerate low temperatures (0–15°C) without damage is chilling tolerance. Several biochemical, molecular, and biochemical processes are associated with chilling tolerance and cold acclimation (Thomashow, 1999; Zhu et al., 2007; Ganeshan et al., 2011). When plants get exposed to low temperature, they sense it and transmit it through molecules like Ca ion at the cellular level, which leads to alterations in expression of various genes. Altered expression of genes leads to an alteration in structure of the membrane, production of compatible osmolytes, antioxidant enzymes, and cold shock proteins etc., which ultimately help plants to ameliorate the harmful effect of low-temperature stress on the growth and development of plant (Maleki and Ghorbanpour, 2018).

A variety of mechanisms is utilized by temperate plants to avoid/reduce damage by low temperature. Growth repression is one of the key mechanisms to cold stress. It allows relocation of resources from growth to enhance tolerance to cold stress. Cold treatment affects the membrane fluidity, which leads to an enhanced rigidity of membrane, which helps plant cells to sense cold stress. Enhanced rigidity of membrane stimulates cold-responsive genes, which are associated with cold acclimation of plants (Denesik, 2007). Alternation in the plasma membrane structure and function is considered as one of the key mechanisms of low-temperature tolerance (Takahashi et al., 2013). This more fluid state of plant cell membranes protects low-temperature stressed cells by preserving the shape of cell and preventing cellular loss of water from cell components. Enzyme oleate desaturase, which is involved in the synthesis of polyunsaturated fatty acids alpha-linolenate and linoleate, is controlled by fatty acid desaturation 2 (fad2) (Okuley et al., 1994). Arabidopsis mutant *fad2* shows rigidification of membrane and sensitivity to cold stress, indicating a relation between cold stress sensing and plasma membrane rigidification (Vaultier et al., 2006). A direct association between the expression of FAD and desaturation of lipid was observed in cold-sensitive genotypes, whereas an inverse relation was observed in cold-resistant genotype of *Olea europaea*, indicating that cold acclimation in these plants needs a fine posttranscriptional regulation of FAD (Matteucci et al., 2011).

Phytohormone JA acts as a key controlling signal in cold tolerance (Zhao et al., 2013). Its biosynthesis is activated upon cold exposure. Hu et al. (2013a) observed that in Arabidopsis, JA is associated with the regulation of CBF/DRE binding factor1 cascade inducer and improvement of freezing tolerance. It is involved in positive upregulation of CBF transcription pathway for the upregulation of downstream cold-responsive (COR) genes associated with cold tolerance. JA interaction with other plant hormone (Aux, gibberellin, and ethylene) signaling to control leaf senescence and cold tolerance has been observed (Hu et al., 2017). In different plant species, SA accumulation under chilling stress has been observed (Kosová et al., 2012; Dong et al., 2014). Accumulation of SA (endogenous) was associated with chilling tolerance of *Cucumis sativus* seedlings (Dong et al., 2014). Also, SA application enhanced cold tolerance of various plant species (Ignatenko et al., 2019). GA associated with the expression of CRT/DRE-binding factor conferred cold, salt, and drought stress tolerance (Niu et al., 2014). GA also helps in SA/JA ratio in the stress response mediated by CBF (Niu et al., 2014). Although CK content was reduced, its application improved cold tolerance in plant. In grapevine, ethylene concentration was also decreased under low temperature (Sun et al., 2016a). BR-biosynthetic genes transcript and concentration level in *Elymus nutans* indicate that BR has an important role in cold tolerance. In *bak1* mutant, proline concentration was low compared to the wild plant, suggesting an association between proline and BR-facilitated cold tolerance. Also, pretreatment of BR enhanced the cold stress-related gene expression. It has been proposed that BR induced cold-related genes, which cause the regulation of antioxidant defense mechanism and proline accumulation in plants (Fu et al., 2019). Brassinazole-resistant 1 (BZR1) gene is known to act upstream of *CBF1* and *CBF2* and involved in the regulation of their expression. It also regulates *COR* genes not coupled with CBFs like *PYL6*, *WKRY6*, *SOC1*, *SAG21*, and *JMT* to control cold stress response. In accordance, *wrky6* mutants displayed lowered freezing tolerance. BZR1 regulate freezing cold tolerance positively through both CBF-independent and CBF-dependent pathways in plants (Li et al., 2017).

Various TFs change downstream genes expression related to stress tolerance. DREB/CBFs belonging to APETALA2/AP2/ERF TF family play a significant role in cold tolerance (Vazquez-Hernandez et al., 2017). The *CdERF1* from *Cynodon dactylon* controls cold tolerance positively (Hu et al., 2020). Plants with less upregulated *CBF* gene transcripts show low ability to acclimate and tolerate low-temperature stress (Rihan et al., 2017). CBF-dependent responsive pathway operating in plants exposed to low temperature is an important regulator of COR genes. Various evidences indicate the key role of Ca^{2+} signaling in the induction of CBF pathways due to cold (Knight and Knight, 2012). Alteration in fluidity of membrane due to cold stress triggers Ca^{2+} influx into the cell and proposed that this influx could mediate *CBF4* induction resulting in cold hardening. It has been proposed that microtubule dynamics determines the sensitivity of cold-induced Ca^{2+} influx, thus playing a role in the induction of CBF4 resulting in cold hardening (Wang et al., 2019b). Microtubule stabilization has been shown to induce cold acclimation (Hiraki et al., 2019). Inhibition of activity of phosphatase 2A by Ca^{2+} is believed to cause phosphorylation of some proteins that may have roles in the activation of genes playing a role in cold tolerance and acclimation (Monroy et al., 1998). A Ca^{2+}-binding protein, calmodulin, binds to a regulatory element present in the promoter of *CBF*2 gene and controls the CBF regulon and freezing tolerance (Doherty et al., 2009). In plants, CBL family proteins are a distinctive group of Ca^{2+} sensors. Ca^{2+} signal is transmitted by CBLs by interaction with and regulation of the CIPKs family. CBL1 interaction with CIPK7 has been proposed to play a role in cold response (Huang et al., 2011).

Inducer of CBF expression (ICE) is involved in CBFs upstream regulation. CBF signaling pathway components can form an association with other transcriptional repressors and activators under cold stress and thus regulate gene expression (Zhou et al., 2011). RAP2.1 and RAP2.6 act as downstream regulators of CBFs. CBF4 participates in ABA-dependent signaling pathways. In *Medicago falcata*, *MfERF1* provides cold tolerance through the promotion of turnover of polyamine, antioxidants, and accumulation of proline (Zhuo et al., 2018). bZIP TF, MdHY5, positively modulates cold tolerance of plants by CBF-dependent and CBF-independent pathways (An et al., 2017). Two R2R3 MYB TFs, MYB 88 and paralogous FLP (MYB 124), positively controlled cold hardiness and expression of cold-responsive genes under cold stress through CBF-dependent and CBF-independent pathways in both apple and Arabidopsis (Xie et al., 2018). CDPKs induce cross-talk among MPKs, ROS, and NO resulting in the induction of ABA signaling in cold stress leading to adaptation in plants (Lv et al., 2018). In rice, OsCPK17 is needed for a suitable cold stress response probably by modulating sugar metabolism and membrane channels (Almadanim et al., 2017).

Some hydrophilic proteins, for example, DHNs or LEA proteins, expressed by COR gene play a role in the stabilization of membrane and counteraction of protein aggregation (Close, 1996; Hundertmark and Hincha, 2008; Janská et al., 2010). Artus et al. (1996) reported that COR15a gene overexpression improved chloroplast freezing tolerance in Arabidopsis by about $2°C$ in nonacclimated plants. Role of ABA in the induction of expression of LEA gene has been shown (Finkelstein et al., 2002; Bies-Etheve et al., 2008). Additionally, expression of HSPs and some PR proteins, including PR1, 2, 5, 10, 11, and 14, is enhanced under cold stress, indicating their involvement in cold tolerance (Timperio et al., 2008; Seo et al., 2010; Zhang and Denlinger, 2010).

Osmolyte accumulation controls osmotic potential in plants under cold stress (Slama et al., 2015). Osmolytes such as proline, total proteins, total soluble sugars, GB, amino acids, and trehalose showed an enhanced accumulation in 30-day-old *Capsella bursa pastoris* seedlings under cold temperature (Wani et al., 2018). Glycometabolism also has a key role in cold tolerance. Differential expression of gene *TATPS11* (trehalose 6-phosphate synthase 11) was observed in high and low freezing-tolerant cultivars of wheat. Expressing *TaTPS11* in *Arabidopsis thaliana* caused lower sucrose, higher starch, and higher activity of enzymes participating in sucrose metabolism such as sucrose synthase, sucrose phosphate synthase, and invertase (Liu et al., 2019). Further, an increased activity of sucrose synthase in *Brachiaria mutica* roots under flooding may have a key role in flooding adaptation. It represents a greater ability to use sucrose in anaerobic tissue (Ram, 2000). It was proposed that NO might act downstream of H_2O_2 to regulate trehalose-mediated cold-induced oxidative stress in tomato (Liu et al., 2020b).

ROS is commonly overproduced in cold stress. De novo synthesis of antioxidants and activity of antioxidative enzymes have been observed to increase under cold stress (Baek and Skinner, 2012). Differences in the ability of GSH and osmolyte accumulation like glucose, proline, fructose, and arginine were observed in cold-sensitive and cold-tolerant seedlings of *Pinus halepensis* (Taïbi et al., 2018). Ribosome is considered as a regulating node for cold responses. STCH4, which is involved in change in ribosome constitution and functions during cold stress, allows protein translation and thus contributes to the growth and survival of plants (Yu et al., 2020).

2.5 Submergence/flood tolerance

Climate models predict an enhancement in the heavy rainfall intensity and frequency, which could add to rises in precipitation-led local flooding (Kundzewicz et al., 2014). Ethylene production enhanced along with a signaling cascade, including a hormone network, and other general secondary signal molecules are typical response against flooding (Dat et al., 2004). Ethylene induces genes encoding enzymes involved in the formation of aerenchyma, fermentation pathway, and glycolysis. Many of the flood tolerance strategies are regulated by members of ERF family of TFs (Voesenek and Bailey-Serres, 2013). Negative correlations were observed across and within species between conductance of stomata and NO concentration (internal) and flux of NO (Copolovici and Niinemets, 2010). Disturbed mitochondrial retrograde signaling causes an enhanced sensitivity to flood stress, and functional analysis of WRKY45 and WRKY40 using overexpressing and knockout lines revealed their involvement in providing submergence tolerance. Therefore, it was proposed that submergence tolerance depends on mitochondrial retrograde signaling, and re-programming of transcription is an adaptation strategy (Meng et al., 2020). In rice seedlings, physiological and molecular evidences demonstrated the important role of MPK3-*SUBMERGENCE1A1* (*SUB1A1*) module in acclimation to harmful effects of submergence (Singh and Sinha, 2016). *SUB1A-1* was observed to be nonessential for submergence tolerance in wild-type rice genotypes having C-genome. It showed SUB1A-independent response to submergence (Niroula et al., 2012). In rice, a calcineurin OsCIPK15 is associated with hypoxia tolerance. Thorough examination showed that in *A. thaliana*, ERFs are targeted by a conserved O_2-sensing protein turnover mechanism (Lekshmy et al., 2015).

Morphological adaptations against flood involve the development of adventitious roots, hypertrophied lenticels initiation, aerenchyma formation, and shoot elongation. During submergence, maintenance of photosynthesis can sustain metabolism (Ashraf, 2012; Colmer and Flowers, 2008). Activating pathways that generate ATP without oxidative phosphorylation (e.g., glycolysis, sucrose catabolism, fermentation) are also flood tolerance mechanism. Plants under anaerobic condition shift from respiration to fermentation, and genes expressed under flooding stress are mostly enzymes of fermentative pathway. No increase in the production of ethanol and no induction in alcohol dehydrogenase (ADH) were observed in plants unharmed by flood. Such activation may lead to the exclusion of plants from wet areas due to the buildup of toxic amounts of ethanol (Crawford, 1967). Increased activity of ADH in flood-tolerant soybean mutant during early stage of flood than in wild type indicated that activating fermentation pathway in the flooding early stages might be the main factor for the attainment of tolerance to flood (Komatsu et al., 2013). Anaerobic condition during flooding led to the accumulation of mRNA and the selective production of around 20 anaerobic proteins (ANPs), which included enzymes involved in glycolysis and connected processes enzymes like aldolase, ADH, enolase, glyceraldehyde-3-phosphate dehydrogenase, glucose-phosphate isomerase, pyruvate decarboxylase, and sucrose synthase. Xyloglucan endotransglycosylase activated by O_2 deprivation has been associated with aerenchyma development during flooding (Sachs et al., 1996).

Less consumption of energy by inhibiting processes that are energetically costly like biogenesis of ribosome and formation of cell wall also helps plants to overcome flooding stress (Piazzi et al., 2019). NO and nonsymbiotic hemoglobin have been proposed as a substitute for fermentation to sustain low ratio of NADH/NAD (low redox potential) (Igamberdiev and Hill, 2004). Oxygen-deficient surroundings during flooding hamper plant survival, growth, and development considerably. Due to the deprivation of light and oxygen, ROS are produced in excess, which can cause serious harm to cells. ROS scavenging ability after de-submergence enables plants to tolerate flood. Increased activity and expression of antioxidant enzymes and content of nonenzymic antioxidants reduce oxidative stress in plants. Flooding tolerance of R5064-5 and MP-29 *Prunus* rootstocks was associated with higher activities of antioxidant enzymes and concentrations of osmolytes (McGee et al., 2021). Osmotic adjustment is considered as a key strategy of stress tolerance, and osmolyte (proline, GB, GABA, and sugars) accumulation is a general reaction witnessed in various plant systems (Suprasanna et al., 2016). Proline was suggested to be associated with stress tolerance irrespective of the stress intensity (Pellegrini et al., 2020). GABA allows plants to ensue H_2O_2 signaling to stimulate a gene cascade that facilitates adaptation of plants to flooding and averting the cell to enter a "suicide program" (Shabala et al., 2014).

2.6 Heavy metal tolerance

Various mechanisms have been evolved in plants, which enable them to tolerate HM stress (Sharma and Dubey, 2010; Kumar et al., 2019; Singh et al., 2019a, b). Cell wall, the plant cells' outmost structure, encounters HMs first. It acts as blockade and restricts HMs transport in the cytoplasm. HM sequestration in cell wall is considered as HM tolerance mechanisms in plants (Torasa et al., 2019). Root of *Athyrium yokoscense* growing naturally on metalliferous habitat showed approximately 70%–90% of total Cd, Zn, and Cu in the cell wall. Cu showed significantly higher affinity for cell wall than Zn and Cd and was restricted to enter the cytoplasm (Nishizono et al., 1987). In *A. thaliana*, binding of Cd to cell wall, hemicellulose, pectin, and cellulose was considerably greater in Cd-tolerant ecotype Tor-1 compared to ecotype Ph2-23, which was sensitive to Cd (Xiao et al., 2020). Concentrations of hemicellulose, cellulose, and pectin were also higher in Tor-1 compared to Ph2-23.

Once HMs make entry in plant cells, they produce signaling molecules like NO, ET, SA, JA etc., which contribute to tolerance against HMs (Di Toppi and Gabbrielli, 1999; Popova et al., 2012). Roles of H_2O_2 in growth responses, HM toxicity, and its cross-talk with other significant plant growth regulators in regulating several processes have been demonstrated (Nazir et al., 2020). Various transcription factors including MAPKs involved in the transduction of signal in plants attenuate HM toxicity (Nazir et al., 2020). Ligands involved in HM binding like GSH, PCs, MTs, chaperons, amino acids, and organic acids help in immobilization and sequestration of HMs. They play a central role in detoxifying HMs (Clemens, 2006; Viehweger, 2014). GSH, MTs, and PCs demonstrate high metal affinity as they are Cys rich (Chaudhary et al.,

2018). PCs that play a crucial role in mediating HM tolerance are synthesized by enzyme phytochelatin synthase (Clemens, 2006; Gupta et al., 2009). Regulation of PC biosynthesis by HMs at the posttranslational level is documented. MTs are small proteins (5–10 kDa) rich in cysteine having various biological functions, including antioxidative defense, detoxification of HM, and metal homeostasis (Hassinen et al., 2011; Singh et al., 2021a). Some organic acids, including citrate, oxalate, malonate, malate, tartrate, and aconitate, can make a strong bond with HMs via chelation with their carboxyl groups. They can participate in extracellular and intracellular HM chelation. Organic acids secreted by the roots of plant form the extracellular complex with HM, leading to a reduction in their bioavailability, whereas intracellular chelation by organic acid enhances plants' tolerance toward HMs (Osmolovskaya et al., 2018).

Vacuolar sequestration regulated mainly by metal chelators, transporters localized on the vacuolar membrane, and interaction between them is implicated in HM tolerance (Peng and Gong, 2014). Various transporter families like natural resistance-associated macrophage proteins (NRAMPs), Ca^{2+} exchangers (CAXs), ATP-binding cassette C proteins (ABCCs), and heavy metal ATPases (HMAs) have been associated with vacuolar HM sequestration (Korenkov et al., 2007; Park et al., 2012; Martinoia, 2018; Sharma et al., 2018; Zhang et al., 2018). Several HM transporters show rather wide substrate specificity, and plants in soils with high concentrations of a specific HM are required to offset the too much uptake of particular HM (Martinoia, 2018).

HMs are known to produce excessive ROS in plants causing oxidative stress. Antioxidants (enzymatic and nonenzymatic) contribute to HM tolerance by scavenging ROS (Sharma et al., 2012, 2014, 2018, 2019; Sharma and Dubey, 2019; Singh et al., 2021b). Enzymatic component comprises many enzymes like SOD, CAT, POD, APX, and GR. SODs catalyze dismutation of superoxide anion to H_2O_2 (Mishra and Sharma, 2019). Enzymes like CAT, APX, and POD can decompose H_2O_2 to produce water. Activities of these enzymes vary according to the plant species and HM content (Sharma et al., 2012). Nonenzymatic antioxidants, including GSH and AsA, protect plants against oxidative stress brought by HMs (Asgher et al., 2017). HSPs also provide HM tolerance to plant by repairing proteins damaged due to HM stress (Neumann et al., 1994). Fig. 2 shows the various heavy metal tolerance mechanisms in plants.

3 Coping against biotic stresses

3.1 Pest resistance

Plants are regularly challenged by various pests, which can significantly affect the plant productivity. Infestation of pest depends on various climatic factors such as precipitations, relative humidity, temperature, CO_2, and solar radiation. Current knowledge suggests that climate change can affect the pest and plant development and influence their interactions (Castex et al., 2018). Increasing concentrations of CO_2 and consequent alteration in climate such as temperature and precipitation extremes will probably lead to an alteration in pest pressures in managed and natural plant communities. Such alterations can be negative (new introductions) or positive

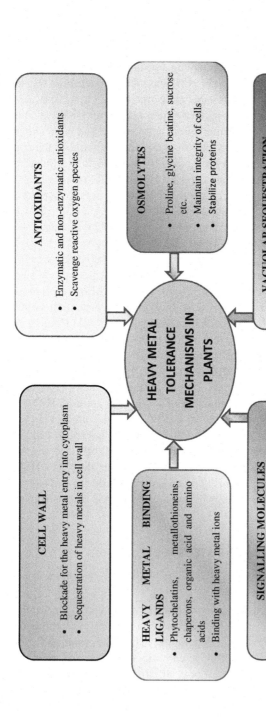

Fig. 2 Heavy metal tolerance mechanisms in plants.

(migration from a region) but probably will be companioned by considerable environmental and economic consequences (Ziska and McConnell, 2016). Also, the performance of pesticide is strongly related to environmental conditions (Taylor et al., 2018). Altered environmental conditions have led to conditional resistance of pest, which is characterized by a decrease in pesticide sensitivity (Matzrafi, 2019).

Plants possess various defense mechanisms to resist insect herbivory. Plant resistance to pests is categorized into (i) antibiosis, (ii) antixenosis, and (iii) tolerance. Antibiosis includes increasing mortality, and reducing longevity, growth, and fecundity of pests by plants (Painter, 1958; Smith, 2005). Antixenosis includes directing the herbivore away from plant due to the presence of unsuitable plant structures or characteristics such as cell wall thickness, surface wax, spines, and trichomes that have the ability to do interference with reproduction, movement, and feeding of insect on or in plants. Both antibiosis and antixenosis lead to a reduction in injury and loss of yield but can put a selection pressure on insect leading to pest resistance. Tolerance recover from injury through herbivores via growing and compensating physiological processes involves only plant responses and hence does not cause pest resistance (Peterson et al., 2017). It maintains insect populations similar to those seen on susceptible plants (Painter, 1958; Panda and Khush, 1995; Smith, 2005). Virulent insect population selection and threat due to evolving biotypes is assumed to be restricted (Koch et al., 2016).

Increased rate of net photosynthesis after injury is one of the key mechanisms through which plants can tolerate herbivory. Plants can tolerate pest by compensating photosynthetically by circumventing feedback inhibition and reduced electron flow via photosystem II, which occurs due to insect feeding. Enhanced photosynthetic activity has been demonstrated in case of tolerance to hemipterans (Luo et al., 2014; Cao et al., 2015). The leaves of tolerant wheat line showed no alteration in chlorophyll fluorescence in uninfested and *Diuraphis noxia* infested leaves, whereas antibiosis and sensitive wheat lines demonstrated reduced chlorophyll fluorescence, indicating that electron transport system was disrupted in susceptible lines decreasing the absorption of light for photosynthesis but not in the wheat lines that were tolerant (Haile et al., 1999).

By delaying the growth and production of fruit and flower, herbivore tolerance is promoted. It helps plants in postponing development of plant until the danger has passed (Tiffin, 2000). In herbivore-tolerant maize, a delay in resource allocation to roots is considered to provide tolerance to western corn rootworm (Robert et al., 2017). Plant vigor has been associated with herbivory tolerance in various plant species (Price, 1991). In red raspberry and sugarcane however, tolerance to aphid herbivory has been associated with an enhanced plant vigor (Karley et al., 2016). Higher abundance and fitness of many insect herbivore groups on vigorous host plants (Cornelissen, 2011) could reveal the capability of vigorous plants to tolerate insect attack. In plant species with multiple meristems, dormant bud activation after elimination or injury to vegetative or flowering meristems is a one kind of mechanism of compensatory growth that permits plant's recovery from herbivore attack (Tiffin, 2000).

ROS is considered as vital early signals that integrate environmental cues and regulate stress tolerance in response to insect attack (Foyer and Noctor, 2005, 2013;

Kerchev et al., 2012). ROS signaling is closely associated with hormonal signaling, and substantial overlap occurs between ROS, SA, and JA pathways (Foyer and Noctor, 2005, 2013; Kwak et al., 2006; Mittler et al., 2011; Kerchev et al., 2012; Santamaria et al., 2013). In the tolerant genotype, constitutive amounts of JA and ABA- and basal JA and ABA-related transcripts were higher (Chapman et al., 2018). CK is involved in initiating the expression of wound-inducible gene and inducing the concentration of antiinsect compounds (Dervinis et al., 2010; Giron et al., 2013). Higher concentrations of it can help in repairing tissue by enhancing division of cells (Giron et al., 2013). CK signaling also influences the availability of nitrogen and micro- and macronutrients, which can critically affect the growth of plants and insects (Dervinis et al., 2010). Both ROS and antioxidant defense system play a key role in SA signaling, PCD regulation, and induction of PR proteins involved in SAR (Foyer and Noctor, 2005, 2016).

Resistance to aphid in *Bemisia tabaci* nymph-infested plants involves enhanced activities of antioxidative enzymes in which CAT could play a key role. This resistance probably acted via interactions with SA-mediated defense responses (Zhao et al., 2016). Higher concentrations of H_2O_2 and ROS in insect harmed Ludwigia species in comparison with unharmed plants indicated that feeding by *Altica cyanea* caused stress in plants. Higher activities of SOD, CAT, APX, and GST in insect harmed Ludwigia species in comparison with unharmed plants indicted that these 4 enzymes were reducing the stress generated in plants because of insect herbivory (Mitra et al., 2019).

3.2 Pathogen stress

Various pathogens such as bacteria, oomycetes, nematodes, and fungi attack naturally grown plants and are responsible for reducing their yield and quality. Climate change can trigger the frequency and severity of pathogen attack on plants (Eastburn et al., 2011; Luck et al., 2011; Burdon and Zhan, 2020). Using different virulence factors, they lead to host defense suppression and the promotion of pathogenicity. Plants have also developed complex defense mechanisms that work against pathogen infections (Lu and Yao, 2018). Resistance, the restriction of pathogen multiplication (constitutive defense mechanism), and tolerance, the reduction of the pathogen multiplication effect on plant (inducible defense mechanism), are the two key mechanisms employed by plants to counteract pathogen infection (Pagán and García-Arenal, 2018). Normally, these two mechanisms coexist. Both constitutive and inducible defense mechanisms act against pathogen in plants. Constitutive defense mechanisms are the first line of defense and help to avert further pathogen invasion (Osbourn, 1996; VanEtten et al., 1994, 2001). These include preformed structural barriers such as lignified cell wall, cutin, waxes and antimicrobial molecules such as phytoanticipins that are nonspecific to any pathogen. Pathogens that are able to cross the first line of defense are counteracted by inducible defense mechanisms. Release of effector proteins or identification of pathogen-associated molecular pattern triggers inducible defense that protects plants by further spreading of pathogen. Hypersensitive response, ROS production, cross-linking of cell wall, and antimicrobial molecules production, such as phytoalexins and PR proteins, constitute inducible defense mechanisms (Van Loon and Van

Strien, 1999; Van Baarlen et al., 2007). Tolerance doesn't exert the selection pressure on pathogen and thus is considered a more stable defense strategy.

A complex network involving signaling pathways including JA, SA, and ET regulates plant defense against microbial attack. The signaling pathways associated with JA and SA are antagonistic and allow plants to adjust their defenses against different kinds of plant pathogens (Kunkel and Brooks, 2002). Their activation cause the accumulation of different PR proteins (chitinase, glucanase, thaumatin, defensin, and thionin). Overexpression of single PR gene or a combination of genes leads to upregulated defense mechanisms in many pathogens. Biotrophic pathogens trigger SA, whereas necrotrophic pathogen triggers JA pathway. Biotrophic pathogen-induced SA pathway activates the transcription of nonexpressor of PR gene 1 (NPR1), which leads to the accumulation of PR1, PR2, and PR5 (SA signature gene products) systematically and locally causing systemic acquired resistance (SAR) to various pathogens. Necrotrophic pathogen-induced JA pathway results in the accumulation of PR3, PR4, and PR12 (JA signature gene products), locally leading to local acquired resistance (LAR) (Ali et al., 2017).

The hypersensitive response (HR) characteristic of higher plants involves a rapid death of cells at the entrance point and is a very effective part of the immune system of the plant. It kills or restricts pathogens causing tissue necrosis, which is linked to pathogen resistance (Balint-Kurti, 2019). HR induces defense mechanism in the distal part of the plant to begin SAR. Pathogen-associated molecular pattern is the prime tier of the innate immunity of plant. HR includes sequential events like transcriptional reprograming, oxidative bursts, phytohormonal signaling, and Ca^{2+} influx. ROS oxidative burst occurs as plants show a hypersensitive response to pathogen (Noman et al., 2020).

Increase in ROS accumulation in response to pathogen infection leads to pathogen resistance (Mittler et al., 1999). During fungal attack, chitin and other elicitors stimulate the plant immune responses that recruit ROS, NO, Ca etc. to counteract pathogen attack and limit its spread to other plant parts (Ali et al., 2018). ROS-mediated defense ensues via cross-talk with other mechanisms (Ali et al., 2018). Alterations in redox status and/or content of GSH and/or AsA and activities of their redox enzymes happen during HR (De Gara et al., 2003). These alterations may not be a simple consequence of oxidative stress induced by ROS, but they appear to be activated as component of the transduction pathways leading to defense responses and PCD (De Gara et al., 2003). Activities of antioxidant enzymes generally increase more in tolerant varieties compared to sensitive varieties. PPO and APX activities were enhanced more in "Vulkan," a fusarium head blight-resistant variety compared to "Kraljica" and "Golubica," which were medium-resistant and medium-susceptible variety, respectively (Spanic et al., 2017). H_2O_2 may show a direct antimicrobial activity at sites where pathogen attacks, trigger PCD that restricts the further spread of infection, lead to phytoalexin synthesis, and cause oxidative cross-linking and lignify proteins rich in hydroxyproline and other polymer present in the cell wall during HR, function as signal in inducing SAR and defense genes (Kuźniak and Urbanek, 2000). As a signaling molecule, H_2O_2 induces genes involved in defense mechanisms like basic and acidic PR proteins, thus causing increased pathogen tolerance (Chamnongpol et al., 1998). However, strong PR protein expression observed during HR happens only when excessive H_2O_2 is companioned by leaf necrosis (Chamnongpol et al., 1998).

4 Plant responses toward climatic stressors in combination with soil heavy metals

Due to climate change, drought, salinity, heavy rainfall (causing flood), temperature extremes etc. are likely to happen more often. Interactions between these stresses and soil heavy metals will influence phytoremediation under changing climatic condition. After perception of stresses that are present simultaneously, interaction between the induced responses is possible and the final responses will be based on the specific combinations of stresses present (Table 1). Combination of stresses can be considered as a new state of stress that generates new type of acclimation responses in plants (Pandey et al., 2017).

Effect of presence of abiotic or biotic stresses along with heavy metals can be antagonistic or synergistic. Combination of heavy metals and drought amplifies the effects of one another (Barceló and Poschenrieder, 1990). In red maple seedlings, an additive effect of these two stresses on the reduction of plant growth was observed. Both drought and heavy metals led to a reduction in xylem tissue proportion, stem conduit density, and root conduit size, leading to reductions in root and stem hydraulic conductance, and leaf-specific and xylem-specific conductivity. Metal stress-specific responses were decreased chlorophyll content and stomatal density. It made plants more susceptible to drought by aggravating water stress (de Silva et al., 2012). In Brachypodium seedling, Cd and combination of Cd and osmotic stress showed more severe effects on the root proteome than osmotic stress alone. 14-3-3 centered subnetwork responded synergistically to cadmium and osmotic stress and their combination (Chen et al., 2018).

Study of the salinity and heavy metals, separately and in combination in plants, revealed that combined effects are more drastic as compared to single stress. In *Conocarpus erectus* plant, biomass of roots and shoots, content of leaf water, and pigment declined greater in the presence of Cd and salinity combination in comparison with Cd individually (Rehman et al., 2019). The Na^+ and Cl^- contents in the roots and shoots remained unaffected by Cd alone but enhanced under Cd + salinity exposure. The decrease in K^+ concentration was observed due to Cd treatment alone as well as Cd + salinity treatment. Uptake of Cd was enhanced due to salinity but Cd translocation to the shoot was not affected. Cd alone and Cd + salinity induced oxidative stress through H_2O_2 overproduction and enhanced lipid peroxidation (Rehman et al., 2019). The antioxidant enzymes activities, including CAT, SOD, APX, and POD, were higher to reduce the oxidative stress. It was suggested that the conocarpus tolerance to Cd was less due to salinity because of enhanced ion uptake and oxidative stress (Rehman et al., 2019). Cu alone and in the presence of salinity considerably reduced shoot length and dry and fresh weights of maize shoot. The salinity as well as Cu and Cd also reduced the leaf chlorophyll contents in leaves. The heavy metals significantly increased the K^+ concentration in the shoots of both cultivars (Gul et al., 2016). Effects of salinity

Table 1 Plant responses toward combination of heavy metal and other environmental stresses.

Heavy metal stress	Other environmental stress	Plant species	Effect of combination of heavy metal and other environmental stress	References
Slag from a Cu-Ni smelter	Drought	*Acer rubrum*	Heavy metals made plants more vulnerable to drought by aggravating water stress. Decrease in hydraulic conductance, xylem and leaf-specific conductivity of plants.	de Silva et al. (2012)
Cd	Osmotic stress	*Brachypodium distachyon*	More severe effects in combined stress. 14-3-3 centered subnetwork responded synergistically to Cd and osmotic stress and their combination.	Chen et al. (2018)
Cd	Salinity	*Conocarpus erectus*	Cd tolerance decreased in due to salinity because of enhanced uptake of ions and oxidative stress.	Rehman et al. (2019)
Cu	Salinity	*Mentha spicata*	Chlorophyll concentration decreased mainly in combined stress. Antioxidants (DPPH, FRAP, ABTS) and polyphenol concentration increased in single stress treatments, but reduced under combined stress.	Chrysargyris et al. (2019)
As	Salinity	*Suaeda maritima*	Tolerate As, high salinity, and their combination without any signs of toxicity.	Panda et al. (2017)
As	Salinity	*Sorghum bicolor*	GPX activity in roots and shoots than single stresses. Salinity and As decreased the CAT activity in root tissues. By increasing As and salinity concentrations in root medium, concentration of soluble carbohydrate increased in leaves.	Talebi et al. (2019)
Cd	Salinity	*Kosteletzkya pentacarpos*	Salinity provides efficient protection from Cd toxicity by improving oxidative stress management and enhancing cytokinin concentration.	Han et al. (2013)
As	Salinity	*Kosteletzkya pentacarpos*	When plants were exposed to Zn stress, the additional NaCl significantly increased the content of both hemicellulose and cellulose, compared to Zn alone. Plant hemicellulose has the potential ability to bind with heavy metal.	Zhou et al. (2020)

Continued

Table 1 Continued

Heavy metal stress	Other environmental stress	Plant species	Effect of combination of heavy metal and other environmental stress	References
Cd	Low and high temperature	*Triticum aestivum* cv. Gerek-79 and Bolal-2973	The total soluble phenolics increase resulting from Cd was enhanced by high temperature. Heavy metal toxicity increased parallelly with temperature. Significant proline accumulation was observed in Gerek-79 variety due to Cd treatment particularly at low temperature, whereas at higher temperature, proline accumulation was decreased. As for the Bolal-2973 variety, Cd stimulated the proline accumulation at low- or high-temperature treatments.	Öncel et al. (2000)
Cd	Heat stress	*Oryza sativa*	Cd treatment and heat stress alone and in combination have different effects on antioxidative defense mechanism. APX and GR appeared important for better survival under combined Cd + heat stress	Nahakpam and Shah (2011)
Cd	Heat stress	*Oryza sativa*	Differential and tissue-specific expression of POD/SOD/CAT/APX and Mn-SOD3/CATR2/APXR4 appeared to be critical constituents of antioxidant defense in the roots of rice plant under Cd + HS, which reduces the harmful impact of low Cd concentration in Cd-tolerant rice cv. Bh-1.	Shah and Nahakpam (2012)
Cd	Heat stress	*Vigna mungo*	One to two POD and SOD transcripts were observed, and signal intensity of them was stronger both in Cd and in HS treatment, and signal intensity was increased more in combination of Cd and heat stress. Inducible as well as constitutive cytosolic HSPs were also involved in combination of stress responses in vivo.	Kochhar and Kochhar (2005)
Cd	Cold	*Populus tremula*	All treatments influenced the plant growth significantly but cold individually and in combined form caused complete arrest of growth and influenced Fv/Fm ratio.	Sergeant et al. (2014)
As	Hypoxia	*Arabidopsis thaliana*	Combination of As and hypoxia treatment led to severe nutrition disorder obvious from deregulated root transcriptome and mineral contents.	Kumar et al. (2020)

Metal	Organism	Plant	Description	Reference
Cd	*Enterobacter* sp. MN17	*Brassica napus*	Inoculation of seed with *Enterobacter* sp. MN17 reversed the harmful effects of Cd in *Brassica napus* plants. Photosynthetic rate, stomatal conductance, transpiration rate, and chlorophyll content of *B. napus* were enhanced.	Saeed et al. (2019)
Zn, Cd	Brassicaceae specialists *Pieris napi Athalia rosae Phaedon cochleariae*	*Arabidopsis halleri*	Feeding of hyperaccumulator *Arabidopsis halleri* by all Brassicaceae specialists was considerably decreased by high concentration of Cd and Zn.	Kazemi-Dinan et al. (2014)
Pb, Cd, As	*Pisum sativum*	*Glomus mosseae*	Growth of plants, contents of photosynthetic pigments, carbohydrates, and total nitrogen were reduced, whereas enzymatic (CAT, SOD, and APX) and nonenzymatic (proline and total phenolics) antioxidants were enhanced upon heavy metal treatment with *G. mosseae*	Chaturvedi et al. (2021)
As, Si	*Magnaporthe oryzae*	*Oryza sativa*	Rice growing under full Si can endure combined stressors of As and *M. oryzae* and, but is highly stressed under Si-deficient condition. As exposure reduced lesion growth, specifically under Si treatments, and reduced the effect of *M. oryzae* on rice.	Griffith et al. (2021)
As	Cucumber mosaic virus	*Solanum lycopersicum*	Arsenic, viral infection, and the combined stresses of both decreased concentrations of chlorophyll *a*, *b* and carotenoids. The viral infection reduced the effect of As at higher doses of As.	Miteva et al. (2005)

and Cu stress singly and in combination on the quality and growth of *Mentha spicata* were studied by Chrysargyris et al. (2019). They observed reduced chlorophyll concentration mainly for the combined stress treatment. Interestingly, concentrations of antioxidants (DPPH, FRAP, ABTS) and polyphenols were increased under high salinity or Cu concentration, but reduced due to the combination of stresses (Chrysargyris et al., 2019). *Suaeda maritima* seedlings can tolerate high salinity, As, and combination of them without any visible signs of toxicity. Accumulation of osmolytes like proline, soluble sugars, and polyphenols up to threshold concentration under salinity and As stress provides osmotic protection under these conditions, thereby maintaining the water status and nutrient uptake. The unchanged chlorophyll content under salinity, As, or salinity + As suggested a minimal harm to the photosynthetic machinery. H_2O_2, $O_2^{\cdot-}$, and MDA concentrations remained low in both NaCl- and As-treated seedlings, suggesting low oxidative stress. It was suggested that *S. maritima* plants can be utilized for phytoremediation/phytostabilization of saline soils contaminated with high arsenic (Panda et al., 2017). The simultaneous application of salinity and As enhanced the GPX activity in roots and shoots than single stresses (Talebi et al., 2019). In halophyte, moderate concentration of salt enhanced plant growth, and diluted the heavy metal concentration, thus decreasing toxicity. It was demonstrated that 50 mM of NaCl delayed senescence in Cd-exposed *Kosteletzkya pentacarpos* by decreasing aminocyclopropane carboxylic acid concentration and enhancing CK concentrations (Han et al., 2013). Salinity enhanced biomass of *K. pentacarpos* and reduced Zn absorption and led to a partial alleviation of the Zn toxicity. Zn stress led to increased crude mucilage content in every plant organ. Excessive Zn caused a modification in the polysaccharides structure and composition, and mucilage might contribute to tolerance to Zn by sequestering toxic ions (Zhou et al., 2020).

Under heat stress, plant cools its leaves via opening of their stomata through transpiration. Combination of heat and heavy metal can enhance transpiration, which could cause an increased uptake of heavy metal or salt (Mittler, 2006). Investigations of interaction between heavy metals and temperature in wheat cvs. Gerek-79 and Bolal-2973 were conducted. High Cd concentrations significantly altered the plant length and concentration of chlorophyll, dry weight, free proline, and total soluble phenolics. No considerable alteration was noted in the examined parameters for the Pb treatments. Results suggested that toxicity of heavy metals increased parallelly with temperature (Öncel et al., 2000). Effects of heat stress and Cd^{2+} toxicity alone and in combination were investigated in Bh-1-tolerant rice cv. and DR-92-sensitive rice cv. This study indicated that in rice plant, Cd^{2+} treatment and heat stress alone and in combination have different effects on antioxidative defense system. GR and APX appeared to be important for the enhanced survival of rice plants growing under combined Cd^{2+} and heat stress (Nahakpam and Shah, 2011). Differential and tissue-specific expression of POD/SOD/CAT/APX and Mn-SOD3/CATR2/APXR4 appeared to be critical constituents of antioxidant defense in the roots of rice plant under Cd^{2+}+HS, which reduced the harmful effect of low Cd^{2+} concentration in tolerant rice cv. Bh-1 (Shah and Nahakpam, 2012). When treated with heat and As, HSP70 expression was observed at the protein and mRNA levels in rice seedlings. The effect was cumulative and increased with the duration of stress for 3 h. During three-hour heat stress recovery at ambient temperature, chaperone exhibited higher expression in plants that were pretreated with As (Goswami et al., 2010). Effects of heat stress and Cd on growth, activity, and isoenzymic profiles of POD, SOD, and their gene expression pattern were studied in seedlings of *Vigna mungo*. Expression of high and low molecular weight HSPs was also studied under these conditions. In the first 12 h, heat stress enhanced germination compared to control. Heat stress also enhanced seedlings growth, and the effect was evident even up to 96 h. Cd^{2+} exposure decreased germination and seedlings length but heat shock before Cd exposure provided considerable cross-protection to harmful effects of Cd^{2+} up to 48 h (Kochhar and Kochhar, 2005). Both POD and SOD activities behaved differently in response to Cd^{2+} stress or heat stress. Cd^{2+} stress increased SOD activity by 5%–10%, whereas heat stress reduced SOD activity insignificantly. Activity of POD increased considerably with heat stress in the presence of as well as in the absence of Cd^{2+}. With time, several bands of POD appeared in all stress treatments, of which some were Cd^{2+} and some were heat stress-specific. One to two POD and SOD transcripts were observed, and signal intensity of them was stronger both in Cd^{2+} and in heat stress treatment, and signal intensity was increased more in combination of Cd^{2+} and heat stress. Inducible as well as constitutive cytosolic HSPs were also involved in combination of stress responses in vivo. The induction and accumulation of these small and high molecular weight HSPs actually showed the outcome of a cross-talk between individual stress response pathways (Kochhar and Kochhar, 2005).

Effects of Cd, cold, and their combination on Poplar clone (*Populus tremula*) at the proteome level were determined. Alteration in protein abundance was classified as general stress, Cd, or cold responsive. All treatments influenced the plant growth significantly but cold individually and in combined form caused a complete arrest of growth and influenced Fv/Fm ratio. Proteins associated with the methionine pathway to activated methyl group were particularly cold-responsive like many HSP and proteins having membrane-stabilizing property. Proteins associated with the import and maturation of mitochondrial protein and proteins associated with nitrogen metabolism were particularly Cd-responsive (Sergeant et al., 2014).

Combination of Cd and As treatment produced higher toxicity to wheat compared to individual metal treatment separately, and As and Cd in combination had an additive influence on the frequency of seed germination and antagonistic influence on biomass and root and shoot elongation of seedlings. Content of MDA, soluble protein, and activity of POD enhanced, whereas SOD and CAT activities reduced with enhanced Cd and As concentrations after an initial enhancement (Liu et al., 2007). The impacts of Cd and As on *Spirodela polyrrhiza* were investigated. Combined treatment reduced fresh biomass and chlorophyll and protein contents; however, nonenzymatic antioxidant did not increase significantly compared to individual treatment (Seth et al., 2007). Plant exposure to As and Cd stress significantly increased ROS, antioxidant activities, and accumulation of osmolytes than control, while a combined treatment of As + Cd was more damaging in decreasing plant biomass of both maize cultivars (Anjum et al., 2016).

Combination of hypoxia and As treatment led to severe nutrition disorder obvious from deregulated transcriptome of root and mineral contents. Both hypoxia and

As posed stress-specific checks on the development of root that caused distinctive root growth phenotype when exposed to combined stress. Besides root, apical meristem activity was inhibited under all treatments. As led to the induction of lateral root growth, whereas density and length of root hairs were enhanced significantly by hypoxia and hypoxia + As treatments. A dual stimulation of phosphate (Pi) starvation response was reported for hypoxia + As-exposed plant roots; however, the responses under hypoxia + As associated more with hypoxia compared to As. Transcriptional and biochemical analyses suggested the association of *PHOSPHATE STARVATION RESPONSE 1*-dependent signaling. Pi metabolism-associated transcripts along with iron homeostasis modulated the development of roots under hypoxia + As. Early alteration in redox potential, alterations in meristematic cells, variable accumulation of ROS in cell layers of root hair, and significant deregulation of NADPH-dependent oxidoreductases, NADPH oxidases, and PODs indicate a role of ROS and redox signaling in remodeling of root architecture under hypoxia + As. Differential expression of aquaporin indicates the transmembrane transport of ROS to control the induction of root hair and growth. Reorganizing energy metabolism via NO-dependent alternate oxidase, phosphofructokinase, and lactate fermentation appeared vital under hypoxia + As. Components of TOR and SnRK signaling network were possibly involved in the regulation of sustainable use of accessible energy reserve for the growth of root hairs under combination of stress and recovery on reaeration (Kumar et al., 2020).

Inoculation of seed with *Enterobacter* sp. MN17 reversed the harmful impacts of Cd in *Brassica napus* plant. Photosynthetic rate, conductance of stomata, transpiration rate, and chlorophyll content were enhanced from 6% to 137% in *B. napus* leaves. Cd content of *B. napus* shoot and root showed that bacterium reduced uptake of Cd from soil (Saeed et al., 2019). Elemental defense theory postulated that high concentration of internal metal can lead to the protection of plant from herbivores. Hyperaccumulation of metal defends plant from herbivory because of toxicity and deterrence to a variety of herbivores. Feeding of hyperaccumulator *Arabidopsis halleri* by all Brassicaceae specialists was considerably decreased by the concentration of metals above 1 mg Zn kg^{-1} DW and 18 ng Cd kg^{-1} DW. However, metals did not influence oviposition. Survival of generalists reduced with increasing individual metal concentration. Combined Zn and Cd treatment showed an additive effect even when applied at the lowest concentration of 2 mg Cd kg^{-1} and 100 ng Zn kg^{-1} (Kazemi-Dinan et al., 2014). Various inorganic and organic compounds can be influenced in plants in various ways by the intricate interaction between herbivores, plants, and heavy metal pollution of soil (Stolpe et al., 2017). The effects of chewing and sucking insects were decreased on *A. halleri* plants growing on soil amended with metals than on plants that were growing in unamended soil. Along with higher concentrations of heavy metal, glucosinolate (GLS) and concentration of elements in plants growing in metal-enriched soil were affected differentially in old vs young leaves, and response was different to herbivory. Most elements and GLS were comparatively lower in old than in young leaves, which support optimal defense theory that predicts improved defense in comparatively valuable tissues (Stolpe et al., 2017). Concentration-dependent enhancement in metal accumulation was observed in plants upon harvest at 60 days. Growth of plants, and contents of photosynthetic pigments, total nitrogen, and carbohydrates were reduced, whereas enzymatic (CAT, SOD, and APX) and nonenzymatic (proline and total phenolics) antioxidants were enhanced upon heavy metal exposure with *Glomus mosseae*. It caused increased growth, photosynthetic pigment concentration, nitrogen, carbohydrates, and antioxidants (proline was reduced) (Chaturvedi et al., 2021). Griffith et al. (2021) suggested that rice growing under full Si can sustain combined stresses of *Magnaporthe oryzae* and As, but can be highly stressed under Si deficiency. Infection of cucumber mosaic virus in tomato plants triggered higher specific peroxidase activity compared to As treatment. The combination of both stresses decreased the positive POD response triggered by viral infection. Arsenic concentrations of 50 and 100 μg g^{-1}, viral infection, and the combined stresses of both at 25 μg g^{-1} decreased the concentrations of chlorophyll *a*, *b* and carotenoids. The viral infection reduced the effect of As at higher As doses. Interaction between the effects of abiotic and biotic stresses was obvious. When As and viral infection were simultaneously applied, they altered the influence of each other on the plants when applied individually (Miteva et al., 2005).

5 Conclusions and future prospects

Climate change is a serious problem due to its adverse impacts on plant, which is predicted to vary in different regions of the world. Physiology, biochemistry, and growth are affected by climate alteration and heavy metal. When abiotic or biotic stresses are applied simultaneously with heavy metal toxicity, they alter the influence of each other on the plants when applied individually. These interactions can affect phytoremediation. Understanding plant responses to abiotic and biotic stresses individually and in combination with soil heavy metal is important for efficient phytoremediation in changing climate. Research focusing on multiple stresses has tried to simulate natural conditions, but conditions are not controlled in the field, and one stress can strongly affect the primary plant defense response. Plants can demonstrate various ranges of sensitivity, which depends on the condition of the field and the stage of plant development. Additional factors that can influence an interaction are the intensity of the stress and the plant species. A better evaluation of effects of climate change on plant growth and phytoremediation and deciphering biochemical and molecular mechanisms which naturally make plants able to withstand various abiotic stresses and heavy metals are important for designing and developing plants with high phytoremediation potential using genetic engineering or breeding.

References

Abrams, M.D., 1990. Adaptations and responses to drought in Quercus species of North America. Tree Physiol. 7, 227–238.

Ahmad, R., Lim, C.J., Kwon, S.Y., 2013. Glycine betaine: a versatile compound with great potential for gene pyramiding to improve crop plant performance against environmental stresses. Plant Biotechnol. Rep. 7, 49–57.

Akcin, T.A., Akcin, A., Yalcin, E., 2015. Anatomical adaptations to salinity in *Spergularia marina* (Caryophyllaceae) from Turkey. Proc. Natl. Acad. Sci. India B Biol. Sci. 85, 625–634.

Alamri, S.A., Siddiqui, M.H., Al-Khaishany, M.Y., Khan, M.N., Ali, H.M., Alakeel, K.A., 2019. Nitric oxide-mediated cross-talk of proline and heat shock proteins induce thermotolerance in *Vicia faba* L. Environ. Exp. Bot. 161, 290–302.

Ali, S., Mir, Z.A., Tyagi, A., Bhat, J.A., Chandrashekar, N., Papolu, P.K., Rawat, S., Grover, A., 2017. Identification and comparative analysis of *Brassica juncea* pathogenesis-related genes in response to hormonal, biotic and abiotic stresses. Acta Physiol. Plant. 39, 1–15.

Ali, M., Cheng, Z., Ahmad, H., Hayat, S., 2018. Reactive oxygen species (ROS) as defenses against a broad range of plant fungal infections and case study on ROS employed by crops against *Verticillium dahliae* wilts. J. Plant Interact. 13, 353–363.

Allakhverdiev, S.I., Los, D.A., Mohanty, P., Nishiyama, Y., Murata, N., 2007. Glycinebetaine alleviates the inhibitory effect of moderate heat stress on the repair of photosystem II during photoinhibition. Biochim. Biophys. Acta 1767, 1363–1371.

Allen, J.A., Chambers, J.L., Stine, M., 1994. Prospects for increasing the salt tolerance of forest trees: a review. Tree Physiol. 14, 843–853.

Almadanim, M.C., Alexandre, B.M., Rosa, M.T., Sapeta, H., Leitão, A.E., Ramalho, J.C., Lam, T.T., Negrão, S., Abreu, I.A., Oliveira, M.M., 2017. Rice calcium-dependent protein kinase OsCPK17 targets plasma membrane intrinsic protein and sucrose-phosphate synthase and is required for a proper cold stress response. Plant Cell Environ. 40, 1197–1213.

Alnsour, M., Ludwig-Müller, J., 2015. Potential effects of climate change on plant primary and secondary metabolism and its influence on plant ecological interactions. J. Endocytobiosis Cell Res. 26, 90–99.

An, J.P., Yao, J.F., Wang, X.N., You, C.X., Wang, X.F., Hao, Y.J., 2017. MdHY5 positively regulates cold tolerance via CBF-dependent and CBF-independent pathways in apple. J. Plant Physiol. 218, 275–281.

Anjum, S.A., Tanveer, M., Hussain, S., Shahzad, B., Ashraf, U., Fahad, S., Hassan, W., Jan, S., Khan, I., Saleem, M.F., Bajwa, A.A., 2016. Osmoregulation and antioxidant production in maize under combined cadmium and arsenic stress. Environ. Sci. Pollut. Res. 23, 11864–11875.

Anjum, S.A., Ashraf, U., Tanveer, M., Khan, I., Hussain, S., Shahzad, B., Zohaib, A., Abbas, F., Saleem, M.F., Ali, I., Wang, L.C., 2017. Drought induced changes in growth, osmolyte accumulation and antioxidant metabolism of three maize hybrids. Front. Plant Sci. 8, 69. https://doi.org/10.3389/fpls.2017.00069.

Araus, J.L., Slafer, G.A., Reynolds, M.P., Royo, C., 2002. Plant breeding and drought in C3 cereals: what should we breed for? Ann. Bot. 89, 925–940.

Artus, N.N., Uemura, M., Steponkus, P.L., Gilmour, S.J., Lin, C., Thomashow, M.F., 1996. Constitutive expression of the cold-regulated *Arabidopsis thaliana* COR15a gene affects both chloroplast and protoplast freezing tolerance. Proc. Natl. Acad. Sci. 93, 13404–13409.

Asgher, M., Per, T.S., Anjum, S., Khan, M.I.R., Masood, A., Verma, S., Khan, N.A., 2017. Contribution of glutathione in heavy metal stress tolerance in plants. In: Reactive Oxygen Species and Antioxidant Systems in Plants: Role and Regulation under Abiotic Stress. Springer, Singapore, pp. 297–313.

Ashraf, M.A., 2012. Waterlogging stress in plants: a review. Afr. J. Agric. Res. 7, 1976–1981.

Atkinson, N., Urwin, P.-E., 2012. The interaction of plant biotic and abiotic stresses: from genes to the field. J. Exp. Bot. 63, 3523–3544.

Baek, K.H., Skinner, D.Z., 2012. Production of reactive oxygen species by freezing stress and the protective roles of antioxidant enzymes in plants. J. Agric. Chem. Environ. 01, 34–40.

Bajguz, A., Hayat, S., 2009. Effects of brassinosteroids on the plant responses to environmental stresses. Plant Physiol. Biochem. 47, 1–8.

Bakhshi, B., Mohseni Fard, E., Nikpay, N., Ebrahimi, M.A., Bihamta, M.R., Mardi, M., Salekdeh, G.H., 2016. MicroRNA signatures of drought signaling in rice root. PLoS One 11, e0156814. https://doi.org/10.1371/journal.pone.0156814.

Bale, J.S., Hayward, S.A.L., 2010. Insect overwintering in a changing climate. J. Exp. Biol. 213, 980–994.

Balint-Kurti, P., 2019. The plant hypersensitive response: concepts, control and consequences. Mol. Plant Pathol. 20, 1163–1178.

Balla, K., Bencze, S., Janda, T., Veisz, O., 2009. Analysis of heat stress tolerance in winter wheat. Acta Agron. Hung. 57, 437–444.

Barceló, J.U.A.N., Poschenrieder, C., 1990. Plant water relations as affected by heavy metal stress: a review. J. Plant Nutr. 13, 1–37.

Begara-Morales, J.C., Sánchez-Calvo, B., Chaki, M., et al., 2014. Dual regulation of cytosolic ascorbate peroxidase (APX) by tyrosine nitration and S-nitrosylation. J. Exp. Bot. 65, 527–538.

Bhattarai, S., Harvey, J.T., Djidonou, D., Leskovar, D.I., 2021. Exploring morpho-physiological variation for heat stress tolerance in tomato. Plan. Theory 10, 347. https://doi.org/10.3390/plants10020347.

Bies-Etheve, N., Gaubier-Comella, P., Debures, A., Lasserre, E., Jobet, E., Raynal, M., Cooke, R., Delseny, M., 2008. Inventory, evolution and expression profiling diversity of the LEA (late embryogenesis abundant) protein gene family in *Arabidopsis thaliana*. Plant Mol. Biol. 67, 107–124.

Bita, C., Gerats, T., 2013. Plant tolerance to high temperature in a changing environment: scientific fundamentals and production of heat stress-tolerant crops. Front. Plant Sci. 4, 273. https://doi.org/10.3389/fpls.2013.00273.

Bloodsworth, A., O'Donnell, V.B., Freeman, B.A., 2000. Nitric oxide regulation of free radical– and enzyme-mediated lipid and lipoprotein oxidation. Arterioscler. Thromb. Vasc. Biol. 20, 1707–1715.

Bohnert, H.J., Nelson, D.E., Jensen, R.G., 1995. Adaptations to environmental stresses. Plant Cell 7, 1099. https://doi.org/10.1105/tpc.7.7.1099.

Bowler, C., Fluhr, R., 2000. The role of calcium and activated oxygens as signals for controlling cross-tolerance. Trends Plant Sci. 5, 241–246.

Burdon, J.J., Zhan, J., 2020. Climate change and disease in plant communities. PLoS Biol. 18, e3000949. https://doi.org/10.1371/journal.pbio.3000949.

Cao, H.H., Pan, M.Z., Liu, H.R., Wang, S.H., Liu, T.X., 2015. Antibiosis and tolerance but not antixenosis to the grain aphid, *Sitobion avenae* (Hemiptera: Aphididae), are essential mechanisms of resistance in a wheat cultivar. Bull. Entomol. Res. 105, 448–455.

Castex, V., Beniston, M., Calanca, P., Fleury, D., Moreau, J., 2018. Pest management under climate change: the importance of understanding tritrophic relations. Sci. Total Environ. 616, 397–407.

Cha-Um, S., Kirdmanee, C., 2010. Effect of glycinebetaine on proline, water use, and photosynthetic efficiencies, and growth of rice seedlings under salt stress. Turk. J. Agric. For. 34, 517–527.

Chamnongpol, S., Willekens, H., Moeder, W., Langebartels, C., Sandermann, H., Van Montagu, M., Inzé, D., Van Camp, W., 1998. Defense activation and enhanced pathogen tolerance induced by H2O2 in transgenic tobacco. Proc. Natl. Acad. Sci. 95, 5818–5823.

Chapman, K.M., Marchi-Werle, L., Hunt, T.E., et al., 2018. Abscisic and jasmonic acids contribute to soybean tolerance to the soybean aphid (*Aphis glycines* Matsumura). Sci. Rep. 8, 15148. https://doi.org/10.1038/s41598-018-33477-w.

Chaturvedi, R., Favas, P.J., Pratas, J., Varun, M., Paul, M.S., 2021. Harnessing *Pisum sativum–Glomus mosseae* symbiosis for phytoremediation of soil contaminated with lead, cadmium, and arsenic. Int. J. Phytoremediation 23, 279–290.

Chaudhary, K., Agarwal, S., Khan, S., 2018. Role of phytochelatins (PCs), metallothioneins (MTs), and heavy metal ATPase (HMA) genes in heavy metal tolerance. In: Mycoremediation and Environmental Sustainability. Springer, Cham, pp. 39–60.

Chen, Z., Zhu, D., Wu, J., Cheng, Z., Yan, X., Deng, X., Yan, Y., 2018. Identification of differentially accumulated proteins involved in regulating independent and combined osmosis and cadmium stress response in Brachypodium seedling roots. Sci. Rep. 8, 1–17.

Chen, D., Shao, Q., Yin, L., Younis, A., Zheng, B., 2019. Polyamine function in plants: metabolism, regulation on development, and roles in abiotic stress responses. Front. Plant Sci. 9, 1945. https://doi.org/10.3389/fpls.2018.01945.

Choi, D.W., Rodriguez, E.M., Close, T.J., 2002. Barley Cbf3 gene identification, expression pattern, and map location. Plant Physiol. 129, 1781–1787.

Chrysargyris, A., Papakyriakou, E., Petropoulos, S.A., Tzortzakis, N., 2019. The combined and single effect of salinity and copper stress on growth and quality of *Mentha spicata* plants. J. Hazard. Mater. 368, 584–593.

Clarke, S.M., Cristescu, S.M., Miersch, O., Harren, F.J., Wasternack, C., Mur, L.A., 2009. Jasmonates act with salicylic acid to confer basal thermotolerance in *Arabidopsis thaliana*. New Phytol. 182, 175–187.

Clemens, S., 2006. Toxic metal accumulation, responses to exposure and mechanisms of tolerance in plants. Biochimie 88, 1707–1719.

Close, T.J., 1996. Dehydrins: emergence of a biochemical role of a family of plant dehydration proteins. Physiol. Plant. 97, 795–803.

Colmenero-Flores, J.M., Rosales, M.A., 2014. Interaction between salt and heat stress: when two wrongs make a right. Plant Cell Environ. 37, 1042–1045.

Colmer, T.D., Flowers, T.J., 2008. Flooding tolerance in halophytes. New Phytol. 179, 964–974.

Cona, A., Rea, G., Angelini, R., Federico, R., Tavladoraki, P., 2006. Functions of amine oxidases in plant development and defence. Trends Plant Sci. 11, 80–88.

Cook, B.I., Mankin, J.S., Anchukaitis, K.J., 2018. Climate change and drought: from past to future. Curr. Clim. Change Rep. 4, 164–179.

Copolovici, L., Niinemets, Ü., 2010. Flooding induced emissions of volatile signalling compounds in three tree species with differing waterlogging tolerance. Plant Cell Environ. 33, 1582–1594.

Cornelissen, T., 2011. Climate change and its effects on terrestrial insects and herbivory patterns. Neotrop. Entomol. 40, 155–163.

Corwin, D.L., 2021. Climate change impacts on soil salinity in agricultural areas. Eur. J. Soil Sci. 72, 842–862.

Crawford, R.M.M., 1967. Alcohol dehydrogenase activity in relation to flooding tolerance in roots. J. Exp. Bot. 18, 458–464.

Cui, F., Liu, L., Zhao, Q., Zhang, Z., Li, Q., Lin, B., Wu, Y., Tang, S., Xie, Q., 2012. Arabidopsis ubiquitin conjugase UBC32 is an ERAD component that functions in brassinosteroid-mediated salt stress tolerance. Plant Cell 24, 233–244.

Dasgupta, S., Hossain, M.M., Huq, M., et al., 2015. Climate change and soil salinity: the case of coastal Bangladesh. Ambio 44, 815–826.

Dat, J.F., Capelli, N., Folzer, H., Bourgeade, P., Badot, P.M., 2004. Sensing and signalling during plant flooding. Plant Physiol. Biochem. 42, 273–282.

Davletova, S., Rizhsky, L., Liang, H.J., Zhong, S.Q., Oliver, D.J., Coutu, J., et al., 2005. Cytosolic ascorbate peroxidase 1 is a central component of the reactive oxygen gene network of Arabidopsis. Plant Cell 17, 268–281.

De Gara, L., de Pinto, M.C., Tommasi, F., 2003. The antioxidant systems vis-à-vis reactive oxygen species during plant–pathogen interaction. Plant Physiol. Biochem. 41, 863–870.

de la Torre, F., Gutiérrez-Beltrán, E., Pareja-Jaime, Y., Chakravarthy, S., Martin, G.B., del Pozo, O., 2013. The tomato calcium sensor Cbl10 and its interacting protein kinase Cipk6 define a signaling pathway in plant immunity. Plant Cell 25, 2748–2764.

de Silva, N.D.G., Cholewa, E., Ryser, P., 2012. Effects of combined drought and heavy metal stresses on xylem structure and hydraulic conductivity in red maple (*Acer rubrum* L.). J. Exp. Bot. 63, 5957–5966.

Denesik, T.J., 2007. Quantitative Expression Analysis of Four Low-Temperature-Tolerance-Associated Genes during Cold Acclimation in Wheat (*Triticum aestivum* L.) (Doctoral dissertation). University of Saskatchewan.

Dervinis, C., Frost, C.J., Lawrence, S.D., Novak, N.G., Davis, J.M., 2010. Cytokinin primes plant responses to wounding and reduces insect performance. J. Plant Growth Regul. 29, 289–296.

Di Toppi, L.S., Gabbrielli, R., 1999. Response to cadmium in higher plants. Environ. Exp. Bot. 41, 105–130.

Doherty, C.J., Van Buskirk, H.A., Myers, S.J., Thomashow, M.F., 2009. Roles for Arabidopsis CAMTA transcription factors in cold-regulated gene expression and freezing tolerance. Plant Cell 21, 972–984.

Dong, C.J., Li, L., Shang, Q.M., et al., 2014. Endogenous salicylic acid accumulation is required for chilling tolerance in cucumber (*Cucumis sativus* L.) seedlings. Planta 240, 687–700.

Drerup, M.M., Schlücking, K., Hashimoto, K., Manishankar, P., Steinhorst, L., Kuchitsu, K., Kudla, J., 2013. The calcineurin B-like calcium sensors CBL1 and CBL9 together with their interacting protein kinase CIPK26 regulate the Arabidopsis NADPH oxidase RBOHF. Mol. Plant 6, 559–569.

Dubey, R.S., Singh, A.K., 1999. Salinity induces accumulation of soluble sugars and alters the activity of sugar metabolising enzymes in rice plants. Biol. Plant. 42, 233–239.

Dusenge, M.E., Duarte, A.G., Way, D.A., 2019. Plant carbon metabolism and climate change: elevated CO_2 and temperature impacts on photosynthesis, photorespiration and respiration. New Phytol. 221, 32–49.

Eastburn, D.M., McElrone, A.J., Bilgin, D.D., 2011. Influence of atmospheric and climatic change on plant–pathogen interactions. Plant Pathol. 60, 54–69.

Egea, I., Pineda, B., Ortíz-Atienza, A., Plasencia, F.A., Drevensek, S., García-Sogo, B., Yuste-Lisbona, F.J., Barrero-Gil, J., Atarés, A., Flores, F.B., Barneche, F., 2018. The SlCBL10 calcineurin B-like protein ensures plant growth under salt stress by regulating Na+ and Ca2+ homeostasis. Plant Physiol. 176, 1676–1693.

Escandón, M., Cañal, M.J., Pascual, J., Pinto, G., Correia, B., Amaral, J., Meijón, M., 2016. Integrated physiological and hormonal profile of heat-induced thermotolerance in *Pinus radiata*. Tree Physiol. 36, 63–77.

Farooq, M., Basra, S.M.A., Wahid, A., Cheema, Z.A., Cheema, M.A., Khaliq, A., 2008. Physiological role of exogenously applied glycinebetaine to improve drought tolerance in fine grain aromatic rice (*Oryza sativa* L.). J. Agron. Crop Sci. 194, 325–333.

Feng, W., Kita, D., Peaucelle, A., Cartwright, H.N., Doan, V., Duan, Q., Liu, M.C., Maman, J., Steinhorst, L., Schmitz-Thom, I., Yvon, R., 2018. The FERONIA receptor kinase maintains cell-wall integrity during salt stress through Ca2+ signaling. Curr. Biol. 28, 666–675.

Finka, A., Cuendet, A.F.H., Maathuis, F.J., Saidi, Y., Goloubinoff, P., 2012. Plasma membrane cyclic nucleotide gated calcium channels control land plant thermal sensing and acquired thermotolerance. Plant Cell 24, 3333–3348.

Finkelstein, R.R., Gampala, S.S., Rock, C.D., 2002. Abscisic acid signaling in seeds and seedlings. Plant Cell 14, S15–S45.

Flowers, T.J., Colmer, T.D., 2008. Salinity tolerance in halophytes. New Phytol. 179, 945–963.

Flowers, T.J., Hajibagheri, M.A., Clipson, N.J.W., 1986. Halophytes. Q. Rev. Biol. 61, 313–337.

Foyer, C.H., Noctor, G., 2005. Redox homeostasis and antioxidant signaling: a metabolic interface between stress perception and physiological responses. Plant Cell 17, 1866–1875.

Foyer, C.H., Noctor, G., 2013. Redox signaling in plants. Antioxid Redox Signal 18, 2087–2090.

Foyer, C.H., Noctor, G., 2016. Stress-triggered redox signalling: what's in pROSpect? Plant Cell Environ. 39, 951–964.

Fraire-Velázquez, S., Rodríguez-Guerra, R., Sánchez-Calderón, L., 2011. In: Shanker, A. (Ed.), Abiotic and Biotic Stress Response Crosstalk in Plants-Physiological, Biochemical and Genetic Perspectives. InTech Open Access Company, Rijeka, pp. 1–26.

Fu, J., Sun, P., Luo, Y., Zhou, H., Gao, J., Zhao, D., Pubu, Z., Liu, J., Hu, T., 2019. Brassinosteroids enhance cold tolerance in *Elymus nutans* via mediating redox homeostasis and proline biosynthesis. Environ. Exp. Bot. 167, 103831.

Fukuda, A., Tanaka, Y., 2006. Effects of ABA, auxin, and gibberellin on the expression of genes for vacuolar H^+-inorganic pyrophosphatase, H^+-ATPase subunit A, and Na^+/H^+ antiporter in barley. Plant Physiol. Biochem. 44, 351–358.

Galani, S., Wahid, A., Arshad, M., 2013. Tissue-specific expression and functional role of dehydrins in heat tolerance of sugarcane (*Saccharum officinarum*). Protoplasma 250, 577–583.

Ganeshan, S., Sharma, P., Young, L., Kumar, A., Fowler, D.B., Chibbar, R.N., 2011. Contrasting cDNA-AFLP profiles between crown and leaf tissues of cold-acclimated wheat plants indicate differing regulatory circuitries. Plant Mol. Biol. 75, 379–398.

Giron, D., Frago, E., Glevarec, G., Pieterse, C.M., Dicke, M., 2013. Cytokinins as key regulators in plant–microbe–insect interactions: connecting plant growth and defence. Funct. Ecol. 27, 599–609.

Gorantla, M., Babu, P.R., Lachagari, V.B.R., Reddy, A.M.M., Wusirika, R., Bennetzen, J.L., Reddy, A.R., 2006. Identification of stress responsive genes in an indica rice (*Oryza sativa* L.) using ESTs generated from drought-stressed seedlings. J. Exp. Bot. 58, 253–265.

Gorham, J., Wyn Jones, R.G., 1990. A physiologists's approach to improve the salt tolerance of wheat. Rachis 9, 20–24.

Goswami, A., Banerjee, R., Raha, S., 2010. Mechanisms of plant adaptation/memory in rice seedlings under arsenic and heat stress: expression of heat-shock protein gene HSP70. AoB Plants 2010, plq023. https://doi.org/10.1093/aobpla/plq023.

Griffith, A., Wise, P., Gill, R., Paukett, M., Donofrio, N., Seyfferth, A.L., 2021. Combined effects of arsenic and *Magnaporthe oryzae* on rice and alleviation by silicon. Sci. Total Environ. 750, 142209.

Gul, S., Nawaz, M.F., Azeem, M., 2016. Interactive effects of salinity and heavy metal stress on ecophysiological responses of two maize (*Zea mays* L.) cultivars. FUUAST J. Biol. 6, 81–87.

Guo, Y.Y., Yu, H.Y., Kong, D.S., Yan, F., Zhang, Y.J., 2016. Effects of drought stress on growth and chlorophyll fluorescence of *Lycium ruthenicum* Murr. seedlings. Photosynthetica 54, 524–531.

Gupta, B., Huang, B., 2014. Mechanism of salinity tolerance in plants: physiological, biochemical, and molecular characterization. Int. J. Genomics 2014. https://doi.org/10.1155/2014/701596.

Gupta, M., Sharma, P., Sarin, N.B., Sinha, A.K., 2009. Differential response of arsenic stress in two varieties of *Brassica juncea* L. Chemosphere 74, 1201–1208.

Guy, C.L., 1990. Cold acclimation and freezing stress tolerance: role of protein metabolism. Annu. Rev. Plant Biol. 41, 187–223.

Haak, D.C., Fukao, T., Grene, R., Hua, Z., Ivanov, R., Perrella, G., Li, S., 2017. Multilevel regulation of abiotic stress responses in plants. Front. Plant Sci. 8, 1564. https://doi.org/10.3389/fpls.2017.01564.

Haile, F.J., Higley, L.G., Ni, X., Quisenberry, S.S., 1999. Physiological and growth tolerance in wheat to Russian wheat aphid (Homoptera: Aphididae) injury. Environ. Entomol. 28, 787–794.

Halfter, U., Ishitani, M., Zhu, J.K., 2000. The Arabidopsis SOS2 protein kinase physically interacts with and is activated by the calcium-binding protein SOS3. Proc. Natl. Acad. Sci. U S A 97, 3735–3740.

Hameed, M., Ashraf, M., Naz, N., 2009. Anatomical adaptations to salinity in cogon grass [*Imperata cylindrica* (L.) Raeuschel] from the Salt Range, Pakistan. Plant and Soil 322, 229–238.

Hameed, M., Ashraf, M., Ahmad, M.S.A., Naz, N., 2010. Structural and functional adaptations in plants for salinity tolerance. In: Plant Adaptation and Phytoremediation. Springer, Dordrecht, pp. 151–170.

Han, R.M., Lefevre, I., Albacete, A., Pérez-Alfocea, F., Barba-Espin, G., Diaz-Vivancos, P., Quinet, M., Ruan, C.J., Hernandez, J.A., Cantero-Navarro, E., Lutts, S., 2013. Antioxidant enzyme activities and hormonal status in response to Cd stress in the wetland halophyte *Kosteletzkya virginica* under saline conditions. Physiol. Plant. 147, 352–368.

Hanif, S., Saleem, M.F., Sarwar, M., Irshad, M., Shakoor, A., Wahid, M.A., Khan, H.Z., 2021. Biochemically triggered heat and drought stress tolerance in rice by proline application. J. Plant Growth Regul. 40, 305–312.

Hasanuzzaman, M., Nahar, K., Alam, M., Roychowdhury, R., Fujita, M., 2013. Physiological, biochemical, and molecular mechanisms of heat stress tolerance in plants. Int. J. Mol. Sci. 14, 9643–9684.

Hassinen, V.H., Tervahauta, A.I., Schat, H., Kärenlampi, S.O., 2011. Plant metallothioneins–metal chelators with ROS scavenging activity? Plant Biol. 13, 225–232.

Hauser, F., Horie, T., 2010. A conserved primary salt tolerance mechanism mediated by HKT transporters: a mechanism for sodium exclusion and maintenance of high K+/Na+ ratio in leaves during salinity stress. Plant Cell Environ. 33, 552–565.

Hernando, C.E., Sanchez, S.E., Mancini, E., Yanovsky, M.J., 2015. Genome wide comparative analysis of the effects of PRMT5 and PRMT4/CARM1 arginine methyltransferases on the *Arabidopsis thaliana* transcriptome. BMC Genomics 16, 1–15.

Hiraki, H., Uemura, M., Kawamura, Y., 2019. Calcium signaling-linked CBF/DREB1 gene expression was induced depending on the temperature fluctuation in the field: views from the natural condition of cold acclimation. Plant Cell Physiol. 60, 303–317.

Hoque, M.A., Banu, M.N.A., Nakamura, Y., Shimoishi, Y., Murata, Y., 2008. Proline and glycinebetaine enhance antioxidant defense and methylglyoxal detoxification systems and reduce NaCl-induced damage in cultured tobacco cells. J. Plant Physiol. 165, 813–824.

Hu, Y., Jiang, L., Wang, F., Yu, D., 2013a. Jasmonate regulates the inducer of CBF expression–crepeat binding factor/DRE binding factor1 cascade and freezing tolerance in Arabidopsis. Plant Cell 25, 2907–2924.

Hu, Y., Chen, L., Wang, H., Zhang, L., Wang, F., Yu, D., 2013b. Arabidopsis transcription factorWRKY8 functions antagonistically with its interacting partner VQ9 to modulate salinity stress tolerance. Plant J. 74, 730–745.

Hu, Y., Jiang, Y., Han, X., Wang, H., Pan, J., Yu, D., 2017. Jasmonate regulates leaf senescence and tolerance to cold stress: crosstalk with other phytohormones. J. Exp. Bot. 68, 1361–1369.

Hu, Z., Huang, X., Amombo, E., Liu, A., Fan, J., Bi, A., Ji, K., Xin, H., Chen, L., Fu, J., 2020. The ethylene responsive factor CdERF1 from bermudagrass (*Cynodon dactylon*) positively regulates cold tolerance. Plant Sci. 294, 110432. https://doi.org/10.1016/j.plantsci.2020.110432.

Huang, B., Xu, C., 2008. Identification and characterization of proteins associated with plant tolerance to heat stress. J. Integr. Plant Biol. 50, 1230–1237.

Huang, C., Ding, S., Zhang, H., Du, H., An, L., 2011. CIPK7 is involved in cold response by interacting with CBL1 in *Arabidopsis thaliana*. Plant Sci. 181, 57–64.

Huang, G.T., Ma, S.L., Bai, L.P., Zhang, L., Ma, H., Jia, P., Liu, J., Zhong, M., Guo, Z.F., 2012. Signal transduction during cold, salt, and drought stresses in plants. Mol. Biol. Rep. 39, 969–987.

Huang, H., Ullah, F., Zhou, D.X., Yi, M., Zhao, Y., 2019. Mechanisms of ROS regulation of plant development and stress responses. Front. Plant Sci. 10, 800. https://doi.org/10.3389/fpls.2019.00800.

Hundertmark, M., Hincha, D.K., 2008. LEA (late embryogenesis abundant) proteins and their encoding genes in *Arabidopsis thaliana*. BMC Genomics 9, 118. https://doi.org/10.1186/1471-2164-9-118.

Husen, A., Iqbal, M., Sohrab, S.S., Ansari, M.K.A., 2018. Salicylic acid alleviates salinity-caused damage to foliar functions, plant growth and antioxidant system in Ethiopian mustard (*Brassica carinata* A. Br.). Agric. Food Secur. 7, 1–14.

Hussain, S., Khalid, M.F., Saqib, M., Ahmad, S., Zafar, W., Rao, M.J., Morillon, R., Anjum, M.A., 2018. Drought tolerance in citrus rootstocks is associated with better antioxidant defense mechanism. Acta Physiol. Plant. 40, 135. https://doi.org/10.1007/s11738-018-2710-z.

Igamberdiev, A.U., Hill, R.D., 2004. Nitrate, NO and haemoglobin in plant adaptation to hypoxia: an alternative to classic fermentation pathways. J. Exp. Bot. 55, 2473–2482.

Ignatenko, A., Talanova, V., Repkina, N., Titov, A., 2019. Exogenous salicylic acid treatment induces cold tolerance in wheat through promotion of antioxidant enzyme activity and proline accumulation. Acta Physiol. Plant. 41, 1–10.

Janská, A., Maršík, P., Zelenková, S., Ovesná, J., 2010. Cold stress and acclimation–what is important for metabolic adjustment? Plant Biol. 12, 395–405.

Jayakannan, M., Bose, J., Babourina, O., Rengel, Z., Shabala, S., 2015. Salicylic acid in plant salinity stress signalling and tolerance. Plant Growth Regul. 76, 25–40.

Jegadeesan, S., Chaturvedi, P., Ghatak, A., Pressman, E., Meir, S., Faigenboim, A., et al., 2018. Proteomics of heat-stress and ethylene-mediated thermotolerance mechanisms in tomato pollen grains. Front. Plant Sci. 9, 1558. https://doi.org/10.3389/fpls.2018.01558.

Jha, A.B., Sharma, P., 2019. Regulation of osmolytes syntheses and improvement of abiotic stress tolerance in plants. In: Hasanuzzaman, M., Nahar, K., Fujita, M., Oku, H., Islam, T. (Eds.), Approaches for Enhancing Abiotic Stress Tolerance in Plants. CRC Press, Boca Raton, FL, pp. 311–338.

Jha, A.B., Misra, A.N., Sharma, P., 2017. Phytoremediation of heavy metal-contaminated soil using bioenergy crops. In: Bauddh, K., Singh, B., Korstad, J. (Eds.), Phytoremediation Potential of Bioenergy Plants. Springer, Singapore, pp. 63–96. Print ISBN 978-981-10-3083-3, Online ISBN 978-981-10-3084-0.

Ji, H., Pardo, J.M., Batelli, G., Van Oosten, M.J., Bressan, R.A., Li, X., 2013. The salt overly sensitive (SOS) pathway: established and emerging roles. Mol. Plant 6, 275–286.

Jiang, D., Zhou, L., Chen, W., Ye, N., Xia, J., Zhuang, C., 2019. Overexpression of a microRNA-targeted NAC transcription factor improves drought and salt tolerance in Rice via ABA-mediated pathways. Rice 12, 1–11.

Joo, J., Lee, Y.H., Song, S.I., 2019. OsbZIP42 is a positive regulator of ABA signaling and confers drought tolerance to rice. Planta 249, 1521–1533.

Karley, A.J., Mitchell, C., Brookes, C., McNicol, J., O'neill, T., Roberts, H., Graham, J., Johnson, S.N., 2016. Exploiting physical defence traits for crop protection: leaf trichomes of *Rubus idaeus* have deterrent effects on spider mites but not aphids. Ann. Appl. Biol. 168, 159–172.

Karwa, S., Bahuguna, R.N., Chaturvedi, A.K., et al., 2020. Phenotyping and characterization of heat stress tolerance at reproductive stage in rice (*Oryza sativa* L.). Acta Physiol. Plant 42, 29. https://doi.org/10.1007/s11738-020-3016-5.

Kasuga, M., Liu, Q., Miura, S., Yamaguchi-Shinozaki, K., Shinozaki, K., 1999. Improving plant drought, salt, and freezing tolerance by gene transfer of a single stress-inducible transcription factor. Nat. Biotechnol. 17, 287–291.

Kaur, H., Kaur, K., Gill, G.K., 2019. Modulation of sucrose and starch metabolism by salicylic acid induces thermotolerance in spring maize. Russ. J. Plant Physiol. 66, 771–777.

Kavar, T., Maras, M., Kidric, M., Sustar-Vozlic, J., Meglic, V., 2007. Identification of genes involved in the response of leaves of *Phaseolus vulgaris* to drought stress. Mol. Breed. 21, 159–172.

Kazemi-Dinan, A., Thomaschky, S., Stein, R.J., Krämer, U., Müller, C., 2014. Zinc and cadmium hyperaccumulation act as deterrents towards specialist herbivores and impede the performance of a generalist herbivore. New Phytol. 202, 628–639.

Kearney, M.R., 2020. How will snow alter exposure of organisms to cold stress under climate warming? Glob. Ecol. Biogeogr. 29, 1246–1256.

Kerchev, P.I., Fenton, B., Foyer, C.H., Hancock, R.D., 2012. Plant responses to insect herbivory: interactions between photosynthesis, reactive oxygen species and hormonal signalling pathways. Plant Cell Environ. 35, 441–453.

Keskin, B.C., Yuksel, B., Memon, A.R., Topal-Sarıkaya, A., 2010. Abscisic acid regulated gene expression in bread wheat ('*Triticum aestivum*' L.). Aust. J. Crop. Sci. 4, 617–625.

Khan, Z., Shahwar, D., 2020. Role of heat shock proteins (HSPs) and heat stress tolerance in crop plants. In: Roychowdhury, R., Choudhury, S., Hasanuzzaman, M., Srivastava, S. (Eds.), Sustainable Agriculture in the Era of Climate Change. Springer, Cham.

Khan, M.I.R., Iqbal, N., Masood, A., Per, T.S., Khan, N.A., 2013. Salicylic acid alleviates adverse effects of heat stress on photosynthesis through changes in proline production and ethylene formation. Plant Signal. Behav. 8, e26374. https://doi.org/10.4161/psb.26374.

Knight, M.R., Knight, H., 2012. Low-temperature perception leading to gene expression and cold tolerance in higher plants. New Phytol. 195, 737–751.

Kobayashi, M., Ohura, I., Kawakita, K., Yokota, N., Fujiwara, M., Shimamoto, K., Doke, N., Yoshioka, H., 2007. Calcium-dependent protein kinases regulate the production of reactive oxygen species by potato NADPH oxidase. Plant Cell 19, 1065–1080.

Koch, K.G., Chapman, K., Louis, J., Heng-Moss, T., Sarath, G., 2016. Plant tolerance: a unique approach to control hemipteran pests. Front. Plant Sci. 7, 1363. https://doi.org/10.3389/fpls.2016.01363.

Kochhar, S., Kochhar, V.K., 2005. Expression of antioxidant enzymes and heat shock proteins in relation to combined stress of cadmium and heat in *Vigna mungo* seedlings. Plant Sci. 168, 921–929.

Komatsu, S., Nanjo, Y., Nishimura, M., 2013. Proteomic analysis of the flooding tolerance mechanism in mutant soybean. J. Proteomics 79, 231–250.

Korenkov, V., Hirschi, K., Crutchfield, J.D., Wagner, G.J., 2007. Enhancing tonoplast Cd/H antiport activity increases Cd, Zn, and Mn tolerance, and impacts root/shoot Cd partitioning in *Nicotiana tabacum* L. Planta 226, 1379–1387.

Kosová, K., Prášil, I.T., Vítámvás, P., Dobrev, P., Motyka, V., Floková, K., Novák, O., et al., 2012. Complex phytohormone responses during the cold acclimation of two wheat cultivars differing in cold tolerance, winter Samanta and spring Sandra. J. Plant Physiol. 169, 567–576.

Kotak, S., Larkindale, J., Lee, U., von Koskull-Döring, P., Vierling, E., Scharf, K.D., 2007. Complexity of the heat stress response in plants. Curr. Opin. Plant Biol. 10, 310–316.

Kuźniak, E., Urbanek, H., 2000. The involvement of hydrogen peroxide in plant responses to stresses. Acta Physiol. Plant. 22, 195–203.

Kulcheski, F.R., de Oliveira, L.F., Molina, L.G., Almerão, M.P., Rodrigues, F.A., Marcolino, J., Barbosa, J.F., Stolf-Moreira, R., Nepomuceno, A.L., Marcelino-Guimarães, F.C., Abdelnoor, R.V., 2011. Identification of novel soybean microRNAs involved in abiotic and biotic stresses. BMC Genomics 12, 1–17.

Kumar, B., Pandey, D.M., Goswami, C.L., Jain, S., 2001. Effect of growth regulators on photosynthesis, transpiration and related parameters in water stressed cotton. Biol. Plant. 44, 475–478.

Kumar, K., Rao, K.P., Sharma, P., Sinha, A.K., 2008. Differential regulation of rice mitogen activated protein kinase kinase (MKK) by abiotic stress. Plant Physiol. Biochem. 46, 891–897.

Kumar, K., Gupta, D., Mosa, K.A., Ramamoorthy, K., Sharma, P., 2019. Arsenic transport, metabolism, and possible mitigation strategies in plants. In: Srivastava, S., Srivastava, A.K., Suprasanna, P. (Eds.), Plant-Metal Interactions. Springer, Cham, pp. 141–168. Print ISBN 978-3-030-20731-1, Online ISBN 978-3-030-20732-8.

Kumar, V., Vogelsang, L., Schmidt, R.R., Sharma, S.S., Seidel, T., Dietz, K.J., 2020. Remodeling of root growth under combined arsenic and hypoxia stress is linked to nutrient deprivation. Front. Plant Sci. 11, 569687. https://doi.org/10.3389/fpls.2020.569687.

Kumar, R.R., Dubey, K., Arora, K., Dalal, M., Rai, G.K., Mishra, D., Chaturvedi, K.K., Rai, A., Kumar, S.N., Singh, B., Chinnusamy, V., 2021. Characterizing the putative mitogen-activated protein kinase (MAPK) and their protective role in oxidative stress tolerance and carbon assimilation in wheat under terminal heat stress. Biotechnol. Rep. 29, e00597. https://doi.org/10.1016/j.btre.2021.e00597.

Kundzewicz, Z.W., Kanae, S., Seneviratne, S.I., Handmer, J., Nicholls, N., Peduzzi, P., Mechler, R., Bouwer, L.M., Arnell, N., Mach, K., Muir-Wood, R., 2014. Flood risk and climate change: global and regional perspectives. Hydrol. Sci. J. 59, 1–28.

Kunkel, B.N., Brooks, D.M., 2002. Cross talk between signaling pathways in pathogen defense. Curr. Opin. Plant Biol. 5, 325–331.

Kwak, J.M., Nguyen, V., Schroeder, J.I., 2006. The role of reactive oxygen species in hormonal responses. Plant Physiol. 141, 323–329.

Laloum, T., Martín, G., Duque, P., 2018. Alternative splicing control of abiotic stress responses. Trends Plant Sci. 23, 140–150.

Lamers, J., Van Der Meer, T., Testerink, C., 2020. How plants sense and respond to stressful environments. Plant Physiol. 182, 1624–1635.

Lang-Mladek, C., Popova, O., Kiok, K., Berlinger, M., Rakic, B., Aufsatz, W., et al., 2010. Transgenerational inheritance and resetting of stress-induced loss of epigenetic gene silencing in Arabidopsis. Mol. Plant 3, 594–602. https://doi.org/10.1093/mp/ssq014.

Larkindale, J., Hall, J.D., Knight, M.R., Vierling, E., 2005. Heat stress phenotypes of Arabidopsis mutants implicate multiple signaling pathways in the acquisition of thermotolerance. Plant Physiol. 138, 882–897.

Lee, U., Rioflorido, I., Hong, S.W., Larkindale, J., Waters, E.R., Vierling, E., 2006. The Arabidopsis ClpB/Hsp100 family of proteins: chaperones for stress and chloroplast development. Plant J. 49 (1), 115–127.

Lekshmy, S., Jha, S.K., Sairam, R.K., 2015. Physiological and molecular mechanisms of flooding tolerance in plants. In: Elucidation of Abiotic Stress Signaling in Plants. Springer, New York, NY, pp. 227–242.

Li, Z., Howell, S.H., 2021. Heat stress responses and thermotolerance in maize. Int. J. Mol. Sci. 22, 948. https://doi.org/10.3390/ijms22020948.

Li, H., Dong, Y., Yin, H., Wang, N., Yang, J., Liu, X., Wang, Y., Wu, J., Li, X., 2011. Characterization of the stress associated microRNAs in *Glycine max* by deep sequencing. BMC Plant Biol. 11, 1–12.

Li, W., Wang, T., Zhang, Y., Li, Y., 2016. Overexpression of soybean miR172c confers tolerance to water deficit and salt stress, but increases ABA sensitivity in transgenic *Arabidopsis thaliana*. J. Exp. Bot. 67, 175–194.

Li, H., Ye, K., Shi, Y., Cheng, J., Zhang, X., Yang, S., 2017. BZR1 positively regulates freezing tolerance via CBF-dependent and CBF-independent pathways in Arabidopsis. Mol. Plant 10, 545–559.

Li, J., Guo, X., Zhang, M., Wang, X., Zhao, Y., Yin, Z., Zhang, Z., Wang, Y., Xiong, H., Zhang, H., 2018. OsERF71 confers drought tolerance via modulating ABA signaling and proline biosynthesis. Plant Sci. 270, 131–139.

Liu, X., Huang, B., 2000. Heat stress injury in relation to membrane lipid peroxidation in creeping bentgrass. Crop. Sci. 40, 503–510.

Liu, J., Ishitani, M., Halfter, U., Kim, C.S., Zhu, J.K., 2000. The *Arabidopsis thaliana* SOS2 gene encodes a protein kinase that is required for salt tolerance. Proc. Natl. Acad. Sci. 97, 3730–3734.

Liu, X., Zhang, S., Shan, X.Q., Christie, P., 2007. Combined toxicity of cadmium and arsenate to wheat seedlings and plant uptake and antioxidative enzyme responses to cadmium and arsenate co-contamination. Ecotoxicol. Environ. Saf. 68, 305–313.

Liu, Y., Song, Q., Li, D., Yang, X., Li, D., 2017. Multifunctional roles of plant dehydrins in response to environmental stresses. Front. Plant Sci. 8, 1018. https://doi.org/10.3389/fpls.2017.01018.

Liu, X., Fu, L., Qin, P., Sun, Y., Liu, J., Wang, X., 2019. Overexpression of the wheat trehalose 6-phosphate synthase 11 gene enhances cold tolerance in *Arabidopsis thaliana*. Gene 710, 210–217.

Liu, H., Yang, Y., Liu, D., Wang, X., Zhang, L., 2020a. Transcription factor TabHLH49 positively regulates dehydrin WZY2 gene expression and enhances drought stress tolerance in wheat. BMC Plant Biol. 20, 1–10.

Liu, T., Ye, X., Li, M., Li, J., Qi, H., Hu, X., 2020b. H_2O_2 and NO are involved in trehalose-regulated oxidative stress tolerance in cold-stressed tomato plants. Environ. Exp. Bot. 171, 103961. https://doi.org/10.1016/j.envexpbot.2019.103961.

Lu, Y., Yao, J., 2018. Chloroplasts at the crossroad of photosynthesis, pathogen infection and plant defense. Int. J. Mol. Sci. 19, 3900. https://doi.org/10.3390/ijms19123900.

Lubovská, Z., Dobrá, J., Štorchová, H., Wilhelmová, N., Vanková, R., 2014. Cytokinin oxidase/dehydrogenase overexpression modifies antioxidant defense against heat, drought and their combination in *Nicotiana tabacum* plants. J. Plant Physiol. 171, 1625–1633.

Luck, J., Spackman, M., Freeman, A., Trębicki, P., Griffiths, W., Finlay, K., Chakraborty, S., 2011. Climate change and diseases of food crops. Plant Pathol. 60, 113–121.

Luo, K., Zhang, G., Wang, C., Ouellet, T., Wu, J., Zhu, Q., Zhao, H., 2014. Candidate genes expressed in tolerant common wheat with resistant to English grain aphid. J. Econ. Entomol. 107, 1977–1984.

Lv, X., Li, H., Chen, X., Xiang, X., Guo, Z., Yu, J., Zhou, Y., 2018. The role of calcium-dependent protein kinase in hydrogen peroxide, nitric oxide and ABA-dependent cold acclimation. J. Exp. Bot. 69, 4127–4139.

Maestri, E., Klueva, N., Perrotta, C., Gulli, M., Nguyen, H.T., Marmiroli, N., 2002. Molecular genetics of heat tolerance and heat shock proteins in cereals. Plant Mol. Biol. 48, 667–681.

Mahajan, S., Pandey, G.K., Tuteja, N., 2008. Calcium-and salt-stress signaling in plants: shedding light on SOS pathway. Arch. Biochem. Biophys. 471, 146–158.

Maleki, M., Ghorbanpour, M., 2018. Cold tolerance in plants: molecular machinery deciphered. In: Biochemical, Physiological and Molecular Avenues for Combating Abiotic Stress Tolerance in Plants. Academic Press, pp. 57–71.

Mansoor, U., Fatima, S., Hameed, M., Naseer, M., Ahmad, M.S.A., Ashraf, M., Ahmad, F., Waseem, M., 2019. Structural modifications for drought tolerance in stem and leaves of *Cenchrus ciliaris* L. ecotypes from the Cholistan Desert. Flora 261, 151485.

Martinoia, E., 2018. Vacuolar transporters–companions on a longtime journey. Plant Physiol. 176, 1384–1407.

Matteucci, M., D'angeli, S., Errico, S., Lamanna, R., Perrotta, G., Altamura, M.M., 2011. Cold affects the transcription of fatty acid desaturases and oil quality in the fruit of *Olea europaea* L. genotypes with different cold hardiness. J. Exp. Bot. 62, 3403–3420.

Matysik, J., Alia, B.B., Mohanty, P., 2002. Molecular mechanisms of quenching of reactive oxygen species by proline under stress in plants. Curr. Sci., 525–532.

Matzrafi, M., 2019. Climate change exacerbates pest damage through reduced pesticide efficacy. Pest Manag. Sci. 75, 9–13.

McGee, T., Shahid, M.A., Beckman, T.G., Chaparro, J.X., Schaffer, B., Sarkhosh, A., 2021. Physiological and biochemical characterization of six Prunus rootstocks in response to flooding. Environ. Exp. Bot. 183, 104368. https://doi.org/10.1016/j.envexpbot.2020.104368.

Meiri, D., Tazat, K., Cohen-Peer, R., Farchi-Pisanty, O., Aviezer-Hagai, K., Avni, A., Breiman, A., 2010. Involvement of Arabidopsis ROF2 (FKBP65) in thermotolerance. Plant Mol. Biol. 72, 191. https://doi.org/10.1007/s11103-009-9561-3.

Meng, X., Li, L., Narsai, R., De Clercq, I., Whelan, J., Berkowitz, O., 2020. Mitochondrial signalling is critical for acclimation and adaptation to flooding in *Arabidopsis thaliana*. Plant J. 103, 227–247.

Miller, G., Schlauch, K., Tam, R., Cortes, D., Torres, M.A., Shulaev, V., Dangl, J.L., Mittler, R., 2009. The plant NADPH oxidase RBOHD mediates rapid systemic signaling in response to diverse stimuli. Sci. Signal. 2, ra45. https://doi.org/10.1126/scisignal.2000448.

Mishra, R.C., Grover, A., 2019. Voyaging around ClpB/Hsp100 proteins and plant heat tolerance. Proc. Indian Natl. Sci. Acad. 85, 791–802.

Mishra, P., Sharma, P., 2019. Superoxide Dismutases (SODs) and their role in regulating abiotic stress induced oxidative stress in plants. In: Hasanuzzaman, M., Fotopoulos, V., Nahar, K., Fujita, M. (Eds.), Reactive Oxygen, Nitrogen and Sulfur Species in Plants: Production, Metabolism, Signaling and Defense Mechanisms. John Wiley & Sons, pp. 53–88. Hard Cover ISBN: 978-1-119-46869-1 EbookISBN: 978-1-119-46864-6.

Miteva, E., Hristova, D., Nenova, V., Maneva, S., 2005. Arsenic as a factor affecting virus infection in tomato plants: changes in plant growth, peroxidase activity and chloroplast pigments. Sci. Hortic. 105, 343–358.

Mitra, S., Mobarak, S.H., Karmakar, A., Barik, A., 2019. Activities of antioxidant enzymes in three species of Ludwigia weeds on feeding by *Altica cyanea*. J. King Saud Univ. Sci. 31, 1522–1527.

Mittler, R., 2006. Abiotic stress, the field environment and stress combination. Trends Plant Sci. 11, 15–19.

Mittler, R., Lam, E., Shulaev, V., Cohen, M., 1999. Signals controlling the expression of cytosolic ascorbate peroxidase during pathogen-induced programmed cell death in tobacco. Plant Mol. Biol. 39, 1025–1035.

Mittler, R., Vanderauwera, S., Gollery, M., Van Breusegem, F., 2004. Reactive oxygen gene network of plants. Trends Plant Sci. 9, 490–498.

Mittler, R., Vanderauwera, S., Suzuki, N., Miller, G.A.D., Tognetti, V.B., Vandepoele, K., Gollery, M., Shulaev, V., Van Breusegem, F., 2011. ROS signaling: the new wave? Trends Plant Sci. 16, 300–309.

Monroy, A.F., Sangwan, V., Dhindsa, R.S., 1998. Low temperature signal transduction during cold acclimation: protein phosphatase 2A as an early target for cold-inactivation. Plant J. 13, 653–660.

Morgan, P.W., He, C.J., De Greef, J.A., De Proft, M.P., 1990. Does water deficit stress promote ethylene synthesis by intact plants? Plant Physiol. 94, 1616–1624.

Mukherjee, K., Choudhury, A.R., Gupta, B., Gupta, S., Sengupta, D.N., 2006. An ABRE-binding factor, OSBZ8, is highly expressed in salt tolerant cultivars than in salt sensitive cultivars of indica rice. BMC Plant Biol. 6, 1–14.

Mukherjee, S., Mishra, A., Trenberth, K.E., 2018. Climate change and drought: a perspective on drought indices. Curr. Clim. Change Rep. 4, 145–163.

Munns, R., 2002. Comparative physiology of salt and water stress. Plant Cell Environ. 25, 239–250.

Munns, R., Tester, M., 2008. Mechanisms of salinity tolerance. Annu. Rev. Plant Biol. 59, 651–681.

Nahakpam, S., Shah, K., 2011. Expression of key antioxidant enzymes under combined effect of heat and cadmium toxicity in growing rice seedlings. Plant Growth Regul. 63, 23–35.

Nakashima, K., Shinwari, Z.K., Sakuma, Y., Seki, M., Miura, S., Shinozaki, K., Yamaguchi-Shinozaki, K., 2000. Organization and expression of two Arabidopsis DREB2 genes encoding DRE-binding proteins involved in dehydration-and high-salinity-responsive gene expression. Plant Mol. Biol. 42, 657–665.

Nakashima, K., Ito, Y., Yamaguchi-Shinozaki, K., 2009. Transcriptional regulatory networks in response to abiotic stresses in Arabidopsis and grasses. Plant Physiol. 149, 88–95.

Narusaka, Y., Nakashima, K., Shinwari, Z.K., Sakuma, Y., Furihata, T., Abe, H., Narusaka, M., Shinozaki, K., Yamaguchi-Shinozaki, K., 2003. Interaction between two cis-acting elements, ABRE and DRE, in ABA-dependent expression of Arabidopsis rd29A gene in response to dehydration and high-salinity stresses. Plant J. 34, 137–148.

Navakoudis, E., Lütz, C., Langebartels, C., Lütz-Meindl, U., Kotzabasis, K., 2003. Ozone impact on the photosynthetic apparatus and the protective role of polyamines. Biochim. Biophys. Acta 1621, 160–169.

Nazir, F., Fariduddin, Q., Khan, T.A., 2020. Hydrogen peroxide as a signalling molecule in plants and its crosstalk with other plant growth regulators under heavy metal stress. Chemosphere 252, 126486. https://doi.org/10.1016/j.chemosphere.2020.126486.

Neumann, D., Lichtenberger, O., Günther, D., et al., 1994. Heat-shock proteins induce heavy-metal tolerance in higher plants. Planta 194, 360–367.

Nguyen, H.T., Babu, R.C., Blum, A., 1997. Breeding for drought resistance in rice: physiology and molecular genetics considerations. Crop. Sci. 37, 1426–1434.

Nilsen, E.T., Orcutte, D.M., 1996. Phytohormones and plant responses to stress. In: Nilsen, E.T., Orcutte, D.M. (Eds.), Physiology of Plant Under Stress: Abiotic Factors. John Wiley and Sons, New York, pp. 183–198.

Niroula, R.K., Pucciariello, C., Ho, V.T., Novi, G., Fukao, T., Perata, P., 2012. SUB1A-dependent and-independent mechanisms are involved in the flooding tolerance of wild rice species. Plant J. 72, 282–293.

Nishizono, H., Ichikawa, H., Suziki, S., et al., 1987. The role of the root cell wall in the heavy metal tolerance of *Athyrium yokoscense*. Plant and Soil 101, 15–20.

Niu, S., Gao, Q., Li, Z., Chen, X., Li, W., 2014. The role of gibberellin in the CBF1-mediated stress-response pathway. Plant Mol. Biol. Report. 32, 852–863.

Noman, A., Aqeel, M., Qari, S.H., Al Surhanee, A.A., Yasin, G., Hashem, M., Al-Saadi, A., 2020. Plant hypersensitive response vs pathogen ingression: death of few gives life to others. Microb. Pathog. 145, 104224. https://doi.org/10.1016/j.micpath.2020.104224.

Ogweno, J.O., Song, X.S., Shi, K., Hu, W.H., Mao, W.H., Zhou, Y.H., Yu, J.Q., Nogués, S., 2008. Brassinosteroids alleviate heat-induced inhibition of photosynthesis by increasing carboxylation efficiency and enhancing antioxidant systems in *Lycopersicon esculentum*. J. Plant Growth Regul. 27, 49–57.

Okuley, J., Lightner, J., Feldmann, K., Yadav, N., Lark, E., 1994. Arabidopsis FAD2 gene encodes the enzyme that is essential for polyunsaturated lipid synthesis. Plant Cell 6, 147–158.

Öncel, I., Keleş, Y., Üstün, A.S., 2000. Interactive effects of temperature and heavy metal stress on the growth and some biochemical compounds in wheat seedlings. Environ. Pollut. 107, 315–320.

Osbourn, A.E., 1996. Preformed antimicrobial compounds and plant defense against fungal attack. Plant Cell 8, 1821–1831.

Osmolovskaya, N., Dung, V.V., Kuchaeva, L., 2018. The role of organic acids in heavy metal tolerance in plants. Biol. Commun. 63, 9–16.

Pagán, I., García-Arenal, F., 2018. Tolerance to plant pathogens: theory and experimental evidence. Int. J. Mol. Sci. 19, 810. https://doi.org/10.3390/ijms19030810.

Painter, R.H., 1958. Resistance of plants to insects. Annu. Rev. Entomol. 3, 267–290.

Pan, Q., Zhan, J., Liu, H., Zhang, J., Chen, J., Wen, P., Huang, W., 2006. Salicylic acid synthesized by benzoic acid 2-hydroxylase participates in the development of thermotolerance in pea plants. Plant Sci. 171, 226–233.

Pan, W.J., Tao, J.J., Cheng, T., Bian, X.H., Wei, W., Zhang, W.K., Ma, B., Chen, S.Y., Zhang, J.S., 2016. Soybean miR172a improves salt tolerance and can function as a long-distance signal. Mol. Plant 9, 1337–1340.

Panda, N., Khush, G.A., 1995. Host Plant Resistance to Insects. CAB International in association with the International Rice Research Institute, Wallingford, Oxon.

Panda, A., Rangani, J., Kumari, A., Parida, A.K., 2017. Efficient regulation of arsenic translocation to shoot tissue and modulation of phytochelatin levels and antioxidative defense system confers salinity and arsenic tolerance in the Halophyte *Suaeda maritima*. Environ. Exp. Bot. 143, 149–171.

Pandey, P., Irulappan, V., Bagavathiannan, M.V., Senthil-Kumar, M., 2017. Impact of combined abiotic and biotic stresses on plant growth and avenues for crop improvement by exploiting physio-morphological traits. Front. Plant Sci. 8, 537. https://doi.org/10.3389/fpls.2017.00537.

Panzade, K.P., Vishwakarma, H., Padaria, J.C., 2020. Heat stress inducible cytoplasmic isoform of ClpB1 from Z. *nummularia* exhibits enhanced thermotolerance in transgenic tobacco. Mol. Biol. Rep. 47, 3821–3831.

Park, J., Song, W.Y., Ko, D., Eom, Y., Hansen, T.H., Schiller, M., Lee, T.G., Martinoia, E., Lee, Y., 2012. The phytochelatin transporters AtABCC1 and AtABCC2 mediate tolerance to cadmium and mercury. Plant J. 69, 278–288.

Park, H.C., Cha, J.Y., Yun, D.J., 2013. Roles of YUCCAs in auxin biosynthesis and drought stress responses in plants. Plant Signal. Behav. 8, 337–349.

Pattanagul, W., Thitisaksakul, M., 2008. Effect of salinity stress on growth and carbohydrate metabolism in three rice (*Oryza sativa* L.) cultivars differing in salinity tolerance. Indian J. Exp. Biol. 46, 736–742.

Pellegrini, E., Forlani, G., Boscutti, F., Casolo, V., 2020. Evidence of non-structural carbohydrates-mediated response to flooding and salinity in *Limonium narbonense* and *Salicornia fruticosa*. Aquat. Bot. 166, 103265. https://doi.org/10.1016/j.aquabot.2020.103265.

Peng, J., Gong, J., 2014. Vacuolar sequestration capacity and long-distance metal transport in plants. Front. Plant Sci. 5, 19. https://doi.org/10.3389/fpls.2014.00019.

Peterson, R.K., Varella, A.C., Higley, L.G., 2017. Tolerance: the forgotten child of plant resistance. PeerJ 5, e3934. https://doi.org/10.7717/peerj.3934.

Piazzi, M., Bavelloni, A., Gallo, A., Faenza, I., Blalock, W.L., 2019. Signal transduction in ribosome biogenesis: a recipe to avoid disaster. Int. J. Mol. Sci. 20, 2718. https://doi.org/10.3390/ijms20112718.

Pierik, R., Sasidharan, R., Voesenek, L.A., 2007. Growth control by ethylene: adjusting phenotypes to the environment. J. Plant Growth Regul. 26, 188–200.

Poljakoff-Mayber, A., 1975. Morphological and anatomical changes in plants as a response to salinity stress. In: Plants in Saline Environments. Springer, Berlin, Heidelberg, pp. 97–117.

Poór, P., Nawaz, K., Gupta, R., et al., 2021. Ethylene involvement in the regulation of heat stress tolerance in plants. Plant Cell Rep. https://doi.org/10.1007/s00299-021-02675-8.

Popova, L.P., Maslenkova, L.T., Ivanova, A., Stoinova, Z., 2012. Role of salicylic acid in alleviating heavy metal stress. In: Environmental Adaptations and Stress Tolerance of Plants in the Era of Climate Change. Springer, New York, NY, pp. 447–466.

Price, P.W., 1991. The plant vigor hypothesis and herbivore attack. Oikos 62, 244–251.

Quintero, F.J., Martinez-Atienza, J., Villalta, I., Jiang, X., Kim, W.Y., Ali, Z., Fujii, H., Mendoza, I., Yun, D.J., Zhu, J.K., et al., 2011. Activation of the plasma membrane Na/H antiporter salt-overly-sensitive 1 (SOS1) by phosphorylation of an auto-inhibitory C-terminal domain. Proc. Natl. Acad. Sci. U. S. A. 108, 2611–2616.

Ram, S., 2000. Role of sucrose hydrolysing enzymes in flooding tolerance in *Brachiaria* species. Indian J. Plant Physiol., 68–72.

Raza, A., Razzaq, A., Mehmood, S.S., Zou, X., Zhang, X., Lv, Y., Xu, J., 2019. Impact of climate change on crops adaptation and strategies to tackle its outcome: a review. Plants 8, 34. https://doi.org/10.3390/plants8020034.

Rehman, S., Abbas, G., Shahid, M., Saqib, M., Farooq, A.B.U., Hussain, M., Murtaza, B., Amjad, M., Naeem, M.A., Farooq, A., 2019. Effect of salinity on cadmium tolerance, ionic homeostasis and oxidative stress responses in conocarpus exposed to cadmium stress: implications for phytoremediation. Ecotoxicol. Environ. Saf. 171, 146–153.

Rejeb, I.B., Pastor, V., Mauch-Mani, B., 2014. Plant responses to simultaneous biotic and abiotic stress: molecular mechanisms. Plan. Theory 3, 458–475.

Ren, J., Sun, L.N., Zhang, Q.Y., Song, X.S., 2016. Drought tolerance is correlated with the activity of antioxidant enzymes in *Cerasus humilis* seedlings. Biomed Res. Int. 2016. https://doi.org/10.1155/2016/9851095.

Rihan, H.Z., Al-Issawi, M., Fuller, M.P., 2017. Advances in physiological and molecular aspects of plant cold tolerance. J. Plant Interact. 12, 143–157.

Rivero, R.M., Ruiz, J.M., Garcıa, P.C., Lopez-Lefebre, L.R., Sánchez, E., Romero, L., 2001. Resistance to cold and heat stress: accumulation of phenolic compounds in tomato and watermelon plants. Plant Sci. 160, 315–321.

Robert, C.A., Zhang, X., Machado, R.A., Schirmer, S., Lori, M., Mateo, P., Erb, M., Gershenzon, J., 2017. Sequestration and activation of plant toxins protect the western corn rootworm from enemies at multiple trophic levels. Elife 6, e29307.

Rodríguez, M., Canales, E., Borrás-Hidalgo, O., 2005. Molecular aspects of abiotic stress in plants. Biotechnol. Appl. 22, 1–10.

Roy, M., Wu, R., 2002. Overexpression of S-adenosylmethionine decarboxylase gene in rice increases polyamine level and enhances sodium chloride-stress tolerance. Plant Sci. 163, 987–992.

Sachs, M.M., Subbaiah, C.C., Saab, I.N., 1996. Anaerobic gene expression and flooding tolerance in maize. J. Exp. Bot. 47, 1–15.

Saeed, Z., Naveed, M., Imran, M., Bashir, M.A., Sattar, A., Mustafa, A., Hussain, A., Xu, M., 2019. Combined use of Enterobacter sp. MN17 and zeolite reverts the adverse effects of cadmium on growth, physiology and antioxidant activity of *Brassica napus*. PLoS One 14, e0213016.

Saidi, Y., Finka, A., Goloubinoff, P., 2011. Heat perception and signalling in plants: a tortuous path to thermotolerance. New Phytol. 190, 556–565.

Sakamoto, A., Murata, N., 2000. Genetic engineering of glycinebetaine synthesis in plants: current status and implications for enhancement of stress tolerance. J. Exp. Bot. 51, 81–88.

Sakamoto, A., Murata, N., 2002. The role of glycine betaine in the protection of plants from stress: clues from transgenic plants. Plant Cell Environ. 25, 163–171.

Santamaria, M.E., Martínez, M., Cambra, I., Grbic, V., Diaz, I., 2013. Understanding plant defence responses against herbivore attacks: an essential first step towards the development of sustainable resistance against pests. Transgenic Res. 22, 697–708.

Sarker, U., Oba, S., 2020. The response of salinity stress-induced A. *tricolor* to growth, anatomy, physiology, non-enzymatic and enzymatic antioxidants. Front. Plant Sci. 11, 559876. https://doi.org/10.3389/fpls.2020.559876.

Scafaro, A.P., Haynes, P.A., Atwell, B.J., 2010. Physiological and molecular changes in *Oryza meridionalis* Ng., a heat-tolerant species of wild rice. J. Exp. Bot. 61, 191–202.

Schmidt, R., Mieulet, D., Hubberten, H.M., Obata, T., Hoefgen, R., Fernie, A.R., Fisahn, J., San Segundo, B., Guiderdoni, E., Schippers, J.H., Mueller-Roeber, B., 2013. SALT-RESPONSIVE ERF1 regulates reactive oxygen species–dependent signaling during the initial response to salt stress in rice. Plant Cell 25, 2115–2131.

Seo, P.J., Kim, M.J., Park, J.Y., Kim, S.Y., Jeon, J., Lee, Y.H., Kim, J., Park, C.M., 2010. Cold activation of a plasma membrane-tethered NAC transcription factor induces a pathogen resistance response in Arabidopsis. Plant J. 61, 661–671.

Sergeant, K., Kieffer, P., Dommes, J., Hausman, J.F., Renaut, J., 2014. Proteomic changes in leaves of poplar exposed to both cadmium and low-temperature. Environ. Exp. Bot. 106, 112–123.

Seth, C.S., Chaturvedi, P.K., Misra, V., 2007. Toxic effect of arsenate and cadmium alone and in combination on giant duckweed (*Spirodela polyrrhiza* L.) in response to its accumulation. Environ. Toxicol. 22, 539–549.

Shabala, S., Shabala, L., Barcelo, J., Poschenrieder, C., 2014. Membrane transporters mediating root signalling and adaptive responses to oxygen deprivation and soil flooding. Plant Cell Environ. 37, 2216–2233.

Shah, K., Nahakpam, S., 2012. Heat exposure alters the expression of SOD, POD, APX and CAT isozymes and mitigates low cadmium toxicity in seedlings of sensitive and tolerant rice cultivars. Plant Physiol. Biochem. 57, 106–113.

Sharma, P., Dubey, R.S., 2005. Drought induces oxidative stress and enhances the activities of antioxidant enzymes in growing rice seedlings. Plant Growth Regul. 46, 209–221.

Sharma, P., Dubey, R.S., 2006. Cadmium uptake and its toxicity in higher plants. In: Samiullah (Ed.), Cadmium Toxicity and Tolerance in Plants. Narosa Publishing House, New Delhi, ISBN: 8173197377, pp. 63–86.

Sharma, P., Dubey, R.S., 2007. Involvement of oxidative stress and role of antioxidative defense system in growing rice seedlings exposed to toxic levels of aluminium. Plant Cell Rep. 27, 2027–2038.

Sharma, P., Dubey, R.S., 2008. Mechanism of aluminium toxicity and tolerance in higher plant. In: Hemantaranjan, A. (Ed.), Advances in Plant Physiology. An International Treatise Series, vol. X. Scientific Publishers, Jodhpur; New Delhi, ISBN: 9788172335243, pp. 145–179.

Sharma, P., Dubey, R.S., 2010. Metal toxicity in plants: uptake of metals, metabolic alterations and tolerance mechanisms. In: Hemantaranjan, A. (Ed.), Advances in Plant Physiology. vol. XI. Scientific Publishers, Jodhpur; New Delhi, pp. 53–106. ISBN: 81-7233-631-8.

Sharma, P., Dubey, R.S., 2011. Abiotic stress induced metabolic alterations in crop plants: strategies for improving stress tolerance. In: Sinha, R.P., Sharma, N.K., Rai, A.K. (Eds.), Botanical Research: The Current Scenario. I K International Publishing House Pvt. Ltd., New Delhi, ISBN: 9789381141045, pp. 1–54.

Sharma, P., Dubey, R.S., 2019. Protein synthesis by plants under stressful conditions. In: Pessarakli, M. (Ed.), Handbook of Plant and Crop Stress, fourth ed. CRC Press, Taylor and Francis, Boca Raton, FL. eISBN 9781351104609.

Sharma, P., Jha, A.B., Dubey, R.S., 2009. Effect of abiotic stresses on growth, metabolic alterations and tolerance mechanisms in rice crop. In: Danforth, A.T. (Ed.), Corn Crops: Growth, Fertilization, and Yield. Agriculture Issues and Policies, Nova Science Publishers, New York, ISBN: 9781607419556.

Sharma, P., Jha, A.B., Dubey, R.S., 2010. Oxidative stress and antioxidative defense system in plants growing under abiotic stresses. In: Pessarakli, M. (Ed.), Handbook of Plant and Crop Stress, third ed. CRC Press, Taylor & Francis Publishing Company, Florida, ISBN: 9781439813966, pp. 89–138. Revised and Expanded.

Sharma, P., Jha, A.B., Dubey, R.S., Pessarakli, M., 2012. Reactive oxygen species, oxidative damage, and antioxidative defense mechanism in plants under stressful conditions. J. Bot. https://doi.org/10.1155/2012/217037, 217037.

Sharma, P., Jha, A.B., Dubey, R.S., 2014. Arsenic toxicity and tolerance mechanisms in crop plants. In: Pessarakli, M. (Ed.), Handbook of Plant and Crop Physiology, third ed. CRC Press, Taylor and Francis Publishing Company, Florida, pp. 733–782.

Sharma, P., Srivastava, V., Kumar, A., et al., 2018. Mechanisms of metalloid uptake, transport, toxicity, and tolerance in plants. In: Abbas, Z., Tiwari, A.K., Kumar, P. (Eds.), Emerging Trends of Plant Physiology for Sustainable Crop Production. Apple Academic Press, Oakville; Waretown, pp. 167–221.

Sharma, P., Jha, A.B., Dubey, R.S., 2019. Oxidative stress and antioxidative defense system in plants growing under abiotic stresses. In: Pessarakli, M. (Ed.), Handbook of Plant and Crop Stress, fourth ed. CRC Press, Taylor and Francis, Boca Raton, FL. eISBN 9781351104609.

Shevyakova, N.I., Musatenko, L.I., Stetsenko, L.A., Vedenicheva, N.P., Voitenko, L.P., Sytnik, K.M., Kuznetsov, V.V., 2013. Effects of abscisic acid on the contents of polyamines and proline in common bean plants under salt stress. Russ. J. Plant Physiol. 60, 200–211.

Singh, P., Sinha, A.K., 2016. A positive feedback loop governed by SUB1A1 interaction with MITOGEN-ACTIVATED PROTEIN KINASE3 imparts submergence tolerance in rice. Plant Cell 28, 1127–1143.

Singh, R., Jha, A.B., Misra, A.N., Sharma, P., 2019a. Differential responses of growth, photosynthesis, oxidative stress, metals accumulation and NRAMP genes in contrasting *Ricinus communis* genotypes under arsenic stress. Environ. Sci. Pollut. Res. 26, 31166–31177.

Singh, R., Jha, A.B., Misra, A.N., Sharma, P., 2019b. Adaption mechanisms in plants under heavy metal stress conditions during phytoremediation. In: Pandey, V., Bauddh, K. (Eds.), Phytomanagement of Polluted Sites. Elsevier, Amsterdam, pp. 329–360.

Singh, R., Jha, A.B., Misra, A.N., Sharma, P., 2020. Entrapment of enzyme in the presence of proline: effective approach to enhance activity and stability of horseradish peroxidase. 3 Biotech, 10. https://doi.org/10.1007/s13205-020-2140-7.

Singh, R., Misra, A.N., Sharma, P., 2021a. Differential responses of thiol metabolism and genes involved in arsenic detoxification in tolerant and sensitive genotypes of bioenergy crop *Ricinus communis*. Protoplasma 258, 391–401.

Singh, R., Misra, A.N., Sharma, P., 2021b. Effect of arsenate toxicity on antioxidant enzymes and expression of nicotianamine synthase in contrasting genotypes of bioenergy crop *Ricinus communis*. Environ. Sci. Pollut. Res. https://doi.org/10.1007/s11356-021-12701-7.

Slama, I., Abdelly, C., Bouchereau, A., Flowers, T., Savouré, A., 2015. Diversity, distribution and roles of osmoprotective compounds accumulated in halophytes under abiotic stress. Ann. Bot. 115 (3), 433–447.

Smith, C.M., 2005. Plant Resistance to Arthropods: Molecular and Conventional Approaches. Springer Science & Business Media, Netherlands.

Somero, G.N., 2010. The physiology of climate change: how potentials for acclimatization and genetic adaptation will determine 'winners' and 'losers'. J. Exp. Biol. 213, 912–920.

Spanic, V., Vuletic, M.V., Abicic, I., Marcek, T., 2017. Early response of wheat antioxidant system with special reference to Fusarium head blight stress. Plant Physiol. Biochem. 115, 34–43.

Spoel, S.H., Dong, X., 2008. Making sense of hormone crosstalk during plant immune responses. Cell Host Microbe 3, 348–351.

Stolpe, C., Krämer, U., Müller, C., 2017. Heavy metal (hyper) accumulation in leaves of *Arabidopsis halleri* is accompanied by a reduced performance of herbivores and shifts in leaf glucosinolate and element concentrations. Environ. Exp. Bot. 133, 78–86.

Strzepek, K., Yohe, G., Neumann, J., Boehlert, B., 2010. Characterizing changes in drought risk for the United States from climate change. Environ. Res. Lett. 5, 044012. Online at stacks. iop.org/ERL/5/044012.

Sugio, A., Dreos, R., Aparicio, F., Maule, A.J., 2009. The cytosolic protein response as a subcomponent of the wider heat shock response in Arabidopsis. Plant Cell 21, 642–654.

Sun, S.J., Guo, S.Q., Yang, X., Bao, Y.M., Tang, H.J., Sun, H., Huang, J., Zhang, H.S., 2010. Functional analysis of a novel Cys2/His2-type zinc finger protein involved in salt tolerance in rice. J. Exp. Bot. 61, 2807–2818.

Sun, X., Zhao, T., Gan, S., Ren, X., Fang, L., Karungo, S.K., Wang, Y., Chen, L., Li, S., Xin, H., 2016a. Ethylene positively regulates cold tolerance in grapevine by modulating the expression of ethylene response factor. Sci. Rep. 6, 24066. https://doi.org/10.1038/srep24066.

Sun, Z., Wang, Y., Mou, F., Tian, Y., Chen, L., Zhang, S., Jiang, Q., Li, X., 2016b. Genome-wide small RNA analysis of soybean reveals auxin-responsive microRNAs that are differentially expressed in response to salt stress in root apex. Front. Plant Sci. 6, 1273. https://doi. org/10.3389/fpls.2015.01273.

Suprasanna, P., Nikalje, G.C., Rai, A.N., 2016. Osmolyte accumulation and implications in plant abiotic stress tolerance. In: Osmolytes and Plants Acclimation to Changing Environment: Emerging Omics Technologies. Springer, New Delhi, pp. 1–12.

Taïbi, K., Del Campo, A.D., Vilagrosa, A., Bellés, J.M., López-Gresa, M.P., López-Nicolás, J.M., Mulet, J.M., 2018. Distinctive physiological and molecular responses to cold stress among cold-tolerant and cold-sensitive *Pinus halepensis* seed sources. BMC Plant Biol. 18. https://doi.org/10.1186/s12870-018-1464-5.

Takahashi, T., Kakehi, J.-I., 2010. Polyamines: ubiquitous polycations with unique roles in growth and stress responses. Ann. Bot. 105, 1–6.

Takahashi, D., Li, B., Nakayama, T., Kawamura, Y., Uemura, M., 2013. Plant plasma membrane proteomics for improving cold tolerance. Front. Plant Sci. 4, 90.

Takasaki, H., Maruyama, K., Kidokoro, S., Ito, Y., Fujita, Y., Shinozaki, K., et al., 2010. The abiotic stress-responsive NAC-type transcription factor OsNAC5 regulates stress-inducible genes and stress tolerance in rice. Mol. Genet. Genomics 284, 173–183.

Talebi, B., Heidari, M., Ghorbani, H., 2019. Arsenic application changed growth, photosynthetic pigments and antioxidant enzymes activity in Sorghum (*Sorghum bicolor* L.) under salinity stress. Sci. Agric. Bohem. 50, 155–163.

Tang, R.J., Yang, Y., Yang, L., Liu, H., Wang, C.T., Yu, M.M., Gao, X.S., Zhang, H.X., 2014. Poplar calcineurin B-like proteins PtCBL10A and PtCBL10B regulate shoot salt tolerance through interaction with PtSOS2 in the vacuolar membrane. Plant Cell Environ. 37, 573–588.

Taylor, R.A.J., Herms, D.A., Cardina, J., Moore, R.H., 2018. Climate change and pest management: unanticipated consequences of trophic dislocation. Agronomy 8, 7. https://doi.org/10.3390/agronomy8010007.

Thomashow, M.F., 1999. Plant cold acclimation: freezing tolerance genes and regulatory mechanisms. Annu. Rev. Plant Biol. 50, 571–599.

Tiffin, P., 2000. Mechanisms of tolerance to herbivore damage: what do we know? Evol. Ecol. 14, 523–536.

Timperio, A.M., Egidi, M.G., Zolla, L., 2008. Proteomics applied on plant abiotic stresses: role of heat shock proteins (HSP). J. Proteomics 71, 391–411.

Tisi, A., Angelini, R., Cona, A., 2008. Wound healing in plants: cooperation of copper amine oxidase and flavin-containing polyamine oxidase. Plant Signal. Behav. 3, 204–206.

Torasa, S., Boonyarat, P., Phongdara, A., Buapet, P., 2019. Tolerance mechanisms to copper and zinc excess in *Rhizophora mucronata* Lam. seedlings involve cell wall sequestration and limited translocation. Bull. Environ. Contam. Toxicol. 102, 573–580.

Tran, D.Q., Konishi, A., Cushman, J.C., Morokuma, M., Toyota, M., Agarie, S., 2020. Ion accumulation and expression of ion homeostasis-related genes associated with halophilism, NaCl-promoted growth in a halophyte *Mesembryanthemum crystallinum* L. Plant Prod. Sci. 23, 91–102.

Turner, N.C., Wright, G.C., Siddique, K.H.M., 2001. Adaptation of grain legumes (pulses) to water-limited environments. Adv. Agron. 71, 193–231. https://doi.org/10.1016/S0065-2113(01)71015-2.

Ullah, A., Manghwar, H., Shaban, M., Khan, A.H., Akbar, A., Ali, U., Ali, E., Fahad, S., 2018. Phytohormones enhanced drought tolerance in plants: a coping strategy. Environ. Sci. Pollut. Res. 25, 33103–33118.

Van Baarlen, P., Van Belkum, A., Summerbell, R.C., Crous, P.W., Thomma, B.P., 2007. Molecular mechanisms of pathogenicity: how do pathogenic microorganisms develop cross-kingdom host jumps? FEMS Microbiol. Rev. 31, 239–277.

Van Loon, L.C., Van Strien, E.A., 1999. The families of pathogenesis-related proteins, their activities, and comparative analysis of PR-1 type proteins. Physiol. Mol. Plant Pathol. 55, 85–97.

VanEtten, H.D., Mansfield, J.W., Bailey, J.A., Farmer, E.E., 1994. Two classes of plant antibiotics: phytoalexins versus phytoanticipins. Plant Cell 6, 1191–1192.

VanEtten, H., Temporini, E., Wasmann, C., 2001. Phytoalexin (and phytoanticipin) tolerance as a virulence trait: why is it not required by all pathogens? Physiol. Mol. Plant Pathol. 59, 83–93.

Vaultier, M.-N., Cantrel, C., Vergnolle, C., Justin, A.-M., Demandre, C., Benhassaine-Kesri, G., Çiçek, D., Zachowski, A., Ruelland, E., 2006. Desaturase mutants reveal that membrane rigidification acts as a cold perception mechanism upstream of the diacylglycerol kinase pathway in Arabidopsis cells. FEBS Lett. 580, 4218–4223.

Vazquez-Hernandez, M., Romero, I., Escribano, M.I., Merodio, C., Sanchez-Ballesta, M.T., 2017. Deciphering the role of CBF/DREB transcription factors and dehydrins in maintaining the quality of table grapes cv. autumn royal treated with high CO2 levels and stored at 0 C. Front. Plant Sci. 8, 1591. https://doi.org/10.3389/fpls.2017.01591.

Viehweger, K., 2014. How plants cope with heavy metals. Bot. Stud. 55, 1–12.

Voesenek, L.A.C.J., Bailey-Serres, J., 2013. Flooding tolerance: O_2 sensing and survival strategies. Curr. Opin. Plant Biol. 16, 647–653.

Wahid, A., 2007. Physiological implications of metabolite biosynthesis for net assimilation and heat-stress tolerance of sugarcane (*Saccharum officinarum*) sprouts. J. Plant Res. 120, 219–228.

Wahid, A., Close, T.J., 2007. Expression of dehydrins under heat stress and their relationship with water relations of sugarcane leaves. Biol. Plant. 51, 104–109.

Wang, L.J., Li, S.H., 2006. Salicylic acid-induced heat or cold tolerance in relation to Ca2+ homeostasis and antioxidant systems in young grape plants. Plant Sci. 170, 685–694.

Wang, G.P., Li, F., Zhang, J., Zhao, M.R., Hui, Z., Wang, W., 2010. Overaccumulation of glycine betaine enhances tolerance of the photosynthetic apparatus to drought and heat stress in wheat. Photosynthetica 48, 30–41.

Wang, J., Yuan, B., Huang, B., 2019a. Differential heat-induced changes in phenolic acids associated with genotypic variations in heat tolerance for hard fescue. Crop. Sci. 59, 667–674.

Wang, L., Sadeghnezhad, E., Riemann, M., Nick, P., 2019b. Microtubule dynamics modulate sensing during cold acclimation in grapevine suspension cells. Plant Sci. 280, 18–30.

Wang, Y., Yu, Y., Huang, M., Gao, P., Chen, H., Liu, M., Chen, Q., Yang, Z., Sun, Q., 2020. Transcriptomic and proteomic profiles of II YOU 838 (*Oryza sativa*) provide insights into heat stress tolerance in hybrid rice. PeerJ 8, e8306. https://doi.org/10.7717/peerj.8306.

Wani, M.A., Jan, N., Qazi, H.A., Andrabi, K.I., John, R., 2018. Cold stress induces biochemical changes, fatty acid profile, antioxidant system and gene expression in *Capsella bursa pastoris* L. Acta Physiol. Plant. 40, 167. https://doi.org/10.1007/s11738-018-2747-z.

Wei, W., Liang, D.W., Bian, X.H., Shen, M., Xiao, J.H., Zhang, W.K., Ma, B., Lin, Q., Lv, J., Chen, X., Chen, S.Y., 2019. GmWRKY54 improves drought tolerance through activating genes in abscisic acid and Ca2+ signaling pathways in transgenic soybean. Plant J. 100, 384–398.

Winicov, I., 1998. New molecular approaches to improving salt tolerance in crop plants. Ann. Bot. 82, 703–710.

Wrzaczek, M., Hirt, H., 2001. Plant MAP kinase pathways: how many and what for? Biol. Cell 93, 81–87.

Wu, H.C., Jinn, T.L., 2012. Oscillation regulation of Ca^{2+}/calmodulin and heat-stress related genes in response to heat stress in rice (*Oryza sativa* L.). Plant Signal. Behav. 7, 1056–1057.

Wu, Y.S., Yang, C.Y., 2019. Ethylene-mediated signaling confers thermotolerance and regulates transcript levels of heat shock factors in rice seedlings under heat stress. Bot. Stud. 60, 1–12.

Xiao, Y., Wu, X., Liu, D., Yao, J., Liang, G., Song, H., Ismail, A.M., Luo, J.S., Zhang, Z., 2020. Cell wall polysaccharide-mediated cadmium tolerance between two *Arabidopsis thaliana* ecotypes. Front. Plant Sci. 11, 473. https://doi.org/10.3389/fpls.2020.00473.

Xie, Y., Chen, P., Yan, Y., Bao, C., Li, X., Wang, L., Shen, X., Li, H., Liu, X., Niu, C., Zhu, C., 2018. An atypical R2R3 MYB transcription factor increases cold hardiness by CBF-dependent and CBF-independent pathways in apple. New Phytol. 218, 201–218.

Xu, Y., Gianfagna, T., Huang, B., 2010. Proteomic changes associated with expression of a gene (ipt) controlling cytokinin synthesis for improving heat tolerance in a perennial grass species. J. Exp. Bot. 61, 3273–3289.

Xuan, Y., Zhou, S., Wang, L., Cheng, Y., Zhao, L., 2010. Nitric oxide functions as a signal and acts upstream of AtCaM3 in thermotolerance in Arabidopsis seedlings. Plant Physiol. 153, 1895–1906.

Yamaguchi, K., Takahashi, Y., Berberich, T., Imai, A., Miyazaki, A., Takahashi, T., Michael, A., Kusano, T., 2006. The polyamine spermine protects against high salt stress in *Arabidopsis thaliana*. FEBS Lett. 580, 6783–6788.

Yamaguchi-Shinozaki, K., Shinozaki, K., 2006. Transcriptional regulatory networks in cellular responses and tolerance to dehydration and cold stresses. Annu. Rev. Plant Biol. 57, 781–803.

Yang, J., Zhang, J., Liu, K., Wang, Z., Liu, L., 2007. Involvement of polyamines in the drought resistance of rice. J. Exp. Bot. 58, 1545–1555.

Yin, Y., Qin, K., Song, X., Zhang, Q., Zhou, Y., Xia, X., Yu, J., 2018. BZR1 transcription factor regulates heat stress tolerance through FERONIA receptor-like kinase-mediated reactive oxygen species signaling in tomato. Plant Cell Physiol. 59, 2239–2254.

Yokoi, S., Bressan, R.A., Hasegawa, P.M., 2002. Salt stress tolerance of plants. JIRCAS Working Rep. 23, 25–33.

Young, T.E., Meeley, R.B., Gallie, D.R., 2004. ACC synthase expression regulates leaf performance and drought tolerance in maize. Plant J. 40, 813–825.

Yu, H., Kong, X., Huang, H., Wu, W., Park, J., Yun, D.J., Lee, B.H., Shi, H., Zhu, J.K., 2020. STCH4/REIL2 confers cold stress tolerance in Arabidopsis by promoting rRNA processing and CBF protein translation. Cell Rep. 30, 229–242.

Zargar, S.M., Nagar, P., Deshmukh, R., Nazir, M., Wani, A.A., Masoodi, K.Z., Agrawal, G.K., Rakwal, R., 2017. Aquaporins as potential drought tolerance inducing proteins: towards instigating stress tolerance. J. Proteomics 169, 233–238.

Zhang, Q., Denlinger, D.L., 2010. Molecular characterization of heat shock protein 90, 70 and 70 cognate cDNAs and their expression patterns during thermal stress and pupal diapause in the corn earworm. J. Insect Physiol. 56, 138–150.

Zhang, F., Wang, Y., Yang, Y., Wu, H.A.O., Wang, D.I., Liu, J., 2007. Involvement of hydrogen peroxide and nitric oxide in salt resistance in the calluses from *Populus euphratica*. Plant Cell Environ. 30, 775–785.

Zhang, J., Martinoia, E., Lee, Y., 2018. Vacuolar transporters for cadmium and arsenic in plants and their applications in phytoremediation and crop development. Plant Cell Physiol. 59, 1317–1325.

Zhang, Y., Li, Y., Hassan, M.J., et al., 2020. Indole-3-acetic acid improves drought tolerance of white clover via activating auxin, abscisic acid and jasmonic acid related genes and inhibiting senescence genes. BMC Plant Biol. 20, 150. https://doi.org/10.1186/s12870-020-02354-y.

Zhang, Y., Lai, X., Yang, S., et al., 2021. Functional analysis of tomato CHIP ubiquitin E3 ligase in heat tolerance. Sci. Rep. 11, 1713. https://doi.org/10.1038/s41598-021-81372-8.

Zhao, J., Ren, W., Zhi, D., Wang, L., Xia, G., 2007. Arabidopsis DREB1A/CBF3 bestowed transgenic tall fescue increased tolerance to drought stress. Plant Cell Rep. 26, 1521–1528.

Zhao, M.L., Wang, J.N., Shan, W., Fan, J.G., Kuang, J.F., Wu, K.Q., et al., 2013. Induction of jasmonate signalling regulators MaMYC2s and their physical interactions with MaICE1 in methyl jasmonate-induced chilling tolerance in banana fruit. Plant Cell Environ. 36, 30–51.

Zhao, H., Sun, X., Xue, M., Zhang, X., Li, Q., 2016. Antioxidant enzyme responses induced by whiteflies in tobacco plants in defense against aphids: catalase may play a dominant role. PLoS One 11, e0165454. https://doi.org/10.1371/journal.pone.0165454.

Zhao, B., Liu, Q., Wang, B., Yuan, F., 2021a. Roles of phytohormones and their signaling pathways in leaf development and stress responses. J. Agric. Food Chem. 69, 3566–3584.

Zhao, J., Lu, Z., Wang, L., Jin, B., 2021b. Plant responses to heat stress: physiology, transcription, noncoding RNAs, and epigenetics. Int. J. Mol. Sci. 22, 117. https://doi.org/10.3390/ijms22010117.

Zhao, Y., Du, H., Wang, Y., Wang, H., Yang, S., Li, C., Chen, N., Yang, H., Zhang, Y., Zhu, Y., Yang, L., 2021c. The calcium-dependent protein kinase ZmCDPK7 functions in heat-stress tolerance in maize. J. Integr. Plant Biol. 63, 510–527.

Zhou, M.Q., Shen, C., Wu, L.H., Tang, K.X., Lin, J., 2011. CBF-dependent signaling pathway: a key responder to low temperature stress in plants. Crit. Rev. Biotechnol. 31, 186–192.

Zhou, S., Duursma, R.A., Medlyn, B.E., Kelly, J.W., Prentice, I.C., 2013. How should we model plant responses to drought? An analysis of stomatal and non-stomatal responses to water stress. Agric. For. Meteorol. 182, 204–214.

Zhou, M.X., Classen, B., Agneessens, R., Godin, B., Lutts, S., 2020. Salinity improves zinc resistance in *Kosteletzkya pentacarpos* in relation to a modification in mucilage and poly-saccharides composition. Int. J. Environ. Res. 14, 323–333.

Zhu, J.K., 2003. Regulation of ion homeostasis under salt stress. Curr. Opin. Plant Biol. 6, 441–445.

Zhu, J., Dong, C.H., Zhu, J.K., 2007. Interplay between cold-responsive gene regulation, metabolism and RNA processing during plant cold acclimation. Curr. Opin. Plant Biol. 10, 290–295.

Zhuo, C., Liang, L., Zhao, Y., Guo, Z., Lu, S., 2018. A cold responsive ethylene responsive factor from *Medicago falcata* confers cold tolerance by up-regulation of polyamine turnover, antioxidant protection, and proline accumulation. Plant Cell Environ. 41, 2021–2032.

Ziska, L.H., McConnell, L.L., 2016. Climate change, carbon dioxide, and pest biology: monitor, mitigate, manage. J. Agric. Food Chem. 64, 6–12.

Structural and functional characteristics of resilient plants for adaptive phytoremediation practices

3

Chapter outline

1 Introduction

1.1 Environmental pollution

The urbanization, industrialization, and traffic in cities worldwide lead to great emissions of pollutants in the environment. Fly ash as a coal residue is generated in thermal power plants for electricity production (Heildrich et al., 2013), whereas mine waste is produced as a result of mining activities (European Commission, 2009). Many inorganic and organic pollutants in urban environment, both from fly ash and mine waste, rock and tailings released in the water, air, and soil cause environmental pollution affecting food security, human health, and life quality (Bell and Treshow, 2002; Maiti, 2013; Gajić et al., 2018a; Pandey, 2020a,b). The main air pollutants are sulfur oxide (SO_2), nitrogen dioxide (NO_2), nitric oxide (NO), carbon dioxide (CO_2), carbon oxide (CO), total volatile organic pollutants (TVOCs), total suspended particulate matter (TSP), and particulate matter (PM_{10} and $PM_{2.5}$) (Bell and Treshow, 2002). The particulate matters (PMs) consist of black carbon, heavy metals, and polycyclic aromatic hydrocarbons (PAHs) (Bell et al., 2011). Furthermore, the urban soils, fly ash, and mine tailings are characterized by high concentrations of As, B, Cd, Cr, Cu, Fe, Hg, Mn, Ni, Pb, Sb, Zn, PAHs, and polychlorinated biphenyls (PCBs) that usually exceed allowed values proposed by national legislation for public health standards (Maiti, 2013; Gajić et al., 2018a; Pandey, 2020b). The pollutants lead to the formation of acid rain and emission of greenhouse gases, which lead to altered relationships between climate, soil, plant, and animal species, thus contaminating the environment and disturbing the balance between ecosystems (Storm et al., 2015; McDonald et al., 2016; O'Connor et al., 2019) (Fig. 1).

Adaptive Phytoremediation Practices. https://doi.org/10.1016/B978-0-12-823831-8.00005-0

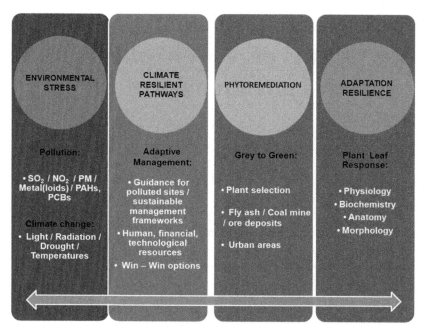

Fig. 1 Trade-off between environmental stresses, climate-resilient pathways, phytoremediation, and plant adaptation/resilience.

Fly ash and mine waste—Coal as a major fuel for production of electricity in thermal power stations generates a large amount of fly ash (580 million tons/year in China; 160 million tons/year in India; 130 million tons/year in the United States, Yao et al., 2015). Furthermore, the opencast coal and metal mining produce a substantial amount of different types of mine waste materials, such as open-pit overburden and waste rock, and tailings (Maiti, 2013). Coal fly ash and mine waste are usually disposed in artificially made lagoons, heaps, and ponds (Reijnders, 2005; European Commission, 2009). Fly ash deposits and mine spoil characterized by unfavorable physicochemical properties: sandy structure with small amounts of colloidal particles and low percentage of hydroscopic water, high/low pH values and electrical conductivity, lack of N and P, and high concentrations of As, B, Cd, Cr, Cu, Hg, Mn, Mo, Ni, Pb, Se, etc., PAH, PCB (Gajić et al., 2018a; Pandey and Singh, 2010; Pandey, 2020b). Fine particles from fly ash and mine waste can be dispersed due to wind in the environment affecting human health by provoking cancer diseases, bronchitis, asthma, and irritation in eyes, skin, nose, and throat (USEPA, 2007).

Urban areas—The main sources in urban areas are products of traffic emission, combustion in households, individual boilers, and waste landfills (Bell and Treshow, 2002). Air pollutant emission from traffic depends on its density, vehicle type (cars trucks, buses), age and degree of vehicle maintenance, fuel type, tire wear and brake wear, and meteorological factors (Colls, 2002; AQEG, 2005; Bukowski et al., 2010; Font and Fuller, 2016). The combustion of 1 L of fuel releases 100 g of CO, 30 g of NO_x, 2.5 kg of CO_2, sulfur compounds, Pb, other heavy metals, and soot (Colls, 2002).

It was estimated that 73% of NO_2 and 42% of PM_{10} are released from traffic in Paris (Airparif, 2016), whereas in London it was 50% for both NO_2 and SO_2 (GLA, 2018). Particles and gases from motor vehicles are released at a height of several tens of centimeters and spread in environment, where the concentrations of pollutants decrease with distance from the emission source. Thus, the concentrations of NO_2 in London are as follows: 2000 ppb along the sidewalk, 80–90 ppb in the central part of street, 50–60 ppb on the sidewalk, and only 40–50 ppb out off the sidewalk (Colls, 2002). According to Karner et al. (2010), air pollutants in cities decreased within 200 m of the source. The clean air requires standards for air quality, human health, and environmental protection. Established standards for primary pollutants are as follows: NO_2—100 ppb for 1 h and 53 ppb for 1 year; SO_2—75 ppb for 1 h; $PM_{2.5}$—12 µg/m^3 for 1 year; PM_{10}—150 µg/m^3 for 24 h; CO—35 ppm for 1 h; and Pb—0.15 µg/m^3 for 3 months (USEPA, 2010). However, despite the ambient air quality standards, long-term exposure to pollutants has been associated with premature deaths (55,000 annually in France and 50,000 in the England) (Guerreiro et al., 2016). NO_2, SO_2, CO, Pb, and PM can adversely affect respiratory/nervous/immune/reproductive/developmental/cardiovascular system, kidney function and cause cancer and brain damage, especially in children and elderly (USEPA, 2010; Pascal et al., 2014).

1.2 Climate-resilient pathways and adaptive management

Climate changes as a result of environmental pollution increase ambient temperature and CO_2 concentration, leading to severe drought, floods, storms, and rainfalls (USEPA, 2011). Climate change resilience is associated with reduction of risk from hazards of extreme events by wide range of assessment, monitoring and sustainable management in order to recognize one/more cleanup solutions in socio-ecological systems by accessing different information, technologies, and economic resources (Denton et al., 2014). A wide range of climate-resilient responses, such as adaptation and mitigation, depend on human, financial, and technological resources and education as well as legal policy frameworks (Denton et al., 2014) (Fig. 1). Adaptive management provides a wide range of sustainable solutions to climate challenges by learning from outcomes of practices (experience, experimenting, and monitoring) and policy frameworks allowing that decision makers, stakeholders, and ecosystem managers implement "win-win" options for nations, business, and civil society (Owens, 2009; CEAA, 2009; Birge et al., 2016) (Fig. 1).

*Healing urban landscapes—Smart Cities—*Cities take action through sustainable projects, adaptation programs, and policy to mitigate and adapt to climate change creating co-benefits for the environment, economies, communities, and human health (Storm et al., 2015). The solutions in green energy sector involve the use of renewable technologies for low-carbon transition in city energy systems: Stockholm—world's first urban carbon sink with biochar from green organic waste (turning park and garden waste into biochar that can replace peat, sand, and clay sequestering carbon and converting it into stable element in the soil lowering in that way the level of CO_2 in the atmosphere); Copenhagen—carbon-neutral district heating system by substituting fossil fuels, such as coal, oil, and natural gas, to sustainable biomass (deploy large-scale heat pumps and underground pipes that run on geothermal energy, i.e., combined heat and power plants that are fired with wood pellets reducing energy consumption in buildings) (Storm et al., 2015). The solutions in solid waste sector include transformation of waste resource into clean energy limited food waste, reducing greenhouse gas emissions, and air and soil pollution. The cities, such as Yokohama, New York, Bogota, Milan, Oakland, and Bengaluru, encourage behavior change by raising awareness and coordinating between residents and city governments in order to increase usage of recyclable materials (ecofriendly products) and food waste that can be used for production of biogas and compost for soil remediation, whereas in Wuhan and Durban landfills can be revitalized by switching to greening that empowers a community (Storm et al., 2015). The solutions in sections, such as adaptation planning and assessment, adaptation implementation, carbon measurement and planning, and building energy efficiency, focus on plans that can make cities more resilient for future climatic changes: Vancouver—building strategy for sea level rise; Columbus—securing local river water supply; Washington—reducing flood risk; New Orleans—coastal storm weather protection; and Melbourne, Sydney, Paris, Rotterdam, and Copenhagen—developing gray to green infrastructure and rooftop gardens to increase ecosystem services to safeguard the city in order to reduce the urban heat island effect (Storm et al., 2015). Smart cities and smart community engagements and sustainable communities include solutions that rely on information technology, data management, and online communications to engage citizens in climate actions: Boston—new media engages residents in climate actions; Melbourne—web tool enables building retrofits; Stockholm—from brownfield to sustainable district; Buenos Aires—improves safety for cyclists and pedestrians; Johannesburg—builds green public transport; Mexico city—park revitalization promotes accessibility; Nanjing—has world's fastest electrical vehicles; Chennai—transforms streets for walking and cycling; Milan—world's first free-floating ride-sharing system; and Heidelberg—world's largest zero-emissions district (Storm et al., 2015).

1.3   Phytoremediation—"Gray to green"

Phytoremediation as a "green" technology uses plants to clean up polluted environment from hazardous inorganic and organic pollutants (Pilon-Smits, 2005) (Fig. 1). The main phytoremediation technologies are as follows: phytostabilization that uses plants in order to reduce the mobility of pollutants in the environment preventing their entry into the soil, the groundwater, and the food web; phytoextraction that uses plants capable of accumulating metal(loid)s in the aboveground and underground plant's parts; rhizodegradation/phytostimulation that uses plant's roots which enhance microbial and fungal activity in the rhizosphere and breakdown pollutants; and phytodegradation/phytotransformation that involves degradation of organic compounds by plant enzymes within roots or leaves (McCutcheon and Schnoor, 2003; Pilon-Smits, 2005).

Revegetation of fly ash, coal, and ore deposits—Establishment of the vegetation cover on fly ash and mine waste deposits is significant for increasing the organic matter and nutrient content, improving the physicochemical and biological characteristics of substrate, retention of moisture, prevention of wind erosion and stabilization of substrate, and reduction of mobility of contaminants in the surrounding environment (Kostić et al., 2012, 2018; Gajić et al., 2016, 2020; Pandey, 2020a,c; Kumar et al., 2021; Maiti and Pandey, 2021; Ahirwal and Pandey, 2021). The selection of native plant species is essential for revegetation and phytoremediation because they are well adapted to drought, high temperatures, and radiation, they possess a high ability to tolerate the toxicity/deficiency of some chemical elements in fly ash and mine tailings, and they are characterized by fast growing, high biomass, and extended root system (Pandey, 2012, 2013, 2021; Gajić et al., 2013, 2019, 2020; Pandey et al., 2015a) (Table 1). Plant species suitable for phytoremediation of fly ash deposits are as follows: *Calamagrostis epigejos, Chenopodium rubrum, Cirsium arvense, Crepis setosa, Cynodon dactylon, Dactylis glomerata, Daucus carota, Echium vulgare, Erigeron canadensis, Medicago sativa, Melilotus officinalis, Oenothera biennis, Phragmites communis, Sorghum halepense, Trifolium repens, Tussilago farfara, Vicia villosa, Verbascum phlomoides, Xanthium strumarium, Robinia pseudoacacia, Rosa canina, Rubus caesius, Populus alba, Salix alba, Tamarix tetrandra, Calotropis procera, Cassia tora, Chenopodium album, Croton bonplandianum, C. dactylon, Ipomea carnea, Parthenium hysterophorus, Saccharum bengalense, Saccharum munja, Saccharum spontaneum, Sida cordifolia, Thelypteris dentata,* and *Typha latifolia* (Pavlović et al., 2004; Djurdjević et al., 2006; Mitrović et al., 2008; Gajić and Pavlović, 2018; Gajić et al., 2018a, 2019; Pandey and Singh, 2014; Pandey, 2015, 2020a; Pandey et al., 2015b, 2016; Kumari et al., 2016). Plant species suitable for phytoremediation of mine wastes are as follows: *Agrostis stolonifera, C. epigejos, Cerastium arvense, Polygonum aviculare, T. farfara, Digitalis purpurea, Mentha suaveolens, Ruscus ulmifolius, Zygophyllum fabago, Helichrysum decumbens, Tamarix, Lygeum spartum, Piptatherum miliaceum, Pinus halepensis, Holcus lanatus, Festuca rubra, D. glomerata, Cytisus striatus, Genista* sp., *Lotus corniculatus, Salsola collina, Festuca elata, M. sativa, Ipomea purpurea, Digitaria sanguinalis, E. canadensis, Phytolacca acinosa, Pteris multifida,* and *C. dactylon* (Conesa et al., 2006; Li et al., 2007; Pratas et al., 2013; Parraga-Aguado et al., 2014; Zhang et al., 2014; Fernandez et al., 2017; Kasowska et al., 2018).

"Planting healthy urban environment"—Phytoremediation in urban areas can be integrated into the transformation processes of public space by cleaning up water, air, and soil by planting trees, shrubs, and grasses creating the sustainable urban environment (McDonald et al., 2016; O'Connor et al., 2019). According to Badach et al. (2020), mitigation of air pollution is achieved by urban greenery where urban vegetation has an essential role in air quality management. Retention, adsorption, and deposition capacity of plants depend on the source of pollution, meteorological factors, deposition rate, canopy geometry, and plant height (Liu et al., 2018; Kwak et al., 2019). Efficiency of green belt depends on vegetation density, plant species, leaf surface area, leaf size and shape, leaf surface characteristics, leaf longevity, and phenology (Shannigrahi et al., 2004; Blanusa et al., 2019; Barwise and Kumar, 2020).

In densely populated urban area, green infrastructure in street canyon offers great benefits by reducing 40% of NO_2 and 60% of PM which mitigates air pollution and creates "filtered avenues" (Pugh et al., 2012). Designing vegetation barriers for air pollution mitigation from traffic emissions is significant because they can limit the exposure of pedestrians (Abhijith et al., 2017; Abhijith and Kumar, 2019). Thus, roadside/street

Table 1 Plant species used in phytoremediation of fly ash, mine waste, and urban areas.

Plant species	References	Plant species	References
Fly ash *Calamagrostis epigejos, Chenopodium rubrum, Cirsium arvense, Crepis setosa, Cynodon dactylon, Dactylis glomerata, Daucus carota, Echium vulgare, Erigeron canadensis, Medicago sativa, Melilotus officinalis, Oenothera biennis, Phragmites communis, Sorghum halepense, Trifolium repens, Tussilago farfara, Vicia villosa, Verbascum phlomoides, Xanthium strumarium, Robinia pseudoacacia, Rosa canina, Rubus caesius, Populus alba, Salix alba, Tamarix tetrandra, Calotropis procera, Cassia tora, Chenopodium album, Croton bonplandianum, Cynodon dactylon, Ipomea carnea, Parthenium hysterophorus, Saccharum bengalense, Saccharum munja, Saccharum, spontaneum, Sida cordifolia, Thelypteris dentate, Typha latifolia*	Pavlović et al. (2004), Djurdjević et al. (2006), Mitrović et al. (2008), Gajić and Pavlović (2018), Gajić et al. (2018a), Pandey and Singh (2014), Pandey (2015), Pandey et al. (2015b), Pandey et al. (2016), Kumari et al. (2016)	**Mine waste** *Agrostis stolonifera, Calamagrostis epigejos, Cerastium arvense, Polygonum aviculare, Tussilago farfara, Digitalis purpurea, Mentha suaveolens, Ruscus ulmifolius, Zygophyllum fabago, Helichrysum decumbens, Tamarix, Lygeum spartum, Piptatherum miliaceum, Pinus halepensis, Holcus lanatus, Festuca rubra, Dactylis glomerata, Cytisus striatus, Genista sp., egionensis, Lotus corniculatus, Salsola collina, Festuca elata, Medicago sativa, Ipomea purpurea, Digitaria sanguinalis, Erigeron canadensis, Phytolacca acinosa, Pteris multifida, Cynodon dactylon*	Conesa et al. (2006), Li et al. (2007), Pratas et al. (2013), Parraga-Aguado et al. (2014), Zhang et al. (2014), Fernandez et al. (2017), Kasowska et al. (2018)

Hedgerows/Green roof		Vegetation barrier	
Berberis thunbergii, Buxus sempervirens, Camellia japonicum, Carpinus betulus, Corylus avellana, Crataegus monogyna, Elaeagnus × ebbingei, Euonymus japonicas, Fagus sylvatica, Forsythia sp., Ilex aquifolium, Ligustrum ovalifolium, Lonicera nitida, Pyracantha coccinea, Rosa rugosa, Symphoricarpos albus, Taxus baccata, Viburnum tinus, Agrostis stolonifera, Festuca rubra, Plantago lanceolata, Sedum album	Librando et al. (2002), Lukasik et al. (2002), Min and Renging (2004), Takahashi et al. (2005), Lei et al. (2006), Gajić et al. (2009), Jim and Chen (2009), Huseyinova et al. (2009), Erdogan and Yazgan (2009), Akguc et al. (2010), Speak et al. (2012), Zolgharnein et al. (2013), Hu et al. (2014), Van Renterghem et al. (2014), Fellet et al. (2016), Fellet et al. (2016), Bianco Mori et al. (2015), O'Sullivan et al. (2017)	*Acer campestre, Acer platanoides, Acer pseudoplatanus, Acer rubrum, Acer tataricum, Ailanthus altissima, Alnus glutinosa, Celtis australis, Celtis occidentalis, Crataegus monogyna, Elaeagnus angustifolia, Fraxinus sp., Ginkgo biloba, Gleditsia triacanthos, Metasequoia sp., Pinus nigra, Pinus strobus, Pinus sylvestris, Platanus sp., Populus nigra, Prunus avium, Rhododendron sp., Quercus coccinea, Quercus frainetto, Quercus ilex, Quercus petraea, Quercus robur, Rhus typhina, Robinia pseudoacacia, Sorbus latifolia, Taxodium distichum, Taxus baccata, Thuja plicata, Tilia cordata, Ulmus sp., Zelkova serrata*	Barwise and Kumar (2020)

hedgerows improve air quality by reduction of pollutant emission (46%–61%) in urban canyons (Gromke et al., 2016). Plant species that are successfully used for urban hedges in ecosystem services are as follows: *Berberis thunbergii* (Hu et al., 2014); *Buxus sempervirens* (Librando et al., 2002); *Camellia japonicum* (Takahashi et al., 2005); *Carpinus betulus* (Zolgharnein et al., 2013); *Corylus avellana* (Huseyinova et al., 2009); *Crataegus monogyna* (Erdogan and Yazgan, 2009); *Elaeagnus × ebbingei* (Fellet et al., 2016); *Euonymus japonicus* (Lei et al., 2006); *Fagus sylvatica* (Van Renterghem et al., 2014); *Forsythia* sp. (Jim and Chen, 2009); *Ilex aquifolium* (Fellet et al., 2016); *Ligustrum ovalifolium* (Gajić et al., 2009); *Lonicera nitida* (Bianco et al., 2017); *Pyracantha coccinea* (Akguc et al., 2010); *Rosa rugosa* (Min and Renging, 2004); *Symphoricarpos albus* (Lukasik et al., 2002); *Taxus baccata* (O'Sullivan et al., 2017); and *Viburnum tinus* (Mori et al., 2015). Plant species in living walls as vertical greenery systems can capture traffic-generated PM (Weerakkody et al., 2018). Green roof vegetation efficiently removes air pollutants (NO_2, SO_2, and PM_{10}) (Yang et al., 2008; Currie and Bass, 2008; Speak et al., 2012), and plants that can be used as a filter of urban air are as follows: *A. stolonifera, F. rubra, Plantago lanceolata,* and *Sedum album* (Speak et al., 2012).

Plant species that are planted in urban areas are characterized by ecophysiological and morphological characteristics which enable them to thrive in environmental stress conditions, such as air pollution, drought, salt, and high temperatures (Gajić et al., 2009; Barwise and Kumar, 2020). In addition, plant species are selected based on air quality improvement potential: evergreen > deciduous; conifer > broadleaf; high leaf area density > low leaf area density; rough texture of leaf structure (trichome, etc.) > smooth texture of leaf structure; and high amount of epicuticular wax > low amount of epicuticular wax (Barwise and Kumar, 2020). Woody plant species that are tolerant to air pollution suitable for planting in temperate regions are as follows: *Acer campestre, Acer platanoides, Acer pseudoplatanus, Acer rubrum, Acer tataricum, Ailanthus altissima, Alnus glutinosa, Celtis australis, Celtis occidentalis, C. monogyna, Elaeagnus angustifolia, Fraxinus* sp., *Ginkgo biloba, Gleditsia triacanthos, Metasequoia* sp., *Pinus nigra, Pinus strobus, Pinus sylvestris, Platanus* sp., *Populus nigra, Prunus avium, Rhododendron* sp., *Quercus coccinea, Quercus frainetto, Quercus ilex, Quercus petraea, Quercus robur, Rhus typhina, R. pseudoacacia, Sorbus latifolia, Taxodium distichum, T. baccata, Thuja plicata, Tilia cordata, Ulmus* sp., and *Zelkova serrata* (Barwise and Kumar, 2020).

2 Adaptive characteristics of resilient plants

2.1 Functional and structural plant response to pollution and climatic change

Adverse climatic and edaphic conditions and emission of pollutants from different sources in urban and industrial environments cause functional and structural changes and damage to plants. Stressful effects of high temperatures, intense radiation, drought, nutrient deficiencies, gaseous pollutants, PM, heavy metals, and organic pollutants affect the water balance and photosynthesis, causing chlorosis and necrosis of leaves (Bell and Treshow, 2002; Gajić et al., 2020). Critical concentrations of pollutants in plants are defined as the highest concentrations above which they become toxic, i.e., cause functional (water balance, photosynthesis, and respiration) and structural (damage of the cuticle, epidermis, stoma, and mesophyll) damage (Bell and Treshow, 2002; Ågreen and Weih, 2012). It was found that inorganic and organic pollutants may affect the metabolism of carbohydrates, proteins, amino acids, and lipids; increase the permeability of cell membranes; inhibit the activity of many enzymes; decrease the amount of chlorophyll in the leaves; cause water deficiency; inhibit the process of photosynthesis and respiration; induce changes in the ultrastructure of chloroplasts and mitochondria; and lead to erosion cuticle, and epidermal and mesophilic cell damage, chlorosis, and necrosis (Viskari, 2000; Bell and Treshow, 2002; Vollenweider et al., 2003).

The analysis of physiological and biochemical responses of plants to the stress effects of natural and anthropogenic environmental factors provides information on plant changes, even when visible symptoms are absent, and allows to better understand the effects of individual stressors on metabolic processes of plants, and selection of stress-resilient species (Bell and Treshow, 2002; Gajić, 2007). Saxe (1996) indicates that the monitoring of the effects of toxic gases, PM, and heavy metals on the structural and functional characteristics of plants is essential for the early diagnosis of damage, selection of resistant species, and their adaptive strategies. Therefore, plant functional and structural changes are used as sensitive diagnostic parameters to indicate and monitor the toxic effects of pollutants on plants (Saxe, 1996). For the purpose of early detection and recognition of plant response to environmental stress, the following factors have been observed: (a) physiological processes, such as photosynthesis, respiration, and transpiration; (b) biochemical changes, such as amount of chlorophylls and carotenoids, activities of enzymes, the content of phenolic compound, and glutathione; (c) (ultra)structural damages of cells, such as leaf cuticles, epidermal and mesophilic cells, and stomata; and (d) visible leaf symptoms, such as chlorosis and necrosis (Bell and Treshow, 2002; Gajić et al., 2009, 2016, 2020) (Fig. 1). However, when these plant responses are analyzed separately, they are not always specific, and the most studies employ several different parameters by which it is possible to detect the effects of pollution and climatic stress on plants (Saxe, 1996).

2.1.1   Functional response of resilient plants to pollution and climatic change

Plant physiology

Water status—Measurements of water status of plants are fundamental for determining the effect of water stress on physiological processes, such as photosynthesis and respiration (Stevanović and Janković, 2001). In polluted areas, there is the appearance of water deficiency in plants as a result of high temperatures, intense insolation, low atmospheric humidity, small amounts of precipitation, and elevated concentrations of pollutants in the air and soil (Bombelli and Gratani, 2003). In the conditions of high air pollution, the supply of leaves with water is 10%–25% lower than in natural environment (Bell and Treshow, 2002). Water deficit and disruption of plant structure and function occur when the amount of transpired exceeds the amount of received water in the plant (Stevanović and Janković, 2001). If the drought lasts longer, the turgor is more difficult to restore, because it disturbs the balance between receiving and giving out water and the plants begin to wither. Lack of water affects the functioning of plants, i.e., the ability of leaves to assimilate CO_2 and roots to absorb essential nutrients. Drought induces closure of stomata, reduces the flow of electrons in PSI and PSII, and leads to inactivation of enzyme activity and reduction in net photosynthesis. Accordingly, the growth of shoots and leaves is reduced and the vitality of the whole plant is reduced. Therefore, plants of low vitality have low photosynthetic activity due to excessive water loss. Under water stress, there is a decrease in the leaf water potential, relative water content, and conductivity of the stoma (Grammatikopoulus, 1999).

Plants that are resistant to drought better tolerate the negative effects of increased concentrations of sulfur compounds in atmosphere (Stevanović and Janković, 2001). Air pollutants have a negative effect on the stoma and epidermal tissue, reducing their ability of regulation of excessive water loss (Wolfenden and Mansfield, 1991). Low concentrations of gaseous pollutants increase stomatal conductivity, and thus their penetration into the mesophile, while its high concentrations cause partial or complete closure of the stomata (Mansfield and Pearson, 1996; Bell and Treshow, 2002). Plant capabilities for water-efficient urban and industrial landscapes are significant for designers, managers, and policymakers (Sun et al., 2012).

Pavlović et al. (2004) found that plant species which are capable of surviving on fly ash deposits are characterized by very high RWC (relative water content) in leaves of *Tamarix gallica* (98.85%), *V. phlomoides* (92.51%), and *C. arvense* (92.51%); moderately high RWC in leaves of *C. setosa* (86.74%), *Epilobium collinum* (84.02%), and *Eupatorium cannabinum* (83.62%); and less favorable RWC in leaves of *Amorpha fruticosa* (75.47%), *P. alba* (75.88%), and *Ambrosia artemisiifolia* (77.83%) (Table 2). According to Skrynetska et al. (2018) in the industrial district of the city Sosnowiec (Poland), relative water content (RWC) ranged as follows: *Plantago major* (46.79%–88.06%), *P. lanceolata* (75.23%–92.19%), *Achillea millefolium* (60.10%–93.45%), and *R. pseudoacacia* (46.83%–91.95%) (Table 2). High RWC in these plants indicates that they were moderately sensitive or resilient to water stress despite elevated Pb and Zn concentration in the soil and leaves. In addition, Kaur and Nagpal (2017) found that in urban area with road traffic, RWC in plant leaves ranged in *Alstonia scholaris* from 72.88% to 87.73%, *Nerium oleander* from 73.13% to 90.37%, *Tabernaemontana coronaria* from 64.60% to 88.51%, and *Thevetia peruviana* from 63.20% to 79.64% (Table 2). Furthermore, during season (May–September) shrub plant species growing in the urban park next to a street with intensive traffic in Belgrade (Serbia) was characterized by a stable water balance: mean RWC in leaves of *L. ovalifolium*, *Syringa vulgaris*, and *Spiraea van-houttei* was 91%, 99.3%, and 85.3%, respectively (Gajić, 2007) (Table 2). The *S. vulgaris* and *L. ovalifolium* as xerophytes functionally better adapted to arid environmental conditions, whereas *S. van-houttei* adapted to a moderate amount of water as mesophyte. In the warm and dry months (May–July), *L. ovalifolium*, *S. vulgaris*, and *S. van-houttei* from the city park had a smaller amount of water in the leaves (86.6%–89.3%, 81.7%–85.1%, and 82.0%–87.3%, respectively)

Table 2 Water status in plant species growing at fly ash deposits, industrial, and urban areas.

Sites/pollutants/climatic factor	Parameters	Plant species	References
Fly ash deposits	RWC	*Tamarix gallica, Verbascum phlomoides, Cirsium arvense, Crepis setosa, Epilobium collinum, Eupatorium cannabinum, Amorpha fruticosa, Populus alba, Ambrosia artemisiifolia*	Pavlović et al. (2004)
Industrial district of the city	RWC	*Plantago major, Plantago lanceolata, Achillea millefolium, Robinia pseudoacacia*	Skrynetska et al. (2018)
Urban area with road traffic	RWC	*Alstonia scholaris, Nerium oleander, Tabernaemontana coronaria, Thevetia peruviana*	Kaur and Nagpal (2017)
Urban park next to the street with intensive traffic	RWC	*Ligustrum ovalifolium, Syringa vulgaris, Spiraea van-houttei*	Gajić (2007)
O_3, drought	Ψ	*Betula* sp.	Pääkkonen et al. (1998)
SO_2, NO_2	g_s	*Hordeum vulgare*	Atkinson et al. (1991)
Urban area	Φ_{PO}	*Acer monspessulanum, Cornus mas, Ginkgo biloba, Koelreuteria paniculata, Pyrus calleryana, Quercus cerris, Quercus frainetto, Syringa reticulata, Acer tataricum, Catalpa speciosa, Celtis occidentalis, Corylus colurna, Ostrya carpinifolia, Prunus sargentii, Tilia tomentosa, Ulmus parvifolia*	Sjöman et al. (2018)
Urban landscape	Ψxylem	*Artemisia tridentata, Holodiscus microphyllus, Ericameria nauseosa*	Martinson et al. (2019)
Water deficit Doubled CO_2	WUE	C3, C4 plant species	Earl (2002), Allen et al. (2003), Ainsworth and Rogers (2007)
Diffuse solar and PAR	WUE	*Zea mays*	Stanhill and Cohen (2001), Gao et al. (2018)

g_s, stomatal conductance; *RWC*, relative water content; *WUE*, water-use efficiency; *ΦPO*, leaf water potential at turgor loss point; *Ψ*, water potential; *Ψxylem*, xylem water potential.

compared to the period (August–September) which was rich in precipitation (91.0%–96.9%, 90.0%–96.26, and 91.8%–92.8%, respectively) (Gajić, 2007). Warm and dry habitat conditions and elevated concentrations of pollutants led to the occurrence of moderate water stress in all three shrub plant species, whereas in September, there was an increase in high relative air humidity, and despite the increased concentrations of air pollutants created a more favorable water balance (Gajić, 2007). In experimental conditions, Pääkkonen et al. (1998) found that in the combined action of O_3 and drought *Betula* sp. is characterized by significant reduction in water potential (Table 2). However, Atkinson et al. (1991) found that leaves of *Hordeum vulgare* are well supplied with water, even if low concentrations of SO_2 and NO_2 lead to stomatal closure (Table 2). According to Sjöman et al. (2018), leaf water potential at turgor loss point (Φ_{PO}) provides insight into plant potential to tolerate warm and dry conditions in the urban area. Thus, *Acer monspessulanum* (− 3.56 MPa), *Cornus mas* (− 3.03 MPa), *G. biloba* (− 3.45 MPa), *Koelreuteria paniculata* (− 3.50 MPa), *Pyrus calleryana* (− 3.03 MPa), *Quercus cerris* (− 3.12 MPa), *Q. frainetto* (− 3.06 MPa), and *Syringa reticulata* (− 3.45 MPa) are drought tolerant whereas *A. tataricum* (− 2.86 MPa), *Catalpa speciosa* (− 2.34 MPa), *C. occidentalis* (− 2.67 MPa), *Corylus colurna* (− 2.56 MPa), *Ostrya carpinifolia* (− 2.88 MPa), *Prunus sargentii* (− 2.86 MP), *Tilia tomentosa* (− 2.73 MPa), and *Ulmus parvifolia* (− 2.69 MPa) are moderately sensitive to drought stress (Table 2). These trees are capable of maintaining leaf gas exchange and growth at low soil water potential, recommending them for roof gardens, green bridges, and street canyons (Sjöman et al., 2018).

Furthermore, plants that are drought resilient and have low water requirements are important for water-efficient urban landscaping. Thus, canopy cover of woody trees and shrubs, and herbaceous perennial plant species is important for water-use landscaping (Sun et al., 2012). According to Martinson et al. (2019), xylem water potential of xerophytic plants in urban landscape, such as *Artemisia tridentata*, *Holodiscus microphyllus*, and *Ericameria nauseosa*, follows seasonal decline in soil moisture whereas in the driest months they show drought tolerance with no visual signs of stress (Table 2). Water-use efficiency (WUE) at the leaf level increased as CO_2 increased until the leaf was exposed to high temperatures, heat, and drought stress because the transpiration rate (E) and leaf temperature increased (Lopes et al., 2011; Hatfield and Dold, 2019). Under water-deficit stress, WUE increases because the reduction in net photosynthetic rate (An) is less than reduction in E and stomatal conductance (g_s) (Earl, 2002) (Table 2). However, with a doubling CO_2, in *Glycine max* and perennial crops g_s decreased which in turn may enable soil water conservation sustaining crop productivity (Allen et al., 2003; Ainsworth and Rogers, 2007) (Table 2). Changing climate increases clouds and aerosols that increase diffuse solar and PAR (photosynthetic active radiation), and this "solar dimming" increases a water vapor in the atmosphere, and in *Zea mays*, WUE can decrease with more direct PAR (Stanhill and Cohen, 2001; Gao et al., 2018) (Table 2).

Photosynthesis—Photosynthesis is a process that primarily determines the growth and productivity of plants and depends on CO_2 concentrations in the atmosphere, humidity, temperature, and mineral regime of plants (Lambers et al., 1998; Blankenship, 2002). The chlorophyll molecule is able to take over and transfer the absorbed energy to another chlorophyll molecule within the light-harvesting complex II (LHCII), which finally reaches the reaction center of photosystem II (PSII) (Blankenship, 2002). Electron flow continues from the excited reaction center PSII to PSI via primary and secondary quinone acceptors (Q_A and Q_B), cytochrome (b6/f), plastocyan (PC), and ferrodoxin (Fd). As a result of electron transport, redox equivalents (NADPH) and ATP synthesis are formed with the help of "coupling factor" (CF) (Trebst, 1995; Blankenship, 2002) (Fig. 2). Chlorophyll is released by excess energy by fluorescence or in the form of heat (Maxwell and Johnson, 2000) (Fig. 2). Photoinhibition of PSII occurs as a result of excessive excitation of the photosynthetic apparatus and can be dynamic and chronic (Osmond and Grace, 1995). Dynamic photoinhibition/photoprotection is associated with dissipation of excess energy as heat and recovery of photosynthetic activity occur through accumulation of zeaxanthin content in PSII antenna whereas chronic photoinhibition/photodamage requires more time for recovery/repair of reaction centers of PSII through the replacement of D1 protein (Osmond and Grace, 1995; Aro et al., 2005). Photoinhibition of PSII at the donor site is associated with dissociation of the LHCPII complex, reduced electron flow through OEC (oxygen-evolving complex), and TyrZ all the way to the PSII reaction center ($P680^+$), resulting in conformational changes in D1 protein and its degradation (Fo), as well as reduced thermal dissipation of excited energy via the xanthophyll cycle (Fv) (Mysliwa-Kurdziel et al., 2002). Photoinhibition at the PSII acceptor site is associated with slowed or blocked electron transfer from Q_A to Q_B (Fm) and a reduced amount of PQ pool ($t_{1/2}$) (Mysliwa-Kurdziel et al., 2002).

Chlorophyll *a* fluorescence emitted from the thylakoid membrane of the chloroplast is a useful indicator of the photochemistry activity (Maxwell and Johnson, 2000;

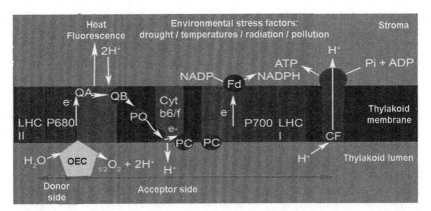

Fig. 2 Effects of environmental stress factors: Drought/temperatures/radiation/pollution on photosynthesis in plant species: excited energy and electron transfer, molecule migration and chemical reactions: P680 (PSII reaction center); P700 (PSI reaction center); LHCII (light-harvesting complex, PSII pigment antenna); LHCI (light-harvesting complex, PSI pigment complex); OEC (oxygen-evolving complex); Q_A, Q_B (electron acceptors, quinones); PQ (plastoquinones); Cytb6/f (cytochrome *b*6/f); PC (plastocyan); Fd (ferrodoxin); CF (coupling factor); redox equivalents (NADPH); and energy equivalents (ATP).

Baker and Rosenqvist, 2004; Baker, 2008). Quenching of the chlorophyll fluorescence can be achieved by combination of photochemical and nonphotochemical quenching. Photochemical quenching (qP) of chlorophyll fluorescence represents the actual fraction of active centers of PSII, increasing the availability of sinks for the electrons which are transported away from PSII (Baker and Rosenqvist, 2004). Nonphotochemical quenching of chlorophyll fluorescence is associated with conversion of violaxanthin to zeaxanthin in the xanthophylls cycle that promotes thermal dissipation in the PSII antenna complexes (NPQ) (Baker and Rosenqvist, 2004).

Chlorophyll fluorescence is used as a nondestructive method for interpreting the mechanisms of action of various environmental stressors, such as intense light, high/low temperatures, drought, air pollutants, and heavy metals on the functioning of PSII activity (Maxwell and Johnson, 2000; Baker and Rosenqvist, 2004; Baker, 2008; Gajić et al., 2018a). The environmental stress factors affect photosynthesis and disrupt the structure of the reaction center and the electron transfer efficiency; i.e., they may alter the structure and function of components of electron transfer chain, oxygen-evolving activity (OEC), decreasing the chlorophyll a fluorescence, CO_2 fixation, and gas exchange (Mysliwa-Kurdziel et al., 2002). Photosynthetic efficiency of plants in urban and industrial areas has proven to reduce as a result of continuous pollution exposure, adverse microclimatic factors, and accumulation of physiological and morphological leaf damages (Gajić, 2007; Gajić et al., 2009).

- *Fly ash deposits.* Photosynthetic efficiency (Fv/Fm) of plants growing on fly ash deposits from the "Nikola Tesla—A" thermal power plant (Serbia) was below optimal range for plants (0.750–0.850, Bjorkman and Demmig, 1987) as the result of elevated/reduced concentrations of metal(loid)s and xerothermic site conditions and indicates high sensitivity of PSII to stress—trees: *P. alba* (0.541–0.625) and *R. pseudoacacia* (0.563–0.675) (Kostić et al., 2012); shrubs: *T. tetrandra* (0.567–0.727) and *A. fruticosa* (0.650–0.684) (Pavlović et al., 2004; Kostić et al., 2012); perennial: *A. artemisiifolia* (0.559), *E. cannabinum* (0.535), *C. setosa* (0.558), *E. collinum* (0.472), *V. phlomoides* (0.620), and *C. arvense* (0.429) (Pavlović et al., 2004); and grasses: *C. epigejos* (0.581–0.585), *F. rubra* (0.543–0.691), and *D. glomerata* (0.633) (Mitrović et al., 2008; Gajić et al., 2016, 2020) (Table 3). However, over time *F. rubra* growing on fly ash deposits showed higher values of Fv/Fm, Fm (maximal chlorophyll fluorescence), and Fv (variable chlorophyll fluorescence) and no changes in Fo (initial chlorophyll fluorescence) and $t_{1/2}$ (half time required to reach maximum fluorescence) indicating an increase in the maximum quantum yield of primary PSII chemistry and number of "open" reaction centers where PSII acceptors are capable of performing photochemistry (Gajić et al., 2016) (Table 3). Similarly, in *D. glomerata* from fly ash deposits, Fo and $t_{1/2}$ were increased under As stress, despite reduced efficiency of photosynthesis (Fv/Fm, Fm, and Fv) indicating that this species possesses some photoprotective mechanisms to maintain the PSII photochemistry activity (Gajić et al., 2020) (Table 3). In addition, Pandey et al. (2020) found that Fv/Fm was 0.66 in *I. carnea* and 0.77 in *Prosopis juliflora* growing on fly ash deposits from thermal power plant (Uttar Pradesh, India), whereas Fv'/Fm' (PSII maximum efficiency from light-adapted leaves) was low in *S. spontaneum* (0.37) which can be associated with high nonphotochemical quenching of chlorophyll fluorescence (qN) (0.47). Furthermore, *I. carnea* showed high photochemical quenching of chlorophyll fluorescence (qP) that is associated with high effective quantum yield of PSII photochemistry (ΦPSII) and high net photosynthetic rate (An); *I. carnea* also showed the

Table 3 Photosynthetic parameters in plants growing at fly ash deposits and urban areas.

Sites/pollutants	Parameters	Plant species	References
Fly ash deposits	Fv/Fm	*Populus alba, Robinia pseudoacacia*	Kostić et al. (2012)
		Tamarix tetrandra, Amorpha fruticosa	Pavlović et al. (2004), Kostić et al. (2012)
		Ambrosia artemisiifolia, Eupatorium cannabinum,	Pavlović et al. (2004)
		Crepis setosa, Epilobium collinum, Verbascum	
		phlomoides, Cirsium arvense, Calamagrostis	
		epigejos	
	Fv/Fm, Fm, Fv, Fo, $t_{1/2}$, Fm/	*Festuca rubra, Dactylis glomerata*	Mitrović et al. (2008), Gajić et al. (2016),
	Fo		Gajić et al. (2020)
	Fv/Fm, ΦPSII, qP, qN, An,	*Ipomea carnea, Prosopis juliflora, Saccharum*	Pandey et al. (2020)
	g_s, E	*spontaneum*	
	Fv/Fm, Fo, qP, NPQ, An, g_s	*Ricinus communis*	Panda et al. (2020)
Urban areas			
CO, NO$_x$, and SO$_2$	Fv/Fm, An	*Fragaria × ananassa*	Muneer et al. (2014)
Roadside	An, g_s, E	*Grevillea robusta, Mangifera indica*	Singh et al. (2020)
PM	Fv/Fm	*Betula pendula, Forsythia × intermedia,*	Popek et al. (2018)
		Physocarpus opulifolius, Platanus × hispanica,	
		Hedera helix, Sorbus intermedia, Spiraea japonica	
Pb	Fv/Fm, Fm, Fv, Fo, $t_{1/2}$, Fm/	*Ligustrum ovalifolium*	Gajić et al. (2009)
	Fo		
	Fv/Fm	*Syringa vulgaris, Spiraea van-houttei*	Gajić (2007)
Pb, Cu, and Zn	Fv/Fm	*Taraxacum* spp., *Sonchus* spp.	Sgardelis et al. (1994)
SO$_2$	Fv/Fm	*Pinus sylvestris, Betula pubescens, Vaccinium*	Odasz-Albrigtsen et al. (2000)
		myrtillus	
O$_3$	Fv/Fm	*Fagus sylvatica*	Gerosa et al. (2003)
O$_3$	Fv/Fm	*Phaseolus vulgaris*	Guidi et al. (2000)
O$_3$	Fv/Fm	*Fraxinus excelsior*	Gerosa et al. (2003)

An, net photosynthetic rate; *E*, transpiration rate; *Fm*, maximum chlorophyll fluorescence; *Fo*, minimal chlorophyll fluorescence; *Fv*, variable chlorophyll fluorescence; *Fv/Fm*, photosynthetic efficiency; *g_s*, stomatal conductance; *NPQ*, nonphotochemical quenching of chlorophyll fluorescence; *qN*, nonphotochemical quenching of chlorophyll fluorescence from a dark-adapted leaf at Fm; *qP*, photochemical quenching of chlorophyll fluorescence; *t1/2*, half time required to reach maximum fluorescence; *ΦPSII*, effective quantum yield of photochemical activity.

highest values of stomatal conductance (g_s) and transpiration rate (E) (Pandey et al., 2020) (Table 3). According to Panda et al. (2020), *Ricinus communis* growing on fly ash showed reduced values of CO_2 photosynthetic rate (10.52 μmol CO_2 m^{-2} s^{-1}), stomatal conductivity (g_s, 46.23.50 mmol H_2O m^{-2} s^{-1}), Fv/Fm (0.639), Fm (1224), and qP (0.871) compared to garden soil (18.51 μmol CO_2 m^{-2} s^{-1}, 77.50 mmol H_2O m^{-2} s^{-1}, 0.789, 1360, and 0.981, respectively), whereas Fo (454), and NPQ (0.054) were increased compared to garden soil (295 and 0.021, respectively) (Table 3).

- *Urban areas.* In urban areas with high pollution level, Muneer et al. (2014) showed that CO, NO_x, and SO_2 led to the reduction of net photosynthetic rate (An) and photosynthetic efficiency (Fv/Fm) in *Fragaria × ananassa* (Table 3). Singh et al. (2020) found that CO_2 assimilation rate (An), stomatal conductance (g_s), and transpiration rate (E) declined in *Grevillea robusta* (9.21 μmol CO_2 m^{-2} s^{-1}, 0.64 mol H_2O m^{-2} s^{-1}, and 0.77 mmol H_2O m^{-2} s^{-1}, respectively) and *Mangifera indica* (9.21 μmol CO_2 m^{-2} s^{-1}, 7.23 μmol CO_2 m^{-2} s^{-1}, and 3.21 mmol H_2O m^{-2} s^{-1}, respectively) growing at urban roadside compared to *G. robusta* (9.66 μmol CO_2 m^{-2} s^{-1}, 0.84 mol H_2O m^{-2} s^{-1}, and 1.33 mmol H_2O m^{-2} s^{-1}, respectively) and *M. indica* (10.81 μmol CO_2 m^{-2} s^{-1}, 0.81 mol H_2O m^{-2} s^{-1}, and 4.78 mmol H_2O m^{-2} s^{-1}, respectively) growing in forest area (Table 3). However, Popek et al. (2018) found that in *Betula pendula*, *Forsythia × intermedia*, *Physocarpus opulifolius*, *Platanus × hispanica*, and *Hedera helix* exposed to high PM deposition in city center (Warsaw, Poland) photosynthetic rate (An) declined, except in *Sorbus intermedia* and *Spiraea japonica*, whereas photosynthetic efficiency (Fv/Fm) did not significantly decrease in all plants, except in *H. helix*, and all values are in the optimal range for plants indicating their resilience to particulate matter (Table 3).

Gajić et al. (2009) showed that Pb derived from traffic can decrease photosynthetic efficiency (Fv/Fm, Fo, Fv, Fm) in *L. ovalifolium* representing negative Pb effect at the donor side of PSII decreasing energy transfer from LHCII to the PSII reaction center through structural changes in antenna pigments and damage of proteins of oxygen-evolving complex (OEC) and D1 protein. Photosynthetic efficiency (Fv/Fm) was in the optimal range for plants (0.788 in June and 0.737 in October), indicating that this plant species tolerates Pb possessing active photoprotective mechanisms, such as relief of electron pressure at the acceptor side of PSII (increase of $t_{1/2}$) that enables electron flow from PSII to PSI (Gajić et al., 2009) (Table 3). Sgardelis et al. (1994) showed that Pb, Zn, and Cu lead to the reduction of photosynthetic efficiency (Fv/Fm) in *Taraxacum* spp. and *Sonchus* spp. on different locations in the center of Thessaloniki (Table 3). Photosynthetic efficiency (Fv/Fm) in *S. vulgaris* growing in the city park next to street with intensive traffic was below optimal range for plants (0.707 in June), indicating the sensitivity of PSII activity to NO_2, SO_2, and PM compared to control site (0.808 in June) (Gajić, 2007) (Table 3). Similarly, in *P. sylvestris*, *Betula pubescens*, and *Vaccinium myrtillus* exposed to SO_2 released from smelter, photosynthetic efficiencies (Fv/Fm) are 0.673, 0.691, and 0.715, respectively (Odasz-Albrigtsen et al., 2000) (Table 3). Gerosa et al. (2003) found that photosynthetic efficiency of (Fv/Fm) *F. sylvatica* exposed to O_3 was lower (0.723) compared to control (0.758) (Table 3). However, photosynthetic efficiency (Fv/Fm) in *S. van-houttei* growing at a city park near the street (0.816 in June and 0.785 in October) and control site (0.824 in June and 0.791 in October) was in the optimal range (Gajić, 2007) (Table 3). Similar photosynthetic activity was measured in experimental conditions between control and leaves treated with high concentrations of O_3 in some varieties of *Phaseolus vulgaris*: the values of Fv/Fm in control plants ranged from 0.817 to 0.823, and in plants exposed to O_3 from 0.807 to 0.813 (Guidi et al., 2000) (Table 3). *Fraxinus excelsior* that was exposed to elevated concentrations of O_3 showed a high photosynthetic activity (Fv/Fm): 0.808 in undamaged leaves, 0.812 in slightly damaged leaves, 0.811 in moderately damaged leaves, and 0.789 in significantly damaged leaves (Gerosa et al., 2003) (Table 3).

Plant biochemistry

Plants are capable of synthesizing metabolites that have a crucial role in plant growth and development as well as in surviving in stressful environmental conditions, such as heat, drought, salinity, flooding, heavy metals, PAHs, and PCBs (Nakabayashi and Saito, 2015; Gajić et al., 2018b). Metabolites of abiotic stress-resilient plants are useful in exploring plant physiology and biochemistry under stressful conditions (Abdelrahman et al., 2018). Stress-resistant metabolites can be used as stress biomarkers to understand metabolic network in maintaining the balance of plant cell with the environment (Nakabayashi and Saito, 2015; Abdelrahman et al., 2018).

Chlorophylls (Chl *a* and Chl *b*) are photosynthetic pigments that are capable of absorbing light as part of LHCP complex (light-harvesting protein complex), and they form the core of photochemical reaction centers (PSI and PSII) (Lambers et al., 1998; Blankenship, 2002). Carotenoids efficiently transfer energy to chlorophylls and have a photoprotective function, because they participate in the nonphotochemical quenching of the excited energy of PSII and the dissipation of excess energy in the form of heat (Smirnoff, 2005). According to Telfer (2002), β-carotene inactivates $^1O_2^*$ in the PSII reaction center, while xanthophylls are involved in the nonphotochemical quenching of excited energy (NPQ) in the LHCP complex (Li et al., 2000). Thus, the components of the xanthophyll cycle (zeaxanthin—Zx, antheraxanthin—Ax, violaxanthin—Vx) divert excess energy from the PSII reaction centers and remove the triplet state of chlorophyll (^{3}Chl *) and singlet oxygen ($^1O_2^*$) (Havaux, 1998). In this way, carotenoids protect chlorophylls from photooxidation, preventing excess energy from generating reactive oxygen species (ROS). Environmental stressors such as intense light, radiation, drought, and pollutants can cause chloroplast membrane destruction and inhibit chlorophyll synthesis and the activity of enzymes involved in the photosynthesis process (Gajić et al., 2009; Mysliwa-Kurdziel et al., 2002). Anthocyanins are pigments that belong to phenolics, i.e., flavonoids (cyanidin glucosides), and they have the protective function of photosynthetic apparatus from stress, reducing membrane lipid peroxidation and can directly remove $O_2^{\bullet-}$ radical, OH•, and H_2O_2 (Yamasaki et al., 1996, 1997; Neill and Gould, 2003) reducing the amount of absorbed light absorbed, thereby reducing photoinhibition of PSII and they are responsible for the redox state of the PQ pool in photosynthetic electron transport (Das et al., 2011).

Malondialdehyde (MDA) is the final product of membrane lipid peroxidation, and it can be used as an indicator of free radicals and membrane damage (Weber et al., 2004). Under "oxidative stress," many functions of the membrane system are impaired, such as electron transport, ATP production, and metabolic compartmenting, because ion pumps, ion channels, and transport proteins are located in them (Rao et al., 2006).

Resilience of plants to environmental stressors is achieved by adaptive antioxidant machinery (Gill and Tuteja, 2010). Maintenance of cellular redox homeostasis is achieved by activating the antioxidant system of plants that consists of a complex of control systems that allow them to cope with stress with minimal damage to cell structure and function (Foyer and Noctor, 2005; Halliwell, 2006). Efficient removal of ROS is enabled by nonenzymatic (ascorbic acid, AsA; carotenoids; α-tocopherol, vitamin E; glutathione, GSH; phytochelatins, PCs; metallothioneins, MTs; phenols, proline) and enzymatic (superoxide dismutase, SOD; ascorbate peroxidase, APX; glutathione reductase, GR; catalase, CAT; glutathione S transferase, GST) components of antioxidant system (Mittler, 2002; Foyer and Noctor, 2005). The balance between ROS production and antioxidant biosynthesis is very important in the redox control of homeostasis in the apoplast and cytoplasm of the plant cell (Halliwell, 2006; Foyer and Shigeoka, 2011). The activities of antioxidants depend on plants species, type of pollutants, exposure time, and environmental condition (Gratão et al., 2005). Therefore, environmental stress activates antioxidant plant defense system; i.e., some plant species activate the redox-linked signaling network and compensation mechanisms (Sharma and Dietz, 2009).

- Tocopherol (α-tocopherol) can scavenge ROS inhibiting lipid peroxidation and protect photoprotection of PSII (Smirnoff, 2005). Proline is an amino acid that can act as osmoprotectant, protein stabilizer, metal chelator, and scavenger of ROS (Alia et al., 2001).
- Phenolic compounds are secondary metabolites that have a structural role in plants building the cell wall, and they can act as active antioxidants (Grace, 2005). Phenolics have the ability to directly remove ROS ($O_2^{\bullet-}$, OH•, H_2O_2, $^1O_2$, peroxyl radical) by donating electrons or H atoms, chelate metal ions, and alter the kinetics of peroxidation by modifying lipid packaging and reducing membrane fluidity (Rice-Evans et al., 1996).
- Ascorbic acid (AsA) can be an antioxidant; a cofactor of the enzyme violaxanthin de-epoxidase, which leads to the formation of zeaxanthin; a cofactor for enzymes that synthesize flavonoids; has a role in the regeneration of α-tocopherol; and can alleviate the inactivation of reactive centers of PSII by acting as an alternative electron donor (Mano et al., 1997; Smirnoff, 2005).
- Glutathione (γ-glutamylcysteinylglycine) is a hub molecule involved in biosynthetic pathways, antioxidant reactions, redox signaling, and redox homeostasis (Noctor et al., 2012). Glutathione can scavenge ROS, and in photosynthesis, it takes part in the control of H_2O_2 levels through AsA-GSH pathway; regenerates AsA; acts as a precursor for the biosynthesis of phytochelatins (PCs) and metallothioneins (MTs); and acts as the substrate for enzymes, such as glutathione peroxidases (GPXs) and glutathione S transferases (GSTs) (Foyer et al., 2005; Foyer and Noctor, 2011; Noctor et al., 2012, 2013).
- *Antioxidant enzymes.* Superoxide dismutase (SOD) is the enzyme which converts the superoxide anion radical ($O_2^{\bullet-}$) by the photoreduction of O_2 at PSI to the H_2O_2 and O_2 (Asada, 1999; Miller, 2012). The ascorbate peroxidase (APX) enzyme is essential in the removal of H_2O_2 together with glutathione reductase (GR), as part of the ascorbate-glutathione Asada-Halliwell pathway (AsaA-GSH cycle) (Mittler and Poulos, 2005). Furthermore, glutathione peroxidase (GPX) catalyzes the reduction of H_2O_2 and lipid peroxide uses GSH and thioredoxin (TRX) as a reducing substrate (Noctor et al., 2002; Anjum et al., 2012). Glutathione-S-Transferases (GST) is an enzyme which can catalyze the conjugation of GSH with toxic organic compounds to form derivatives that can be sequestered in the vacuole or catabolized (Anjum et al., 2012).

Fly ash deposits—According to Gajić et al. (2018a), *F. rubra* sown on fly ash deposits in the condition of elevated concentrations of As and B and low concentrations of Cu, Zn, and Mn in leaves had high content of MDA (1.25–1.87 nmol/g) and reduced content of chlorophylls (Chl *a*—2.46–3.21 mg/kg; Chl *b*—0.90–1.22 mg/g; Chl *a* + *b*—3.36–4.43 mg/g) and carotenoids (0.78–0.87 mg/g) compared to control site (0.79 nmol/g; 3.60 mg/g; 1.26 mg/g; 4.86 mg/g; 1.01 mg/g, respectively) indicating oxidative stress, whereas concentrations of anthocyanins (0.563–0.620 mg/g), phenolics (Free Ph—12.85–13.36 mg/g; Bound Ph—26.61—32.97 mg/g; Tot Ph—39.46–45.74 mg/g), ascorbic acid (1.076–1.142 mg/g), and total antioxidant activity (0.254) increased compared to control site (0.501 mg/g; 7.85 mg/g; 17.96 mg/g; 25.89 mg/g; 0.595 mg/g; 0.288 mg/mL, respectively) indicating high adaptive/resilient potential of this species to stress (Table 4). Similarly, excess As in leaves of *D. glomerata* sown on fly ash deposits decreases chlorophyll (Chl *a*—4.30 mg/g; Chl *b*—1.12 mg/g; Chl *a* + *b*—5.42 mg/g), carotenoid (1.33 mg/g), and anthocyanin contents (0.905 mg/g) compared to control site (6.49 mg/g; 2.09 mg/g; 8.58 mg/g; 1.77 mg/g; 1.270 mg/g, respectively), and high content of MDA (0.44 nmol/g) can be a signal for activation of antioxidant machinery, such as increased content of phenolics (Free Ph—18.16 mg/g; Bound Ph—16.04 mg/g; Tot Ph—34.20 mg/g), AsA (1.07 mg/g), and total scavenging activity (14.75%) compared to control site

Table 4 Metabolites in plants growing at fly ash deposits, mine waste, and urban areas.

Sites/pollutants	Parameters	Plant species	References
Fly ash deposits	Chl *a*, Chl *b*, Chl *a* + *b*, Tot Carotenoids, Anthocyanins, MDA, Free Phenolics, Bound Phenolics, Total Phenolics, AsA, DPPH total antioxidant capacity	*Festuca rubra* *Dactylis glomerata*	Gajić et al. (2016) Gajić et al. (2020)
	Chlorophylls, Carotenoids, MDA, proline, sugar, SOD, APX, POD, CAT	*Pithecellobium dulce*	Qadir et al. (2019)
	Chlorophylls, Carotenoids, AsA, Phenolics	*Beta vulgaris*	Singh et al. (2008)
Fly ash-amended soil	Chlorophylls, Carotenoids, MDA, SOD, GR, POD, CAT, GST	*Cicer arietinum*	Pandey et al. (2010)
	Chlorophylls, Carotenoids, MDA, cysteine, NPSH	*Thelypteris dentata*	Kumari et al. (2013)
	Chlorophylls, Carotenoids, MDA, protein, AsA, Proline, Cysteine, NPSH	*Sesbania cannabina*	Sinha and Gupta (2005)
	Chlorophylls, Carotenoids, Cysteine, NPSH	*Oryza sativa*	Dwivedi et al. (2007)

Continued

Table 4   Continued

Sites/pollutants	Parameters	Plant species	References
Mine waste	Chlorophylls, Carotenoids, MDA, Free Phenolics, Bound Phenolics, Total Phenolics	*Epilobium dodonaei*	Randjelović et al. (2016)
	ROS, Chlorophylls, Carotenoids, Proteins, Starch, Soluble sugars, Soluble Phenolics, Bound Phenolics, GSH, PAL, PRX	*Dittrichia viscosa*	Lopez-Orenes et al. (2018)
	MDA $O_2{}^{\cdot-}$, H_2O_2, Chlorophylls, Proteins, Proline, AsA, GSH, NPSH, Phenolics, Soluble sugars, PAL, sPRX, SOD, DPPH	*Zygophyllum fabago*	Lopez-Orenes et al. (2017)
Metallurgy and Mining activities	Proline, GSH, NPSH, SOD, POD	*Cardaminopsis arenosa, Plantago lanceolata*	Nadgórska-Socha et al. (2013)
Mine tailings	Chlorophylls, MDA, SOD, POD, CAT, APX	*Arrhenatherum elatius, Sonchus transcapicus*	Lu et al. (2013)
Urban areas			
CO, NO*x*, and SO$_2$	H_2O_2, $^1O_2$, Total Carbohydrate, Sucrose, SOD, CAT, GR, APX	*Fragaria* × *ananassa*	Muneer et al. (2014)
SO$_2$	H_2O_2, MDA, AsA, GSH, GSH/GSSG, APX, GPX, GR, GST	*Tagetes erecta*	Wei et al. (2015)
NO$_2$	Chl *a*, Chl *b*, Carotenoids, MDA, POD	41 plant species with different functional groups	Shun and Zhu (2019)
PM	Total Chlorophyll, AsA	*Azadirachta indica, Mangifera indica, Delonix regia, Cassia fistula*	Thawale et al. (2011)
Pb	Chlorophylls, Carotenoids, Free Phenolics, Bound Phenolics, Total Phenolics	*Ligustrum ovalifolium*	Gajić et al. (2009)

APX, ascorbate peroxidase; *AsA*, ascorbic acid; *CAT*, catalase; *Chl a*, chlorophyll *a*; *Chl a + b*, total chlorophyll; *Chl b*, chlorophyll *b*; *DPPH*, total scavenging activity; *GPX*, glutathione peroxidase; *GR*, glutathione reductase; *GSH*, glutathione; *GSH/GSSG*, ratio of reduced glutathione and oxidized glutathione; *GST*, glutathione S transferase; *H2O2*, hydrogen dioxide; *MDA*, malondialdehyde; *NPSH*, nonprotein glutathione; *1O2*, singlet oxygen; *PAL*, phenylalanine-ammonia-lyase; *POD*, peroxidase; *PRX*, peroxidase; *ROS*, reactive oxygen species; *SOD*, superoxide dismutase; *sPRX*, class III peroxidases.

(13.56 mg/g; 11.45 mg/g, 25.01 mg/g; 0.49 mg/g; 24.22%, respectively) indicating high adaptive capacity to As stress (Table 4). Qadir et al. (2019) found that in tree leaves of *Pithecellobium dulce* (Fabaceae) growing on fly ash dumps, chlorophylls and carotenoids decreased while the content of MDA, proline, and sugar content as well as SOD, APX, POD, and CAT activities increased compared to control site indicating the tolerance of this tree to stress (Table 4). Singh et al. (2008) showed that in leaves of *Beta vulgaris* grown on fly ash with high levels of Cu, Cd, and Zn, chlorophyll and carotenoid contents decreased whereas the content of ascorbic acids (AsA) and phenols increased (Table 4).

In the experimental conditions, *Cicer arietinum* grown in fly ash-treated soil showed reduced content of chlorophylls whereas carotenoid content, MDA content, and SOD, GR, POD, CAT and GST activities increased (Pandey et al., 2010) (Table 4). In addition, in fern *T. dentata* grown on fly ash-amended soils, concentrations of chlorophylls increased significantly in treatment (75% fly ash and 25% garden soil) and then dropped in treatment with 100% of fly ash whereas carotenoid content and MDA increased with increased fly ash (Kumari et al., 2013) (Table 4). However, this plant showed tolerance to stress due to elevated levels of cysteine and nonprotein thiols (NPSH) (Kumari et al., 2013). Similarly, Sinha and Gupta (2005) showed that in *Sesbania cannabina* grown on different amendments of fly ash, MDA content, protein, AsA, free proline, cysteine, and NPSH content increased in roots whereas in leaves, chlorophyll and carotenoid contents also increased together with MDA content and antioxidants indicating the resilience of this species to elevated levels of Fe, Mn, Cu, Zn, Pb, and Ni in fly ash. In *Oryza sativa* grown in fly ash-amended soil with high concentrations of Fe, Si, Cu, Zn, Mn, Ni, Cd, and As in leaves, concentrations of chlorophylls and carotenoids decreased, but the content of cysteine and NPSH increased (Dwivedi et al., 2007) (Table 4).

Mine waste—Epilobium dodonaei grown on mine waste with elevated As and Cu concentrations were characterized by high content of MDA in roots and leaves, reduced levels of chlorophylls and carotenoids in leaves, and enhanced levels of phenolics and total scavenging activity in leaves and roots compared to control site which indicates that oxidative stress promotes antioxidant protection and its resilience to stress (Randjelović et al., 2016) (Table 4). *Dittrichia viscosa* which grows at mine tailings showed that leaf content of ROS, chlorophylls, proteins, starch and soluble sugars, and GSH was not significantly different between spring and summer populations whereas carotenoids, soluble and cell wall phenolics, phenylalanine-ammonia-lyase (PAL), and peroxidase (PRX) activity were significantly higher in summer than in spring populations indicating the activation of antioxidants and redox buffers that maintained redox homeostasis in stress conditions (Lopez-Orenes et al., 2018) (Table 4). Similarly, oxidative/antioxidative profile of *Z. fabago* at mine tailings showed a decrease in chlorophylls, protein, and proline content and an increase in AsA, GSH, NPSH, phenolics, and soluble sugars content as well as PAL, sPRX (class III peroxidases), and SOD activities together with high antioxidant capacity (DPPH, FRAP, ABTS) that decreased MDA level and $O_2^{\bullet-}$ and H_2O_2 content (Lopez-Orenes et al., 2017) (Table 4). Nadgórska-Socha et al. (2013) found that in leaves of *Cardaminopsis arenosa* grown at metalliferous sites (metallurgy and mining activity), proline content was positively correlated with Fe, GSH content with Cd, and SOD and POD activities with Fe and Mn whereas in leaves of *P. lanceolata*, proline content was negatively correlated with Zn, GSH positively correlated with Pb, NPSH content with Pb and Cu, SOD activity with Mn and Fe, and POD activity positively correlated with Cd and Pb (Table 4). Furthermore, in roots of *Arrhenatherum elatius* and *Sonchus transcapicus* growing in metal polluted site (mine tailings) MDA content, SOD, POD, CAT, and APX activities increased whereas in their leaves chlorophyll content decreased, and SOD (except of *A. elatius*), POD, CAT, and APX increased (except of *A. elatius*) indicating that *S. transcapicus* had more efficient antioxidant system in stress conditions (Lu et al., 2013) (Table 4).

Urban areas—In *Fragaria* × *ananassa* exposed to high levels of CO, NO_x, and SO_2, concentrations of H_2O_2 and $^1O_2$ were increased, total carbohydrate and sucrose decreased, but SOD, CAT, GR and APX activities indicating that gaseous pollutants can induce antioxidative response (Muneer et al., 2014) (Table 4). Wei et al. (2015) found that SO_2 in *Tagetes erecta* induce oxidative stress by increased H_2O_2 and MDA content that activates antioxidative machinery, such as enhanced content of AsA, GSH, ratio of GSH/GSSG, and activity of APX, GPX, GR, and GST (Table 4). (Shun and Zhu (2019)) researched the effect of NO_2 (72 h) to 41 plant species from different functional groups to biochemical leaf traits and compared to natural recovery (NR, after 30 days). Thus, content of Chl *a*, Chl *b*, and carotenoids in different life forms (herb, shrub, tree, evergreen, deciduous, broadleaf, needle-like), phylogeny (Gymnosperm, Angiosperm, Monocotyledon, Dicotyledon), and photosynthetic pathway (C3 and C4 herbs) was lower in NO_2 treatment in comparison with recovery treatment (except carotenoid content in evergreen plant species) and that can be associated with higher MDA concentrations (especially in shrubs, deciduous, needle-like, Gymnosperm, Dicotyledon plant species) and higher protein content and POD activity in all plants (except POD in deciduous plant species) indicating the oxidative stress and activation of protection enzyme (Table 4). Furthermore, in leaves of *Azadirachta indica*, *M. indica*, *Delonix regia*, and *Cassia fistula* total chlorophyll and AsA contents were higher at city sites (India, commercial/residential with automobile) that are exposed to high concentrations of particulate matter (PM), except chlorophyll content in *A. indica* indicating high tolerance capacity and their resilience to pollution stress (Thawale et al., 2011) (Table 4). Gajić et al. (2009) showed that *L. ovalifolium* sown as hedgerow near street with intensive traffic is a plant species resilient to traffic-generated Pb because it reduces chlorophylls and carotenoids content, but it can maintain optimal photosynthesis increasing the content of soluble and cell wall phenolics (Table 4). Soluble phenolics in plant cells, such as sinapate esters, can protect PSII reaction center preventing degradation of D1/D proteins (Booij-James et al., 2000) or mesophyll ortho-dihydroxylated B-ring flavonoids that can scavenge ROS and efficiently dissipate excess energy (Rice-Evans et al., 1996; Tattini et al., 2005).

Lipidomics-based understanding of adaptation between plants and climate change/pollution

Lipidomics is a prominent field of metabolomics that addresses lipids as biomolecules and their response to environmental stressors, such as climate change and pollution

(Koelmel et al., 2020). Lipids have essential structural and signaling role in plants and participate in energy storage (Quartacci et al., 1995). The thylakoid membrane in chloroplast harbors photosynthetic machinery and consists of 50%–80% monogalactosyldiacylglycerol (MGDG) and 12%–26% digalactosyldiacylglycerol (DGDG), phosphatidylglycerol (PG), and sulfoquinovosyldiacylglycerol (SQDG), and according to Dörmann (2013), lipids in photosynthetic membranes have bilayer functions stabilizing membrane architecture and formation of grana thylakoid (grana stacks). The maintenance of membrane fluidity and permeability in stress conditions is achieved by lipid remodeling and changes in the degree of fatty acid saturation (Zheng et al., 2011). Stress factors may lead to the excessive fluidization, disrupting lipid bilayer, causing membrane damages, and affecting photosynthetic activity (chloroplast light energy capture and electron transport) (Los and Zinchenko, 2009). However, plants under stress condition are capable of limiting functional and structural damages due to enhanced biosynthesis of lipid soluble molecules, such as carotenoids (β-carotenoids, lutein, neoxanthin, zeaxanthin, violaxanthin, and antheraxanthin) and phenylquinones (tocopherol, plastoquinone) that act as membrane-protective antioxidant molecules, and they together with activation of plastoglobule-localized enzymes are involved in the lipid metabolic pathways (Larkindale and Huang, 2004). Furthermore, under stress conditions, plants are capable of activating lipid enzymes (phospholipases, acyl hydrolases, phytosphingosine kinases, diacylglycerol kinases, fatty acid amide hydrolases) that take a part in biosynthesis of signaling lipids (lysophospholipids, fatty acids, phosphatidic acid, inositol phosphate, diacylglycerols, oxylipins, sphingolipids, *N*-acylethanolamine) (Hubac et al., 1989; Kaoua et al., 2006). Under high/low temperatures, heat, and drought stress, chloroplast membrane lipids are being remodeled through synthesis/modification of fatty acids (FA) changing its unsaturation degree, content, and composition as well as the balance of lipid classes in order to maintain membrane stability and optimization of photosynthetic activity (Falcone et al., 2004; Scotti-Campos et al., 2014). The reduction in trienoic fatty acid (TFA) and the increase in saturated FAs are important to maintain membrane fluidity. If PUFAs (polyunsaturated fatty acids) increase membrane lipoperoxidation, damage can be compensated by reinforcing the antioxidative system which prevents further ROS production contributing to high photosynthetic efficiency (Martins et al., 2016).

Heat stress—*Coffea* spp. showed resilience to heat and elevated CO_2 concentrations and this plant was capable of maintaining membrane fluidity and its integrity changing lipid dynamics (Scotti-Campos ct al., 2019). A strong reduction was noted in TFA and their double bond index (DBI) under heat and elevated CO_2 stress; i.e., strong declining of TFA unsaturation may be related to the upregulation of lipoxygenase genes *LOX5A* and *LOX5B* and downregulation of fatty acid desaturase *FAD3* gene. Furthermore, elevated CO_2 promoted the enrichment in galactolipids (GLs), such as MGDG and DGSG, and phospholipids (PLs), such as PC (phospatidylcholine) and PG (phospatidylglycerol), and these modifications did not impair the metabolic functioning and photosynthetic performance (Scotti-Campos et al., 2019) (Table 5).

Higashi et al. (2015) showed that under heat stress in *Arabidopsis thaliana*, glycolipids MGDG, DGDG, PC (phosphatydilcholine), and PE (phosphatidylethanolamine) increased, but during recovery they decreased (Table 5). In leaves of *A. thaliana*, triacylglycerol (TAG) levels were increased functioning as an intermediary of lipid turnover resulting in a decrease in polysaturated fatty acids. TAG is accumulated in plastoglobules in chloroplasts, and expression of fibrillin 1 (*FIB1a* and *FIB1b*) genes increased under heat stress (Youssef et al., 2010) (Table 5). In resilient *G. max* under heat stress, Narayanan et al. (2020) observed decrease of lipid unsaturated fatty acids (linolenic acid) and an increase in the amount of less unsaturated fatty acids (linoleic, oleic, and palmitoleic acids) and saturated fatty acids (palmitic and stearic acids) that contribute to overall decreasing in unsaturation index. All that is related to the reduction in the expression levels of the fatty acid desaturase (*FAD3A*, *FAD3B*) genes that are responsible for enzymes which introduce double bonds (unsaturation) in the fatty acyl chains in lipids and contribute to membrane functionality and heat resilience (Narayanan et al., 2020) (Table 5).

Temperature stress—Lipidomic research in *Solanum lycopersicum* showed that fatty acid saturation decreases at high temperatures whereas it increases at low temperatures (Spicher et al., 2016) (Table 5). Unsaturated PE, MGDG, and DGDG decrease at high temperature and increase at low temperature. Furthermore, at high temperatures lipid antioxidant remodeling was observed; i.e., concentrations of tocopherol and plastoquinone increased in order to protect photosynthetic machinery from photoinhibition (Spicher et al., 2016). Narayanan et al. (2016) showed lipid remodeling in high temperatures of resilient *Triticum aestivum*; i.e., levels of lipid unsaturation (MGDG, DGDG, PG, PC, and PE) were decreased whereas levels of sterol lipids (SG, sterol glycoside and ASG, acylated stero, glycoside), triacylgylcerols (TAG), and ox-lipids were increased (Table 5).

Plant resilience to chilling stress is associated with desaturase genes that desaturate palmitate to palmitoleate/modifying the biosynthetic pathway to phosphatidylglycerol (Gao et al., 2015; Barrero-Sicilia et al., 2017). In addition, decline in PC rather than in PE can be related to phospholipase D activity (PLD) which increases biosynthesis of signaling molecule PA (phosphatidic acid) (Ruelland et al., 2002). Increasing number of cold-inducible genes leads to plant resilience to cold: COR (COLD-REGULATED), CIN (COLD-INDUCIBLE), LTI (LOW-TEMPERATURE-INDUCED), and RD (RESPONSE TO DESSICATION) (Thomashow, 2010) (Table 5). Sphingolipids (phyto-SP1) that derived from LCB kinase and phosphorylated ceramides (Cer-P) that derived from ceramide-kinase have active role in cold transduction pathways as signal molecules (Worall et al., 2008; Liang et al., 2003) (Table 5). Zheng et al. (2011) noted that plants growing in the environment with frequent/nonfrequent temperature alterations have different strategies of remodeling the membrane lipids related to maintenance of the degree of unsaturation.

Drought stress—Moradi et al. (2017) have noted the effects of drought on lipid profile and physiology/biochemistry of *Thymus serpyllum*. Drought decreases MGDG, DGDG, PG, PE, and PA levels in sensitive plants indicating that low level of galactolipids may affect the structure of stroma lamella, grana, and chloroplast membrane and leads to the membrane damage and inhibition of photosynthesis. Similar was found for *Brassica* (Dakhma et al., 1995), *Triticum* (Quartacci et al., 1995), and *Lupinus albus* (Hubac et al., 1989) (Table 5). However, sphingolipid (sphinganine 1-phosphate) levels increased in the leaves of *T. serpyllum* under water stress (Moradi et al., 2017)

Table 5 Lipidomics remodeling in plants/microbe/plant-microbe system under climate change and pollution.

Stress factor	Parameters	Plant species	References
Heat, [CO₂]	TFA, DBI, MGDG, DGDG, PL, PC, PG, *LOX5A, LOX5B, FAD3*	*Coffea* spp.	Scotti-Campos et al. (2019)
Heat	MGDG, DGDG, PC, PE TAG, *FIB1a, FIB1b*	*Arabidopsis thaliana*	Higashi et al. (2015) Youssef et al. (2010)
Heat	Linolenic acid, linoleic acid, oleic acid, palmitoleic acid, palmitic acid, stearic acid, *FAD3A, FAD3B*	*Glycine max*	Narayanan et al. (2020)
High temperature	MGDG, DGDG, PE, PG, PC, TAG, SG, ASG, ox-lipids, Tocopherol, plastoquinone	*Solanum lycopersicum*	Spicher et al. (2016)
High temperature	MGDG, DGDG, PG, PC, PE, TAG, SG, ASG, ox-lipids	*Triticum aestivum*	Narayanan et al. (2016)
Low temperatures	Phyto-SP1, Cer-P, COR, CIN, LTI, RD	*Arabidopsis*	Thomashow (2010), Liang et al. (2003)
Drought	MGDG, DGDG, PG, PE, sphinganine 1-phosphate, PC, PI, ox-lipids	*Thymus serpyllum*	Moradi et al. (2017)
	MGDG, DGDG, PG, PE	*Brassica, Triticum, Lupinus albus*	Hubac et al. (1989), Dakhma et al. (1995), Quartacci et al. (1995)
Pollution/ Plastics	Membrane reconfiguration, membrane fluidity	*Rhodococcus*	Gravouil et al. (2017)
Pollution/ Obesogens	Alteration of lipid level	Engineer plant-microbe system	(Lusting (2006))

ASG, acylated stereoglycoside; *Cer-P*, phosphorylated ceramides; *CIN*, COLD-INDUCIBLE gene; *COR*, COLD-REGULATED gene; *DBI*, double bond index; *DGDG*, digalactosyldiacylglycerol; *FAD3*, fatty acid desaturase gene; *FIB1a, FIB1b*, fibrillin 1 genes; *LOX5A, LOX5B*, lipoxygenase genes; *LTI*, LOW-TEMPERATURE-INDUCED gene; *MGDG*, monogalactosyldiacylglycerol; *PC*, phosphatidylcholine; *PE*, phosphatidylcholine; *PG*, phosphatidylglycerol; *phyto-SP1*, sphingolipids; *PI*, phosphatidylinositol; *PL*, phospholipids; *RD*, RESPONSE TO DESSICATION gene; *SG*, sterol glycoside; *TAG*, triacylglycerol; *TFA*, trienoic fatty acid.

(Table 5). Furthermore, phospholipids (PI, phosphatidylinositol, PS, PC) and oxylipins decreased. In drought-resilient plants, fatty acid degree of unsaturation was increased, indicating reduced damage to cell lipids and efficient scavenging of ROS (Moradi et al., 2017). Increased content of antioxidant compounds in water-stressed plants reduces fatty acid oxidation and oxylipin production protecting plants from oxidative damages (Moradi et al., 2017).

Pollution stress—Lipidomics can be used to understand the complex interaction between pollutants in the environment, plants, and microbial communities, usually in order to engineer plant-microbe systems (Koelmel et al., 2020). Environmental lipidomics such as obesogens (pharmaceuticals, pesticides, industrial chemicals) that are present in water, soil, and sediments gain new insight into mechanisms of environmental and human exposure, because these pollutants induce perturbations in lipid pathways and homeostasis altering lipid level or upregulated gene leading to the lifelong implications on human health (Lusting, 2006). Furthermore, in bioremediation of contaminated sites, lipidomics can be used to develop transgenic plants with improved traits for remediation, such as high growth rate, deep root system, and tolerance to pollutants (Koelmel et al., 2020). In lipidomics, a gene cytochrome P450 2E1 in mammalian system can be translated to the transgene engineered plant system because it increases plant's ability to degrade organic pollutants (Greenen et al., 2013). Lipidomics can be used for the determination of mechanisms of microbial degradation of plastics and microplastics (polyethylene). It is possible to employ bioengineered tolerant microbes which can affect the membrane reconfiguration and fluidity (*Rhodococcus*) (Gravouil et al., 2017) (Table 5).

2.1.2 Structural response of resilient plants to pollution and climatic change

Plant leaf anatomy

Drought, UV radiation, O_3, SO_2, NO_2, PM, and metal(loid)s lead to changes in the anatomical structure of plant leaves, and they are a consequence of metabolic disorders in stress conditions (Turunen and Huttunen, 1990; Bussotti et al., 1995; Bell and Treshow, 2002). Leaf damage caused by pollutants depends on vitality of plants, pollutant toxicity, rate, mobility, and localization in tissues (Kovacs, 1993; Bell and Treshow, 2002). Pollutants penetrate the leaf through the stomata and cuticle leading to cuticle erosion and collapse of epidermal cells and stomata (Balaganskaya and Kurdjavtseva, 1998; Bell and Treshow, 2002). They dissolve in a moist film on the cell walls of mesophiles, react with plasmalemma, pass through it and come into contact with cellular organelles, damage their structure and function, and lead to the collapse/ changing of mesophyll cells' shape, size, and content (Kukkola and Huttunen, 1998). According to Bargagli (1998), mesophilic cells are the most sensitive to NO_2 and mesophilic and epidermal cells to SO_2.

Bussotti et al. (1995) considered that the thickening of the epidermis in *F. sylvatica* was associated with drought stress. In conditions of air pollution, it comes to the thickening of mesophilic cell walls (Pääkkonen et al., 1995; Moss et al., 1998), as well as thickening of the cuticle and leaf epidermis (Bussotti et al., 2002). In the urban environment, plants are able to avoid, compensate, and/or repair cell damage caused by toxic effects of pollutants (Bell and Treshow, 2002). The cell walls act as a mechanical barrier preventing the penetration of pollutants into cells, and detoxification processes also take place in them (Bell and Treshow, 2002). According to Evans et al. (1996), plant species that have a palisade parenchyma composed of two or more cell layers and tiny intercellular spaces between cells have greater resistance of mesophiles to the diffusion of gaseous pollutants. Eleftheriou (1987) indicates that reducing intercellular leaf space decreases circulation of toxic gases, and thus intoxication of internal tissues. Similarly, Syversten et al. (1995) found that leaves with small intercellular space show greater resistance to diffusion gases through tissues. According to Niinemets (1999), thicker leaves have larger photosynthetic capacity; i.e., the structure of palisade tissue is associated with efficient process photosynthesis, because the increase in the internal free surface is achieved through magnification of the whole tissue volume. Between compact, elongated palisade cells achieve minimal contact allowing a large free area for gas exchange. However, chronic exposure of plants to urban environmental stress factors decreases antioxidant protection capacity, which further increases sensitivity to oxidative stress, and physiological and morphological damage occurs (Polle et al., 2001).

Urban areas—In the city park close to street with intensive traffic (Belgrade, Serbia) in June, thicker leaf, palisade tissue, spongy tissue, and upper epidermis in *S. vulgaris* (173.6 μm, 86.9 μm, and 56.1 μm, respectively) and *L. ovalifolium* (256.6 μm, 112.9 μm, and 102.3 μm, respectively) were found compared to control site (*S. vulgaris*—143.0 μm, 65.0 μm, and 40.3 μm, respectively; *L. ovalifolium*—210.8 μm, 96.7 μm, 77.5 μm, and 21.5 μm, respectively), except for upper epidermis in *S. vulgaris* (16.0 μm in city park and 19.0 μm from control site) (Gajić, 2007) (Table 6). *Olea europaea* grown in the city center (Thessaloniki, Greece) had thicker leaves compared to the control site that is 15 km from the city center, due to thickened spongy tissue rather than palisade tissue which is achieved through reducing intercellular space; thus, it reduced the circulation of toxic gases and intoxication of internal tissues (Eleftheriou, 1987). Similarly, Syversten et al. (1995) found that leaves with small intercellular space show greater resistance to diffusion of gases through tissues. However, in *S. van-houttei* from the city park, values of leaf thickness (107.4 μm), palisade (45.0 μm) and spongy tissues (42.5 μm μm), upper (9.0 μm) and lower epidermis (7.3 μm), stomata number (155), stomata width (10.3 μm), and stomata length (14.8 μm) were higher compared to the control site (129.3 μm, 56.0 μm, 56.0 μm, 14.3 μm, 10.3 μm, 111.2, 12.3 μm, and 19.3 μm, respectively) (Table 6). Bennett et al. (1992) found that O_3-sensitive individuals of the *Fraxinus pennsylvanica* and *Prunus serotina* had a thinner leaf compared to the tolerant. Furthermore, Gellini et al. (1987) found a larger number of stoma in damaged leaves of *Fagus* sp. in relation to undamaged leaves from the control site.

In *S. vulgaris* from the city park, the first symptoms of damage of the leaf structure appeared in June, on fully formed leaves, in the form of an increased volume of palisade and spongy cells, granulation, or absence of cellular content (Gajić, 2007) (Fig. 3A and B). The upper leaf surfaces in *S. vulgaris* are mostly preserved, and pollutants enter into the leaf through the stoma located on the lower epidermis (Gajić, 2007). Leaf anatomical damage first occurs in spongy tissue, then on the lower epidermis, palisade tissue, and upper epidermis (Gajić, 2007). An increase in cell volume was also observed in *C. avellana*, *Sorbus aria*, and *F. sylvatica*, and cell wall damages in *R. pseudoacacia* (Vollenweider et al., 2003), destruction of cell structure and condensation of cell content in *Pinus cembra* (Fett and Jones, 1995). In June, anatomical changes in *L. ovalifolium* from the city park were expressed in the form of erosion

Table 6 Anatomical traits in plants growing at urban areas, mine waste, and under laboratory conditions.

Sites/pollutants	Parameters	Plant species	References
Urban areas	LT, UE, LE, PT, ST, SN, SW, SL	*Syringa vulgaris, Ligustrum ovalifolium, Spiraea van-houttei*	Gajić (2007)
Mine waste	UC, LC, UE, LE, PT, ST, ECN, TFAbax, TFAdax, SF, SS	*Chromolaena odorata, Tithonia diversifolia, Waltheria indica, Emilia coccinea, Trema orientalis, Aspilia africana, Hyptis suaveolens*	Ogundare et al. (2018)
	LT, PT	*Taraxacum officinale*	Bini et al. (2012)
Serpentine, calamine, nonmetallicolous sites	LT, PT, ST, SNAbax, SNAdax	*Silene vulgaris*	Muszynska et al. (2019)
Laboratory			
Pb	CT, ET	*Cymodocea serrulata*	Rosalina et al. (2019)
Ag, Cu	LT, UE, LE,	*Potamogeton crispus, Potamogeton perfoliatus*	Al-Saadi et al. (2013)
Hg, Pb	UE, LE, AT, VBT, ECN	*Vallisneria spiralis*	Al-Saadi and Qader (2016)
Zn	ET, PT	*Brassica juncea*	Sridhar et al. (2005)
Cr	UE, LE, M, X, P	*Triticum aestivum*	Akcin et al. (2018)
Cr	LT, ET, M	*Pteris vittata*	Sridhar et al. (2011)
Cu, Cd, Pb	UE, LE, ECN, SN, BCS, SCT, P, LT, AME, SD	*Brachiaria decumbens*	Gomes et al. (2011)

AME, area of metaxylem elements; *AT*, aerenchyma thickness; *BCS*, bulliform cell size; *CT*, cuticle thickness; *ECN*, epidermal cell number; *ET*, epidermal tissue; *ET*, epidermis thickness; *LC*, lower cuticle; *LE*, lower epidermis; *LT*, leaf thickness; *M*, mesophyll; *P*, pericycle; *P*, phloem; *PT*, palisade tissue; *SCT*, sclerenchyma thickness; *SD*, stomatal density; *SF*, stomatal frequencies; *SL*, stomatal length; *SN*, stomatal number; *SNAbax*, stomatal number on abaxial side; *SNAdax*, stomatal number on adaxial side; *SS*, stomatal size; *ST*, spongy tissue; *SW*, stomatal width; *TFAbax*, trichome frequency on abaxial side; *TFAdax*, trichome frequencies on adaxial side; *UC*, upper cuticle; *UE*, upper epidermis; *VBT*, vascular bundle thickness; *X*, xylem.

of upper epidermal cells that were filled with black content, damage of mesophiles that was in the form of altered shape and size of palisade cells, and shrinkage of intercellular space (Gajić, 2007) (Fig. 3C and D). Tissue damage was high in *S. van-houttei* from the city park in June and expressed in the form of epidermal erosions and complete mesophile necrosis; i.e., cell damage was in the form of plasmolysis, loss of chloroplasts, dark staining of the cellular content of epidermal tissue, granulation cytoplasmic content, loss of cellular content in mesophilic tissue, and the presence of

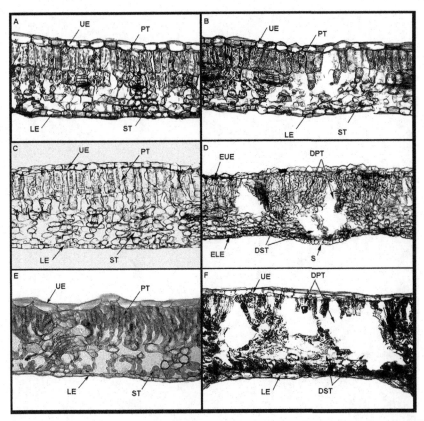

Fig. 3 Cross section through the plant leaves at the control and polluted sites in June (× 400): *Syringa vulgaris* (A, B), *Ligustrum ovalifolium* (C, D), and *Spiraea van-houttei* (E, F). *DPT,* damaged palisade tissue; *DST,* damaged spongy tissue; *ELE,* erosion of lower epidermis; *EUP,* erosion of upper epidermis; *LE,* lower epidermis; *PT,* palisade tissue; *S,* stomata; *ST,* spongy tissue; *UE,* upper epidermis.

large cavity within the mesophile (Gajić, 2007) (Fig. 3E and F). Moss et al. (1998) were observed anatomical leaf changes in *Picea rubens*, as a consequence of the action of O_3 and low pH.

The number of stomata, their width, and length in *S. vulgaris* and *L. ovalifolium* did not differ significantly in June between the polluted and the control sites whereas in *S. van-houttei* the number of stomata was higher, but their width and length were lower at the polluted site compared to the control site (Gajić, 2007). However, atmospheric particles were observed on the stomata of all three shrub species at the polluted site compared to the control site (Fig. 4) (Gajić, 2007). Atmospheric particles, according to Bacic et al. (1999) can cause changes in physiology and morphology of the leaves by increasing the temperature of the leaves, passing through the stomata, or stomata can be mechanically blocked (Fluckiger et al., 1979). Also, particles cause erosion of cuticular waxes (Eveling and Bataille, 1984), which can result in disturbances in the

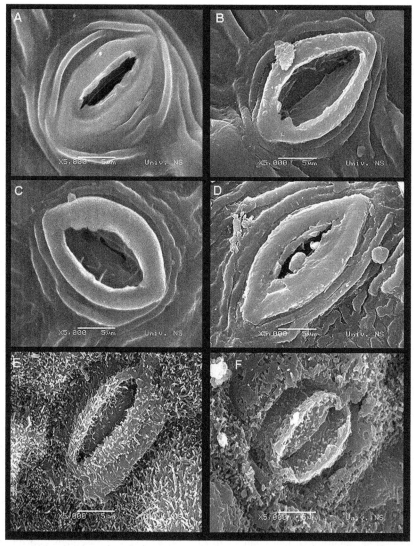

Fig. 4 SEM micrograph of the stomata on the lower surface of the plant leaves at the control and polluted sites in June (× 5000): *Syringa vulgaris* (A, B), *Ligustrum ovalifolium* (C, D), and *Spiraea van-houttei* (E, F).

photosynthetic and water regime of plants (Stevanović and Janković, 2001). Deposited particles can reduce the stomata conductivity and uptake of CO_2. Particles that block the stomata may prevent their closure, and thus, constantly open stomata allow continuous and increased uptake of pollutants which can lead to the continuous water release and water stress (Fluckiger et al., 1979).

The leaf damage was similar in *S. vulgaris* and *L. ovalifolium*, while the most intense leaf tissue damage was in *S. van-houttei* (Gajić, 2007). *S. vulgaris* and *L. ovalifolium* have xeromorphic leaf structure, characterized by a well-developed structure of peripheral protections, i.e., thick cuticle and thickened cell walls of epidermal cells, as well as small intercellular spaces between mesophilic cells, thus preventing high absorption and diffusion of gaseous pollutants and heavy metals. The leaves of *S. van-houttei* are characterized by typical mesomorphic structure, i.e., thin cuticle, thin cell walls of epidermal cells and mesophiles, as well as large intercellular spaces, which resulted in ineffective protection of internal tissues and high sensitivity of mesophile cells to the toxic effects of pollutants (Gajić, 2007).

Mine waste—According to Ogundare et al. (2018), plants grown on gemstone mine site in Nigeria showed higher values of thickness of upper and lower cuticle (*Chromolaena odorata, Tithonia diversifolia, Waltheria indica, Emilia coccinea*) and thickness of upper and lower epidermis (*T. diversifolia, W. indica, Trema orientalis, E. coccinea, Aspilia africana,* and *Hyptis suaveolens*) compared to unmined site. Thickness of palisade tissue was higher in *C. odorata, E. coccinea,* and *A. africana* and lower in *T. diversifolia, W. indica,* and *H. suaveolens* compared to unmined site, whereas thickness of spongy tissue was higher in *C. odorata, T. diversifolia, W. indica, E. coccinea,* and *H. suaveolens* compared to unmined site (Ogundare et al., 2018) (Table 6). Furthermore, in most of the plant species a significant increase in stomatal frequencies, epidermal cell number, and trichome frequencies on abaxial and adaxial leaf surfaces has been found (Table 6). Stomatal size was significantly reduced and that can be the result of avoidance mechanisms in order to control the entry of pollutants and their inhibitory effects to photosynthesis and/or control transpiration for water conservation (Gomes et al., 2011) indicating that these plants can adapt to mining pollution.

Anatomical leaf changes in *Taraxacum officinale* growing on sulfide mine waste (Italy) with high concentrations of Cd, Cr, Cu, Pb, Zn, and Fe were expressed in high reduction of leaf thickness, lacking palisade tissue, and large intercellular spaces with few rounded cells compared to the unpolluted site (Bini et al., 2012) (Table 6). However, anatomical leaf structure in *Silene vulgaris* was different between serpentine, calamine, and nonmetallicolous ecotypes (Muszynska et al., 2019). Total leaf thickness and thickness of palisade and spongy parenchyma were higher, and stomata number on adaxial and abaxial sides was lower in serpentine and calamine ecotypes compared to nonmetallicolous ecotypes (Table 6) and that can be associated with plant adaptation to high insolation, water deficiency, strong wind, and excess heavy metals, regulating uptake and translocation of heavy metals and reducing evapotranspiration, but it can limit CO_2 flux to mesophyll cells which partially decreases photosynthetic efficiency (Wierzbicka and Panufnik, 1998; Karnaukhova, 2016; Pereira et al., 2016; Muszynska et al., 2019).

Laboratory conditions—Rosalina et al. (2019) have found that leaf anatomical structure of *Cymodocea serrulata* (seagrass) was altered in the condition of high Pb concentrations; i.e., cuticle and epidermis were thickened whereas air space was irregulated (Table 6). Cuticle is able to bind Pb in cuticle cavity (Lakrum et al., 2006), whereas thicker epidermis may prevent Pb transportation inside mesophyll and chloroplasts (Rosalina et al., 2019). In the condition of high Ag and Cu concentrations, anatomical leaf changes in *Potamogeton crispus* and *Potamogeton perfoliatus* were expressed in the form of thickened upper and lower epidermis which can bind metals to cell wall preventing their transport to mesophyll whereas leaf blade thickness of both plants was reduced due to decreasing cellular size and intercellular spaces of aerenchyma (Al-Saadi et al., 2013) (Table 6). Similarly, in *Vallisneria spiralis* exposed to Hg and Pb, upper and lower epidermis were thickened and aerenchyma thickness and vascular bundle thickness were lower whereas the increase in the number of epidermal cells indicates that plants activate compensatory mechanisms for the reduced photosynthetic tissue indicating plant resilience to excess metal concentrations (Al-Saadi and Qader, 2016) (Table 6).

Brassica juncea exposed to high Zn concentrations in the laboratory conditions showed changes in leaf anatomical structure in the form of reduced epidermal and palisade cells, disintegration of palisade and spongy area that is followed by loss of cell shape and decreasing of intercellular spaces compared to the control site all of which is associated with reduction in RWC (Sridhar et al., 2005) (Table 6). Furthermore, *T. aestivum* exposed to high Cr concentrations showed thickened upper and lower epidermis and thinner mesophyll, xylem, and phloem (Akcin et al., 2018) (Table 6). Similarly, a reduction in leaf thickness and shrinkage of mesophyll and epidermis exposed to high levels of Cr was shown in *Pteris vittata* (Sridhar et al., 2011) (Table 6). According to Gomes et al. (2011) in the conditions of high levels of Cu, Cd, and Pb, *Brachiaria decumbens* showed leaf anatomical changes in the form of thicker upper and lower epidermis, a high number of epidermal cells and the number of stomata, a larger size of bulliform cells, sclerenchyma, and pericycle whereas leaf blade thickness and area of metaxylem elements were decreased (Table 6). The thickened epidermis and larger bulliform cells can be strategy to minimize transpiration and water loss as well as acquisition of metal ions whereas thickened sclerenchyma and pericycle can be related to metal binding for cell walls preventing their translocation to the photosynthetic tissue. Increased stomatal density in *B. decumbens* can be compensatory mechanisms ensuring CO_2 flux maintaining photosynthesis and indicating the plant resilience to pollution stress (Gomes et al., 2011).

Plant leaf morphology

Visible symptoms of plant leaf damages in polluted environment are the result of complex effects of stress abiotic and biotic factors, such as drought, high temperatures, radiation, lack/excess of essential chemical elements, pathogens, air pollutants, and heavy metals (Bell and Treshow, 2002). The sensitivity of plants to pollutants and visible leaf damages depend on the type, concentrations and length of exposure to pollutants, distance from pollutant emission sources, meteorological factors, and characteristics of plants (genetic predispositions, general physiological condition, stage of development) (Kozlowski and Constandiniou, 1986; Larcher, 1995). Plants exposed to pollution show acute and chronic symptoms of leaf damages (Bargagli, 1998; Bell and Treshow, 2002). Acute damages occurs due to the absorption and accumulation of lethal amounts of toxic pollutants in leaf tissues in a short period of time and manifest themselves in the form of chlorosis, necrosis, drying, and premature leaf fall. Chronic damages are associated with the action of low concentrations of pollutants in the long-run time period and result in mild chlorosis, reduced production, premature aging, and drying of the leaves. Simultaneous presence of several different pollutants in the soil and air causes more damages than expected based on their individual toxicity, and such an effect is synergistic (Bell and Treshow, 2002).

Diagnosis of visible symptoms of plant leaf damage caused by toxic effects of pollutants is aggravated by the fact that the same or similar symptoms occur when abiotic and biotic environmental stressors act together (Saxe, 1996). However, Vollenweider and Gunthardt-Goerg (2005) indicate that visible symptoms of leaf damages can be one of the diagnostic parameters for the identification of various environmental stressors. According to the same authors, it is possible to distinguish visible leaf damages caused by pollutants from natural leaf aging. Kozlowski and Constandiniou (1986) indicate that in a polluted environment morphological damage of plant leaves is caused by the action of SO_2, whereby its low concentrations cause chlorosis of leaves that are light green, yellow, white, brown, red, or black color. The high concentrations of SO_2 and NO_2 lead to the appearance of marginal and interventional leaf necrosis (Freer-Smith, 1984), whereas SO_2, Cu, Cd, Pb, and Zn can cause functional-structural changes in plants (Dueck, 1986).

The cuticle of plants is the main barrier between the external environment and internal tissues. The structure of epicuticular waxes can be damaged by the mechanical action of wind, rain, snow (Jeffree et al., 1994), heavy metals, and air pollution (Huttunen, 1994). The main physiological function of the cuticle is protection against uncontrolled water loss by transpiration (Burghardt and Riederer, 2006), but also reduced excretion of ions and polar organic solutions from cells (Schönherr, 2001). It plays an important role in the structural stabilization of epidermal tissue (Matas et al., 2004). Particle deposition on plant leaf surfaces depends on climatic conditions (wind direction and speed, relative humidity, precipitation, and temperature), chemical properties of elements, particle size, and morphological characteristics of leaves (Beckett et al., 1998; Sawidis et al., 2011). Therefore, heavy rains and intense wind can lead to washing of deposited particles from the leaf surface. Periods with increased humidity, dew, fog, light rainfall, and wind of lower intensity are suitable for retention and efficient binding of particles to leaf surfaces (Bell and Treshow, 2002). Then, the deposited particles dissolve more easily and the metal(loid)s are more easily absorbed through the cuticle or stoma into the internal tissues, causing functional damages (Bell and Treshow, 2002).

Cuticle damages caused by toxic pollutants depend on pollutant concentration, stage of development, and genetic predisposition of plants (Cape, 1994). Micromorphological changes of the cuticle are nonspecific for the effects of pollutants, and they are often similar to changes related to natural aging or changes caused by others environmental stressors, such as high temperature and drought. Nevertheless, Berg (1989) considers that the cuticle can be used as a general diagnostic marker of stress. Also, numerous studies show that erosive cuticle damage can be used as an early indicator of toxic effects of pollutants derived from traffic (Viskari, 2000; Viskari et al., 2000). The use of epicuticular waxes as potential indicators of air pollution is based on direct observations of physical changes in the cuticle using transmission (Turunen et al., 1995) and electron microscopy scanning (Cape and Fowler, 1981) and indirect methods, such as estimating surface moisture by measuring contact angle of water droplets with the surface or measuring water retention capacity (Cape, 1983; Schreuder et al., 2001). Mannien et al. (1996) developed the SEM-EDS method for precise and detailed assessment of changes in epicuticular waxes of plant leaves.

The cuticle damages and the amount of deposited particles on the surface of plant leaves depend on the morphological characteristics of leaf peripheral structures and distance from pollutant emission sources (Pal et al., 2002). Godzik et al. (1979) indicate that smooth leaves with a thick layer of epicuticular waxes retain a smaller amount of particles on the surface unlike hairy leaves or leaves with uneven surface. Fauser (1999) found significantly more particles caused by mechanical wear of car tires on hairy leaves of *Ulmus carpinifolia* and *Salix aurita*, in relation to the smooth leaves of *Ligustrum vulgare* and *F. sylvatica*. Sawidis et al. (1995) recommend *Ligustrum japonicum* for planting along busy streets and roads due to the smooth surfaces and lower particle deposition. According to Birbaum et al. (2010) and Uzu et al. (2010), agglomerates of particles are retained on the surface of epicuticular waxes and trichomes.

Fly ash deposits—*F. rubra* sown on fly ash deposits showed significant reduction in plant growth (root/shoot length, plant height, belowground/underground biomass) compared to control site pointing to adverse conditions on fly ash, such as drought, high temperatures, and high concentrations of As, B, and deficiency of Cu and Zn (Gajić et al., 2016) (Table 7). However, over time (after 11 years of biorecultivation), this grass showed enhanced growth due to favorable physicochemical characteristics of fly ash. Morphological damages of leaves of *F. rubra* on fly ash deposits were chlorosis in the form of green, yellow, and red color, and necrosis of red, brown, and black color (Gajić et al., 2016) (Table 7). SEM analysis of surface structure of leaves showed that damage of cuticle and epidermis was not large although particles of fly ash were retained between slightly protruding epidermal cells (Gajić et al., 2020) (Table 7). Silicate body on the leaf surface of *F. rubra* may increase resilience of this species to environmental stresses (Gajić et al., 2020). In addition, SEM (EDS) analysis of deposited particles on the leaf surface of *F. rubra* sown on fly ash deposits showed that particles had more concentrations of Al, Fe, Si, Cu, Ti, Mg, Ca, and K compared to the control site (Gajić et al., 2020) (Table 8). Furthermore, visible leaf symptoms of *D. glomerata* sown on fly ash deposits were expressed in the form of chlorosis light green and yellow color and necrosis violet and brown color that can be the result of high As concentrations in the leaves (Gajić et al., 2020) (Table 7). SEM analysis of the surface structure of leaves *D. glomerata* sown on fly ash deposits showed no significant cuticle damages despite large particles and aggregates that are retained between convex, elongated epidermal cells and sharp short trichomes (Gajić, 2014) (Table 7). SEM (EDS) analysis of deposited particles on the leaf surface of *D. glomerata* sown on fly ash deposits showed that particles contain the high concentrations of O, Mg, and Ca compared to the control site whereas Al, Fe, and Si were present only in particles from fly ash (Gajić, 2014) (Table 8). Jamil et al. (2009) noted that a large number of plant species growing in the immediate vicinity of a thermal power plant in India are capable of depositing fly ash particles on the leaf surface depending on their surface structure (smooth, uneven, presence of trichomes, bulges). Similarly, Sawidis et al. (2011) showed that *V. phlomoide's* leaves, which have a large number

Table 7 Morphological leaf changes in plants growing at fly ash deposits, mine waste, and urban areas.

Sites/pot experiment	Parameters/ leaf traits/leaf changes	Plant species	References
Fly ash deposits	RL, SL, PH, BGB, UGB Leaf chlorosis, Leaf necrosis Smooth cuticle, protruding epidermal cells, silica body	*Festuca rubra*	Gajić et al. (2016) Gajić et al. (2020)
	Long trichomes	*Verbascum phlomoides*	Sawidis et al. (2011)
	Smooth cuticle	*Rumex acetosa*	
Fly ash amendments	RL, SL, PL, TLA, NoN, PB	*Cicer arietinum*	Pandey et al. (2010)
	PG	*Ricinus communis*	Panda et al. (2020)
	PH, RW, GW, SW, NoT	*Oryza sativa*	Dwivedi et al. (2007)
	RL, SL, B	*Sesbania cannabina*	Sinha and Gupta (2005)
	RL, SL, PL, NoL, B, Y	*Beta vulgaris*	Singh et al. (2008)
Mine waste	TLA, PL	*Chromolaena odorata, Waltheria diversifolia, Hyptis suaveolens, Trema orientalis, Tithonia diversifolia, Emilia coccinea, Aspilia africana*	Ogundare et al. (2018)
	PH, SB, BD	*Arrhenatherum elatius, Sonchus transcapicus*	Lu et al. (2013)
Urban areas	Leaf chlorosis, Leaf necrosis Deep lesions of epidermis, erosion of cuticles	*Syringa vulgaris, Ligustrum ovalifolium, Spiraea van-houttei*	Gajić (2007)

BD, basal diameter; *BGB*, belowground biomass; *GW*, grain weight; *NoL*, number of leaves; *NoN*, number of nodules per plant; *NoT*, number of tillers; *PB*, plant biomass; *PG*, plant growth; *PH*, plant height; *PL*, petiole length; *RL*, root length; *RW*, root weight; *SB*, shoot biomass; *SL*, shoot length; *SW*, straw weight; *TLA*, total leaf area; *UGB*, underground biomass; *Y*, yield.

of long trichomes, deposited a larger amount of particles originating from a thermal power plant in Greece, in contrast to *Rumex acetosa*, which is characterized by a smooth cuticle which retains less particles (Table 7). SEM (EDS) analysis of fly ash particles that were deposited on the surface of the leaves of *V. phlomoides* and *R. acetosa* showed the following chemical elements: Mg, Al, Si, P, Cl, K, Ca, Cr, Mn, Fe,

Table 8 SEM (EDS) spectral analysis of particles from fly ash deposits and urban areas.

Sites	Chemical elements	Plant species	References
Fly ash deposits	Al, Fe, Si, Cu, Ti, Mg, Ca, K	*Festuca rubra*	Gajić et al. (2020)
	Mg, Ca, O, Al, Fe, Si	*Dactylis glomerata*	Gajić (2014)
	Mg, Al, Si, P, Cl, K, Ca, Cr, Mn, Fe, Ni, Cu, Zn, Pb	*Verbascum phlomoides Rumex acetosa*	Sawidis et al. (2011)
Urban areas	O, Mg, Al, Si, Cl, K, Ca, Na, Fe, Cu, Zn	*Syringa vulgaris, Ligustrum ovalifolium, Spiraea van-houttei*	Gajić (2007)
	Na, Mg, Al, Si, P, S, Cl, K, Ca, Fe	*Platanus hispanica, Tilia europaea*	Beckett et al. (1998)
	K, Na, Ca, Mg, Cl	*Sorbus aria, Populus deltoides, Pinus nigra, Cupressocyparis leylandii*	Beckett et al. (2000)

Ni, Cu, Zn, and Pb, where the concentrations of Al, Si, Ca, Mg, K, and Fe were the highest (Sawidis et al., 2011) (Table 8). The concentrations of Fe, Al, and Cu in the fly ash particles in plants may lead to the reduced permeability of the aqueous solutions in which these elements or various salts are present. It was found that at low winds and about 100% relative humidity, Fe (III) chelates dissolve and pass through the plant cuticle (Schönherr et al., 2005). However, at a relative humidity of less than 90%, Cu^{2+}, Fe^{3+}, and Al^{3+} ions form the insoluble hydroxide complexes on the leaf surface which reduce the permeability of the cuticle to water and dissolved salts (Schönherr et al., 2005). The same authors showed that with increasing concentration of Fe chelate, the permeability of the cuticle to $CaCl_2$ and Fe chelate itself decreases. According to them, these complexes reduce the size of aquatic pores by forming hydrogen bonds between the hydroxide complex of Fe(III) (HOH) and the dipole of the cell wall pores, which leads to slower diffusion of Fe or salt within aquatic pores. Aquatic pores are formed only in the presence of water, and in heavy rains or fogs, the cuticle of plant leaves is permeable to aqueous solutions, while in dry air these pores disappear or shrink, and water permeability decreases (Schönherr, 2006). Therefore, according to Schönherr (2006) the surface structure of leaves may be less damaged or even undamaged due to the presence of large ionic molecules and ion-dipole interactions.

Pandey et al. (2010) showed that in *C. arietinum* grown in pot with 100% fly ash, plant/root/shoot length, total leaf area, number of nodules per plant, and biomass were reduced compared to pots with 100% garden soil, 50% fly ash + 50% garden soil, and 25% fly ash and 75% garden soil indicating that this species can successfully survive in different treatments of fly ash (Table 7). Similarly, Panda et al. (2020) showed that in *R. communis* grown in pots with 50% of fly ash and 50% of garden soil, plant growth was higher compared to pots with 100% of garden soil and pots with 100%

of fly ash (Table 7). Dwivedi et al. (2007) have found that plant height, root/grain weight, number of tillers, and straw weight were reduced in different cultivars of *O. sativa* at 50%, 75%, and 100% amendments of fly ash whereas at 10% and 25% of fly ash amendments plant growth increased which can be related to higher availability of essential chemical elements required for physiological and metabolic processes (Table 7). In addition, Sinha and Gupta (2005) showed that *S. cannabina* had the highest root length at 10% fly ash treatment and highest shoot length and biomass at 25% fly ash treatment compared to garden soil (Table 7). Furthermore, Singh et al. (2008) observed that in *B. vulgaris* grown in different fly ash amendments, plant/root/shoot length and number of leaves/total biomass of plant/yields were reduced compared to nonfly ash–amended soil (Table 7).

Mine waste—Ogundare et al. (2018) showed different growth response in plants that grow at mined sites (Nigeria) compared with unmined site: leaf area and petiole length increased in *C. odorata, Waltheria diversifolia, H. suaveolens,* and *T. orientalis* and decreased in *T. diversifolia, E. coccinea,* and *A. africana* indicating their resilience to pollutants from mining activities (Table 6). Plant height, shoot biomass, and basal diameter were significantly reduced in *A. elatius* and *S. transcapicus* grown at mine tailings (China) (Lu et al., 2013) (Table 7).

Urban areas—Leaf damages in the form of chlorosis light green and olive green color, marginal and intravenous necrosis from light brown to dark brown and black color were observed in June in *S. vulgaris, L. ovalifolium,* and *S. van-houttei* grown in a city park along the street with intense traffic (Gajić, 2007) (Fig. 5, Table 7). These symptoms are often associated with damage action of high concentrations of SO_2, NO_2, and soot (Bell and Treshow, 2002). Leaf cuticle of *S. vulgaris* and *L. ovalifolium* from the control site was smooth, whereas in *S. van-houttei* it was covered with small crystals without symptoms of damage (Figs. 6A, 7A, and 8A). However, SEM microscopy revealed deep lesions in the leaf's upper surface epidermis in *S. vulgaris* and cuticle erosion in *L. ovalifolium* and *S. van-houttei* (Gajić, 2007) (Figs. 6B, 7B, and 8B; Table 7). The largest amount of particles on the leaf surface and the greatest damages of the leaf surface structures were registered in the *S. van-houttei* (Gajić, 2007). This species is characterized by a thin cuticle which is highly permeable and very sensitive to the stress effects of light, temperature, wind, rain, and air pollution (Hansen-Bristow, 1986). Crystals on the upper surface of the leaves of *S. van-houttei* were lost through gradual erosion of the cuticle, and according to Neinhuis and Barthlott (1998), leaf surface is suitable for the deposition of particles from the atmosphere. Grill et al. (1987) found that the cuticle of plants from unpolluted sites was well preserved compared with plants from polluted sites where it looks much older than expected based on chronological age. Turunen et al. (1995) indicate that SO_2, NO_2, and heavy metals lead to erosion and faster aging of plant cuticles at contaminated sites. Similarly, in plants that grow along street with intense traffic there is a rapid erosion of epicuticular waxes, while the cuticle of plants from the control site remains unchanged and without deposited dust particles (Pal et al., 2002).

Erosive damage to the cuticle occurs as a result of chemical effects of gaseous pollutants, mechanical (abrasive) and chemical effects of atmospheric dust particles (Riding and Percy, 1985; Viskari et al., 2000). Pb, Cu, and Zn in particles are in the

Fig. 5 Leaves of *Syringa vulgaris* at the control site (A) and polluted sites (B); Leaves of *Ligustrum ovalifolium* at the control site (C) and polluted sites (D); and Leaves of *Spiraea van-houttei* at the control site (E) and polluted sites (F).

soluble form and can be easily absorbed through the cuticle leading to leaf damages (Bell and Treshow, 2002). Mäkelä and Huttunen (1987) found that even low concentrations of SO_2, from 10 to 20 $\mu g/m^3$, can cause destruction of waxy leaf structures. Sauter et al. (1987) have shown in experimental conditions that pollutants originate from car exhaust gases (NOx, SO_2, CO, and hydrocarbons) which can cause erosive damages of cuticle within 15 min. Similarly, pollutants originate from exhaust gases from cars (NOx, PAH, and carbon black) which can cause erosive cuticle damages in *Picea abies* (Viskari et al., 2000).

SEM (EDS) analysis of particles that were deposited on the leaf surface in *S. vulgaris*, *L. ovalifolium*, and *S. van-houttei* in the city park showed a presence of following chemical elements: O, Mg, Al, Si, Cl, K, Ca, Na, Fe, Cu, and Zn where concentrations of Al, Fe, Si, Cu, Zn, Ca, and C were higher at polluted sites compared to the control site (Gajić, 2007) (Figs. 6–8; Table 8). It was found that soot particles originating from car exhaust gases include Al, Fe, and Si (Celis et al., 2003), Cu and Zn (Heal et al., 2005), and Ca and K (Li et al., 2004). Motor vehicles are

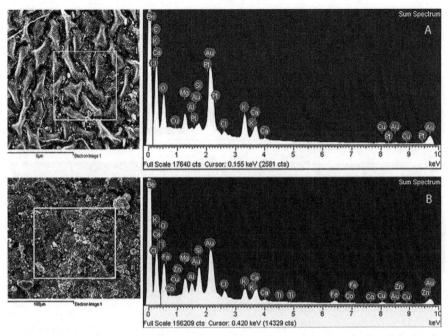

Fig. 6 SEM (EDS) spectral analysis of chemical composition of the deposited particles on the leaf surface of *Syringa vulgaris* the control site (A) and the polluted sites (B).

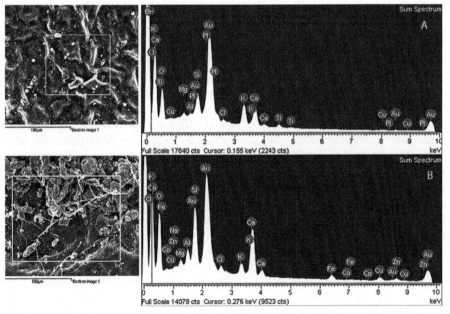

Fig. 7 SEM (EDS) spectral analysis of chemical composition of the deposited particles on the leaf surface of *Ligustrum ovalifolium* at the control site (A) and the polluted sites (B).

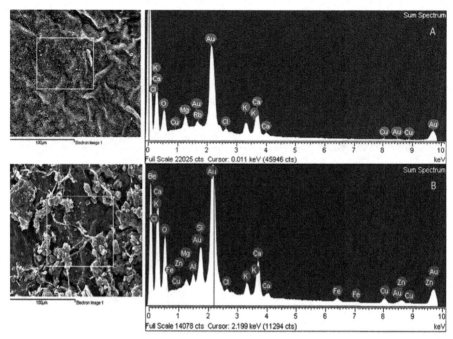

Fig. 8 SEM (EDS) spectral analysis of chemical composition of the deposited particles on the leaf surface of *Spiraea van-houttei* at the control site (A) and the polluted sites (B).

the dominant sources of organic (OC) and elemental (EC) carbon in $PM_{2.5}$ particles (Liu et al., 2005). Kinney et al. (2000) found that high concentrations of EC in the particles near the sidewalk edge in New York can be linked to a large number of buses and trucks in the surrounding streets as well as nearby bus stops. Götschi et al. (2005) analyzed the elementary composition of fine particles in 21 cities in Europe and found the presence of the following elements: Al, As, Br, Ca, Cl, Cu, Fe, K, Mn, Pb, S, Si, Ti, V, and Zn. Similarly, in the chemical composition of PM_{10} particles in the city air (Chile) there were Si, K, Fe, Ca, Al, Cl, Zn, Ti and Cu, EC, nitrates, and ammonium (Celis et al., 2003). Similar elemental composition of particles (Na, Mg, Al, Si, P, S, Cl, K, Ca, and Fe) was registered on the leaf surface in *P. hispanica* and *Tilia europaea* which grow in the park located next to a busy street (Beckett et al., 1998) (Table 8). Also, Beckett et al. (2000) detected the presence of K, Na, Ca, Mg, and Cl in fine particles on the leaf surface of *S. aria*, *Populus deltoides*, *P. nigra*, and *Cupressocyparis leylandii* in a city park located nearby the road in Brighton (United Kingdom) (Table 8). The chemical composition of ultrafine atmospheric particles in the south California (United States) indicated that particles consist of 32%–67% organic compounds, 3.5%–17.5% EC, 1%–18% sulfate ion, 0%–19% nitrate ion, 0%–9% ammonium ion, 1%–26% metal and metal oxide, 0%–2% Na, and 0%–2% chloride (Cass et al., 2000). The same authors found that in very busy streets the proportion of particles originating from traffic exceeds 43% and that has a negative effect on human health and plants.

3 Conclusions

Human activities lead to the pollution of air, soil, and water worldwide having negative impacts to the environment and human health. Industrial and mining activities, waste disposal, cities, and agricultural practice released a huge amount of contaminants. Climate is changed due to pollution increasing CO_2 concentrations, global temperatures, and precipitations leading to the serious drought, rainfalls, floods, etc. Climate-resilient pathways include mitigation and adaptation which provide sustainable green solutions and policy frameworks that can clean up contaminated area having "win-win" ecological, economic, and human benefits. Phytoremediation as environmentally friendly technology was used for healing fly ash, mine, and urban areas transforming them from "gray" to "green." Plant species that are used in the phytoremediation and ecosystem services are capable of improving physicochemical properties of soil, mitigate air pollution, and increase carbon sequestration and biodiversity.

Plants growing at the contaminated sites are capable of coping with the effects of climate change and pollutants developing various functional and structural traits. Plant species that are growing at polluted sites, such as fly ash deposits, mine wastes, and urban areas, show decreased leaf water potential, relative water content, transpiration rate, stomatal conductance, net photosynthetic rate, photosynthetic efficiency, and chlorophylls content which altogether may lead to the chlorosis and necrosis of leaves. Adverse climatic factors, NO_2, SO_2, PM, metal(loid)s, and organic pollutants generate reactive oxygen species (ROS) which can lead to the oxidative stress. Plants resilient to environmental stresses possess adaptive mechanisms to compensate changes in photosynthesis, water balance, metabolites' content, leaf anatomical tissues, and leaf surface. They activate antioxidant machinery in order to maintain the cellular redox homeostasis. An efficient quenching of ROS is enabled by enhanced biosynthesis of carotenoids, anthocyanins, phenolics, proline, tocopherol, AsA, GSH, NPSH, cysteine, and PCs, as well as enhanced activities of enzymes, such as SOD, POD, CAT, APX, GR, and GST. Membrane lipid remodeling and lipodome changes in the conditions of climate change and pollution are significant for the maintenance of membrane functionality, optimal photosynthesis, and antioxidant capacity in resilient plants. However, high/moderate environmental stresses lead to the physiological and biochemical disorders which negatively affect leaf anatomical structure (reduced total leaf thickness, cuticle and epidermis thickness, palisade and spongy thickness, stomatal number and size, etc.). Visible leaf damages in the form of reduced plant growth, leaf chlorosis and necrosis, and erosion of epicuticular waxes can be consequences of drought, high temperature, excess metal(loid)s, air pollutants, and PM. SEM (EDS) analysis of particulates from fly ash and urban areas showed high concentrations of Al, Fe, Si, Cu, Zn, Ca, Mg, O, K, Na, C, and Ti indicating that these elements in PM deposited on leaf surface may have mechanical (abrasive) and chemical effects on plant physiology and morphology. Generally, plants resilient to stress had xeromorphic leaf structure, i.e., thick cuticle, thick cell wall of epidermal cells, well-developed palisade and spongy tissues, small intercellular species, and a large number of small size stomata in order to control entry of pollutants, decrease transpiration, and enable enough water conservation and CO_2 concentrations for optimal photosynthetic and metabolic activity.

4 Future perspectives

The future directions of the research of plant functional and structural response to pollution stress and climate change require introduction of biochemistry, biophysics, molecular biology, metabolomics, proteomics, and genomics. Engineered nanoparticles (ENPs) can use as "smart" delivery agents in a number of industrial and agricultural applications (herbicides/fertilizers). Research of chemistry of NPs and their effects on plant growth, photosynthesis and biochemistry, and oxidative/antioxidative response are needed to be further explored. Nowadays, synthetic biology (SB) as a new platform for industrial biotechnology represents a technology for engineering new traits for plant improvements; i.e., it is focusing on design of new metabolic pathways, physiological traits and developmental strategies, and engineering all that at the cellular and subcellular level, frequently using microbes. Photosynthesis in C3 and C4 plants represents the most basic target input for SB-re-engineering metabolic pathways and enhanced carbon fixation by re-engineering carboxylating enzymes (RuBisCO) that drive photosynthesis. Formation of N_2-fixing nodules by using *Rhizobium* species can use as a tool for engineering N-fixing pathways. Nutrient input traits can also be used for SB engineering of new traits in plants which can provide optimal fertilizer usage and plant resilience to environmental stresses, such as climate change and pollution (synthetic signaling networks and antioxidant system). In order to design plant resilience to contamination (herbicides), synthetic site-specific nucleases (zinc-finger nuclease, ZFNs; transcription activator-like nuclease, TALENs; regularly interspaced short palindromic repeats, CRISPR) may be used to engineer plant genomes and metabolics. Output plant traits can be used for crops that generate high biomass yield—intermediates of secondary metabolism which can use in chemical industry, such as lignocellulose production pathways, carbohydrate, fatty acid, phenolics and terpenoid biosynthetic pathways, and heme-containing cytochrome P450s in photosynthesis. Furthermore, SB can use plants for detection of pollutants (explosives); i.e., detector plants have the potential to detect chemicals in the air and soil by using redesign signal transduction systems based on recognition of small molecules forming new sensor/reporting system. However, applications of gene editing and SB technology are needed to be regulated by frameworks, performing strength, weakness, opportunities, and threats having societal, economical, and political trade-off alongside with safety, security, and ethics in large-scale perspectives.

References

Abdelrahman, M., Burritt, D.J., Tran, L.-S.P., 2018. The use of metabolomics quantitative traits locus mapping and osmotic adjustment traits for the improvement of crop yields under environmental stresses. Semin. Cell Dev. Biol. 83, 86–94.

Abhijith, K., Kumar, P., 2019. Field investigations for evaluating green infrastructure effects on air quality in open-road conditions. Atmos. Environ. 201, 132–147.

Abhijith, K., Kumar, P., Gallagher, J., McNabola, A., Baldauf, R., Pilla, F., Broderick, B., Di Sabatino, S., Pulvirenti, B., 2017. Air pollution abatement performances of green infrastructure in open road and built-up street canyon environments—a review. Atmos. Environ. 162, 71–86.

Ågreen, G.I., Weih, M., 2012. Plant stochiometry at different scales: element concentration patterns reflect environmental more than genotype. New Phytol. 194, 944–952.

Ahirwal, J., Pandey, V.C., 2021. Ecological rehabilitation of mine-degraded land for sustainable environmental development in emerging nations. Restor. Ecol. https://doi.org/10.1111/rec.13268.

Ainsworth, E.A., Rogers, A., 2007. The response of photosynthesis and stomatal conductance to rising (CO_2): mechanisms and environmental interaction. Plant Cell Environ. 30, 258–270.

Airparif, 2016. Inventaire regional des emissions en ille-de-France Annee de reference 2012—Elements synthetiques Edition mai 2016., p. 32.

Akcin, T.A., Akcin, A., Yildirim, C., 2018. Effects of chromium on anatomical characteristics of bread wheat (*Triticum aestivum* L. cv. 'Ekiz'). J. Int. Environ. Appl. Sci. 13 (1), 27–32.

Akguc, N., Ozyigit, I., Yasar, U., Leblebici, Z., Yarci, C., 2010. Use of *Pyracantha coccinea* Roem. as possible biomonitor for the selected heavy metals. Int. J. Environ. Sci. Technol. 7 (3), 427–434.

Alia, Mohanty, P., Matysik, J., 2001. Effect of proline on the production of singlet oxygen. Amino Acids 21, 195–200.

Allen, I.H., Pan, D., Boote, K.J., Pickering, N.B., Jones, J.W., 2003. Carbon dioxide and temperature effects on evapotranspiration and water use efficiency of soybean. Agron. J. 95, 1071–1081.

Al-Saadi, S.A.A.M., Qader, K.O., 2016. The effect of some heavy metals accumulation on anatomical and physiological characteristic of the submerged macrophyte Vallisneria plant. In: Proceedings of 40th IASTEM International Conference, Kuala Lumpur, Malaysia, 1st–2nd December 2016.

Al-Saadi, S.A.A.M., Al-Saadi, W.M., Al-Waheeb, A.N.H., 2013. The effect of some heavy metals accumulation on physiological and anatomical characteristics of some *Potamogeton* L. plant. J. Ecol. Environ. Sci. 4 (1), 100–108.

Anjum, N.A., Ahmad, I., Mohmood, I., Pacheco, M., Duarte, A.C., Pereira, E., Umar, S., Ahmad, A., Khan, N.A., Iqbal, M., Prasad, M.N.V., 2012. Modulation of glutathione and its related enzymes in plant's responses to toxic metals and metalloids—a review. Environ. Exp. Bot. 75, 307–324.

AQEG, 2005. Particulate Matter in the UK. Defra, London.

Aro, E.M., Suorsa, M., Rokka, A., Allahverdiyeva, Y., Paakkarinen, Saleem, A., Battchikova, N., Rintamäki, E., 2005. Dynamics of photosystem II: a proteomic approach to thylakoid protein complexes. J. Exp. Bot. 56 (411), 347–356.

Asada, K., 1999. The water-water cycle in chloroplasts: scavenging of active oxygen and dissipation of excess photons. Annu. Rev. Plant. Physiol. Plant. Mol. Biol. 50, 601–639.

Atkinson, C.J., Wookey, P.A., Mansfield, T.A., 1991. Atmospheric pollution and sensitivity of stomata on barley leaves to abscisic acid and carbon dioxide. New Phytol. 117, 535–541.

Bacic, T., Lynch, A.H., Cutler, D., 1999. Reactions to cement factory dust contamination by *Pinus halepensis* needles. Environ. Exp. Bot. 41, 155–166.

Badach, J., Dymnicka, M., Baranowski, A., 2020. Urban vegetation in air quality management: a review and policy framework. Sustainability 12, 1258.

Baker, N.R., 2008. Chlorophyll fluorescence: a probe of photosynthesis *in vivo*. Annu. Rev. Plant Biol. 59, 659–668.

Baker, N.R., Rosenqvist, E., 2004. Applications of chlorophyll fluorescence can improve crop production strategies: an examination of future possibilities. J. Exp. Bot. 55 (403), 1607–1621.

Balaganskaya, E.D., Kurdjavtseva, O.V., 1998. Change of the morphological structure of leaves of *Vaccinium vitis-idaea* caused by heavy metal pollution. Chemosphere 36, 721–726.

Bargagli, R., 1998. Trace Elements in Terrestrial Plants. Springer Verlag, R. G. Landes Company, p. 344.

Barrero-Sicilia, C., Silvestre, S., Haslam, R.P., Michaelson, L.W., 2017. Lipid remodeling: unravelling the response to cold stress in *Arabidopsis* and its extremophile relative *Eutrema salsugineum*. Plant Sci. 263, 194–200.

Barwise, Y., Kumar, P., 2020. Designing vegetation barriers for urban air pollution abatement: a practical review for appropriate plant species selection. npj Clim. Atmos. Sci. 3, 1–19.

Bell, J.N.B., Treshow, M., 2002. Air Pollution and Plant Life, second ed. John Wiley & Sons, Ltd., England, p. 465.

Beckett, K.P., Freer-Smith, P.H., Taylor, G., 1998. Urban woodlands: their role in reducing the effects of particulate pollution. Environ. Pollut. 99 (3), 347–360.

Beckett, P., Freer-Smith, P., Taylor, G., 2000. Effective tree species for local air-quality management. J. Arboric. 26 (1), 12–18.

Bell, M.L., Morgenstern, R.D., Harrington, W., 2011. Quantifying the human health benefits of air pollution policies: a review of recent studies and new directions in accountability research. Environ. Sci. Policy 14, 357–368.

Bennett, J.P., Rassat, P., Berrang, P., Karnosky, D.F., 1992. Relationship between leaf anatomy and ozone sensitivity of *Fraxinus pennsylvanica* Marsh. and *Prunus serotina* Ehrh. Environ. Exp. Bot. 32, 33–41.

Berg, V.S., 1989. Leaf cuticles as potential markers of air pollutants exposure in trees. In: National Research Council (Ed.), Committee on Biological Markers of Air Pollution Damage in Trees, Board on Environmental Studies and Toxicology, Commission on Life Science. Biologic Markers of Air Pollution Stress and Damage in Forests. National Academy Press, Washington, DC, pp. 333–338.

Bianco, L., Serra, V., Larcher, F., Perino, M., 2017. Thermal behavior assessment of a novel vertical greenery module system: first results of a long-term monitoring campagn in an outdoor test cell. Energ. Effic. 10 (3), 625–638.

Bini, C., Washa, M., Fontana, S., Maleci, L., 2012. Effects of heavy metals on morphological characteristics of *Taraxacum officinale* Web. growing on mine soils in NE Italy. J. Geochem. Explor. 123, 101–108.

Birbaum, K., Brogiolo, R., Schellenberg, M., Martinoia, E., Stark, W.J., Günther, D., Limbach, L., 2010. No evidence for cerium dioxide nanoparticle translocation in maize plants. Environ. Sci. Technol. 44 (22), 8718–8723.

Birge, H.E., Allen, C.R., Garmestani, A.S., Pope, K.L., 2016. Adaptive management for ecosystem services. J. Environ. Manage. 183, 343–352.

Bjorkman, O., Demmig, B., 1987. Photon yield of O_2 evolution and chlorophyll fluorescence characteristics at 77 K among vascular plants of diverse origins. Planta 170, 489–504.

Blankenship, R.E., 2002. Molecular Mechanisms of Photosynthesis. Blackwell Science Publishers Ltd., Oxford, p. 321.

Blanusa, T., Garratt, M., Cathcart-James, M., Hunt, L., Cameron, R.F., 2019. Urban hedges: a review of plant species and cultivars for ecosystem service delivery in north-west Europe. Urban For. Urban Green. 44, 126391.

Bombelli, A., Gratani, L., 2003. Interspecific differences of leaf gas exchange and water relations of three evergreen Mediterranean shrub species. Photosynthetica 41 (4), 619–625.

Booij-James, I.S., Dube, S.K., Jansen, M.A.K., Edelman, M., Mattoo, A.K., 2000. Ultraviolet-B radiation impacts light-mediated turnover of the photosystem II reaction center heterodimer in *Arabidopsis* mutants altered in phenolic metabolism. Plant Physiol. 124, 1275–1283.

Bukowski, N., Lienemann, P., Hill, M., Furger, M., Richard, A., Amato, F., Prevot, A.S.H., Baltensperger, U., Buchmann, B., Gehring, R., 2010. PM10 emission factors for non-exhaust particles generated by road traffic in an urban street canyon and along a freeway in Switzerland. Atmos. Environ. 44, 2330–2340.

Burghardt, M., Riederer, M., 2006. Cuticular transpiration. In: Riederer, M., Müller, C. (Eds.), Biology of the Plant Cuticle. Blackwell Publishing, Oxford, pp. 291–310.

Bussotti, F., Bottacci, A., Bartolesi, A., Grossoni, P., Tani, C., 1995. Morpho-anatomical alterations in leaves collected from beech trees (*Fagus sylvatica* L.) in conditions of natural water stress. Environ. Exp. Bot. 35, 201–213.

Bussotti, F., Gravano, E., Grossoni, P., Tani, C., Mori, B., 2002. Response of a Mediterranean evergreen shrub (*Arbutus unedo* L.) to ozone fumigation. In: Karnosky, D.F., Percy, K.E., Chapelka, A.H., Simpson, C.J. (Eds.), Air Pollution, Global Change and Forests in the New Millennium. Development in Environmental Science. Elsevier, Oxford, pp. 259–268.

Cape, J.N., 1983. Contact angles of water droplets on needles of Scots pine (*Pinus sylvestris*) growing in polluted atmospheres. New Phytol. 93 (2), 293–299.

Cape, J.N., 1994. Evaluation of pollutant critical levelsfrom leaf surfaces characteristics. In: Pearcy, K.E., Cape, J.N., Jagels, R., Simpson, C.D. (Eds.), Air Plutants and the Leaf Cuticle. Springer-Verlag, Berlin, pp. 123–138.

Cape, J.N., Fowler, D., 1981. Changes in epicuticular wax of *Pinus sylvestris* exposed to polluted air. Silva Fenn. 15 (4), 457–458.

Cass, G., Hughes, L., Kleeman, M., Allen, J., Salmon, L., 2000. The chemical composition of atmospheric ultrafine particles. Philos. Trans. R. Soc. Lond. 358 (1775), 2581–2592.

CEAA, 2009. Operational Policy Statement: Adaptive Management Measures under the Canadian Environmental Assessment Act. Canadian Environmental Assessment Agency, Ottawa, ON.

Celis, J., Morales, R., Zaror, C., Inzunza, J., Flocchini, R., Carvacho, O., 2003. Chemical characterization of the inhalable particulate matter in city of Chillan, Chile. J. Chil. Chem. Soc. 48 (2), 49–55.

Colls, J., 2002. Air Pollution. Spon Press, London and New York, p. 560.

Conesa, H., Faz, A., Arnaldos, R., 2006. Heavy metal accumulation and tolerance in plants from mine tailings of the semiarid Cartagena-La Union mining district (SE Spain). Sci. Total Environ. 366, 1–11.

Currie, B.A., Bass, B., 2008. Estimates of air pollution mitigation with green plants and green roofs using the UFORE model. Urban Ecosyst. 11, 409–422.

Dakhma, W.S., Zarrouk, M., Cherif, A., 1995. Effects of drought-stress on lipids in rape leaves. Phytochemistry 40, 1383–1386.

Das, P.K., Geul, B., Choi, S.-B., Yoo, S.-D., Park, Y., 2011. Photosynthesis-dependent anthocyanin pigmentation in *Arabidopsis*. Plant Signal. Behav. 6 (1), 23–25.

Denton, F., Wilbanks, T.J., Abeysinghe, A.C., Burton, I., Gao, Q., Lemos, M.C., Masui, T., O'Brien, K.L., Warner, K., 2014. Climate-resilient pathways: adaptation, mitigation, and sustainable development. In: Field, C.B., Barros, V.R., Dokken, D.J., Mach, K.J., Mastrandea, M.D., Bilir, T.E., White, L.L. (Eds.), Climate Change 2014: Impacts, Adaptation, and Vulnerability. Part A: Global and Sectoral Aspects. Contribution of Working Group II to the Fifth Assessment Report of the Intergovermental Panel on Climate Change. Cambridge University Press, Cambridge and New York, NY, pp. 1101–1131.

Djurdjević, L., Mitrović, M., Pavlović, P., Gajić, G., Kostić, O., 2006. Phenolic acids as bioindicators of fly ash deposit revegetation. Arch. Environ. Contam. Toxicol. 50, 488–495.

Dörmann, P., 2013. Galactolipids in plant membranes. In: eLS. John Willey and Sons, Ltd, Chichester, pp. 1–6.

Dueck, T.A., 1986. The combined effect of sulphur dioxide and copper on two populations of *Trifolium repens* and *Lolium perenne*. In: Impact of Heavy Metals and Air Pollutants on Plants. Academisch Proef-schrift, Free University Press, Amsterdam, pp. 102–114.

Dwivedi, S., Tripathi, R.D., Srivastava, S., Mishra, S., Shukla, M.K., Tiwari, K.K., Singh, R., Rai, U.N., 2007. Growth performance and biochemical responses of three rice (*Oryza sativa* L.) cultivars grown in fly-ash amended soil. Chemosphere 67, 140–151.

Earl, H.J., 2002. Stomatal and non-stomatal restrictions to carbon assimilation in soybean (*Glycine max*) lines differing in water use efficiency. Environ. Exp. Bot. 48, 237–246.

Eleftheriou, E.P., 1987. A comparative study of the leaf anatomy of olive trees growing in the city and the country. Environ. Exp. Bot. 27 (1), 103–107.

Erdogan, E., Yazgan, M.E., 2009. Landscaping in reducing traffic noise problem in cities: Ankara case. Afr. J. Agric. Res. 4 (10), 1015–1022.

European Commission, 2009. Reference Document on Best-Available Techniques for Management of Tailings and Waste–Rock in Mining Activities. http://eippcb.jrc.ec.europa.eu.

Evans, L.S., Adamski, I.I.J.H., Renfro, J.R., 1996. Relationship between cellular injury, visible injury of leaves, and ozone exposure levels for several dicotyledonous plant species at great smoky mountains National Park. Environ. Exp. Bot. 36 (2), 229–237.

Eveling, D.W., Bataille, A., 1984. The effects of deposits of small particles on the resistance of leaves and petals to water loss. Environ. Pollut. A 36, 229–238.

Falcone, D.L., Ogas, J.P., Somerville, C.R., 2004. Regulation of membrane fatty acid composition by temperature in mutants of Arabidopsis with alterations in membrane lipid composition. BMC Plant Biol. 4, 17.

Fauser, P., 1999. Particulate air pollution, with emphasis on traffic generated aerosols. Risø National Laboratory, Technical University of Denmark, Roskilde, Denmark, p. 151.

Fellet, G., Pošćić, F., Licen, S., Marchiol, L., Musetti, R., Tolloi, A., Barbieri, P., Zerbi, G., 2016. PAHs accumulation on leaves of six evergreen urban shrubs: a field experiment. Atmos. Pollut. Res. 7 (5), 915–924.

Fernandez, S., Poschenrieder, C., Marceno, C., Gallego, J.R., Jimenez-Gamez, D., Bueno, A., Afif, A., 2017. Phytoremediation capability of native plant species living on Pb-Zn and Hg-As mining wastes in the Cantabrian range, north Spain. J. Geochem. Explor. 174, 10–20.

Fett, W.F., Jones, S.B., 1995. Microscopy of the interaction of hrp mutants of *Pseudomonas syringae* pv. *phaseolica* with nonhost plant. Plant Sci. 107, 27–39.

Fluckiger, W., Oertli, J.J., Fluckiger, H., 1979. Relationship between stomatal diffusive resistance and various applied particle sizes on leaf surfaces. Z Pflanzenphysiol 91, 173–175.

Font, A., Fuller, G.W., 2016. Did policies to abate atmospheric emissions from traffic have a positive effect in London. Environ. Pollut. 218, 1–12.

Foyer, C.H., Noctor, G., 2005. Redox homeostasis and antioxidant signaling: a metabolic interface between stress perception and physiological responses. Plant Cell 17, 1866–1875.

Foyer, C.H., Noctor, G., 2011. Ascorbate and glutathione: the heart of the redox hub. Plant Physiol. 155, 2–18.

Foyer, C., Shigeoka, S., 2011. Understanding oxidative stress and antioxidant functions to enhance photosynthesis. Plant Physiol. 155, 93–100.

Foyer, C.H., Gomez, L.D., van Heerden, P.D.R., 2005. Glutathione. In: Smirnoff, N. (Ed.), Antioxidants and Reactive Oxygen Species in Plants. Blackwell Publishing Oxford, UK, pp. 1–24.

Freer-Smith, P.H., 1984. The responses of six broadleaved trees during long-term exposure to SO_2 and NO_2. New Phytol. 97, 49–61.

Gajić, G., 2007. The Impact of Air Pollution on Functional—Structural Characteristics of Cultivated Shrub Species *Syringa vulgaris* L., *Ligustrum ovalifolium* Hassk. and *Spiraea van-houttei* (Briot.) Zab. in Belgrade. thesis, Magister of Sciences, of Biological Sciences, Faculty of Biology, University of Belgrade, Belgrade, p. 186.

Gajić, G., Pavlović, P., 2018. The role of vascular plants in the phytoremediation of fly ash deposits. In: Matichenkov, V. (Ed.), Phytoremediation: Methods, Management and Assessment. Nova Science Publishers, Inc., New York, pp. 151–236.

Gajić, G., Mitrović, M., Pavlović, P., Stevanović, B., Djurdjević, L., Kostić, O., 2009. An assessment of the tolerance of *Ligustrum ovalifolium* Hassk.to traffic-generated Pb using physiological and biochemical markers. Ecotoxicol. Environ. Saf. 72, 1090–1101.

Gajić, G., Pavlović, P., Kostić, O., Jarić, S., Djurdjević, L., Pavlović, D., Mitrović, M., Pavlović, P., 2013. Ecophysiological and biochemical traits of three herbaceous plants growing of the disposed coal combustion fly ash of different weathering stage. Arch. Biol. Sci. 65 (1), 1651–1667.

Gajić, G., 2014. Ecophysiological adaptations of selected species of herbaceous plants on the fly ash deposits at the thermal power plant "Nikola Tesla-A" in Obrenovac. Biological Faculty, University of Belgrade, Belgrade, Serbia, p. 406.

Gajić, G., Djurdjević, L., Kostić, O., Jarić, S., Mitrović, M., Stevanović, B., Mitrović, M., Pavlović, P., 2016. Assessment of the phytoremediation potential and an adaptive response of *Festuca rubra* L. sown on fly ash deposits: native grass has a pivotal role in ecorestoration management. Ecol. Eng. 93, 250–261.

Gajić, G., Đurđević, L., Kostić, O., Jarić, S., Mitrović, M., Pavlović, P., 2018a. Ecological potential of plants for phytoremediation and ecorestoration of fly ash deposits and mine wastes. Front. Environ. Sci. 6, 124.

Gajić, G., Stamenković, M., Pavlović, P., 2018b. Plant photosynthetic response to metal(loid) stress. In: Singh, V.P., Singh, S., Singh, R., Prasad, S.M. (Eds.), Environment and Photosynthesis. A Future Prospect. Studium Press (India), Pvt. Ltd, pp. 145–209.

Gajić, G., Mitrović, M., Pavlović, P., 2019. Ecorestoration of fly ash deposits by native plant species at thermal power stations in Serbia. In: Pandey, V.C., Bauddh, K. (Eds.), Phytomanagement of Polluted Sites: Market Opportunities in Sustainable Phytoremediation. Elsevier, Amsterdam, pp. 113–177.

Gajić, G., Djurdjević, L., Kostić, O., Jarić, S., Mitrović, M., Stevanović, B., Mitrović, M., Pavlović, P., 2020. Phytoremediation potential, photosynthetic and antioxidant response to arsenic-induced stress of *Dactylis glomerata* L. sown on the fly ash deposits. Plants 9 (5), 657.

Gao, J., Wallis, J.G., Browse, J., 2015. Mutation in the prokaryotic pathway rescue the fatty acid biosynthesis1 mutant in the cold. Plant Physiol. 169 (1), 442–452.

Gao, X., Gu, F., Mei, X., Hao, W., Li, H., Gong, D., Li, X., 2018. Light and water use efficiency as influenced by clouds and/or aerosols in a rainfed spring maize cropland on the loess plateau. Crop. Sci. 58, 853–862.

Gellini, R., Grossoni, P., Bussoti, F., 1987. Danni di nuovo tipo in *Fagus sylvatica* L. Osservazioni al SEM delle superfici fogliari. Giornalle Bot. Ital. 121, 337–351.

Gerosa, G., Marzuoli, R., Bussotti, F., Pancrazi, M., Ballarin-Denti, A., 2003. Ozone sensitivity of *Fagus sylvatica* and *Fraxinus excelsior* young trees in relation to leaf structure and foliar ozone uptake. Environ. Pollut. 125, 91–98.

Gill, S.S., Tuteja, N., 2010. Reactive oxygen species and antioxidant machinery in abiotic stress tolerance in crop species. Plant Physiol. Biochem. 48, 909–930.

GLA, 2018. London Environmental Strategy. Greater London Authority, p. 442.

Godzik, S., Florkowski, T., Piorek, S., Sassen, M.A., 1979. An attempt to determine the tissue contamination of *Quercus robur* L. and *Pinus sylvestris* L. foliageby particulates from zinc and lead smelters. Environ. Pollut. 18 (2), 97–106.

Gomes, M.P., de Sae Melo Marqes, T.C.L., de Oliviera Goncalves Nogueira, M., de Castro, E.M., Soares, A.M., 2011. Ecophysiological and anatomical changes due to uptake and accumulation of heavy metal in *Brachiaria decumbens*. Sci. Agric. 68 (5), 566–573.

Götschi, T., Hazenkamp-von Arx, M.E., Heinrich, J., Bono, R., Burney, P., Forsberg, B., Jarvis, D., Maldonaldo, J., Norback, D., Stern, W.B., Sunyer, J., Toren, K., Verlato, G., Villani, S., Kunzli, N., 2005. Elemental composition and reflectance of ambient fine particles at 21 Europian locations. Atmos. Environ. 39 (32), 5947–5958.

Grace, S.C., 2005. Phenolics as antioxidants. In: Smirnoff, N. (Ed.), Antioxidants and Reactive Oxygen Species in Plants. Blackwell Publishing Ltd, Oxford, pp. 141–168.

Grammatikopoulus, G., 1999. Mechanisms for drought tolerance in two Mediterranean seasonal dimorphic shrubs. Aust. J. Plant Physiol. 26, 587–593.

Gratão, P.L., Polle, A., Lea, P.J., Azevedo, R.A., 2005. Making the life of heavy metal stressed plants a little easier. Funct. Plant Biol. 32, 481–494.

Gravouil, K., Ferru-Clement, R., Colas, S., Helye, R., Kadri, L., Bourdeau, L., Moumen, B., Mercier, A., Ferreira, T., 2017. Transcriptomics and Lipidomics of the environmental strain Rhodococcus ruber point out consumption pathways and potential metabolic bottlenecks for polyethylene degradation. Environ. Sci. Technol. 51, 5172–5181.

Greenen, S., Cojocariu, C., Gethings, L., Isaac, G., Fernandes, L., Tonge, R., Vissers, J., Langrige, J., Wilson, I., Martin, L., 2013. Qualitative and quantitative characterization of the metabolome, lipidome, and proteome of human hepatocytes stable transfected with cytochrome P450 2E1 using data independent LC-MS. J. Biomol. Tech. 24, S61–S62.

Grill, D., Pfeifhofer, G., Halbwachs, G., Waltinger, W., 1987. Investigations on epicuticular waxes of differently damaged spruce needles. Eur. J. For. Pathol. 17 (4–5), 246–255.

Gromke, C., Jamarkattel, N., Ruck, B., 2016. Influence of roadside hedgerows on air quality in urban street canyons. Atmos. Environ. 139, 75–86.

Guerreiro, C., Gonzales Ortiz, A., de Leeuw, F., Viana, M., Horalek, J., 2016. Air quality in Europe—2016 report.

Guidi, L., Di Cagno, R., Soldatini, G.F., 2000. Screening of bean cultivars for their response to ozone as evaluated by visible symptoms and leaf chlorophyll fluorescence. Environ. Pollut. 107, 349–355.

Halliwell, B., 2006. Reactive species and antioxidants. Redox biology is a fundamental theme of aerobic life. Plant Physiol. 141, 312–322.

Hansen-Bristow, K., 1986. Influence of increasing elevation on growth characteristics at timberline. Can. J. Bot. 64 (11), 2517–2523.

Hatfield, J.L., Dold, C., 2019. Water-use efficiency: advances and challenges in a changing climate. Front. Plant Sci. 10, 103.

Havaux, M., 1998. Carotenoids as membrane stabilizers in chloroplast. Trends Plant Sci. 3, 147–151.

Heal, M.R., Hibbs, L.R., Agius, R.M., Beverland, I.J., 2005. Total and water-soluble trace metal content of urban background PM_{10}, $PM_{2.5}$ and black smoke in Edinburgh, UK. Atmos. Environ. 39 (8), 1417–1430.

Heildrich, C., Feueborn, H.J., Weir, A., 2013. Coal combustion products: a global perspective. In: Proceedings World of Coal Ash (WOCA) Conference, Lexington, KY, USA, April 22–25.

Higashi, Y., Okazaki, Y., Myouga, F., Shinozaki, K., Saito, K., 2015. Landscape of the lipidome and transcriptome under heat stress in *Arabidopsis thaliana*. Sci. Rep. 5, 10533.

Hu, Y., Wang, D., Wei, L., Zhang, X., Song, B., 2014. Bioaccumulation of heavy metals in plant leaves from Yan'an city of the Loess Plateau, China. Ecotoxicol. Environ. Saf. 110, 82–88.

Hubac, C., Guerrier, D., Ferran, J., Tremoliers, A., 1989. Change of leaf lipid composition during water stress in two genotypes of *Lupinus albus* resistant or susceptible to drought. Plant Physiol. Biochem. 27, 737–744.

Huseyinova, R., Kutbay, H.G., Bilgin, A., Kilic, D., Horuz, A., Kirmanglu, C., 2009. Sulphur and some heavy metal contents in foliage of *Corylus avellana* and some roadside native plants in Ordu Province, Turkey. Ekol. Derg. 18, 70.

Huttunen, S., 1994. Effects of air pollutants on epicuticular wax structure. In: Percy, K.E., Cape, J.N., Jagels, R., Simpson, C.D. (Eds.), Air Pollutants and the Leaf Cuticle. Springer-Verlag, Berlin, pp. 123–138.

Jamil, S., Abhilash, P.C., Singh, A., Singh, N., Behl, H.M., 2009. Fly ash trapping and metal accumulating capacity of plants: implication for green belt around thermal power plants. Landsc. Urban Plan. 92 (2), 136–147.

Jeffree, C.E., Grace, J., Hoad, S.P., 1994. Spatial distribution of sulphate uptake by wind-damaged beech leaves. In: Percy, K.E., Cape, J.N., Jagels, R., Simpson, C.D. (Eds.), Air Pollutants and the Leaf Cuticle. Springer-Verlag, Berlin, pp. 183–193.

Jim, C., Chen, W.Y., 2009. Ecosystem services and valuation of urban forests in China. Cities 26 (4), 187–194.

Kaoua, M., Serraj, R., Benichou, M., Hsissou, D., 2006. Comparative sensitivity of two Moroccan wheat varieties to water stress: the relationship between fatty acids and proline accumulation. Bot. Stud. 47, 51–60.

Karnaukhova, N.A., 2016. Anatomo-morphological features of the leaves of *Hedysarum theinum* (Fabaceae) in Western Altai. Contemp. Probl. Ecol. 9, 349–354.

Karner, A.A., Eisinger, D.S., Niemeier, D.A., 2010. Near-roadway air quality: synthesizing the findings from real-world data. Environ. Sci. Technol. 44 (14), 5334–5344.

Kasowska, D., Gediga, K., Spiak, Z., 2018. Heavy metal and nutrient uptake in plants colonizing post-flotation copper tailings. Environ. Sci. Pollut. Res. 25, 824–835.

Kaur, M., Nagpal, A.K., 2017. Evaluation of ait pollution tolerance index and anticipated performance index of plants and their application in development of green space along the urban areas. Environ. Sci. Pollut. Res. 24, 18881–18895.

Kinney, P.L., Aggrawal, M., Northridge, M.E., Janssen, N.A., Shepard, P., 2000. Airborne concentrations of $PM_{2.5}$ and diesel exhaust particles on Harlem sidewalks: a comminity-based pilot study. Environ. Health Persp. 108 (3), 213–218.

Koelmel, J.P., Napolitano, M.P., Ulmer, C.Z., Vasiliou, V., Garrett, T.J., Yost, R.A., Prasad, M.N.V., Godro Pollitt, K.J., Bowden, J.A., 2020. Environmental lipidomics: understanding the response of organisms and ecosystems to a changing world. Metabolomics 16, 56.

Kostić, O., Mitrović, M., Knežević, M., Jarić, S., Gajić, G., Djurdjević, L., Pavlović, P., 2012. The potential of four woody species for the revegetation of fly ash deposits from the 'Nikola Tesla–A' thermoelectric plant (Obenovac, Serbia). Arch. Biol. Sci. 64 (1), 145–158.

Kostić, O., Jarić, S., Gajić, G., Pavlovć, D., Pavlović, M., Mitrović, M., Pavlović, P., 2018. Pedological properties and ecological implications of substrates derived 3 and 11 years after the revegetation of lignite fly ash disposal sites in Serbia. Catena 163, 78–88.

Kovacs, M., 1993. Biological Indicators of Environmental Protection. Elis Harwood, Ltd, New York, p. 207.

Kozlowski, T.T., Constandiniou, A.H., 1986. Environmental pollution and tree growth. For. Abst. 47, 105–125.

Kukkola, E., Huttunen, S., 1998. Structural observations on needles exposed to elevated levels of copper and nickel. Chemosphere 36 (4–5), 727–732.

Kumar, R., Thangaraju, M.M., Kumar, M., Thul, S.T., Pandey, V.C., Yadav, S., Singh, L., Kumar, S., 2021. Ecological restoration of coal fly ash–dumped area through bamboo plantation. Environ. Sci. Pollut. Res. https://doi.org/10.1007/s11356-021-12995-7.

Kumari, A., Pandey, V.C., Rai, U.N., 2013. Feasibility of fern *Thelypteris dentata* for revegetation of coal fly ash landfills. J. Geochem. Explor. 128, 147–152.

Kumari, A., Lal, B., Rai, U.N., 2016. Assessment of native plant species for phytoremediation of heavy metals growing in the vicinity of NTPC sites, Kahagon, India. Int. J. Phytoremediation 18 (6), 592–597.

Kwak, M.J., Lee, J., Kim, H., Park, S., Lim, Y., Kim, J.E., Baek, S.G., Seo, S.M., Kim, K.N., Woo, S.Y., 2019. The removal efficiency of several temperate tree species at adsorbing airborne particulate matter in urban forests and roadsides. Forests 10, 960.

Lakrum, A.E.W.D., Orth, R.J., Duarte, C.M., 2006. Seagrasses: Biology, Ecology and Conservation. Springer, Netherlands.

Lambers, H., Chapin, F.S., Pons, T.L., 1998. Plant Physiological Ecology. Springer-Verlag, New York, p. 540.

Larcher, W., 1995. Physiological Plant Ecology. Springer-Verlag, Berlin Heidelberg, p. 506.

Larkindale, I., Huang, B., 2004. Changes in lipid composition and saturation level in leaves and roots for heat-stressed and heat-acclimated creeping bentgrass (*Agrostis stolonifera*). Environ. Exp. Bot. 51, 57–67.

Lei, W., Liu, L.-Y., Gao, S.-Y., Eerdum, H., Zhi, W., 2006. Physicochemical characteristics of ambient particles settling upon leaf surfaces of urban plants in Beijing. J. Environ. Sci. 18 (5), 921–926.

Li, X.P., Björkman, O., Shih, C., Grossman, A.R., Rosenquist, M., Jansson, S., Niyogi, K.K., 2000. A pigment binding protein essential for regulation of photosynthetic light harvesting. Nature 403, 391–395.

Li, Z., Hopke, P.K., Husain, L., Quereshi, S., Dutkewitz, V.A., Schwab, J.J., Drewnick, F., Demerijan, K., 2004. Sources of fine particles composition in New York city. Atmos. Environ. 38 (38), 6521–6529.

Li, M.S., Luo, Y.P., Su, Z.Y., 2007. Heavy metal concentrations in soils and plant accumulation in a restored manganese mineland in Guangxi, South China. Environ. Pollut. 147, 168–175.

Liang, H., Yao, N., Song, J.T., Luo, S., Lu, H., Greenberg, J.T., 2003. Ceramides modulate programmed cell death in plants. Genes Dev. 17 (21), 2636–2641.

Librando, V., Perrini, G., Tomasello, M., 2002. Biomonitoring of atmospheric PAHs by evergreen plants: correlations and applicability. Polycycl. Aromat. 33 (3–4), 549–559.

Liu, J., Cao, Z., Zou, S., Liu, H., Hai, X., Wang, S., Duan, J., Xi, B., Yan, G., Zhang, S., Jia, Z., 2018. An investigation of the leaf retention capacity, efficiency and mechanisms for atmospheric particulate matter of five greening tree species in Beijing, China. Sci. Total Environ. 616–617, 417–426.

Liu, W., Wang, Y., Russel, A., Edgerton, E.C., 2005. Atmospheric aerosol over two urban-rural pairs in the southeastern United States: chemical composition and possible sources. Atmos. Environ. 39 (25), 4453–4470.

Lopes, M.S., Araus, J.I., van Heerden, P.D.R., Foyer, C.H., 2011. Enhancing drought tolerance in C4 crops. J. Exp. Bot. 62, 3135–3153.

Lopez-Orenes, A., Bueso, M.C., Conesa, H.M., Calderon, A., Ferer, M., 2017. Seasonal changes in antioxidative/oxidative profile of mining and non-mining populations of Syrian beancaper as determined by soil conditions. Sci. Total Environ. 575, 437–447.

Lopez-Orenes, A., Bueso, M., Parraga-Aguado, I.M., Calderon, A., Ferrer, M.A., 2018. Coordinated role of soluble and cell wall bound phenols is a key feature of the metabolic adjustment in mining woody fleabane (*Dittrichia viscosa* L.) population under semi-arid conditions. Sci. Total Environ. 618, 1139–1151.

Los, D.A., Zinchenko, V.V., 2009. Regulatory role of membrane fluidity in gene expression. In: Wada, H., Murata, N. (Eds.), Lipids in Photosynthesis: Essential and Regulatory Functions. Series Advances in Photosynthesis and Respiration, vol. 30. Springer Verlag, pp. 329–348.

Lu, Y., Li, X., He, M., Zeng, F., 2013. Behavior of native species *Arrhenatherum elatius* (Poaceae) and *Sonchus transcapicus* (Asteraceae) exposed to a heavy metal-polluted field: plant metal concentration, phytotoxicity, and detoxification responses. Int. J. Phytoremediation 15, 924–937.

Lukasik, L., Palowski, B., Ciepal, R., 2002. Lead, cadmium, and zinc contents in soil and in leaves of selected tree and shrub species grown in urban parks of Upper Silesia. Chem. Inzymeria Ekol. 9 (4), 431–439.

Lusting, R.H., 2006. Childhood obesity: behavioral aberration or biochemical drive? Reinterpreting the first law of thermodynamics. Nat. Clin. Pract. Endocrinol. Metab. 2, 447.

Maiti, S.K., 2013. Ecorestoration of the Coalmine Degraded Lands. Springer, New Delhi.

Maiti, D., Pandey, V.C., 2021. Metal remediation potential of naturally occurring plants growing on barren fly ash dumps. Environ. Geochem. Health 43, 1415–1426. https://doi.org/10.1007/s10653-020-00679-z.

Mäkelä, A., Huttunen, S., 1987. Cuticular Needle Erosion and Winter Drought in Polluted Environments—A Model Analysis. The International Institute for Applied Systems Analysis, Laxenburg, Austria, p. 33.

Mannien, S., Peura, R., Huttunen, R., 1996. A new SEM method for scoring the wax tube distribution of needles. J. Exp. Bot. 47 (Supplem), 60.

Mano, S., Yamaguchi, K., Hayashi, M., Nishimura, M., 1997. Stromal and thylakoid-bound ascorbate peroxidases are produced by alternative splicing in pumpkin. FEBS Lett. 413, 21–26.

Mansfield, T.A., Pearson, M., 1996. Disturbances in stomatal behavior in plants exposed to air pollution. In: Yunus, M., Iqbal, M. (Eds.), Plant Response to Air Pollution. John Wiley & Sons, Chichester, pp. 179–194.

Martins, M.Q., Rodrigues, W.P., Fortunato, A.S., Leitao, A.E., Rofrigues, A.P., Pais, I.P., Martins, L.D., Silva, M.J., Reboredo, F.H., Partelli, F.I., Campostrini, E., Tomaz, M.A., Scotti-Campos, P., Ribeiro-Barros, A.I., Lidon, F.J., DaMatta, F.M., Ramalho, J.C., 2016. Protective response mechanisms to heat stress in interaction with high [CO2] conditions in *Coffea* spp. Front. Plant Sci. 7, 947.

Martinson, R., Lambrinos, J., Mata-Gonzales, R., 2019. Water stress patterns of xerophytic plants in an urban landscaping. Hort. Sci. 54 (5), 818–823.

Matas, A.J., Cuartero, J., Herèdia, A., 2004. Phase transitions in the biopolyester cutin isolated from tomato fruit cuticles. Thermochim. Acta 409, 165–168.

Maxwell, K., Johnson, G.N., 2000. Chlorophyll fluorescence—a practical guide. J. Exp. Bot. 51 (345), 659–668.

McCutcheon, S.C., Schnoor, J.L., 2003. Phytoremediation. In: Transformation and Control of Contaminants. John Wiley and Sons, Inc., Hoboken, NJ.

McDonald, R., Kroeger, T., Boucher, T., Longzhu, W., Salem, R., Adams, J., Bassett, S., Edgecomb, M., Garg, S., 2016. Planting Healthy Air. A Global Analysis of the Role of Urban Trees in Addressing Particulate Matter Pollution and Extreme Heat. The Nature Conservancy, p. 128.

Miller, A.-F., 2012. Superoxide dismutases: ancient enzymes and new insight. FEBS Lett. 586 (5), 585–595.

Min, L., Renging, W., 2004. Acute responses of some plants upon combined pollution of sulfur dioxide and lead. Shandong Da Xue xue bao Yi xue ban 39 (5), 116–121.

Mitrović, M., Pavlović, P., Lakušić, D., Stevanović, B., Djurdjevic, L., Kostić, O., Gajić, G., 2008. The potential of *Festuca rubra* and *Calamagrostis epigejos* for the revegetation on fly ash deposits. Sci. Total Environ. 72, 1090–1101.

Mittler, R., 2002. Oxidative stress, antioxidants and stress tolerance. Trends Plant Sci. 7 (9), 405–410.

Mittler, R., Poulos, T.L., 2005. Ascorbate peroxidase. In: Smirnoff, N. (Ed.), Antioxidant and Reactive Oxygen Species in Plants. Blackwell Publishing, Oxford, pp. 87–100.

Moradi, P., Mahdavi, A., Khoshkam, M., Iriti, M., 2017. Lipidomics unravels the role of leaf lipids in thyme plant response to drought stress. Int. J. Mol. Sci. 18, 2067.

Mori, J., Saeba, A., Hanslin, H.M., Teani, A., Ferrini, F., Fini, A., Burchi, G., 2015. Deposition of traffic-related air pollutants on leaves of six evergreen shrub species during a Mediterranean summer season. Urban For. Urban Green. 14 (2), 264–273.

Moss, D.M., Barrett, N.R., Bogle, A.L., Bilkova, J., 1998. Anatomical evidence of the development of damage symptoms across a growing season in needles of red spruce from central New Hampshire. Environ. Exp. Bot. 39, 247–262.

Muneer, S., Kim, T.H., Choi, B.C., Lee, B.S., Lee, J.H., 2014. Effect of CO, NOx and SO2 on ROS production, photosynthesis and ascorbate-glutathione pathway to induce *Fragaria x ananassa* as a hyperaccumulator. Redox Biol. 2, 91–98.

Muszynska, E., Labudda, M., Rozanska, E., Hanus-Fajerska, E., Koszelnik-Leszek, A., 2019. Structural, physiological and genetic diversification of *Silene vulgaris* ecotypes from heavy metal-contaminated areas and their synchronous in vitro cultivation. Planta 249, 1761–1778.

Mysliwa-Kurdziel, B., Prasad, M.N.V., Strzalka, K., 2002. Heavy metal influence on the light phase of photosynthesis. In: Prasad, M.N.V., Strzalka, K. (Eds.), Physiology and Biochemistry of Metal Toxicity and Tolerance in Plants. Kluwer Academic Publishers, Dordrecht, pp. 229–255.

Nadgórska-Socha, A., Ptasiñski, B., Kita, A., 2013. Heavy metal bioaccumulation and antioxidative responses in *Cardaminopsis arenosa* and *Plantago lanceolata* leaves from metalliferous and non-metalliferous sites: a field study. Ecotoxicology 22, 1422–1434.

Nakabayashi, R., Saito, K., 2015. Integrated metabolomics for abiotic stress responses in plants. Curr. Opin. Plant Biol. 24, 10–16.

Narayanan, S., Tamura, P.J., Roth, M.R., Prasad, P.V., Welti, R., 2016. Wheat leaf lipids during heat stress: I. High day and night temperatures result in major lipid alterations. Plant Cell Environ. 39, 787–803.

Narayanan, S., Zoong-Lwe, Z.S., Gandhi, N., Welti, R., Fallen, B., Smith, J.R., Rustgi, S., 2020. Comparative lipidomic analysis reveals heat stress responses of two soybean genotypes differing in temperature sensitivity. Plan. Theory 9, 457.

Neill, S.O., Gould, K.S., 2003. Anthocyanins in leaves: light attenuators or antioxidants? Funct. Plant Biol. 30, 865–873.

Neinhuis, C., Barthlott, W., 1998. Seasonal changes of leaf surface contamination in beech, oak, and ginkgo in relation to leaf micromorphology and wettability. New Phytol. 138, 91–98.

Niinemets, U., 1999. Components of leaf dry mass per area-thickness and density—alter leaf photosynthetic capacity in reverse directions in woody plants. New Phytol. 144, 35–47.

Noctor, G., Gomez, L.A., Vanacker, H., Foyer, C., 2002. Interactions between biosynthesis, compartmentation and transport in the control of glutathione homeostasis and signaling. J. Exp. Bot. 53, 1283–1304.

Noctor, G., Mhamdi, A., Chaouch, S., Han, Y., Neukermans, J., Marquez-Garcia, B., Queval, G., Foyer, C.H., 2012. Glutathione in plants: an integrated overview. Plant Cell Environ. 35, 454–484.

Noctor, G., Mhamdi, A., Queval, G., Foyer, C.H., 2013. Regulation the redox gatekeeper: vacuolar sequestration puts glutathione disulfide in its place. Plant Physiol. 163, 665–671.

O'Connor, D., Zheng, X., Hou, D., Shen, Z., Li, G., Miao, G., O'Connell, S., Guo, M., 2019. Phytoremediation: climate change resilience and sustainability assessment at a coastal brownfield redevelopment. Environ. Int. 130, 104945.

O'Sullivan, O.S., Holt, A.R., Warren, P.H., Evans, K.L., 2017. Optimising UK urban road verge contribution to biodiversity and ecosystem services with cost-effective management. J. Environ. Manage. 191, 162–171.

Odasz-Albrigtsen, A., Tommervik, H., Murphy, P., 2000. Decreased photosynthetic efficiency in plant species exposed to multiple airborne pollutants along the Russian-Norwegian border. Can. J. Bot. 78, 1021–1033.

Ogundare, C.S., Jimoh, M., Saheed, S.A., 2018. Changes in leaf morphological and anatomical characters of some plant species in response to gemstone mining in southwestern Nigeria. IFE J. Sci. 20 (3), 475–486.

Osmond, C.B., Grace, S.C., 1995. Perspectives on photoinhibition and photorespiration in the field: quintessential inefficiencies of the light and dark reactions of photosynthesis. J. Exp. Bot. 46, 1351–1362.

Owens, P.N., 2009. Adaptive management frameworks for natural resource management at the landscape scale: implications and applications for sediment resources. J. Soils Sediments 9, 578–593.

Pääkkonen, E., Metsarinne, S., Holopainen, T., Karenlampi, L., 1995. The ozone sensitivity of birch (Betula pendula) in relation to the developmental stage of leaves. New Phytol. 132, 145–154.

Pääkkonen, E., Gunthardt-Goerg, M.S., Holopainen, T., 1998. Responses of leaf processes in sensitive birch (Betula pendula Roth) clone to ozone combined with drought. Ann. Bot. 82, 49–59.

Pal, A., Kulshreshtha, K., Ahmad, K.J., Behl, H.M., 2002. Do leaf surface characters play a role in plant resistance to auto-exhaust pollution? Flora 197 (1), 47–55.

Panda, D., Mandal, L., Barik, J., Padhan, B., Bisoi, S.S., 2020. Physiological response of metal tolerance and detoxification in castor (Ricinus communis L.) under fly-ash amended soil. Helyon 6, e04567.

Pandey, V.C., 2012. Invasive species based efficient green technology for phytoremediation of fly ash deposits. J. Geochem. Explor. 123, 13–18.

Pandey, V.C., 2013. Suitability of Ricinus communis L. cultivation for phytoremediation of fly ash disposal sites. Ecol. Eng. 57, 336–341.

Pandey, V.C., 2015. Assisted phytoremediation of fly ash dumps through naturally colonized plants. Ecol. Eng. 82, 1–5.

Pandey, V.C., 2020a. Phytomanagement of Fly Ash. Elsevier, Amsterdam, p. 352.

Pandey, V.C., 2020b. Fly ash properties, multiple uses, threats, and management: an introduction. In: Phytomanagement of Fly Ash. Elsevier, Amsterdam, pp. 1–34, https://doi.org/10.1016/B978-0-12-818544-5.00001-8.

Pandey, V.C., 2020c. Afforestation on fly ash catena: an adaptive fly ash management. In: Phytomanagement of fly ash. Elsevier, Amsterdam, pp. 195–234, https://doi.org/10.1016/B978-0-12-818544-5.00007-9.

Pandey, V.C., 2021. Direct seeding offers affordable restoration for fly ash deposits. Energy Ecol. Environ. https://doi.org/10.1007/s40974-021-00212-7.

Pandey, V.C., Singh, N., 2010. Impact of fly ash incorporation in soil systems. Agric. Ecosyst. Environ. 136 (1), 16–27.

Pandey, V.C., Singh, N., 2014. Fast green capping on coal fly ash basins through ecological engineering. Ecol. Eng. 73, 671–675.

Pandey, V.C., Singh, J.S., Kumar, A., Tewari, D.D., 2010. Accumulation of heavy metals by chickpea grown in fly ash treated soil: effects on antioxidants. CLEAN Soil Air Water 38 (12), 1116–1123.

Pandey, V.C., Pandey, D.N., Singh, N., 2015a. Sustainable phytoremediation based on naturally colonizing and economically valuable plants. J. Clean. Prod. 86, 37–39.

Pandey, V.C., Prakash, P., Bajpai, O., Kumar, A., Sing, N., 2015b. Phytodiversity on fly ash deposits: evaluation of naturally colonized species for sustainable phytorestoration. Environ. Sci. Pollut. Res. 22 (4), 2776–2787.

Pandey, V.C., Bajpai, O., Sinhg, N., 2016. Plant regeneration potential in fly ash ecosystem. Urban For. Urban Green. 15, 40–44.

Pandey, V.C., Sahu, N., Singh, D.P., 2020. Physiological profiling of invasive plant species for ecological restoration of fly ash deposits. Urban For. Urban Green. 54, 126773.

Parraga-Aguado, I., Querejeta, J.-I., Gonzales-Alcaraz, M.-N., Jimenez-Carceles, F.J., Conesa, H.M., 2014. Usefulness of pioneer vegetation for the phytomanagement of metal(lod)s enriched tailings: grasses vs. shrubs vs. trees. J. Environ. Manage. 133, 51–58.

Pascal, M., Falq, G., Wagner, V., Chatignoux, E., Corso, M., Blanchard, M., Host, S., Pascal, L., Larrieu, S., 2014. Short-term impacts of particulate matter (PM10, PM10-2.5, PM2.5) on mortality in nine French cities. Atmos. Environ. 95, 175–184.

Pavlović, P., Mitrović, M., Djurdjevic, L., 2004. An ecophysiological study of plants growing on the fly ash deposits from the "Nikola Tesla—A" thermal power station in Serbia. Environ. Manag. 33, 654–663.

Pereira, M.P., de Almeida, R.L.C., Correa, F.F., de Castro, E.M., Ribeiro, V.E., Pereira, F.J., 2016. Cadmium tolerance in Schinus molle trees is modulated by enhanced leaf anatomy and photosynthesis. Trees 30, 807–814.

Pilon-Smits, E., 2005. Phytoremediation. Annu. Rev. Plant Biol. 56, 15–39.

Polle, A., Schwanz, P., Rudolf, C., 2001. Developmental and seasonal changes of stress responsiveness in beech leaves (Fagus sylvatica L.). Plant Cell Environ. 24, 821–829.

Popek, R., Przybysz, A., Gawronska, H., Klamkowski, K., Gawronsky, S.W., 2018. Impact of particulate matter accumulation on the photosynthetic apparatus of roadside woody plants growing in the urban conditions. Ecotoxicol. Environ. Saf. 163, 56–62.

Pratas, J., Favas, P.J.C., D'Souza, R., Varun, M., Pau, M.S., 2013. Phytoremediation assessment of flora tolerant to heavy metals in the contaminated soils of an abandoned Pb mine in Central Portugal. Chemosphere 90, 2216–2225.

Pugh, T.A.M., Mackenzie, A.R., Whyatt, J.D., Hewitt, C.N., 2012. Effectiveness of green infrastructure for improvement of air quality in urban street canyons. Environ. Sci. Technol. 46, 7692–7699.

Qadir, S.U., Raja, V., Sidiqqi, W.A., Mahmooduzzafar, Allah, E.F.A., Hashem, A., Alam, P., Ahmad, P., 2019. Fly-ash pollution modulates growth, biochemical attributes, antioxidant activity and gene expression in Pithecellobium dulce (Roxb) Benth. Plan. Theory 8, 528.

Quartacci, M.F., Pinzino, C., Sgherri, C.L., Navari-Izzo, F., 1995. Lipid composition and protein dynamics in thylakoids of two wheat cultivars differently sensitive to drought. Plant Physiol. 108, 191–197.

Randjelović, D., Gajić, G., Mutić, J., Pavlović, P., Mihailović, N., Jovanović, S., 2016. Ecological potential of Epilobium dodonaei Vill. for restoration of metalliferous mine waste. Ecol. Eng. 95, 800–810.

Rao, K.V.M., Raghavendra, A.S., Reddy, K.J., 2006. Physiology and Molecular Biology of Stress Tolerance in Plants. Springer, Dordrecht, p. 345.

Reijnders, L., 2005. Disposal, uses and treatments of combustion ashes: a review. Resour. Conserv. Recycl. 43, 313–336.

Rice-Evans, C.A., Miller, J.M., Paganga, G., 1996. Structure-antioxidant activity relationship of flavonoids and phenolic acids. Free Radic. Biol. Med. 20, 933–956.

Riding, R.T., Percy, K.E., 1985. Effects of SO_2 and other air pollutants on the morphology of epicuticular waxes on needles of *Pinus strobus* and *Pinus banksiana*. New Phytol. 99 (4), 555–563.

Rosalina, D., Herawati, E.Y., Musa, M., Sofarini, D., Amin, M., Risjani, Y., 2019. Lead accumulation and its histological impact on *Cymodocea serrulata* seagrass in the laboratory. Sains Malaysiana 48 (4), 813–822.

Ruelland, E., Cantrel, C., Gawerm, M., Kader, J.-C., Zachowski, A., 2002. Activationof phospholipases C and D is an early response to a cold exposure in *Arabidopsis* suspension cells. Plant Physiol. 130 (2), 999–1007.

Sauter, J.J., Kammerbauer, H., Pambor, L., Hock, B., 1987. Evidence for the accelerated micromorphological degradation of epistomatal waxes in Norway spruce by motor vehicle emissions. Eur. J. For. Pathol. 17 (7), 444–448.

Sawidis, T., Marnasidis, A., Zachariadis, G., Stratis, J., 1995. A study of air pollution with heavy metals in Thessaloniki City (Greece) using trees as biological indicators. Arch. Environ. Contam. Toxicol. 28, 118–124.

Sawidis, T., Metentzoglou, E., Mitrakas, M., Vasara, E., 2011. A study of chromium, copper, and lead distribution from lignite fuels using cultivated and non-cultivated plants as biological monitors. Water Air Soil Pollut. 220 (1), 339–352.

Saxe, H., 1996. Physiological and biochemical tools in diagnosis of forest decline and air pollution injury to plants. In: Yunus, M., Iqbal, M. (Eds.), Plant Response to Air Pollution. John Wiley & Sons, Chichester, pp. 449–487.

Schönherr, J., 2001. Cuticular penetration of calcium salts: effects of humidity, anions, and adjuvants. J. Plant Nutr. Soil Sci. 164, 225–231.

Schönherr, J., 2006. Characterization of aqueous pores in plant cuticles and permeation of ionic solutes. J. Exp. Bot. 57 (11), 2471–2491.

Schönherr, J., Fernandez, V., Schreiber, L., 2005. Rates of cuticular penetration of chelated Fe(III): role of humidity, concentration, adjuvants, temperature and type of chelate. J. Agr. Food Chem. 53 (11), 4484–4492.

Schreuder, M.D., Van Hove, L.W.A., Brewer, C.A., 2001. Ozone exposure affects leaf wettabilty and tree water balance. New Phytol. 152 (3), 443–454.

Scotti-Campos, P., Pais, I., Partelli, F.I., Batista-Santos, P., Ramalho, J.C., 2014. Phospholipids profile in chloroplasts of *Coffea* spp. genotypes differing in cold acclimation ability. J. Plant Physiol. 171 (3–4), 243–249.

Scotti-Campos, P., Pais, I.P., Ribeiro-Barros, A.I., Martins, L.D., Tomaz, M.A., Rodriques, W.P., Campostrini, E., Semedo, J.N., Fortunato, A.S., Martins, M.Q., Partelli, F.L., Lidon, F.C., DeMatta, F.D., Ramalho, J.C., 2019. Lipid profile adjustments may contribute to warming acclimation and to heat impact mitigation by elevated $[CO_2]$ in *Coffea* spp. Environ. Exp. Bot. 167, 103856.

Sgardelis, S., Cook, C.M., Pantis, J.D., Lanaras, T., 1994. Comparison of chlorophyll fluorescence and some heavy metal concentrations in *Sonchus* spp. and *Taraxacum* spp. along an urban pollution gradient. Sci. Total Environ. 158, 157–164.

Shannigrahi, A.S., Fukushima, T., Sharma, R.C., 2004. Anticipated air pollution tolerance of some plant species considered for green belt development in and around an industrial/urban area in India: an overview. Int. J. Environ. Sci. 61 (2), 125–137.

Sharma, S.S., Dietz, K.-J., 2009. The relationship between metal toxicity and cellular redox imbalance. Trends Plant Sci. 14 (1), 43–50.

Shun, Q., Zhu, Z., 2019. Effects of nitrogen dioxide on biochemical responses in 41 garden plants. Plants 8, 45.

Singh, A., Sharma, R.K., Agrawal, S.B., 2008. Effects of fly ash incorporation on heavy metal accumulation, growth and yield responses of Beta vulgaris plants. Bioresour. Technol. 99, 7200–7207.

Singh, H., Yadav, M., Kumar, N., Kumar, A., Kumar, M., 2020. Assessing adaptation and mitigation potential of roadside trees under the influence of vehicular emissions: a case study of Grevillea robusta and Mangifera indica planted in an urban city of India. PLoS One 15 (1), e0227380.

Sinha, S., Gupta, A.K., 2005. Translocation of metals from fly ash amended soil in the plant of Sesbania cannabina L. Ritz.: effect on antioxidants. Chemosphere 61, 1204–1214.

Sjöman, H., Hirons, A.D., Bassuk, N.L., 2018. Improving confidence in tree species selection for challenging urban sites: a role for leaf turgor loss. Urban Ecosyst. 21, 1171–1188.

Skrynetska, I., Ciepal, R., Kandizora-Ciupa, M., Barczyk, G., Nadgorska-Socha, A., 2018. Ecophysiological responses to environmental pollution of selected plant species in industrial urban area. Int. J. Environ. Res. 12, 255–267.

Smirnoff, N., 2005. Ascorbate, tocopherol and carotenoids: metabolism, pathway engineering and function. In: Smirnoff, N. (Ed.), Antioxidants and Reactive Oxygen Species in Plants. Blackwell Publishing Ltd, Oxford, pp. 54–86.

Speak, A.F., Rothwell, J.J., Lindley, S.J., Smith, C.L., 2012. Urban particulate pollution reduction by four species of green roof vegetation in UK city. Atmos. Environ. 61, 283–293.

Spicher, L., Glauser, G., Kessler, F., 2016. Lipid antioxidant and galactolipid remodeling under temperature stress in tomato plants. Front. Plant Sci. 7, 167.

Sridhar, B.B.M., Diehl, S.V., Han, F.X., Monts, D.L., Su, Y., 2005. Anatomical changes due to uptake and accumulation of Zn and Cd in Indian mustard (Brassica juncea). Environ. Exp. Bot. 54, 131–141.

Sridhar, B.B.M., Han, F.X., Diehl, S.V., Monts, D.L., Su, Y., 2011. Effect of phytoaccumulation of arsenic and chromium on structural and ultrastructural changes of brake fern (Pteris vittata). Braz. J. Plant Physiol. 23 (4), 285–293.

Stanhill, G., Cohen, S., 2001. Global dimming: a review of the evidence for a widespread and significant reduction in global radiation with discussion of its probable causes and possible agricultural consequences. Agric. For. Meteorol. 107, 255–278.

Stevanović, B., Janković, M., 2001. Plant Ecology With Basis in Plant Ecophysiology. NNK International, Belgrade, p. 514.

Storm, L., Watts, M., Austin, K., 2015. 100 Solutions for Climate Action in Cities. Sustainia, C40 Cities Climate Leadership Group.

Sun, H., Kopp, K., Kjwlgren, R., 2012. Water-efficient urban landscapes: integrating different water use categorizations and plant types. Hort. Sci. 47 (2), 254–263.

Syversten, J.P., Lloyd, J., McConchie, C., Kriedemann, P.E., Farquhar, G.D., 1995. On the relationship between leaf anatomy and CO2 diffusion through the mesophyll of hypostomatous leaves. Plant Cell Environ. 1, 149–157.

Takahashi, M., Higaki, A., Nohno, M., Kamada, M., Okamura, Y., Matsui, K., Kitani, S., Morikawa, H., 2005. Differential assimilation of nitrogen dioxide by 70 taxa of roadside trees at an urban pollution level. Chemosphere 61 (5), 633–639.

Tattini, M., Guidi, L., Morassi-Bonzi, L., Pinelli, P., Remorini, D., Innocenti, E., Giordano, C., Massai, R., Agati, G., 2005. On the role of flavonoids in the integrated mechanisms of response of Ligustrum vulgare and Phillyrea latifolia to high solar radiation. New Phytol. 167, 457–470.

Telfer, A., 2002. What is carotene doing in the photosystem II reaction centre. Philos. Trans. R. Soc. Lond. B Biol. Sci. 357, 1431–1440.

Thawale, P.R., Babu, S.S., Wakode, R.R., Singh, S.K., Kumar, S., Juwarkar, A.A., 2011. Biochemical changes in plant leaves as a biomarker of pollution due to anthropogenic activity. Environ. Monit. Assess. 177, 527–535.

Thomashow, M.F., 2010. Molecular basis of plant cold acclimation: insights gained from studying the CBF cold response pathway. Plant Physiol. 154 (2), 571–577.

Trebst, A., 1995. Dynamics in photosystem II. Structure and function. In: Schulze, E.-D., Caldwell, M. (Eds.), Ecophysiology of Photosynthesis. Springer-Verlag, Berlin, Heidelberg, New York, pp. 3–16.

Turunen, M., Huttunen, S., 1990. A review of the response of epicuticular wax conifer needles to air pollution. J. Environ. Qual. 19, 35–45.

Turunen, M., Huttunen, S., Back, J., Lamppu, J., 1995. Acid rain induced in cuticles and Ca distribution in Scot pine and Norway spruce seedlings. Can. J. For. Res. 25 (8), 1313–1325.

USEPA (US Environmental Protection Agency), 2007. EPA's 2007 Report on the Environment: Science Report. Office of Research and Development, Washington, DC, p. 539.

USEPA (US Environmental Protection Agency), 2010. Criteria Air Pollutants. National Ambient Air Quality Standards (NAAQS). Federal Register, Washington DC.

USEPA (US Environmental Protection Agency), 2011. Agriculture and Food Supply: Climate Change, Health, and Environmental Effects.

Uzu, G., Sobanska, S., Sarret, G., Munoz, M., Dumat, C., 2010. Foliar lead uptake by lettuce exposed to atmospheric fallouts. Environ. Sci. Technol. 44 (3), 1036–1042.

Van Renterghem, T., Attenborough, K., Maennel, M., Defrance, J., Horoshenkov, K., Kang, J., Bashir, I., Taherzadeh, S., Altreuther, B., Khan, A., 2014. Measured light vehicle noise reduction by hedges. Appl. Acoust. 78, 19–27.

Viskari, E.L., 2000. Epicuticular wax of Norway spruce needles as indicator of traffic pollutant deposition. Water Air Soil Pollut. 121, 327–337.

Viskari, E.L., Holopainen, T., Karenlampi, L., 2000. Responses of spruce seedlings (*Picea abies*) to exhaust gas under laboratory conditions-II ultrastructural changes and stomatal behavior. Environ. Pollut. 107 (1), 99–107.

Vollenweider, P., Gunthardt-Goerg, M.S., 2005. Diagnosis of abiotic and biotic stress factors using the visible symptoms in foliage. Environ. Pollut. 137, 455–465.

Vollenweider, P., Ottiger, M., Gunthardt-Goerg, M.S., 2003. Validation of leaf ozone symptoms in natural vegetation using microscopical methods. Environ. Pollut. 124, 101–118.

Weber, H., Chetelat, A., Reymond, P., Farmer, E.E., 2004. Selective and powerful stress gene expression in *Arabidopsis* in response to malondialdehyde. Plant J. 37, 877–888.

Weerakkody, U., Dover, J.W., Mitchell, P., Reiling, K., 2018. Quantification of the traffic-generated particulate matter capture by plant species in a living wall and evaluation of the important leaf characteristics. Sci. Total Environ. 635, 1012–1024.

Wei, A., Fu, B., Wang, Y., Li, R., Zhang, C., Cao, D., Zhang, X., Duan, J., 2015. The defence potential of glutathione-ascorbate dependent detoxification pathway to sulfur dioxide exposure in *Tagetes erecta*. Ecotoxicol. Environ. Saf. 111, 117–122.

Wierzbicka, M., Panufnik, D., 1998. The adaptation of *Silene vulgaris* to grown on a calamine waste heap (S. Poland). Environ. Pollut. 101, 415–426.

Wolfenden, L., Mansfield, T.A., 1991. Physiological disturbances in plants caused by air pollutants. Proc. R. Soc. Edinb. 97B, 117–138.

Worall, D., Liang, Y.-K., Alvarez, S., Holroyd, G.H., Spigel, S., Panagopulos, M., Gray, J.E., Hetherington, A.M., 2008. Involvement of sphingosine kinase in plant cell signaling. Plant J. 56 (1), 64–72.

Yamasaki, H., Uefuji, H., Sakihama, Y., 1996. Bleaching of the red anthocyanin induced by superoxide radical. Arch. Biochem. Biophys. 332, 183–186.

Yamasaki, H., Sakihama, Y., Ikehara, N., 1997. Flavonoid peroxidase reaction as a detoxification mechanism of plant cells against H_2O_2. Plant Physiol. 115, 1405–1412.

Yang, J., Yu, Q., Gong, P., 2008. Quantifying air pollution removal by green roofs in Chicago. Atmos. Environ. 42, 7266–7273.

Yao, Z.T., Ji, X.S., Sarker, P.K., Tang, J.H., Ge, L.Q., Xia, M.S., Xi, Y.Q., 2015. A comprehensive review on the applications of coal fly ash. Earth Sci. Rev. 141, 105–121.

Youssef, A., Laizet, Y., Block, M.A., Marechal, E., Alcaraz, J.-P., Larson, T.R., Pontier, D., Gaffe, J., Kuntz, M., 2010. Plant lipid associated fibrillin proteins condition jasmonate production under photosynthetic stress. Plant J. 61, 1616–1626.

Zhang, Y., Yang, J., Wu, H., Shi, C., Zhang, C., Li, D., Feng, M., 2014. Dynamic changes in soil and vegetation during varying ecological-recovery conditions of abandoned mines in Beijing. Ecol. Eng. 73, 676–683.

Zheng, G., Tian, B., Zhang, F., Tao, F., Li, W., 2011. Plant adaptation to frequent alterations between high and low temperatures: remodeling of membrane lipids and maintenance of unsaturation levels. Plant Cell Environ. 34, 1431–1442.

Zolgharnein, J., Asanarani, N., Shariatmanesh, T., 2013. Taguchi L16 orthogonal array optimization for Cd (II) removal using *Carpinus betulus* tree leaves adsorption characterization. Int. Biodeter. Biodegr. 85, 66–77.

Soil and phytomanagement for adaptive phytoremediation practices

Chapter outline

1 Introduction

1.1 Soil management

Soil is the main medium for plant growth, source of raw materials in industry and agriculture, and a place for the foundation of civil structures. The main functions of soil are also carbon sequestration of terrestrial ecosystems and a disposal site for urban/industrial/agricultural waste (Smith and Powlson, 2007; Lal, 2008). However, the rapid industrialization and urbanization are increasing the world population as well as the demand for civil/engineering structures, water/energy/wood products, food production, and land area for waste disposal (Lal, 2008). Climate change and contamination lead to a great extent to the degradation and desertification of soils and soil fertility loss. Decline in soil structure (compaction and crusting) leads to the accelerated erosion by water/wind, depletion of soil organic matter and negative nutrient balance, decline in soil quality (food shortage), whereas soil degradation leads to poverty, hunger, malnutrition, and ultimately to the political instability and civil turbulence (Lal, 2008). According to McCarthy and Dickson (2000) trading between climate change, pollution, soil degradation, and loss of biodiversity can be an entry point for environmental and human vulnerability as well as for low yield, poverty, and hunger.

Soil sustainability and management improve soil conservation and increase soil organic carbon pool through restoration of degraded soils by using organic amendments, mulching, green manure, mixed farming, cultivating cover, and biofuel crops that improve food security and biodiversity (Smith and Powlson, 2007; Lal, 2008). Developing farming/cropping systems may improve water use efficiency and supply of essential nutrients for plant growth. Soil management provides the use of crop residues (lignocellulosic biomass) as a source of energy, use of plant cover canopy,

Adaptive Phytoremediation Practices. https://doi.org/10.1016/B978-0-12-823831-8.00002-5

and technologies for maintaining microbial activity in rhizosphere for organic matter turnover and nutrient cycling (Lal, 2006). Soil conservation can minimize the runoff, control erosion, decrease evaporation, and reduce contamination, whereas sequestration of soil carbon, remediation of air quality, and increasing the land biodiversity improve the environment and mitigate climate change (IPCC, 2000; Lal, 2006, 2008). The key environmental and soil factors that affect the soil management and recovery are: (a) topography and rainfall that are important for soil erosion; (b) acidity/alkalinity/sodicity/salinity; (c) cation exchange capacity (CEC); (d) phosphorous fixation; (e) cracking properties, soil depth, soil texture, and porosity; and (f) nutrient content and organic matter (FAO, 2000). Therefore, according to FAO (2000) sustainable soil management practice brings multiple benefits:

- It increases the soil cover, reduces water and wind erosion, increases the rainfall infiltration, reduces the moisture loss by evaporation, increases the moisture availability, reduces the temperature, improves the germination conditions, increases the organic matter content on the surface soil layer, increases the capacity of soil for nutrient retention, improves the structural stability of the surface aggregates, increases porosity, stimulates the biological activity of the soil, suppresses the weed growth, improves the rooting development and growth; increases the yield and plant biomass, reduces production cost, protects the field, and reduces the pollution of the soil and the environment (Fig. 1A).

Green manure is usually used for conserving/restoring the productivity of degraded soils protecting the surface layer from rainstorms, sun, and wind; reducing the soil temperature variations; increasing the moisture availability; reducing the runoff velocity; increasing the organic matter; increasing the recycling of nutrients; reducing the leaching of nutrients; fixing nitrogen; and stimulating microorganisms (FAO, 2000). The main plant species used as green manure and soil cover are *Avena sativa, Cajanus cajan, Crotalaria juncea, Lathyrus sativus, Lolium multiflorum, Lupinus albus, Melilotus officinalis, Pisum sativum, Raphanus sativus, Secale cereale, Sesbania speciosa, Trifolium* sp., *Vicia faba, Vicia villosa,* etc. (FAO, 2000).

Selection of soil management technologies depends on the characteristics of soil, climate, land use potentials, financial resources, markets, prices, i.e., selections include climatic, environmental, and socioeconomic factors (FAO, 2000). General framework of the production systems of farmers includes land preparation, tillage, crop rotation, seeding, possible applications of fertilization, harvest, yield storage, residue management, machinery and equipment, irrigation, drainage system, and soil conservation. Socioeconomic limitations for soil management involve land/farm size and production levels, financial resources, prices, costs, availability of inputs, marketing, transport, crop storage, manual labor, farmer organization, land tenure, and technical assistance (FAO, 2000). However, soil technologies should be suitable and must not pose a threat to the environment, human, and animal life avoiding any contamination, developing "no regrets" strategy for sustainable soil management that will increase the resilience of the soil/ecosystem to the environmental stresses (Smith and Powlson, 2007).

1.2 *Phytomanagement*

Phytomanagement of polluted sites (fly ash deposit, coal mine, cities, agricultural fields, etc.) is based on phytoremediation, sustainable site management, and

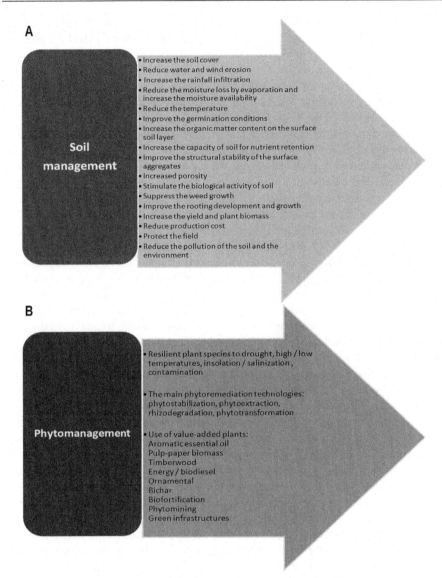

A

Soil management

- Increase the soil cover
- Reduce water and wind erosion
- Increase the rainfall infiltration
- Reduce the moisture loss by evaporation and increase the moisture availability
- Reduce the temperature
- Improve the germination conditions
- Increase the organic matter content on the surface soil layer
- Increase the capacity of soil for nutrient retention
- Improve the structural stability of the surface aggregates
- Increased porosity
- Stimulate the biological activity of soil
- Suppress the weed growth
- Improve the rooting development and growth
- Increase the yield and plant biomass
- Reduce production cost
- Protect the field
- Reduce the pollution of the soil and the environment

B

Phytomanagement

- Resilient plant species to drought, high / low temperatures, insolation / salinization, contamination

- The main phytoremediation technologies: phytostabilization, phytoextraction, rhizodegradation, phytotransformation

- Use of value-added plants:
  Aromatic essential oil
  Pulp-paper biomass
  Timberwood
  Energy / biodiesel
  Ornamental
  Bichar
  Biofortification
  Phytomining
  Green infrastructures

Fig. 1 Soil and phytomanagement practice for adaptive phytoremediation.

ecosystem services (Pandey and Bauddh, 2019; Pandey, 2020a). Phytoremediation as a green solution uses plants for the cleanup of contaminated sites (Gajić and Pavlović, 2018; Gajić et al., 2018; Pandey and Bauddh, 2019). Trade-off between pollutants, climatic conditions, characteristics of soil/fly ash/coal mine, and the selection of plant species have already been researched (Pavlović et al., 2004, 2016; Djurdjević et al., 2006; Mitrović et al., 2008; Pandey, 2012, 2015, 2020b; Gajić et al., 2013; Maiti and Ahirwal, 2019; Grbović et al., 2019, 2020; Pandey et al., 2020). Phytostabilization, phytoextraction, rhizodegradation, phytotransformation, and phytovolatilization are the main phytoremediation technologies that can be used in decontamination of inorganic (metal(loid)s) and organic (herbicides, explosives, etc.) pollutants (McCutcheon and Schnoor, 2003; Pilon-Smits, 2005; Pandey and Bajpai, 2019). Phytoremediation and revitalization of contaminated sites have been achieved by selecting native plant species that are best adapted to environmental stresses, such as drought, high/low temperatures, insolation/salinization, and contamination (Pandey et al., 2015a,b, 2016a,b; Gajić et al., 2016, 2019, 2020a,b) (Fig. 1B).

Plants used in phytoremediation can be excluders and/or accumulators. Excluders are commonly used in phytostabilization, and they are capable to limit the entry of excess concentrations of pollutants inside the plants, i.e., they avoid/exclude the pollutants from further transport at the root level by mycorrhizal fungi and/or by root rhizosphere exudates binding them to the cell wall and/or preventing their transport through the plasma membrane (Gajić and Pavlović, 2018; Gajić et al., 2018). Accumulators are commonly used in phytoextraction and they are capable of accumulating pollutants in the plant aboveground tissues that are then harvested (Gajić and Pavlović, 2018; Gajić et al., 2018; Pandey and Bajpai, 2019). Rhizodegradation is a technology in which plants break down the organic pollutants by stimulating the microbial and fungal activity that helps to release exudes (Pilon-Smits, 2005; Ma et al., 2011). Therefore, plant-growth-promoting bacteria (PGPB) are beneficial for plants because they stimulate plant growth, decrease metal(loid) toxicity, and degrade organic compounds enhancing the phytoremediation (Glick, 2012), whereas arbuscular mycorrhizal fungi (AMF) improve plant nutrient and water uptake and resilience to salt stress, drought, and excess of metal(loid)s and organic pollutants in soils (Gonzáles-Chávez et al., 2006). Phytotransformation is a technology in which plants use enzymes in roots and leaves to degrade organic compounds to CO_2 and H_2O and the cascade of metabolic reactions in which plants transform, conjugate, and store/eliminate organic pollutants is called the "green liver model" (Sandermann, 1994; Burken, 2003; Pilon-Smits, 2005).

Phytomanagement of polluted sites provides: (a) minimum maintenance and low input by sowing native plant species on sites or spontaneous native plants growing on the contaminated sites; (b) restoration and revegetation time is reduced by using fast-growing plant species (leguminous, grasses, *Populus* sp., *Salix* sp.); (c) low risk from entry of pollutants into food web; and (d) use of value-added plants (Pandey et al., 2015a; Pandey and Bajpai, 2019). Application of plants with multiple benefits has ecological and socioeconomic importance providing biodiversity conservation, pleasant landscape, soil quality improvement, carbon sequestration, biomass, fiber, energy, biodiesel, essential oils, mineral elements for animal fodder/human edible yield, pulpwood, timber wood, floriculture (Pandey et al., 2015a,b; Pandey and Bauddh, 2019; Pandey and Bajpai, 2019) (Fig. 1B). Therefore, according to Pandey and Souza-Alonso (2019), economically valuable plants in commercial phytoremediation can be used for:

- Aromatic essential oils—*Vetiveria zizanioides, Cymbopogon citratus, Chrysopogon zizanioides, Mentha arvensis, Mentha piperita, Anethum graveolens, Lavandula angustifolia, Ocimum basilicum*
- Pulp-paper biomass—*Dendrocalamus strictus, Populus, Leucaena leucocephala*

- Timber wood products—*Eucalyptus tereticornis, Tectona grandis, Salix, Azadirachta indica*
- Energy/Biodiesel—*Ricinus communis, Jatropha curcas, Miscanthus*
- Ornamental purpose—*Jasminum* sp., *Nerium oleander, Chrysanthemum indicum, Gladiolus grandiflorus, Calendula officinalis*
- Biochar
- Biofortification
- Phytomining
- Green roofs and green infrastructures in smart cities

2 Soil management

2.1 Arbuscular mycorrhizae against drought stress

Drought stress is one of the abiotic stress factors that limit plant growth, development, and productivity jeopardizing food security worldwide (Wang et al., 2003). Drought lowers soil water potential inducing cell dehydration that results in the inhibition of root proliferation, disturbed plant water, and nutrient uptake and reduced the size of stem and leaf (Wu et al., 2013). Arbuscular mycorrhizal fungi (AMF) that belong to the phylum *Glomeromycota* establish a symbiosis with plants enabling them to grow efficiently even under stress (Porcel and Ruiz-Lozano, 2004). AMF use photosynthetic products from plants, while they provide the host plant with water and nutrients (N, P, K), thus increasing water use efficiency (WUE) and stomatal regulation, reducing oxidative stress, and activating antioxidant machinery under drought stress (Chittara et al., 2016).

Fungal hyphae enter epidermal and cortical root cell layer and inner cortex, whereas hyphae may extend the root/soil interface, establish hyphal networks, and secrete glomalin that facilitates nutrient and water absorption and improves soil structure (Pagano, 2014). In soils with water deficit, extraradical hyphae penetrate into soil pores giving mycorrhizal roots access to available soil water (Wu et al., 2013). Glomalin as glycoprotein is released from fungi hyphae into the soil (glomalin-related soil protein—GRSP), and it is insoluble in water (Rilling, 2004). Soil with AMF shows high distribution of water-stable aggregates in rhizosphere which is related to GRSP (Rilling, 2004; Wu et al., 2008). GRSP is more active under water stress, it coats on hyphae and forms a hydrophobic layer in the aggregate surface in *Glomus mosseae*, *Glomus diaphanum*, and *Glomus versiforme*, conferring drought tolerance of the host plant (Wu et al., 2008; Nichols, 2008).

Plants inoculated with AMF (*Glomus intraradices*, *Laccaria bicolor*, and *Terfezia claveryi*), possess aquaporins (pores/water channel) in cell membranes and they are involved in water transport (Li et al., 2013a; Xu et al., 2013). In roots of *Zea mays* that are inoculated by AMF, enhanced levels of aquaporin genes, *GintAQPF1* and *GintAQPF2*, have been found in arbuscule-enriched cortical cells and extraradical mycelia (Li et al., 2013a). However, ABA with AMF may have an inhibitory effect on the expression of *PIP* (plasma membrane intrinsic protein) aquaporin gene, and its reduction in expression can be compensated by changing the activity of other aquaporins

(Ruiz-Lozano et al., 2009). Quiroga et al. (2017) observed the key aquaporin genes involved in drought tolerance in *Z. mays* colonized by AMF, i.e., *ZmPIP1;6*, *ZmPIP2;2*, and *ZmTIP4;1* genes were upregulated by AMF symbiosis. In mycorrhizal plants under drought stress, AMF symbiosis upregulates a gene from *G. intraradices* that encodes a binding protein (GiBiP, molecular chaperone) (Porcel et al., 2007) or increases the expression of *Mn-sods II* and *Fe-sods II* genes (Ruiz-Lozano et al., 2001a,b). However, downregulation of key proline synthetase Δ^1-pyrroline-5-carboxylate synthetase (p5cs) was found in *Glycine max* and *lettuce* (Porcel et al., 2004). Thus, AMF symbiosis with plants facilitates water transport and hydraulic conductivity demanding upregulation of AQPs genes and enhancement of drought resilience in the host plant (Sanchez-Romera et al., 2016).

Arbuscular mycorrhizal fungi (AMF) colonization under drought stress induces root system architecture (RSA) changes in the host plant that depends on plant/fungal species and nutrient/water status (Li et al., 2013b). Plant inoculation with AMF increased the absorption areas of root systems under drought stress, such as root length, number of lateral roots, root diameter, and root branching density (Wu et al., 2012). AMF-mediated modifications in root morphology facilitate nutrient uptake and maintain water balance in plants under water stress (Wu et al., 2013). Mycorrhizal effect on nutrient uptake (N, P, K, Ca, Fe, Mn, Zn) is more prominent under drought stress, i.e., AMF enhance nutrient acquisition and absorption into host plant (Wu and Zou, 2009).

Osmotic stress induced by drought, plants may tolerate by enhanced biosynthesis of some metabolites/osmoprotectants, such as K^+, Ca^{2+}, Mg^{2+}, sugars, glycinbetain, and proline, which decrease osmotic potential and sustain high water status (Martinez et al., 2004). Plants that are inoculated with AMF have greater potential for osmotic adjustment due to under drought stress-enhanced content of osmoprotectants in root and leaves sustaining cell hydration and turgor level and maintaining photosynthetic efficiency and plant growth (Wu et al., 2007). High proline accumulation was observed in AMF-inoculated *Lactuca sativa*, *Oryza sativa*, and *Macadamia tetraphylla* under drought stress providing a greater capacity of plants to combat water stress (Azcon et al., 1996; Ruiz-Sanchez et al., 2011; Yooyongwech et al., 2013). However, in some plants, AMF did not induce the expression of the proline synthases (Δ^1-pyrroline-5-carboxylate synthetase) and low proline content was the consequence of increased proline catabolic enzyme activity (proline oxidase and dehydrogenase) providing less damage due to avoiding water stress (Auge and Moore, 2005).

Drought stress induces a generation of reactive oxygen species (ROS) that lead to lipid peroxidation, chlorophyll bleaching, protein oxidation affecting photosynthesis, and plant yield (Anjum et al., 2012). Plants activate nonenzymatic (carotenoids, anthocyanins, ascorbic acid, phenolics, GSH—glutathione, and tocopherol) and enzymatic (superoxide dismutase (SOD), peroxidase (POD), catalase (CAT), glutathione reductase (GR), ascorbate peroxidase (APX), and guaiacol peroxidase (GPOD)) antioxidant system to cope with stress. Therefore, AMF symbiosis with plants confers antioxidant defense system under drought stress by lowering oxidative stress and producing antioxidants at a high level (Wu and Xia, 2006). In plants colonized by AMF, accumulation of ROS in roots and leaves is lower due to enhanced activation of antioxidants (Wu et al., 2013). Plants inoculated by AMF possess the *Sod1* genes that confer drought resilience (Corradi et al., 2009). Furthermore, plants colonized with AMF increase the production of calmodulin-binding protein that regulates Cu/Zn-SOD, Mn-SOD, and CAT activity under drought stress (Huang et al., 2014).

Haslem et al. (2019) showed that drought stress in *Cicer arietinum* was alleviated by inoculation with AMF and biochar amendments, i.e., AMF and biochar improve N fixation and P uptake in root and leaves, increased root and shoot length, leaf area, number of branches/plant, chlorophylls and carotenoids content in leaves stomatal pore aperture, stomatal density, photosynthetic rate, relative water content (RWC), and membrane stability (Table 1). Inoculation of *Onobrychis viciifolia* by AMF improved its resistance to drought stress because AMF enhanced N and P uptake accelerated plant growth (plant height, root length, and root/shoot biomass), increased RWC and soluble proteins, reduced MDA (malondialdehyde) content, and increased activity of SOD and CAT (Jing et al., 2014) (Table 1). According to Barros et al. (2018), AMF colonization of woody evergreen species *Cynophalla flexuosa* enhanced leaf RWC, increased specific leaf area (SLA) lowering leaf construction cost (CC), and increased photosynthetic energy use efficiency (PEUE) promoting its resilience to drought stress (Table 1). Inoculation of *Phoenix dactylifera* by AMF, *Glomus monosporus*, *Glomus clarum*, and *Glomus deserticola* improved tolerance to drought stress due to increased growth parameters (number of produced leaves, height of aerial part, and leaf surface), ionic content in leaves (P, Ca, Mg, Na, K, Cu, and Mn), RWC, water potential, stomatal resistance, phenolic content, and activity of peroxidase and polyphenol oxidase (Meddich et al., 2015) (Table 1). Asensio et al. (2012) showed that AMF root colonization increases essential isoprenoides (abscisic acid, chlorophylls, and carotenoids) under drought stress in *Solanum lycopersicum* (Table 1). AMF symbiosis can confer tolerance to water stress in *Cupressus atlantica* having a positive effect on shoot height, stem diameter, biomass, and growth rate; improve uptake of minerals N, P, K, Ca, and Mg; and increase content of proline, soluble sugars, and activity of SOD, POD, and CAT maintained RWC and water potential (Zarik et al., 2016) (Table 1).

In addition, Huang et al. (2020) showed that *Malus hupehensis* inoculated with AMF improved drought tolerance; enhanced the plant growth (plant height, root length, root/shoot biomass); increased chlorophyll content, total sugar content, proline, lowered MDA content, and relative electrolyte leakage; and increased activity of SOD, POD, CAT, raised net photosynthetic rate, stomatal conductance, transpiration rate, WUE (water use efficiency), Fv/Fm (photosynthetic efficiency), ΦPSII (actual photosystem II efficiency), electron transport rate, qP (photochemical quenching of chlorophyll fluorescence), and NPQ (nonphotochemical quenching of chlorophyll fluorescence) together with the upregulation of MAPK (mitogen-activated protein kinase) gene that is involved in signal transduction pathway and enhanced expression of *PIP1–3*, *PIP1–4*, *Rir-AQP1*, and *Rir-AQP2* genes of aquaporins that provide high water capacity in plants (Table 1). In *Ceratonia siliqua* exposed to drought stress, AMF mediated in its tolerance: increased shoot height root length, shoot/root biomass, improved mineral absorption efficiency (P, K, Na, and Ca), increased total sugar content and proteins, decreased MDA content and H_2O_2 concentrations, and enhanced stomatal conductance and photosynthetic efficiency (Fv/Fm) (Boutasknit et al., 2020) (Table 1). Furthermore, Li et al. (2019) showed that AMF-alleviated drought stress

Table 1 Morphological, physiological, and biochemical response of plants inoculated with AMF (arbuscular mycorrhizal fungi) to drought stress.

Plant species	Parameter	References
Cicer arietinum	N, P, root/shoot length, leaf area, number of branches/per plant, chlorophylls, carotenoids, stomatal pore aperture, stomatal density, A, RWC	Haslem et al. (2019)
Onobrychis viciifolia	N, P, plant height, root length, root/shoot biomass, RWC, soluble proteins, MDA, SOD, CAT	Jing et al. (2014)
Cynophalla flexuosa	RWC, SLA, CC, PEUE	Barros et al. (2018)
Phoenix dactylifera	Number of produced leaves, height of aerial part, P, Ca, Mg, Na, K, Cu, Mn, RWC, water potential, stomatal resistance, phenolics, POD, PPO	Meddich et al. (2015)
Solanum lycopersicum	Isoprenoides: abscisic acid, chlorophylls, carotenoids	Asensio et al. (2012)
Cupressus atlantica	Shoot height, stem diameter, biomass, growth rate, N, P, K, Ca, Mg, proline, soluble sugars, SOD, POD, CAT RWC, water potential	Zarik et al. (2016)
Malus hupehensis	Plant height, root length, root/shoot biomass, chlorophyll, total sugar, proline, MDA, relative electrolyte leakage, SOD, POD, CAT, A, gs, E, WUE, Fv/Fm, ΦPSII, ETR, qP, NPQ, MAPK, *PIP1–3, PIP1–4, Rir-AQP1, Rir-AQP2*	Huang et al. (2020)
Ceratonia siliqua	Shoot height, root length, shoot/root biomass, P, K, Na, Ca, sugar, proteins, MDA, H_2O_2, gs, Fv/Fm	Boutasknit et al. (2020)
Leymus chinensis Hemarthria altissima	Biomass, A, gs, WUE, SOD, CAT, MDA	Li et al. (2019)

A, photosynthetic rate; *CAT*, catalase; *CC*, leaf construction cost; *E*, transpiration rate; *ETR*, electron transport rate; *Fv/Fm*, photosynthetic efficiency; *gs*, stomatal conductance; *MAPK*, mitogen-activated protein kinase gene; *MDA*, malondi-aldehyde; *NPQ*, nonphotochemical quenching of chlorophyll fluorescence; *PEUE*, photosynthetic energy use efficiency; *PIP1–3, PIP1–4, Rir-AQP1, Rir-AQP2*, aquaporin genes; *POD*, peroxidase; *PPO*, polyphenol oxidase; *qP*, photochemical quenching of chlorophyll fluorescence; *RWC*, relative water content; *SLA*, specific leaf area; *SOD*, superoxide dismutase; *WUE*, water use efficiency; Φ*PSII*, actual photosystem II efficiency.

in C3 (*Leymus chinensis*) and C4 grasses (*Hemarthria altissima*) enhanced biomass, net photosynthetic rate (A), stomatal conductance (gs), intrinsic water use efficiency (WUE), SOD and CAT activity, and reduced MDA content (Table 1).

2.2 Arbuscular mycorrhizae against salinity stress

Salinization of soils presents one of the major environmental threats worldwide limiting crop productivity due to deleterious effect of NaCl on plants (Evelin et al., 2019).

According to Ruiz-Lozano et al. (2001a, b), saline soils occupy about 7% of the Earth's land due to the rise of the sea levels, inappropriate irrigation practice, use of salts on the roads, etc. Salt stress negatively affects plant growth by reducing osmotic potential of soil solution due to the accumulation of Na and Cl ions and by reducing the water availability from soil that causes physiological drought to plants (Jahromi et al., 2008). Under salt stress, water and nutrient uptake is decreased leading to the ion toxicity (Na^+, Cl^-); nutrient imbalance; and disruption of photosynthesis, respiration, and protein synthesis (Feng et al., 2002).

Arbuscular mycorrhizal fungi (AMF) colonization in plant roots enhances plant tolerance to salt stress (Evelin et al., 2009, 2019). AMF, such as *G. intraradices, G. versiforme, Glomus etunicatum, Glomus geosporum*, were found in saline soils, and they colonize some halophytes (Aliasgharzadeh et al., 2001). AMF occasionally colonize nonmycorrhizal halophytes, such as *Salicornia europaea, Sagina maritima, Armeria maritima*, and *Lepidium crassifolium*, but they can strongly inoculate other mycorrhizal halophytes, such as *Puccinellia* sp., *Aster tripolium, Plantago maritima, Artemisia maritima, Glaux maritima, Oenanthe lachenalii*, and *Inula crithmoides* (Bothe, 2012).

Plants which inoculated with AMF maintain high RWC have improved hydraulic conductivity of roots that are related to longer roots and altered root architecture (Kapoor et al., 2008; Ruiz-Lozano et al., 2012). Salt stress enhances the expression of some important genes, such as Na^+/H^+ antiporters and aquaporins (MIP, major intrinsic protein family of transmembrane channels), that permit selective membrane passage of Na^+, H^+, and water, respectively (Hill et al., 2004; Munns, 2005). Overexpression of genes (*LeNHX1, LeNHX2*) that code Na^+/H^+ transporters from NHX family (catalyze transport of Na^+ from cytoplasm into vacuole/across the plasmalemma to the outside in exchange of H^+) provides plant tolerance to salt stress (Wu et al., 2005). Aquaporins from cytoplasm membrane (*LePIP1* and *LePIP2*) and tonoplast aquaporin (*LeTIP*) are involved in the regulation of water permeability and osmotic stress, i.e., they provide salt tolerance in AMF colonized plants (Ouziad et al., 2006).

Arbuscular mycorrhizal fungi (AMF) regulate nutrient uptake (N, P, K, Ca, Mg, Na^+, and Cl^-) (Evelin et al., 2009). AMF inoculation can increase P uptake in plants by external hyphae reducing the negative effect of Na^+ and Cl^- maintaining vacuolar membrane integrity (Feng et al., 2002). Similarly, extraradical mycelia from AMF take up N from the soil, and this process is connected with increased expression of enzyme involved in N fixation; improved N nutrition can indirectly maintain chlorophyll content and reduce toxic effects of Na ions (Evelin et al., 2009). AMF enhance uptake of K ions, activate a number of enzymes, and have a key role in stomatal movements and protein biosynthesis, while high $K^+:Na^+$ ratio provides ionic balance of the cytoplasm or Na^+ efflux from plant cells (Colla et al., 2008). AMF enhance uptake of Ca^{2+} that can act as a transducer signal and a second messenger (Cantrell and Linderman, 2001) and improve uptake of Mg^{2+} supporting higher chlorophyll concentrations in plant leaves (Giri et al., 2003).

Arbuscular mycorrhizal fungi (AMF) plants accumulate osmoprotectants, such as proline and glycine betaine in order to maintain osmotic balance under low water potential stabilizing the structures and activities of enzyme and protein complex that maintains membrane integrity (Evelin et al., 2009). AMF inoculation activates antioxidant defense system in plants, and it particularly increases the activity of SOD,

POD, CAT, GST (glutathione S transferase), and APX as a response to salinity stress (Alguacil et al., 2003). Effects of AMF on carotenoids, tocopherol, and ascorbic acid in the host plant have not been reported. However, Klinger et al. (1995) noted in AMF yellow pigment (carotenoid with ^{14}C atoms), mycoradicin that protects cells against ROS, and it can be also found in the vacuole of root cells. In addition, Schröder et al. (2001) reported carotenoid corticrocin deposited on the outer surface of hyphae of *Piloderma croceum* surrounding the roots. Mycorrhizal-inoculated plants showed high photosynthetic efficiency (Fv/Fm), photochemical quenching of chlorophyll fluorescence (qP), and nonphotochemical quenching of chlorophyll fluorescence (NPQ) (Sheng et al., 2008).

Chandrasekaran et al. (2016) showed that mycorrhizal inoculation under saline stress increased root and shoot biomass in C3 and C4 plants—*Rhizophagus irregularis* and *Funneliformis mosseae* were more effective with C3 plants; *F. mosseae*, *Rhizophagus fasciculatus*, and *R. irregularis* were more effective with C4 plants. Furthermore, mycorrhizal C4 plants had higher N, P, and K uptake than C3 plants (Table 2). C4 plants had a higher increase in P uptake efficiency because preferential

Table 2 Morphological, physiological, and biochemical response of plants inoculated with AMF (arbuscular mycorrhizal fungi) to salinity stress.

Plant species	AMF	Parameter	References
C3, C4	*Rhizophagus irregularis, F. mosseae, R. fasciculatus,* Glomus sp.	N, P, K, K^+/Na^+, proline	Chandrasekaran et al. (2016)
Panicum turgidum	AMF	Na^+, K^+, Ca^{2+}, MDA, H_2O_2, SOD, POD, CAT, GR, total phenolics, NADP-malic enzyme, *Rubisco*, proline, glycine betaine	Hashem et al. (2015)
Triticum aestivum	Glomus mosseae, G. etunicatum, G. intraradices	N, P, K, Ca, Mg, Mn, Cu, Fe, Zn, Na^+, Cl^-	Mardukhi et al. (2011)
Ocimum basilicum	Glomus deserticola	K, P, Ca^{2+}, Mg^{2+}, Cl^-, K/Na, Ca/Na, root, shoot biomass, shoot length, leaf number, leaf area, inflorescence length, chlorophylls, proline, A, E, gs, WUE	Elhindi et al. (2017)
Euonymus maackii	AMF	Plant growth, N, P, K in roots and leaves Na^+, Cl^-, A, gs, Ci, E, ΦPSII, qN, SOD, POD, CAT	Li et al. (2020)

A, net photosynthetic rate; *Ci*, intercellular CO_2 concentrations; *E*, evapotranspiration; *GR*, glutathione reductase; *gs*, stomatal conductance; *MDA*, malondialdehyde; *POD*, peroxidase; CAT, catalase; *qN*, nonphotochemical quenching of chlorophyll fluorescence; *SOD*, superoxide dismutase; *WUE*, water use efficiency; *ΦPSII*, actual quantum yield.

P allocation for photosynthesis required de novo biosynthesis of phosphoenolpyruvate carboxylase (PEPC) that needed more P for ATP requirement (Tang et al., 2006). C4 plants had more K uptake that is associated with more exclusion of Na^+ and higher K^+/Na^+ ratio that enhanced total plant growth under saline stress. Proline accumulation is increased in plants inoculated with *R. fasciculatus* and *Glomus* sp., whereas inoculation with *R. irregularis* and *F. mosseae* significantly decreased proline accumulation that is related to Na uptake, i.e., proline content in plants is decreased with more efflux of Na^+ (Chandrasekaran et al., 2016) (Table 2). Hashem et al. (2015) found that AMF enhanced salinity tolerance in *Panicum turgidum* altering nutrient status, oxidative/antioxidative system: content of K^+ and Ca^{2+} was increased, whereas Na^+ decreased; MDA content and H_2O_2 concentrations were decreased; activity of SOD, POD, CAT, GR, and total phenolics was increased; NADP-malic enzyme and *Rubisco* activity were increased and had enhanced photosynthetic efficiency; and decreased content of proline and glycine betaine can be related to lower content of Na^+ mitigating salinity stress and maintaining osmotic and water balance (Table 2). According to Mardukhi et al. (2011), *G. mosseae*, *G. etunicatum*, and *G. intraradices* enhanced nutrient uptake, such as N, P, K, Ca, Mg, Mn, Cu, Fe, and Zn in *Triticum aestivum*, whereas AMF had an inhibitory effect of Na^+ and Cl^- (Table 2). It was suggested that Na^+ might be retained in vacuoles of root cells and intraradical fungi hyphae (Cantrell and Linderman, 2001). *G. deserticola* (AMF) mitigates salt-induced adverse effects in *O. basilicum* by increasing leaf K, P, and Ca^{2+}; decreasing Mg^{2+} and Cl^-; increasing K/Na and Ca/Na ratio; increasing root and shoot biomass, shoot length, leaf number, leaf area, and inflorescence length; enhancing biosynthesis of chlorophyll and proline; raising photosynthetic activity, such as net photosynthetic rate (A), evapotranspiration (E), and stomatal conductivity (gs); and improving water use efficiency (WUE) (Elhindi et al., 2017) (Table 2). Li et al. (2020) reported that AMF enhanced the tolerance of *Euonymus maackii* to moderate level of salinity stress: plant growth increased, N, P, K content in roots and leaves increased, Na^+ and Cl^- decreased, net photosynthetic rate (A), stomatal conductivity (gs), intercellular CO_2 concentrations (Ci), evapotranspiration (E), actual quantum yield of photosystem II (ΦPSII), non-photochemical quenching of chlorophyll fluorescence (qN) increased together with increased activity of SOD, POD, and CAT enzymes in roots and leaves (Table 2).

2.3 Using rhizobacteria in management of polluted soils

The main sources of soil pollution are agrochemicals (pesticides and fertilizers), organic phosphate products (insecticides), urban by-products (municipal waste, construction related contaminants, and trash cans), domestic and industrial waste (polyethylene bags, bottles), heavy metals, etc. (Chitara et al., 2021). Chemicals and products that are not safely disposed of lead to soil contamination. Plant-growth-promoting rhizobacteria (PGPR) are used in soil pollution management alleviating the contamination. PGPR that are used in bioremediation of soil are *Azotobacter*, *Achromobacter*, *Azospirillum*, *Enterobacter*, *Serratia*, *Pseudomonas*, *Streptomyces*, *Klebsiella*, *Lysobacter*, *Rhizobium*, *Burkholderia*, *Mesorhizobium*, *Arthrobacter*, *Psychrobacillus*, *Rhodococcus*, *Alcaligenes*, *Arthrobacter*, *Bacillus*, *Flavobacterium*,

Sphingomonas, Frankia, etc. (Khan et al., 2009; Singh et al., 2011; Srivastava et al., 2020; Chitara et al., 2021). Bacteria obtain food for host plants and solubilize/degrade pollutants, i.e., they enhance the plant growth, facilitate nitrogen fixation and phosphorous solubilization, increase the nutrient availability in the rhizosphere, increase root surface area, decrease heavy metal toxicity, and degrade organic compounds accelerating phytoremediation (Ma et al., 2011; Glick, 2012; Praveen et al., 2019; Guo et al., 2020). Rhizospheric PGPR assist in phytoremediation through different mechanisms, such as immobilization (phytostabilization)/bioaccumulation (phytoextraction)/biostimulation/biotransformation (biodegradation) (Roy et al., 2015; Asad et al., 2019; Praveen et al., 2019).

Siderophores produced by PGPR (*Rhizobium, Bradyrhizobium, Pseudomonas, Azotobacter,* etc.) bind heavy metals (Cu, Ni, Cr, Cd, Co, Pb, Mn, and Zn) and reduce bioavailability of metal ions (Dimpka et al., 2008). According to Dimpka et al. (2009), siderophores produced by *Streptomycetes* improve Fe uptake, but reduce Cd^{2+} absorption. Retamal-Morales et al. (2018) reported that siderophores produced by *Rhodococcus, Arthrobacter,* and *Kocuria* were able to chelate As (III), whereas siderophores from *Kluyvera ascorbata* bind Ni, Pb, and Zn protecting *Brassica juncea, Brassica napus,* and *Lycopersicum esculentum* from toxicity (Burd et al., 2000). PGPB can modify phytohormones levels in plants (indole-3-acetic acid—IAA, auxin, cytokinins, gibberellins, and ethylene) altering plant growth (Khan et al., 2009; Glick, 2012; Guo et al., 2020). They have the potential to produce ACC-deaminase (1-aminocyclopropane 1-carboxylate deaminase) that reduces ethylene production and improves plant growth in stressful conditions, such as pollution with heavy metals/organic compounds (Reed and Glick, 2005; Farwell et al., 2006).

Plant-growth-promoting rhizobacteria (PGPR) alleviate the heavy metal accumulation in plants preventing them to enter inside the plant tissues, or they restrict them in roots (Khan et al., 2009; Ma et al., 2011). According to Guo et al. (2020), phosphate-solubilizing bacteria (PSB) provide phosphorous to plants, form phosphate complex with metals, form insoluble phosphate precipitate with metals, induce the activation of heavy metal resistance genes, and affect the biosynthesis of antioxidative enzymes; potassium-solubilizing bacteria (KSB) provide K to plants, induce biosynthesis of antioxidants, and induce the expression of resistance genes; siderophore-producing bacteria (SPB) provide Fe to plants, chelate metals with siderophores, induce IAA biosynthesis, induce the activation of heavy metal resistant genes, and decrease generation of free radicals in the rhizosphere; nitrogen-fixing bacteria (NFB) provide N to plants, convert N_2 to NH_4, and induce the activation of metal resistance genes; organic acid-producing bacteria (OA-PB) form chelates with metals (metal oxalate), facilitate the acquisition of nutrients, induce biosynthesis of antioxidants, and activate heavy metal resistance genes; IAA-PB (indol-3-acetic acid-producing bacteria) promote nutrient availability (Fe, P, Cu, and Zn) by increasing root area and they activate heavy metal resistant genes; ACCD-PB (1-aminocyclopropane 1-carboxylate deaminase-producing bacteria) increase root exudates, reduce ethylene by hydrolyzing ACC to NH_3 and they induce activation of resistant genes; BE-PB (biosurfactant/exopolymer-producing bacteria) promote binding heavy metals and induce activation of resistant genes.

Plant-growth-promoting rhizobacteria (PGPR) induce systemic tolerance (IST) to drought, salinity, flooding, nutrient stress, and metal(loid)s (Sarma et al., 2012). Therefore, lipopolysaccharides of external membrane of bacteria, biosurfactants (surface-active, amphiphilic substances), siderophores, organic acids (oxalic, glucaric, succinic, and citric acids) from bacteria are associated with the induction of ISR. Metal(loid) resistance of PGPR and their immobilization of plants also involve intracellular/extra-cellular sequestration by protein/chelator binding, active transport of metal(loid)s, and enzyme detoxification of metal(loid)s (Guo et al., 2020).

Plant-growth-promoting rhizobacteria (PGPR) can be used as an alternative to chemical fertilizers having a huge potential in agriculture sustainability. Therefore, *Paenibacillus*, *Enterobacter*, *Klebsiella*, *Bacillus*, *Azotobacter*, *Micrococcus*, and *Pseudomonas* can be used as biofertilizers to clean up metal(loid)s, such as Cd, Cr, As, Hg, Pb, Ni, and Zn, in rhizosphere soils of *Gossypium hirsutum*, *O. sativa*, *C. arietinum*, *T. aestivum*, and *Vitis vinifera*, thus improving plant growth, seed germination, increasing N and P uptake, and increasing chlorophyll content and amylase and protease activity (Islam et al., 2014; Pramanik et al., 2017, 2018; Kumari and Thakur, 2018; Ghosh et al., 2018).

Metagenomics is used for creation of soil-based metagenomics library that includes phylogenetic analysis of soil microbial flora as a source of pollutant-degrading genes for transgenic microbes/plants (Daniel, 2005). Metagenomics facilitates the production of degrading enzymes from uncultivable bacteria for accelerated enzymatic remediation via gene/enzyme products (Rayu et al., 2012). Furthermore, a pyrethroid hydrolyzing enzyme coding by a novel gene could be used for the degradation of pyrethroids (3,5,6-trichloro-2-pyridinol) (Fan et al., 2012). Also, it is possible that through metabolic engineering cellular activities are improved by manipulation of metabolic pathways of transportation/regulation by using recombinant DNA (Nielsen, 2001). According to Rayu et al. (2012), combination of metabolic pathways in *Pseudomonas putida* can increase the degradation rate of benzene, toluene, and *p*-xylene.

Metal(loid)s. Gullap et al. (2014) found that the application of PGPR (*Bacillus megaterium* var. *phosphaticum*) together with P fertilizer significantly increases Pb, Ni, B, Zn, and Mn availability in soil and their uptake by *Trifolium hybridum*, *Alopecurus pratensis*, *Poa pratensis*, *Hordeum violaceum*, *Ranunculus kotschyi*, and *Cerastium* sp. without diminishing their yield indicating that meadow plants are tolerant to high heavy metal concentrations in the soil and do not present a risk for animal feeding (Table 3). Plant-growth-promoting rhizobacteria *Bacillus* improve soil quality, enhance enzymatic activity in rhizosphere (acid phosphatase), alleviate uptake of Fe and P, increase the biomass of roots and shoots of *Solanum nigrum*, and increase Pb and Cd in aerial parts of plants (He et al., 2020) (Table 3). PGPR-assisted phytoremediation of heavy metals showed the reduction of their toxicity and improved plant growth: *Azotobacter chroococcum*, *B. megaterium*, and *Bacillus mucilaginosus* stimulate the growth of *B. juncea* in the Pb and Zn stress conditions (Wu et al., 2006); *Bacillus subtilis* facilitate Ni accumulation in *B. juncea* (Zaidi et al., 2006); *Xanthomonas* sp., *Azomonas* sp., *Pseudomonas* sp., and *Bacillus* sp. enhance growth of *B. napus* and increase Cd accumulation (Sheng and Xia, 2006); *Pseudomonas* sp. and *Bacillus* sp. stimulate the growth of *Brassica* sp. and decrease Cr content (Rajkumar et al., 2006);

Pseudomonas fluorescens promote the growth of *G. max* in the condition of high Hg concentrations (Gupta et al., 2005); *Pseudomonas brassicacearum* and *Rhizobium leguminosarum* increase the growth of *B. juncea* and enhance root exudation of Zn chelates (Adediran et al., 2016); *Enterobacter ludwigii* and *Klebsiella pneumoniae* improve the growth of *T. aestivum*, enhance the ACC deaminase activity and IAA production, facilitate the uptake of P, K, and Zn, and reduce proline accumulation and Hg biosorption (Gontia-Mishra et al., 2017) (Table 3).

Organic compounds. Sampaio et al. (2019) reported that *Pseudomonas aeruginosa* and *Bacillus* sp. can promote the development of the propagule of mangrove plants *Rhizophora mangle* that grow in 55% of sediment polluted with diesel (polycyclic aromatic hydrocarbons, PAHs), i.e., these PGPR had the ability to degrade dibenzo(a,h)anthracene (90%), naphthalene (80%), acenaphthylene and benzo(a)anthracene (60%), anthracene and benzo(a)pyrene (50%); *Mesorhizobium* sp. is capable to degrade the chlorpyrifos (CP) in rhizosphere of *L. multiflorum* (Jabeen et al., 2016); *Acinetobacter* sp. can promote the solubilization of phosphate and siderophore activity in the presence of PAHs (naphthalene, fluorene, phenanthrene, anthracene, and dibenzo(a,h)anthracene in *O. sativa* rhizosphere) (Kotoky et al., 2017); *Klebsiella* sp., *Pseudomonas* sp., *Lysobacter* sp., *Pseudoxanthomonas* sp., and *Planctomyces* sp. increase root biomass of *Festuca arundinacea*, enhance mineral solubilization, biosurfactant production, and phytohormones biosynthesis under petroleum hydrocarbons (Hou et al., 2015); *Pseudomonas* sp. assists *Withania somnifera* to increase the production of biosurfactants and efficiently degrades petroleum oil as a carbon source (Das and Kumar, 2016); *Pseudomonas rhizophila* is capable to synthetize ACC deaminase, putative dioxygenase, auxin, pyroverdin, exopolysaccharide, and rhamnolipid biosurfactant in order to reduce pesticide content in *Cynara scolymus* rhizosphere (Hassen et al., 2018); *Burkholderia* sp. can promote the solubilization of phosphate and biosynthesis of ACC deaminase and siderophores in *Brassica chinensis* and *Ipomea aquatica* in order to degrade phenols (Chen et al., 2017); *Flavobacterium*, *Serratia*, *Pasteurella*, and *Azotobacter* increase seed germination of *J. curcas*, solubilize the phosphate, and produce siderophores and IAA in order to degrade the 1,4-dichlorbenzene (Pant et al., 2016); *Rhizobacteria* is capable to promote the growth of *Vigna radiata* and degrade the organophosphate pesticides (OPP) utilizing OPP as carbon/nitrogen source (Pratibha and Krishna, 2015) (Table 3).

2.4 Using industrial waste in management of degraded soils

Waste is generated by human activities due to rapid industrialization, urbanization, technological developments, and population growth. According to EA (2012), waste presents any substance/material/object that the holder is required to discard. However, some waste can be considered as a resource that may be useful in the production of economic valuable products. There are different types of waste: agricultural, municipal, construction and demolition waste, medical waste, industrial waste, nuclear waste, and electronic waste (Millati et al., 2019). Waste can be solids, liquids, and gases and can cause contamination of water, air, and soil having a negative impact on the environment and human health (Artiola, 2019). Sustainable waste management services

Table 3 Plant-growth-promoting rhizobacteria (PGPR) in phytoremediation of metal(loid)s and organic compounds.

Pollutant	Plant species	PGPR	Parameter	References
Metal(loid)s				
Pb, Ni, B, Zn, and Mn	Trifolium hybridum, Alopecurus pratensis, Poa pratensis, Hordeum violaceum, Ranunculus kotschyi, Cerastium sp.	Bacillus megaterium var. phosphaticum	Pb, Ni, B, Zn, and Mn availability in soil and uptake by plants	Gullap et al. (2014)
Pb and Cd	Solanum nigrum	Bacillus	Improve soil quality, enhance enzymatic activity in rhizosphere (acid phosphatase), alleviate uptake of Fe and P, increase biomass of roots and shoots, increase Pb and Cd in aerial parts of plants	He et al. (2020)
Pb and Zn	Brassica juncea	Azotobacter chroococcum, Bacillus megaterium, Bacillus mucilaginosus	Stimulate plant growth	Wu et al. (2006)
Ni	Brassica juncea	Bacillus subtilis	Facilitate Ni accumulation	Zaidi et al. (2006)
Cd	Brassica napus	Xanthomonas sp. Azomonas sp., Pseudomonas sp., Bacillus sp.	Increased plant growth and Cd accumulation	Sheng and Xia (2006)
Cr	Brassica sp.	Pseudomonas sp., Bacillus sp.	Stimulate plant growth and decrease Cr content	Rajkumar et al. (2006)
Hg	Glycine max	Pseudomonas fluorescens	Promoted plant growth	Gupta et al. (2005)
Zn	Brassica juncea	Pseudomonas brassicacearum, Rhizobium leguminosarum	Increased plant growth, enhance root exudation of Zn chelates	Adediran et al. (2016)
Hg	Triticum aestivum	Enterobacter ludwigii, Klebsiella pneumoniae	Enhance the ACC deaminase activity, IAA production, facilitate the uptake of P, K, and Zn, reduce proline accumulation and Hg biosorption	Gontia-Mishra et al. (2017)

Continued

Table 3 Continued

Pollutant	Plant species	PGPR	Parameter	References
Organic compounds				
PAHs (polycyclic aromatic hydrocarbons)	*Rhizophora mangle*	*Pseudomonas aeruginosa, Bacillus* sp.	Promote the development of the plant propagule, degradation of dibenzo(*a,h*) anthracene, naphthalene, acenaphthylene and benzo(*a*)anthracene, anthracene and benzo(*a*)pyrene	Sampaio et al. (2019)
Chlorpyrifos (CP) PAHs	*Lolium multiflorum* *Oryza sativa*	*Mesorhizobium* sp. *Acinetobacter* sp.	Degradation of CP Promote solubilization of phosphate and siderophore activity and degradation of naphthalene, fluorene, phenanthrene, anthracene, dibenzo(*a,h*)anthracene	Jabeen et al. (2016) Kotoky et al. (2017)
Petroleum hydrocarbons	*Festuca arundinacea*	*Klebsiella* sp., *Pseudomonas* sp., *Lysobacter* sp., *Pseudoxanthomonas* sp., *Planctomyces* sp.	Increase root biomass, enhance mineral solubilization, biosurfactant production, and phytohormones biosynthesis	Hou et al. (2015)
Petroleum oil	*Withania somnifera*	*Pseudomonas* sp.	Increase the production of biosurfactants and efficiently degraded petroleum oil	Das and Kumar (2016)
Pesticide	*Cynara scolymus*	*Pseudomonas rhizophila*	Synthetize of ACC deaminase, putative dioxygenase, auxin, pyroverdin, exopolysaccharide, and rhamnolipid biosurfactants	Hassen et al. (2018)
Phenols	*Brassica chinensis, Ipomea aquatica*	*Burkholderia* sp.	Promote solubilization of phosphate and biosynthesis of ACC deaminase and siderophores	Chen et al. (2017)
1,4-Dichlorbenzene	*Jatropha curcas*	*Flavobacterium, Serratia Pasteurella, Azotobacter*	Solubilize the phosphate, and produce siderophores and IAA	Pant et al. (2016)
Organophosphate pesticides (OPP)	*Vigna radiata*	*Rhizobacteria*	Promote the growth of *Vigna radiata* and degraded OPP	Pratibha and Krishna (2015)

include separating, sorting, profiling, packaging, classifying, inventory, storage, logistic management, valorization, recycling, reusing, and traceability (JeyaSundar et al., 2020). Any type of waste should be stored (tanks and containers), collected (properly packed before transportation), and disposed (treatment for biodegradable and nonbiodegradable substances, landfilling, stockpiling, tailings, mud) (Artiola, 2019; JeyaSundar et al., 2020).

Industrial waste can be classified into two types: hazardous and nonhazardous (Chertow and Park, 2019). Hazardous waste is the residue that is flammable, corrosive, active, poisonous/toxic, infectious (pharmachemicals, dyes, paints, synthetic resins, naphthalene, petrochemical waste, polyethanolamine, phthalates, agrochemicals, pesticides, acrylates, etc.) and can harm public health/environment, whereas nonhazardous waste does not pose a threat to human health and environment (carton, glass, rock, organic waste, etc.) (Millati et al., 2019). Industrial waste is generated from various industrial sectors: (a) mining, quarrying, oil and gas extraction; (b) energy (electricity, gas, transmission, and distribution); (c) manufacturing (chemical, food, textile, paper, plastic and rubber, and computer and electronic equipment); and (d) construction (Millati et al., 2019) (Table 4). According to Chertow and Park (2019), industrial waste categories are:

- Chemical waste (solvents, acids/alkali, sulfur, activated carbon, and gases),
- Metallic waste (metal scrap and slag with steel, Fe, Pb, and Zn; spent lead-acid batteries),
- Nonmetallic waste (synthetic gypsum, coal mine overburden, lime dust, concrete, asphalt),
- Ash (fly ash, bottom ash, burnt residue, sludge),
- Sludge (sewage sludge, refinery sludge, paper sludge, fiber mud),
- Paper and wood waste (mixed paper and wood dust),
- Plastics and rubber (polystyrene and rubber scrap),
- Waste oil (oil from chemical processes, edible oil from food manufacturing) (Table 4).

Industrial waste management includes source reduction, recycling/reuse/recovery, and treatment/disposal (Artiola, 2019) (Table 4). Source reduction involves methods that reduce/eliminate the amount of waste and decrease the risk to public health: technological modification (upgrading equipment), clean and well-organized facility with good inventory control, reformulating and redesigning materials (using alternate materials with low risk). Recycling/reuse/recovery methods include processes that promote alternative strategy and optimization of some waste in order to mitigate contamination. Metal recovery refers to the physical separation of useful materials and transforming them to new products, such as economically valuable elements (silver, gold, platinum, cobalt, antimony, and tungsten) or metals that can react with synthetic organic chelates and zeolite, etc. (JeyaSundar et al., 2020). Coal-burning waste, such as fly ash has been used as soil amendment or it can be used as construction material (cement) due to pozzolanic nature and high content of Si, Fe, Al, and Ca/sulfur (Artiola, 2019). Technologies for waste treatment can be physical, chemical, and biological (Table 4). Physical waste treatment includes immobilization, carbon absorption, distillation, filtration, evaporation/volatilization, grinding, and compacting. Chemical waste treatment includes neutralization, oxidation/reduction, precipitation, acid leaching, ion exchange, incineration, and thermal desorption. Biological treatment involves the use

Table 4 Industrial waste management—Waste category, technologies, and processes.

Industrial sector	Waste category	Waste material	Waste management	Technologies for waste management	Processes
Mining, quarrying, oil, gas extraction	Chemical	Solvents, acids, alkali, sulfur, activated carbon, gases	Reduction	Physical	Immobilization, carbon absorption, distillation, filtration, volatilization, grinding, compacting
Energy	Metallic	Fe, Pb, Zn, metal scrap, slag, and steel	Recycling	Chemical	Neutralization, oxidation/reduction, precipitation, acid leaching, ion exchange, incineration, thermal desorption
Manufacturing	Nonmetallic	Synthetic gypsum, coal mine overburden, lime dust, concrete, asphalt	Reuse	Biological	Aerobic and anaerobic decomposition organic compounds by plant/microbe system
Construction	Ash	Fly ash, bottom ash, burnt residue, sludge	Recovery		
	Sludge	Sewage sludge, refinery sludge, paper sludge, fiber muds	Treatment		
	Paper and wood	Mixed paper, wood dust	Disposal		
	Plastics and rubber	Polystyrene, rubber scrap			
	Waste oil	Oil from chemical processes, edible oil from food manufacturing			

of aerobic and anaerobic organisms in order to decompose organic compounds, i.e., these processes are microbial/plant mediated (JeyaSundar et al., 2020).

Proper waste treatment and disposal create challenges that can positively affect the sustainability of industry and society. Legal and institutional waste management programs and services are in the function of governments, private stakeholders, regulatory agencies, etc. The legal framework regarding the environment is associated with economic growth and people's welfare (Artiola, 2019; JeyaSundar et al., 2020).

2.5 Using organic waste in management of polluted soils

Organic waste presents biodegradable food residues from food industry, municipal waste, sewage sludge, garden industry, and private household, i.e., a huge amount of organic waste is generated from plants, animals, and industrial activities where the considerable part remains unutilized; they are burnt or dumped creating pollution and threat for the environment and human health (Chatterjee et al., 2017; Pandey and Singh, 2019). According to Chatterjee et al. (2017), types of organic waste are:

- Plant waste (crop residue, kitchen waste, green market waste, forest biomass, roadside vegetation, and aquatic plant biomass),
- Animal waste (urine, poultry excrete, fish meal, and waste),
- Industrial and municipal solid waste (city garbage, biogas slurry, sewage, sugar industry, paper mill waste) (Table 5).

Recycling organic waste has benefits, such as utilization of nutrients, conservation of energy, reduction of fertilizers cost, reduction of soil pollution, and sustainability of plant growth (Kowalska et al., 2020). Reutilization of organic waste as soil conditioner, energy composting, mulching, green manure covering, and vermicomposting are ecologically sound recycling methods that contribute to the objectives of the EU's "Zero Waste Policy," "End of Waste" Policy, and "Circular Economy Strategy" (Saveyn and Eder, 2014; Kacprzak et al., 2017) (Table 5).

Recycling of organic waste is important for waste management services combating soil degradation and contamination (Gomez-Sagasti et al., 2018; Dwibedi et al., 2021). Organic amendments are suitable for soil restoration and remediation and they can be added to soil through crop residues, rotations of cover crops or as compost, manures, and biosolid (Larkin, 2015). Organic amendment application improves physicochemical properties of degraded soils: decreases soil crusting and bulk density, increases soil water content, increases a stable C stock recovery and content of humic acids, elevates soil cation exchange capacity (CEC), increases soil pH, increases the availability of nutrients (N, P, K, Ca, and Mg), improves the conditions for microbial activity, immobilizes metal(loid)s, and degrades organic pollutants in the soil (Chatterjee et al., 2017; Gomez-Sagasti et al., 2018; Kowalska et al., 2020) (Table 5). Application of organic waste to degraded soils stimulates carbon sequestration due to soil's high potential for stable and safe carbon storage that supports ecosystem balance (Torri et al., 2014). Organic waste application improves plant productivity: increases root/shoot length, leaf area, number of leaves/nodules, total biomass in *V. radiata* (Singh and Agrawal, 2010); increases root growth in *Lepidium sativum*, *Sinapis alba*, and

Table 5 Organic waste management—Category, reutilization, and impact on soil.

Waste category	Waste material	Reutilization	Impact
Plant	Crop residue, kitchen waste, green market waste, forest biomass, roadside vegetation	Soil conditioner	Decrease soil crusting and bulk density Increase soil water content Increase a stable C stock recovery and content of humic acids Elevate soil cation exchange capacity (CEC) Increase soil pH Increase the availability of nutrients (N, P, K, Ca, and Mg) Improve the conditions for microbial activity Immobilize metal(loid)s Degrade organic pollutants
Animal	Urine, fish meal, and waste	Green manuring	Improve soil health, fertility, structure, and water holding capacity Enhance nutrient uptake efficiency Prevent soil erosion Reduce weed growth
Industrial and municipal solid waste	City garbage, biogas slurry, sewage, sugar industry, paper mill waste	Mulching	Cover root rhizosphere Accelerate the decomposition of crop residues Increase nutrient recycling Suppress weed growth Increase water infiltration and retention
		Composting	Improve physicochemical properties of soils Prevent erosion Increase soil microbial biomass
		Vermicomposting	Enhance the rate of decomposition of organic matter, mineralization, humification, and microbial activity

Sorghum saccharatum (Urbaniak et al., 2017); increases biomass yield in *Dactylis glomerata, F. arundinacea, Festuca rubra, Lolium perenne* (Kacprzak et al., 2013); increases root/stem/leaf/total biomass in *Eucalyptus, Populus, Salix,* and *Helianthus* (Nissim et al., 2018); increases leaf length/width, chlorophyll content in *S. lycopersicum* and *Brassica rapa* (Yoo et al., 2018); increases yield at tillering stage, flowering stage, and maturity stage in *T. aestivum* (Nisanth and Biswas, 2008).

Green manuring is one of the best solutions to improve soil health, fertility, structure, water holding capacity, by enhancing nutrient uptake efficiency, preventing soil erosion, and reducing weed growth (Gill et al., 2020) (Table 5). Plants used as green manuring are forage and leguminous crops with deep rooting system, low nutrient and water requirement, fast growing, ability to fix nitrogen, and high biomass (Gill et al., 2020). Green manuring crops are *C. cajan, V. radiata, G. max, C. juncea, Sesbania* sp., *Trifolium* sp., *V. faba, Vicia sativa, Melilotus* sp., *L. albus, Lotus* sp. *Medicago* sp., *S. cereale, A. sativa, Hordeum vulgare, L. perenne, L. multiflorum, D. glomerata, S. alba, Brassica* sp., *R. sativus, L. leucocephala, Cassia* sp., *Hibiscus viscosa, Panicum maximum, Pennisetum purpureum, Eichhornia crassipes, Ipomea carnea, Calotropis gigantea*, etc. (Maitra et al., 2018).

Organic mulches cover root rhizosphere, accelerate the decomposition of crop residues and increase nutrient recycling, suppress weed growth, and increase water infiltration and retention (Chatterjee et al., 2017) (Table 5). In addition, composting is the most common option to recover material from organic waste and it consists of mixture of decayed organomineral materials from food scraps, leaves, sewage sludge, and animal manure (Cesaro et al., 2015). Compost improves physicochemical properties of degraded soils, prevents erosion, and increases soil microbial biomass (Table 5). Composting converts lignin, cellulose, hemicellulose, polysaccharides, and proteins into simple available nutrients; microorganisms break down organic matter and use C as a source of energy and N for cell structure producing CO_2, water, and energy transformed in humus (Chatterjee et al., 2017; Gomez-Sagasti et al., 2018). Vermicompost is organic waste obtained from ingested biomass of earthworms (Ali et al., 2015). It consists of N, P, K, humic acids, auxins, gibberellins, cytokinins, N-fixing and phosphate-solubilizing bacteria, vitamins, etc. During ingestion, the earthworms fragment waste, enhancing the rate of decomposition of organic matter, mineralization, humification, and microbial activity (Chatterjee et al., 2017) (Table 5). Temperature, moisture, pH, C/N ratio, and feeding are critical factors for this process. Nutrient-rich vermicompost can be used for biogas generation as a result of anaerobic digestion of organic waste (Suthar, 2010).

According to Chatterjee et al. (2017), factors that affect recycling of organic waste are availability of organic waste (soil properties, crop type, and crop agronomic practice), quality of organic waste (nutrient composition, C/N ratio, and heavy metals), and sociotechnological factors (proper technology and preparation for recycling of waste and cost).

3 Phytomanagement

3.1 Climate-resilient economic crops

3.1.1 Biofuel crop production—Role in phytoremediation

Biofuels are produced from living plant materials or by-products from agricultural production and can be grouped in biodiesel, bioethanol, biogas, and biomass (Kocar and Civas, 2013; Pandey et al., 2012, 2016a; Pandey, 2017). Energy crops can grow on energy plantations at contaminated lands, such as coal mine spoil, fly ash deposits, fly ash-amended soil, phosphatic clay, red mud deposits, chromite-asbestos deposits where they can also be used as excellent phytoremediators (Pandey et al., 2015a,b, 2016a,b; Pandey, 2017). Biofuels are renewable, generated from sugar plants, oleaginous plants, forest biomass, and algae biomass, and majority of these plants can be used for food and/or energy (Kocar and Civas, 2013).

According to Ziolkowska (2020), "first-generation biofuels" (ethanol and biodiesel) are produced from food crops (oil plants, sugar-containing plants), whereas "advanced biofuels" can be second-generation biofuels that are produced from cellulosic biomass (energy crops specifically for that purpose, crop waste, green waste, wood, and forest/park residue), third-generation biofuels that are produced from algae biomass, and fourth-generation biofuels that are produced from genetically engineered plants. Energy crops suitable for biofuels are grasses/perennial herbaceous plants/starch and sugar-containing plants/oilseed plants, and they have high yield, low energy input, low cost, and low nutrient requirements (McKendry, 2002).

According to Kocar and Civas (2013) and Ziolkowska (2020), energy crops that are used for the production of bioethanol are cereals with starchy materials (*H. vulgare, T. aestivum, Z. mays, A. sativa,* and *S. cereale*); cereals with sucrose materials (*Ipomea batatas, Beta vulgaris, Saccharum officinarum, Sorghum bicolor, Helianthus tuberosum,* and *Manihot esculenta*); oil crops (*B. napus, B. rapa, Cannabis sativa, Linum usitatissimum, Helianthus annuus, Carthamus tinctorius, R. communis, Olea europaea, Elaeis guineensis,* and *Cocos nucifera*); cellulose crops with lignocelluloses biomass (wood, straw, grasses, and rotation crops); and solid energy crops that can be utilized as whole plants (*Cynara cardunculus, S. bicolor, Z. mays, Hibiscus cannabinus, Phalaris arundinacea, Miscanthus* sp., *Panicum virgatum, Salix* sp., *Populus* sp., and *Eucalyptus* sp.) (Fig. 2).

According to Demirbas (2008), energy crops that can be used for the production of biodiesel and have high potential annual yield are palm oil (400–650 gal/acre), coconut (250–300 gal/acre), Jatropha (140–200 gal per acre), canola (110–1045 gal/acre), rapeseed (110–130 gal/acre), mustard (60–140 gal/acre), sunflower (75–105 gal/acre), camelina (60–65 gal/acre), soybean (40–45 gal acre), cotton (35–45 gal/acre), corn (18–20 gal/acre); and other plants with sources of edible and nut oils: *B. napus, B. juncea, C. tinctorius, G. max, Arachis hypogea, R. communis, O. europaea,* etc. (Fig. 2).

Biogas is generated by anaerobic digestions of organic material by microbes converting plant biomass, crop residues, and energy crops to methane-rich biogas, and it does not contribute to greenhouse gases in the atmosphere (Lehtomaki et al., 2007). Optimization of energy crop production, harvesting time, nutrient composition, conservation, and pretreatment is required for maximizing biogas yield (Amon et al., 2007). Plants used for the production of biogas are maize, wheat (*T. aestivum*), oat (*A. sativa*), ryegrass (*L. perenne*), clover (*Trifolium* sp.), flax (*L. usitatissimum*), sunflower (*H. annuus*), oilseed rape (*B. napus*), Jerusalem artichoke (*H. tuberosum*), pea (*P. sativum*), potato (*Solanum tuberosum*), sugar beet (*B. vulgaris*), hemp (*C. sativa*), barley (*H. vulgare*), sorghum (*Sorghum* sp.), alfalfa (*Medicago sativa*), reed canary grass (*P. arundinacea*), miscanthus (*Miscanthus* sp.), etc. (Braun et al., 2008) (Fig. 2).

Energy plants that grow at energy plantations on waste dumpsite have environmental and socioeconomic benefits. Therefore, energy plants that can grow on fly ash

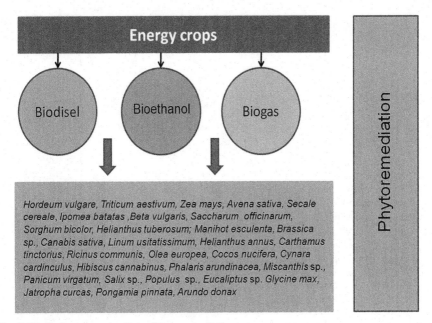

Fig. 2 Biofuel crop production—Phytoremediation.

deposits/amended soil/coal mine spoil/sewage sludge/metalliferous sites are *J. curcas*, *R. communis*, *Pongamia pinnata*, *Saccharum spontaneum*, *L. usitatissimum*, *H. cannabinus*, *C. cardunculus*, *Miscanthus giganteus*, *Arundo donax*, *P. arundinacea*, *P. virgatum*, *H. annuus*, *A. indica*, *Populus*, *Salix*, *Eucalyptus*, and *Linum* (Pandey et al., 2016a; Pandey, 2017) (Fig. 2). These plants decrease soil fly ash/soil erosion contributing substrate stabilization and leaching control, reduce dust pollution, increase soil fertility and quality of waste dumpsite, combat metal(loid)s, sequester atmospheric CO_2 in substrate, enhance biodiversity, and provide a pleasant landscape. In addition, energy plantation reduces input cost and maintenance of waste dumpsite, with no competition between food and bioenergy for land, it supports local communities, employees, and livelihoods (Pandey et al., 2015a, 2016a; Pandey, 2017).

3.1.2   Fiber crop production—Role in phytoremediation

Natural plant fibers provide stiffness, i.e., strength to the composite; they are recyclable, have low density, low cost, biodegradable and nonabrasive, ecofriendly with no health hazards (Rana et al., 2014). Useful fibers are obtained from plant leaves, stems (bast fibers), fruits, and seeds. They have heavily lignified cell walls where sclerenchyma gives mechanical strength and rigidity to the plant (Sfilgoj-Smole et al., 2013). Plant fibers are composed from structural polymers (cellulose, hemicelluloses, and lignin) together with proteins and minerals (Marques et al., 2010). Natural fibers produced from seeds are *G. hirsutum* (cotton) and *Ceiba pentandra* (kapok), usually used as textile fibers; bast fibers are *L. usitatissimum* (flax), *Corchorus capsularis* (jute), *C. sativa* (hemp), *H. cannabinus* (kenaf), and *Boehmeria nivea* (ramie) that

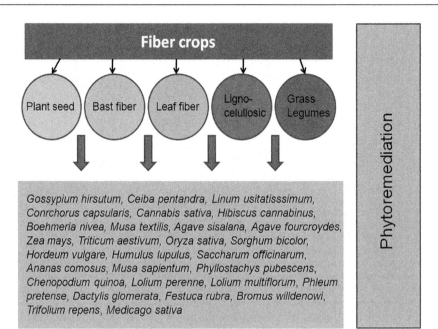

Fig. 3 Fiber crop production—Phytoremediation.

can be used as textile, paper products, composites, etc.; leaf fibers are hard fibers that have stiffer and coarser texture than others and are used for cordage and ropes: *Musa textilis* (abaca), *Agave sisalana* (sisal), *Agave fourcroydes* (henequen), etc. (Sfilgoj-Smole et al., 2013) (Fig. 3). However, lignocellulosic nonconventional plant fibers and their by-products are beneficial for cellulose fibers due to their physicochemical properties and composition, they can be used in textile and paper industry for fibers: *Z. mays* (corn), *T. aestivum* (wheat), *O. sativa* (rice), *S. bicolor* (sorghum), *H. vulgare* (barley), *Humulus lupulus* (hop), *S. officinarum* (sugarcane), *Ananas comosus* (pineapple), *Musa sapientum* (banana), *Phyllostachys pubescens* (bamboo), *Chenopodium quinoa* (quinoa) (Sfilgoj-Smole et al., 2013) (Fig. 3). Grasses and legume fibers have fiber cells bound by pectin middle lamellae and have potential for textile applications: *L. perenne* (ryegrass), *L. multiflorum* (Italian ryegrass), *Phleum pratense* (timoty), *D. glomerata* (cockfoot), *F. rubra* (red fescue), *Bromus willdenowii* (bromes), *Trifolium repens* (clover), and *M. sativa* (lucerne) (Sfilgoj-Smole et al., 2004, 2005) (Fig. 3).

Fiber crops grown over diverse range of climatic conditions and soil qualities. They are resilient to high temperatures, drought, salinity, and metal(loid)s, and they have a high phytoremediation potential (Ludvikova and Griga, 2018; Saleem et al., 2020; Pandey et al., 2022). Fiber plants can accumulate heavy metals in roots and leaves and have a strong antioxidant system that makes them tolerant to contamination stress (Saleem et al., 2020). Therefore, *G. hirsutum* is tolerant to Cd, accumulated in roots and leaves (Angelova et al., 2004). According to Linger et al. (2005), *C. sativa* is capable to accumulate about 800 mg/kg Cd in roots and 50–100 mg/kg Cd in stems and leaves indicating that under moderate Cd concentrations, this fiber plant can grow well and preserve photosynthetic machinery from heavy metal stress. *H. cannabinus* showed a high phytoremediation potential for hydrocarbons in soil (lubricating oil) (Abioye et al., 2010). This fiber plant is capable to accumulate about 140 mg/kg Zn and 4.5 mg/kg Cd in its tissues (Arbaoui et al., 2016). Furthermore, *B. nivea* has a high potential for accumulation and tolerance to Cd with no decrease in shoot biomass and fiber yield, i.e., it retains Cd in roots binding a large amount of Cd to cell walls indicating that this plant is suitable for phytoremediation of Cd-contaminated soils (Zhu et al., 2013). *L. usitatissimum* (flax) is considered as a suitable phytoremediator of Cd, Pb, and Zn (Najmanova et al., 2012; Hosman et al., 2017). This fiber can sequester Cd through phytochelatins (PC), nonenzymatic antioxidant, whereas the highest Cd concentrations are retained in roots (Najmanova et al., 2012). Smykalova et al. (2010) noted that different varieties of flax showed different tolerance mechanisms to Cd, i.e., they can restrict the uptake of Cd inside the plant or can sequester Cd in root cytoplasm and vacuole forming chelates. Hosman et al. (2017) found that flax retained the highest concentrations of Cd in roots and this plant can be considered as Cd excluder, whereas Pb and Zn are retained in the capsule and it can be considered as Pb and Zn hyperaccumulator. *C. capsularis* (jute) showed a high phytoremediation potential to Cu with no significant decrease in growth/biomass at 300 mg/kg Cd in substrate retaining the highest concentrations in roots in the earlier stage of growth (Saleem et al., 2019a). Also, jute can tolerate Cu stress by activating a strong antioxidative system, and because of that, this fiber plant can be considered as a good Cu phytoremediator (Saleem et al., 2019b, 2020).

3.1.3 Aromatic essential oil crop production—Role in phytoremediation

Aromatic plant species that yield essential oils are widely produced on the globe due to their flavor and fragrance. They are used in cosmetics, perfumes, soaps, detergents, repellents, and food processing industries, and they are used as antiseptics, antivirals, bactericides, etc. (Lubbe and Verpoorte, 2011). Essential oils are volatile mixtures of liposoluble organic substances, aldehydes, ketones, terpenes, and aromatic polypropanoids that are synthetized via shikimate and mevalonate pathways (Sangwan et al., 2001). Each essential oil retains the organoleptic character of the original plant (flavor and taste) and unique aroma pattern from its aromatic components. Market and commercial data reveal that there are 400 plant species from 67 plant families that are used for the production of essential oils (Duke et al., 2002). However, the most significant plant species for essential oil cultivation belong to plant families, such as *Asteraceae*, *Lamiaceae, Apiaceae, Rosacea, Fabaceae, Rutaceae, Pinaceae, Myrtaceae*, and *Piperaceae* (Kalia, 2005). Cultivation of essential oil crops and quality of their yield and composition depend on environmental factors (temperature, radiation, long-day conditions, altitude, soil and moisture, wind, and nutrient levels), harvesting-related factors (season, geographic provenance, and age), crop management (soil nutrients, fertilization, irrigation, and mechanization) (Carrubba and Catalano, 2009; Bhattacharya, 2016). Furthermore, planting aromatic essential oil crops includes population density, arrangement in space, and time of sowing (Carrubba and Catalano,

2009; Bhattacharya, 2016). According to Carrubba and Catalano (2009), every plant extract must retain characteristics "quality-security-effectiveness," i.e., the product should meet all these standards. Generally, essential oils are processed from plant distillation that can be undertaken using on-site facilities or mobile units from farms (Carrubba and Catalano, 2009).

Aromatic essential oil plants are adaptable to degraded soils; they improve soil properties; prevent erosion and dust flow; improve soil fertility; enhance carbon sequestration; and tolerate water deficit, salinity, and sodicity; and mitigate contamination; thus, phytomanagement of aromatic crops has ecological and economical benefits for environment and society (Gupta et al., 2013; Verma et al., 2014; Pandey and Singh, 2015; Pandey et al., 2019a,b; Pandey and Praveen, 2020). Cultivation of aromatic crops on contaminated land has been suggested as a profitable and "safe" option because they avoid pollutants' entry into food web and they are unpalatable to livestock (Lal et al., 2008; Verma et al., 2014; Pandey and Singh, 2015; Pandey et al., 2019a; Pandey and Praveen, 2020). However, before using essential oils produced from plants grown on contaminated soils, proper chemical analysis is required: qualitative and quantitative analysis of essential oils, and risk assessment should be done before the products have begun usage on a large scale (Pandey and Singh, 2015; Pandey et al., 2019a; Pandey and Praveen, 2020). According to Zheljakov and Jekov (1996), low concentrations of heavy metals have been found in essential oils of *Rosa*, *Lavandula*, *Mentha*, *Salvia*, *Ocimum*, *Foeniculum*, *Coriandrum*, *Anethum*, and *Rhus* (Fig. 4). Oil extraction by steam distillation process extracts oils, but heavy metals remain in plant residues during the process (Zheljakov et al., 2006; Pandey et al., 2019a; Pandey and Praveen, 2020).

Promising plants for phytoremediation of metalliferous sites are the following: (a) aromatic grasses (*Poaceae*): *Chrysopogon/V. zizanioides* (vetiver), *Cymbopogon flexuosus* (lemongrass), *Cymbopogon winterianus* (citronella), *Cymbopogon martini* (palmarosa); (b) aromatic plants from *Lamiaceae*: *Mentha* sp. (mint), *Ocimum* sp. (basil), *Rosmarinus officinalis* (rosemary), *Salvia* sp. (sage), and *Lavandula* sp. (lavandula); aromatic plants from *Asteraceae*: *Matricaria* sp. (chamomile); and aromatic plants from *Geraniaceae*: *Pelargonium* sp. (geranium) (Verma et al., 2014; Pandey and Singh, 2015; Pandey et al., 2019a,b; Pandey and Praveen, 2020) (Fig. 4).

According to Verma et al. (2014), proposed aromatic grasses that can grow on fly ash deposits are *C. flexuosus*, *C. winterianus*, *C. martini*, and *V. zizanioides*, and they can obtain high biomass. Vetiver (*V. zizanioides*) is a good phytoremediator of As (Pandey and Singh, 2015); it accumulates more As in roots (268 mg/kg) than in leaves (11.2 mg/kg) (Truong, 1999). This species is a phytostabilizer of Pb and Zn (Shu et al., 2002); Pb, Zn, and Cu (Chiu et al., 2006); Cr (Sinha et al., 2013); Cd and Ni (Gunwal et al., 2014), while it is a suitable phytoextractor for Pb, Cu, Zn, and Cd (Chen et al., 2004). Lemongrass (*C. flexuosus*) can accumulate As and Cr (Jha and Kumar, 2017); Pb, Cd, and Zn (Hassan, 2016); and Cu (Das and Maiti, 2009), whereas citronella (*C. winterianus*) has a great potential for phytostabilization of Cr and Cd (Sinha et al., 2013). Mint (*Mentha* sp.) is a good phytostabilizer of Pb, Cr, and Cu (Prasad et al., 2010; Anwar et al., 2016). Geranium (*Pelargonium* sp.) showed a high translocation and phytoextraction of Pb, Cd, Cr, and Ni (Manshadi et al., 2013). Basilicum

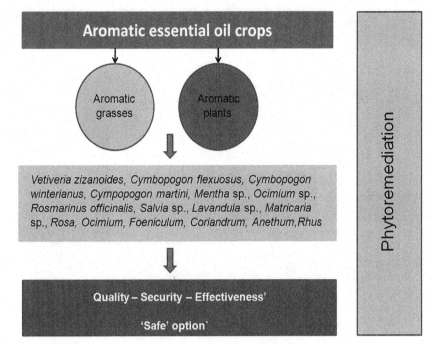

Fig. 4 Aromatic essential oil crop production—Phytoremediation.

(*O. basilicum*) can be used as a phytostabilizer of Cd, Cr, Pb, and Ni (Prasad et al., 2010), whereas chamomile (*Matricaria* sp.) can act as Pb, Cd, Cr, and Cu excluder (Kovacik and Klejdus, 2008; Kovacik, 2013; Farzadfar et al., 2013). Rosemary (*R. officinalis*) is a good phytostabilizer of As, Pb, Sb, Zn, and Cu (Affholder et al., 2014). Sage (*Salvia* sp.) accumulates more Cu, Pb, and Cd in roots than in leaves (Stancheva et al., 2014), whereas lavender (*Lavandula* sp.) efficiently translocates Pb, Cd, and Zn from roots to leaves (Angelova et al., 2015).

3.1.4 Fortified crop production—Role in phytoremediation

Biofortification is a cost-effective and sustainable technique that provides mineral micronutrients to the human population with limited access to diets (Garg et al., 2018). According to McGuire (2015), about 792.5 million people around the world are malnourished where the majority of them live in developing countries (South Asia and Africa). "Hidden hungry" refers to an inadequate intake of essential micronutrients in the daily diet and has negative consequences for human health resulting in sickness, increased morbidity, and disability, stunted mental and physical growth, and cognitive development (Chizuru et al., 2004). Micronutrient deficiencies negatively affect 38% of pregnant women and 43% of children around the globe (Garg et al., 2018). Humans require around essential 40 micro/macronutrients for good health: macrominerals (K, Ca, Mg, S, P, Na, and Cl); microminerals (Fe, Zn, Cu, Mn, I, Se, Mo, Co, and Ni); vitamins (A, D, E, K, C, B1, B2, B3, B5, B6, B7, and B9), amino acids, and fatty acids

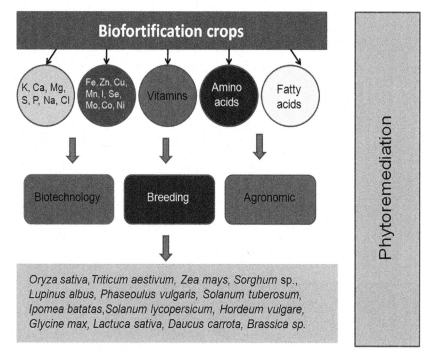

Fig. 5 Fortified crop production—Phytoremediation.

(linoleic and linolenic acids) (Prasanth et al., 2015) (Fig. 5). The main diet of humans is based on cereals, such as rice, wheat, cassava, and maize and they have insufficient amounts of Fe, Zn, Ca, Mg, Co, iodine, Se, and vitamin A. Biofortification of crop plants provides enough nutrients and vitamins in edible portions of plants to meet the energy for sound health. According to WHO (2000), vitamin A, C, D, B, and Fe, Zn, and I deficiencies cause the most serious health constraints.

Biofortification approaches include transgenic (biotechnology), convectional (crop breeding), and agronomic (fertilizers) strategies, and they increase the bioavailable mineral content in food crops, such as cereals, vegetables, legumes, oilseeds, fruits, fodder rice (*O. sativa*), wheat (*T. aestivum*), maize (*Z. mays*), sorghum (*Sorghum* sp.), lupine (*L. albus*), common bean (*Phaseolus vulgaris*), potato (*S. tuberosum*), sweet potato (*I. batatas*), tomato (*S. lycopersicum*), barley (*H. vulgare*), soybean (*G. max*), lettuce (*L. sativa*), carrot (*Daucus carota*), canola (*B. napus*), and mustard (*Brassica* sp.) (Mayer et al., 2008; Garg et al., 2018; Freire et al., 2020) (Fig. 5).

Genetic engineering employs transgenic crops that are developed to improve abilities to accumulate minerals in edible plant parts and it is used when there is limited or no genetic variation in nutrient content between plant varieties or for the redistribution of micronutrients between tissues; it increases their concentrations in the edible portions and increases the efficiency of biochemical pathways (Garg et al., 2018). The main targets for biofortification through transgenic approach are PSY, carotene desaturase, lycopene β-cyclase for vitamins, ferritin, and nicotinamide synthase, phytic acid for minerals, albumin, lysine, methionine, tryptophan for amino acids, Δ6 desaturase, oleic acids, and linoleic acids for fatty acids (Garg et al., 2018). Biofortification through agronomic methods by application of fertilizers (N, P, K, Se, Fe, and Zn) increases their solubilization/mobilization from soil to edible part of plants (Erisman et al., 2008). Biofortification through breeding is the most accepted method that utilizes genotypic variations in the target trait to improve the amount of minerals and vitamins in plants. Generally, implementing biofortification of crops for farmers and consumers involves: (a) identification of target population and staple food consumption profile; setting nutrient target levels, applied technology; (b) crop improvement, nutrient retention and bioavailability, nutritional efficiency studies on humans; and (c) releasing biofortified crops, facilitation of dissemination, marketing and public acceptance (Bouis et al., 2011).

According to Yin et al. (2012), plants generated from phytoremediation can be used as supplementary sources for mineral elements to produce food, i.e., phytoremediation and biofortification can be "two sides of one coin" because they are based on phytoextraction process that involves plant acquisition, uptake, accumulation, transport and homeostasis of nutrient elements (Zhao and McGrath, 2009). Both processes enhance the efficiency of element uptake, by plants, their translocation, distribution, and transformation in soil-plant system. New research efforts have been made to integrate phytoremediation and biofortification processes (Lin et al., 2014; Pandey, 2013). Plant species are capable to secrete organic compounds to chelate heavy metals in rhizosphere in order to increase solubility and availability. Secretion of phytosiderophores (PS) as a response to Fe/Zn deficiency induces Fe/Zn uptake creating Fe/Zn-PS complex supplying the plants with these nutrients (Zhao and McGrath, 2009). Furthermore, synthetic chelators (EDTA) can be used to enhance the mobilization of heavy metals from soil to plant, but this method has a high risk of leaching and a high cost of chelators (Nowack et al., 2006).

Integration of phytoremediation and biofortification requires the selection of appropriate plant species with edible parts that should accumulate safe concentrations of nutrient elements, especially not accumulate toxic metals (broccoli, Rodrigo et al., 2014). *B. juncea* is usually used in phytoremediation of Se-contaminated soils, and according to Banuelos et al. (2009), after harvesting, this plant can be used as a biofortified Se supplement for animals. In addition, Zhu et al. (2009) proposed biofortification of multiple micronutrients, such as Se and Zn in crops, i.e., the applications of Se/Zn enriched plant materials as green manures can increase bioavailability, uptake, and accumulation in edible parts of biofortified crops. Se-enriched crops, such as rice, maize, and wheat can be safely used as Se-supplement in Se-deficient soils. According to Yuan et al. (2013), Se-hyperaccumulator *Cardamine hupingshanesis* can be used as a source of Se fertilizers for Se-biofortification practice.

Phytoremediation and biofortification of nutrient elements involve gene expression, protein modifications, i.e., transcriptomics and proteomics research can help to better understand nutrient element-targeted pathways for both processes. However, assessment of the availability of any nutrient element to humans in biofortified crops and potential accumulation of toxic metal(loid)s, such as As, Hg, and Cd, through phytoextraction on contaminated soils are significant, particularly for toxic elements that share the same transporters as essential elements (Fe, Zn, and Ca) (Zhao and McGrath, 2009).

4   Pros and cons of adaptive phytoremediation practice

Pros of adaptive phytoremediation practice:

- Green, ecofriendly technology that uses plants to clean up contaminated soils;
- Applied to a wide range of inorganic and organic pollutants;
- Multiple contaminants can be removed from sites;
- Minimal soil disturbance;
- Prevention of soil erosion and spreading of contaminants;
- Simple to use and maintain;
- Low cost;
- Technologically simple method;
- No need for expensive equipment/highly qualified stuff;
- Ecologically safe, aesthetically pleasing, and publicly acceptable;
- Phytomining.

Cons of adaptive phytoremediation practice:

- Suitable for low/medium contaminated sites,
- Limited remediation depth,
- Slow rate of remediation (several years),
- Depends on climatic and seasonal conditions, geology,
- Agronomic practice can affect the mobility of contaminants,
- Possible spread of contamination through falling leaves,
- Contaminants can enter trophic chains,
- Contaminated biomass after phytoextraction can be treated as dangerous waste.

5   Conclusion

Global climate change and pollution lead to the degradation of soils, depletion of soil organic matter, lack of nutrients, and contamination of environment that negatively affects human health and causes malnutrition and hunger. Appropriate soil management improves the quality of soil and crop productivity contributing to soil conservation and sustainability. Phytomanagement of polluted sites involves green solutions in revitalization with multiple ecological and socioeconomic benefits. The main phytoremediation technologies are used for removing pollutants from the environment. Selective plant species are usually the best adapted to local climate and contamination. Plants inoculated with arbuscular mycorrhizal fungi (AMF) are tolerant to drought and salinity stress as well as to pollutants. These plants have a great potential to regulate water-deficit and osmotic stress increasing osmoprotectants in roots and leaves sustaining cell turgor and hydrations and they activate the antioxidant system to combat oxidative stress. Under drought and salinity stress, AMF increase nutrient uptake in plants, enhance photosynthetic activity, and increase plant yields. Plant-growth-promoting bacteria (PGPR) are used in soil management and accelerate the phytoremediation of polluted sites. PGPR improve soil quality, enhance enzymatic activity in rhizosphere, increase nutrient solubilization, produce ACC-deaminase that reduce ethylene level, improve plant growth, reduce uptake of metal(loid)s, and degrade organic pollutants.

Industrial waste management includes source reduction, recycling/reuse/recovery of hazardous and nonhazardous wastes that are generated from industrial sectors (fly ash, mining) and manufacturing. Organic waste as biodegradable food residues can be used as soil conditioner, green manure, compost, and vermicompost contributing as organic amendments to sustainable soil restoration decreasing soil contamination pollution and improving plant growth. Recommendations have been made for the application of economically valuable plants in phytoremediation. Selected plants for phytoremediation of contaminated sites contribute to substrate stabilization preventing erosion and dust spread, increase soil fertility, reduce levels of pollutants, but at the same time, they can be used for biofuel production (biodiesel, bioethanol, and biogas), fiber production, aromatic essential oil production, and fortified crop production. Commercial crop production on the contaminated soils should have "safe" options avoiding pollutants into trophic chain. Further research should be focused on combined physiology, biochemistry of plants genomics, proteomics, agronomic, and engineering strategies for crop improvement in order to integrate phytoremediation and commercial/market plant production. Adaptive phytoremediation practice and soil management positively affect the environmental protection and human health providing sustainability of industry, economy, and society building community relations and people's welfare worldwide.

References

Abioye, O.P., Agamuthu, P., Abdul, A.A., 2010. Phytoremediation potential of kenaf (*Hibiscus cannabinus* L.) in soil contaminated with lubricant oil. In: Linnaeus ECO-TECH 10, Kalmar, Sweden.

Adediran, G.A., Ngwenya, B.T., Mosselmans, J.F.W., Heal, K.V., Harvie, B.A., 2016. Mixed planting with a leguminous plant outperforms bacteria in promoting growth of a metal remediating plant through histidine synthesis. Int. J. Phytoremediation 18 (7), 720–729.

Affholder, M.C., Pricop, A.D., Laffont-Schwob, I., Coulomb, B., Rabier, J., Borla, A., Demelas, C., Prudent, P., 2014. As, Pb, Sb, and Zn transfer from soil to root of wild rosemary: do native symbionts matter. Plant Soil 382 (1–2), 219–236.

Alguacil, M.M., Hernandez, J.A., Caravaca, F., Portillo, B., Roldan, A., 2003. Antioxidant enzyme activities in shoots from three mycorrhizal shrub species afforested in a degraded semi-arid soil. Physiol. Plant. 118, 562–570.

Ali, U., Sajid, N., Khalid, A., Riaz, L., Rabbani, M.M., Syed, J.H., Malik, R.N., 2015. A review on vermicomposting of organic wastes. Environ. Progr. Sustain. Energy 34 (4), 1050–1062.

Aliasgharzadeh, N., Saleh, R., Towfighi, H., Alizadeh, A., 2001. Occurrence of arbuscular mycorrhizal fungi in saline soils of the Tabriz Plain of Iran in relation to some physical and chemical properties of soil. Mycorrhiza 11, 119–122.

Amon, T., Amon, B., Kryvoruchko, V., Machmuller, A., Hopfner-Sixt, K., Bodiroza, V., Friedel, J.K., Potsch, E.M., Wagentristi, H., Scheiner, M., Zollitsch, W., 2007. Methane production through anaerobic digestion of various energy crops grown in sustainable crop rotations. Bioresour. Technol. 98 (17), 3204–3212.

Angelova, V., Ivanova, R., Delibaltova, V., Ivanov, K., 2004. Bioaccumulation and distribution of heavy metals in fibre crops. (flax, cotton and hemp). Ind. Crop. Prod. 19, 197–205.

Angelova, V.R., Grekov, D.F., Kisyov, V.K., Ivanov, K.I., 2015. Potential of lavender (*Lavandula vera* L.) for phytoremediation of soils contaminated with heavy metals. Int. J. Biol. Biomol. Agric. Food Biotechnol. Eng. 9, 465–472.

Anjum, S.A., Farooq, M., Xie, X.Y., Liu, X.J., Ijaz, M.F., 2012. Antioxidant defence system and proline accumulation enables hot pepper to perform better under drought. Sci. Hortic. 140, 66–73.

Anwar, S., Nawaz, M.F., Gu, S., Rizwa, M., Ali, S., Kareem, A., 2016. Uptake and distribution of minerals and heavy metals in commonly grown leafy vegetable species irrigated with sewage water. Environ. Monit. Assess. 188 (9), 541.

Arbaoui, S., Soufi, S., Roger, P., Bettaieb, T., 2016. Phytoremediation of trace metal polluted soil with fiber crop: kenaf (*Hibiscus cannabinus* L). Int. Adv. Agric. Environ. Eng. 3, 2.

Artiola, J.F., 2019. Industrial waste and municipal solid waste treatment and disposal. In: Brusseau, M.L., Pepper, I.C., Gerba, C.P. (Eds.), Environmental and Pollution Science, third ed. Academic Press, pp. 377–391.

Asad, S.A., Farooq, M., Afzal, A., West, H., 2019. Integrated phytobial heavy metal remediation strategies for a sustainable clean environment—a review. Chemosphere 217, 925–941.

Asensio, D., Rapparini, F., Penuelas, J., 2012. AM fungi root colonization increases the production of essential isoprenoids vs. nonessential isoprenoids especially under drought stress conditions or after jasmonic acid application. Phytochemistry 77, 149–161.

Auge, R.M., Moore, J.L., 2005. Arbuscular mycorrhizal symbiosis and plant drought tolerance. In: Mehotra, V.S. (Ed.), Mycorrhiza: Role and Applications. Allied Publishers, Hagpur, pp. 136–162.

Azcon, R., Gomez, M., Tobar, R., 1996. Physiological and nutritional responses by *Lactuca sativa* L., to nitrogen sources and mycorrhizal fungi under drought conditions. Biol. Fertil. Soils 22, 156–161.

Banuelos, G.S., Robinson, J., Da Roche, J., 2009. Developing selenium-enriched animal feed and biofuel from canola planted for managing Se-laden drainage waters in the Westside of Central California. Int. J. Phytoremediation 12, 243–254.

Barros, V., Frosi, G., Santos, M., Ramos, D.G., Falcao, H.M., Santos, M.G., 2018. Arbuscular mycorrhizal fungi improve photosynthetic energy use efficiency and decrease foliar construction cost under recurrent water deficit in woody evergreen species. Plant Physiol. Biochem. 127, 469–477.

Bhattacharya, S., 2016. Cultivation of essential oils. In: Preedy, V. (Ed.), Essential Oils in Food Preservation, Flavor, and Safety. Elsevier, Inc, pp. 19–29.

Bothe, H., 2012. Arbuscular mycorrhizae and salt tolerance of plants. Symbiosis 58, 7–16.

Bouis, H.E., Hotz, C., McClafferty, B., Meenakshi, J.V., Pfeiffer, W.H., 2011. Biofortification: a new tool to reduce micronutrient malnutrition. Food Nutr. Bull. 32 (1), S31–S40.

Boutasknit, A., Baslam, M., Ait-El-Mokhtar, M., Anli, M., Ben-Laouane, R., Douira, A., El Modaraf, C., Mitsui, T., Wahbi, S., Meddich, A., 2020. Arbuscular mycorrhizal fungi mediate drought tolerance and recovery in two contrasting carob (*Ceratonia siliqua* L.) ecotypes by regulating stomatal, water relations, and (in) organic adjustment. Plants 9, 80.

Braun, R., Weiland, P., Werllinger, A., 2008. A Biogas from Energy Crops Digestion. IEA Bioenergy Task, p. 37.

Burd, G.I., Dixon, D.G., Glick, B.R., 2000. Plant growth-promoting bacteria that decrease heavy metal toxicity in plants. Can. J. Microbiol. 46, 237–245.

Burken, J.G., 2003. Uptake and metabolism of organic compounds: green liver model. In: McCutcheon, S.C., Schnoor, J.L. (Eds.), Phytoremediation. Transformation and Control of Contaminants. John Wiley and Sons, Inc., Publications, Hoboken, NJ, pp. 59–84.

Cantrell, I.C., Linderman, R.G., 2001. Preinoculation of lettuce and onion with VA mycorrhizal fungi reduces deleterious effects of soil fertility. Plant Soil 233, 269–281.

Carrubba, A., Catalano, C., 2009. Essential oil crops for sustainable agriculture—a review. In: Lichtfouse, E. (Ed.), Sustainable Agriculture Reviews: Climate Change, Intercoping, Pest Control and Beneficial Microorganisms, Sustainable Agriculture Reviews. vol. 2. Springer Science + Business Media B.V, pp. 137–187.

Cesaro, A., Belgiorno, V., Guida, M., 2015. Compost from organic solid waste: quality assessment and European regulations for its sustainable use. Resour. Conserv. Recycl. 94, 72–79.

Chandrasekaran, M., Kim, K., Krishnamoorthy, R., Walitang, D., Sundaram, S., Joe, M.M., Selvakumar, G., Hu, S., Oh, S.-H., Sa, T., 2016. Mycorrhizal symbiotic efficiency on C3 and C4 plants under salinity stress—a meta-analysis. Front. Microbiol. 7, 1246.

Chatterjee, R., Gajjela, S., Thirumdasu, R.K., 2017. Recycling of organic wastes for sustainable soil health and crop growth. Int. J. Waste Resour. 7 (3), 1000296.

Chen, Y., Shen, Z., Li, X., 2004. The use of vetiver grass (Vetiveria zizanioides) in the phytoremediation of soils contaminated with heavy metals. Appl. Geochem. 19 (10), 1553–1565.

Chen, J., Li, S., Xu, B., Su, C., Jiang, Q., Zhou, C., Jin, Q., Zhao, Y., Xiao, M., 2017. Characterization of Burkholderia sp. XTB-5 for phenol degradation and plant growth promotion and its application in bioremediation of contaminated soil. Land Degrad. Dev. 28 (3), 1091–1099.

Chertow, M., Park, J., 2019. Reusing nonhazardous industrial waste across business clusters. In: Letcher, T.M., Vallero, D.A. (Eds.), Waste. A Handbook for Management. Elsevier, Inc, pp. 353–363.

Chitara, M.K., Chauhan, S., Singh, R.P., 2021. Bioremediation of polluted soil by using plant growth—promoting rhizobacteria. In: Panpatte, D.G., Jhala, Y.K. (Eds.), Microbial Rejuvenation of Polluted Environment, Microorganisms for Sustainability. Springer Nature Singapore, Ltd, pp. 204–228.

Chittara, W., Pagliarani, C., Maserati, B., Lumini, E., Siciliano, I., Cascone, P., Schubert, A., Gambino, G., Balestrini, R., Guerrieri, E., 2016. Insight on the impact of arbuscular mycorrhizal symbiosis on tomato tolerance to water stress. Plant Physiol. 171, 1009–1023.

Chiu, K.K., Ye, Z.H., Wong, M.H., 2006. Growth of Vetiveria zizanioides and Phragmites australis on Pb/Zn and Cu mine tailings amended with manure compost and sewage sludge: a greenhouse study. Bioresour. Technol. 97 (1), 158–170.

Chizuru, N., Ricardo, U., Shiriki, K., Prakash, S., 2004. The joint WHO/FAO expert consultation on diet, nutrition and the prevention of chronic diseases: process, product, and policy implications. Public Health Nutr. 7 (1a), 245–250.

Colla, G., Rouphael, Y., Cardarelli, M., Tullio, M., Rivera, C.M., Rea, E., 2008. Alleviation of salt stress by arbuscular mycorrhizal in zucchini plants grown at low and high phosphorous concentrations. Biol. Fertil. Soils 44, 501–509.

Corradi, N., Buffner, B., Croll, D., Colard, A., Horak, A., Sanders, I.R., 2009. High-level molecular diversity of copper-zinc superoxide dismutase genes among ad within species of arbuscular mycorrhizal fungi. Appl. Environ. Microbiol. 75, 1970–1978.

Daniel, R., 2005. The metagenomics of soil. Nat. Rev. Microbiol. 3 (6), 470.

Das, A.J., Kumar, R., 2016. Bioremediation of petroleum contaminated soil to combat toxicity on Withania somnifera through seed priming with biosurfactant producing plant growth promoting rhizobacteria. J. Environ. Manag. 174, 79–86.

Das, M., Maiti, S.K., 2009. Growth of Cymbopogon citratus and Vetiveria zizanioides on Cu mine tailings amended with chicken manure and manure-soil mixture: a pot scale study. Int. J. Phytoremediation 11 (8), 651–663.

Demirbas, A., 2008. A new liquid biofuels from vegetables oils via catalytic pyrolysis. Energy Educ. Sci. Technol. 21, 1–59.

Dimpka, C., Svatoš, A., Merten, D., Buchel, G., Kothe, E., 2008. Hydroxamate siderophores produced by *Streptomyces acidiscabies* E13, bind nickel and promote growth in cowpea (*Vigna unguiculata* L.) under nickel stress. Can. J. Microbiol. 54, 163–172.

Dimpka, C.O., Merten, D., Svatos, A., Buchel, G., Kothe, E., 2009. Siderophores mediate reduced and increased uptake of cadmium by *Streptomyces tendae* F4 and sunflower (*Helianthus annuus*), respectively. J. Appl. Microbiol. 107 (5), 1687–1696.

Djurdjević, L., Mitrović, M., Pavlović, P., Gajić, G., Kostić, O., 2006. Phenolic acids as bioindicators of fly ash deposit revegetation. Arch. Environ. Contam. Toxicol. 50, 488–495.

Duke, J.A., Bogenschuz-Godwin, M.J., duCellier, J., Duke, P.K., 2002. Handbook of Medicinal Plants, second ed. CRC Press, Boca Raton, FL.

Dwibedi, S.K., Mohanty, M.K., Pandey, V.C., Divyasree, D., 2021. Sustainable biowaste management in cereal systems: a review (Edited Book). In: Goyal, A. (Ed.), Cereal Grains. IntechOpen, https://doi.org/10.5772/intechopen.97308.

EA (Environment Agency), 2012. In: Guidance on the Legal Definition of Waste and Its Application. Waste Regulation Policy Branch, Waste & Resource Efficiency Division, Department for Environment and Sustainable Development, Welsh Government, Cardiff, p. 9.

Elhindi, K.M., El-Din, A.S., Elgorban, A.M., 2017. The impact of arbuscular mycorrhizal fungi in mitigating salt-induced adverse effects in sweet basil (*Ocimum basilicum* L.). Saudi J. Biol. Sci. 24, 170–179.

Erisman, J.W., Sutton, M.A., Galloway, J.N., Klimont, Z., Winiwarter, W., 2008. How a century of ammonia synthesis changed the world. Nat. Geosci. 1, 636–639.

Evelin, H., Kapoor, R., Giri, B., 2009. Arbuscular mycorrhizal fungi alleviation of salt stress: a review. Ann. Bot. 104, 1263–1280.

Evelin, H., Devi, T.S., Gupta, S., Kapoor, R., 2019. Mitigation of salinity stress in plants by arbuscular mycorrhizal symbiosis: current understanding and new challenges. Front. Plant Sci. 10, 470.

Fan, X., Liu, X., Huang, R., Liu, Y., 2012. Identification and characterization of a novel thermostable pyrethroid-hydrolyzing enzyme isolated through metagenomic approach. Microb. Cell Fact. 11, 33–37.

FAO (Food and Agricultural Organization of the United Nations), 2000. Manual on Integrated Soil Management and Conservation Practices. FAO Land and Water Bulletin 8.

Farwell, A.J., Vesely, S., Nero, V., Rodriguez, H., Shah, S., Dixon, D.G., Glick, B.R., 2006. The use of 877 transgenic canola (*Brassica napus*) and plant growth-promoting bacteria to enhance plant 878 biomass at a nickel-contaminated field site. Plant Soil 288, 309–318.

Farzadfar, S., Zarinkamar, F., Modarres-Sanavy, S.A.M., Hojati, M., 2013. Exogenously applied calcium alleviates cadmium toxicity in *Matricaria chamomilla* L. plants. Environ. Sci. Pollut. Res. 20 (3), 1413–1422.

Feng, G., Zhang, F.S., Li, X.T., Tian, C.Y., Rengel, Z., 2002. Improved tolerance of maize plants to salt stress by arbuscular mycorrhiza is related to higher accumulation of soluble sugars in roots. Mycorrhiza 12, 185–190.

Freire, B.M., Pereira, R.M., Lange, C.N., Batista, B.L., 2020. Biofortification of crop plants: a practical solution to tackle elemental deficiency. In: Mushra, K., Tandon, P.K., Srivastava, S. (Eds.), Sustainable Solutions for Elemental Deficiency and Excess in Crop Plants. Springer Nature, Singapore Pte Ltd, pp. 135–182.

Gajić, G., Pavlović, P., 2018. The role of vascular plants in the phytoremediation of fly ash deposits. In: Matichenkov, V. (Ed.), Phytoremediation: Methods, Management and Assessment. Nova Science Publishers Inc., New York, pp. 151–236.

Gajić, G., Pavlović, P., Kostić, O., Jarić, S., Djurdjević, L., Pavlović, D., Mitrović, M., 2013. Ecophysiological and biochemical traits of three herbaceous plants growing of the disposed coal combustion fly ash of different weathering stage. Arch. Biol. Sci. 65 (1), 1651–1667.

Gajić, G., Djurdjević, L., Kostić, O., Jarić, S., Mitrović, M., Stevanović, B., Pavlović, P., 2016. Assessment of the phytoremediation potential and an adaptive response of *Festuca rubra* L. sown on fly ash deposits: native grass has a pivotal role in ecorestoration management. Ecol. Eng. 93, 250–261.

Gajić, G., Đurđević, L., Kostić, O., Jarić, S., Mitrović, M., Pavlović, P., 2018. Ecological potential of plants for phytoremediation and ecorestoration of fly ash deposits and mine wastes. Front. Environ. Sci. 6, 124.

Gajić, G., Mitrović, M., Pavlović, P., 2019. Ecorestoration of fly ash deposits by native plant species at thermal power stations in Serbia. In: Pandey, V.C., Bauddh, K. (Eds.), Phytomanagement of Polluted Sites: Market Opportunities in Sustainable Phytoremediation. Elsevier, Amsterdam, pp. 113–177.

Gajić, G., Mitrović, M., Pavlović, P., 2020a. Feasibility of *Festuca rubra* L. native grass in phytoremediation. In: Pandey, V.C., Singh, D.P. (Eds.), Phytoremediation Potential of Perennial Grasses. Elsevier, Amsterdam, pp. 115–164.

Gajić, G., Djurdjević, L., Kostić, O., Jarić, S., Mitrović, M., Stevanović, B., Mitrović, M., Pavlović, P., 2020b. Phytoremediation potential, photosynthetic and antioxidant response to arsenic-induced stress of *Dactylis glomerata* L. sown on the fly ash deposits. Plants 9 (5), 657.

Garg, M., Sharma, N., Sharma, S., Kapoor, P., Kumar, A., Chundri, V., Arora, P., 2018. Biofortified crops generated by breeding, agronomy, and transgenic approaches are improving lives of millions of people around the world. Front. Nutr. 5, 12.

Ghosh, P.K., Maiti, T.K., Pramanik, K., Ghosh, S.K., Mitra, S., De, T.K., 2018. The role of arsenic resistant *Bacillus aryabhattai* MCC3374 in promotion of rice seedlings growth and alleviation of arsenic phytotoxicity. Chemosphere 211, 407–419.

Gill, K., Sandhu, S., Mor, M., Kalmodiya, T., Singh, M., 2020. Role of green manuring in sustainable agriculture: a review. Eur. J. Mol. Clin. Med. 7 (7), 2361–2366.

Giri, B., Kapoor, R., Mukerji, K.G., 2003. Influence of arbuscular mycorrhizal fungi and salinity on growth, biomass, and mineral nutrition of *Acacia auriculiformis*. Biol. Fertil. Soils 38, 170–175.

Glick, B.R., 2012. Plant growth-promoting bacteria: mechanisms and applications. Scientifica 2012, 963401.

Gomez-Sagasti, M.T., Hernandez, A., Artetxe, U., Garbisu, C., Becereli, J.M., 2018. How valuable are organic amendments as tools for the phytomanagement of degraded soils? The knows, known, unknown, and unknowns. Front. Sustain. Food Syst. 2, 68.

Gontia-Mishra, I., Sapre, S., Kachare, S., Tiwari, S., 2017. Molecular diversity of 1-aminocyclopropane-1-carboxylate (ACC) deaminase producing PGPR from wheat (*Triticum aestivum* L.) rhizosphere. Plant Soil 414 (1–2), 213–227.

Gonzáles-Chávez, M.C.A., Vangronsveld, J., Colpaert, J., Leyval, C., 2006. Arbuscular mycorrhizal fungi and heavy metals: tolerance mechanisms and potential use in bioremediation. In: Prasad, M.N.V., Sajwan, K.S., Naidu, R. (Eds.), Trace Elements in the Environment (Biogeochemistry, Biotechnology, and Bioremediation). CRC Taylor and Francis, Boca Raton, FL, pp. 211–234.

Grbović, F., Gajić, G., Branković, S., Simić, Z., Ćirić, A., Rakonjac, L., Pavlović, P., Topuzović, M., 2019. Allelopathic potential of selected woody species growing on fly-ash deposits. Arch. Biol. Sci. 71 (1), 83–94.

Grbović, F., Gajić, G., Branković, S., Simić, Z., Vuković, N., Pavlović, P., Topuzović, M., 2020. Complex effect of *Robinia pseudoacacia* L. and *Ailanthus altissima* (Mill.) Swingle growing on asbestos deposits: allelopathy and biogeochemistry. J. Serb. Chem. Soc. 85 (1), 141–153.

Gullap, M.K., Dasci, M., Erkovan, H.I., Koc, A., Turan, M., 2014. Plant growth-promoting rhizobacteria (PGPR) and phosphorous fertilizer-assisted phytoextraction of toxic heavy metals from contaminated soils. Commun. Soil Sci. Plant Anal. 45, 2593.

Gunwal, I., Singh, L., Mago, P., 2014. Comparison of phytoremediation of cadmium and nickel from contaminated soil by *Vetiveria zizanioides* L. Int. J. Sci. Res. Publ. 4 (10), 1–7.

Guo, J.K., Muhammad, H., Lv, X., Wei, T., Ren, X.H., Jia, H.L., Atif, S., Hua, L., 2020. Prospects and applications of plant growth promoting rhizobacteria to mitigate soil metal contamination. Chemosphere 246, 125823.

Gupta, A., Rai, V., Bagdwal, N., Goel, R., 2005. In situ characterization of mercury resistant growth promoting fluorescent pseudomonads. Microbiol. Res. 160, 385–388.

Gupta, A.K., Verma, S.K., Khan, K., Verma, R.K., 2013. Phytoremediation using aromatic plants: a sustainable approach for remediation of heavy metals polluted sites. Environ. Sci. Technol. 47, 10115–10116.

Hashem, A., Abd_Allah, E.F., Alqarawi, A.A., Aldubise, A., Egaberdieva, D., 2015. Arbuscular mycorrhizal fungi enhances salinity tolerance of *Panicum turgidum* Forssk by altering photosynthetic and antioxidant pathways. J. Plant Interact. 10 (1), 230–242.

Haslem, A., Kumar, A., Al-Dbass, A.M., Alqarawi, A.A., Al-Arjani, A.-B.F., Singh, G., Farooq, M., Abd_Allah, E.F., 2019. Arbuscular mycorrhizal fungi and biochar drought tolerance in chickpea. Saudi J. Biol. Sci. 26, 614–624.

Hassan, E., 2016. Comparative study on the biosorption of Pb (II), Cd (II), and Zn (II) using Lemon grass (*Cymbopogon citratus*): kinetics, isotherms, and thermodynamics. Chem. Int. 2 (2), 89–102.

Hassen, W., Neifar, M., Cherif, H., Najjari, A., Chouchane, H., Driouchi, R.C., Salah, A., Naili, F., Mosbah, A., Soussi, Y., Raddadi, N., 2018. *Pseudomonas rhizophila* S211, a new plant growth-promoting rhizobacterium with potential in pesticide-bioremediation. Front. Microbiol. 9, 34.

He, X., Xu, M., Wei, Q., Tang, M., Guan, L., Lou, L., Xu, X., Hu, Z., Chen, Y., Shen, Z., Xia, Y., 2020. Promotion of growth and phytoextraction of cadmium and lead in *Solanum nigrum* L., mediated by plant-growth-promoting rhizobacteria. Ecotoxicol. Environ. Saf. 205, 111333.

Hill, A.E., Scachar-Hill, B., Schachar-Hill, Y., 2004. What are aquaporins for? J. Membr. Biol. 197, 1–32.

Hosman, M.E., El-Frky, S.S., Elshahawy, M.I., Shaker, E.M., 2017. Mechanism of phytoremediation potential of flax (*Linum usitatissimum* L.) to Pb, Cd, and Zn. Asian J. Plant Sci. Res. 7 (4), 30–40.

Hou, J., Liu, W., Wang, B., Wang, Q., Luo, Y., Franks, A.E., 2015. PGPR enhanced phytoremediation of petroleum contaminated soil and rhizosphere microbial community response. Chemosphere 138, 592–598.

Huang, Y.M., Srivastava, A.K., Zou, Y.N., et al., 2014. Mycorrhizal-induced calmodulin mediated changes in antioxidant enzymes and growth response of drought-stressed trifoliate orange. Front. Microbiol. 5, 682.

Huang, D., Ma, M., Wang, Q., Zhang, M., Jing, G., Li, C., Ma, F., 2020. Arbuscular mycorrhizal fungi enhanced drought resistance in apple by regulating genes in the MAPK pathway. Plant Physiol. Biochem. 149, 245–255.

IPCC, 2000. Special Report on Land Use; Land-Use Change and Forestry. Cambridge University Press, Cambridge, p. 377.

Islam, F., Yasmeen, T., Ali, Q., Ali, S., Saleem, M., Hussain, S., Rizvi, H., 2014. Ecotoxicology and environmental safety influence of *Pseudomonas aeruginosa* as PGPR on oxidative stress tolerance in wheat under Zn stress. Ecotoxicol. Environ. Saf. 104, 285–293.

Jabeen, H., Iqbal, S., Ahmad, F., Afzal, M., Firdous, S., 2016. Enhanced remediation of chlorpyrifos by ryegrass (*Lolium multiflorum*) and a chlorpyrifos degrading bacterial endophyte *Mesorhizobium* sp. HN3. Int. J. Phytoremediation 18 (2), 126–133.

Jahromi, F., Aroca, R., Porcel, P., Ruiz-Lozano, J.M., 2008. Influence of salinity on the in vitro development of *Glomus intraradices* and on the in vitro physiological and molecular responses of mycorrhizal lettuce. Mol. Ecol. 55, 45–53.

JeyaSundar, P.G.S.A., Ali, A., Guo, D., Zhang, Z., 2020. Waste treatment approaches for environmental sustainability. In: Chowdhary, P., Raj, A., Verma, D., Akhter, J. (Eds.), Mechanisms for Sustainable Environment and Health. Elsevier, Inc, pp. 119–135.

Jha, A.K., Kumar, U., 2017. Studies on removal of heavy metals by *Cymbopogon flexuosus*. Int. J. Agric. Environ. Biotechnol. 10 (1), 89.

Jing, K., Zongping, P., Min, D., Gan, S., Xin, Z., 2014. Effects of arbuscular mycorrhizal fungi on the drought resistance of the mining area repair plant Sainfoin. Int. J. Min. Sci. Technol. 24, 485–489.

Kacprzak, M., Grobelak, A., Grosser, A., Prasad, M.N.V., 2013. Efficiency of biosolids in assisted phytostabilization of metalliferous acidic sandy soils with five grass species. Int. J. Phytoremediation 16, 593–608.

Kacprzak, M., Neczaj, E., Fijalkowski, K., Grobelak, A., Grosser, A., Worwag, M., Rorat, A., Brattebo, H., Almas, A., Singh, B.R., 2017. Sewage sludge disposal strategies for sustainable development. Environ. Res. 156, 39–46.

Kalia, A.N., 2005. Text Book of Industrial Pharmacognosy, first ed. CBS Publishers and Distribution, New Delhi.

Kapoor, R., Sharma, D., Bhatnagar, A.K., 2008. Arbuscular mycorrhizae in micropropagation system and their potential applications. Sci. Hortic. 116, 227–239.

Khan, M.S., Zaidi, A., Wani, P.A., Oves, M., 2009. Role of plant growth promoting rhizobacteria in the remediation of metal contaminated soils. Environ. Chem. Lett. 7, 1–19.

Klinger, A., Bothe, H., Wray, V., Marner, F.J., 1995. Identification of a yellow pigment formed in maize roots upon mycorrhizal colonization. Phytochemistry 38, 53–55.

Kocar, G., Civas, N., 2013. An overview of biofuels from energy crops: current status and future prospects. Renew. Sustain. Energy Rev. 28, 900–916.

Kotoky, R., Das, S., Singha, L.P., Pandey, P., Singha, K.M., 2017. Biodegradation of benzo(a) pyrene by biofilm forming and plant growth promoting *Acinetobacter* sp. strain PDB4. Environ. Technol. Innov. 8, 256–268.

Kovacik, J., 2013. Hyperaccumulation of cadmium in *Matricaria chamomilla*: a never-ending story. Acta Physiol. Plant. 35 (5), 1721–1725.

Kovacik, J., Klejdus, B., 2008. Dynamics of phenolic acids and lignin accumulation in metal-treated *Matricaria chamomilla* roots. Plant Cell Rep. 27 (3), 605–615.

Kowalska, A., Grobelak, A., Almas, A.R., Singh, B.R., 2020. Effect of biowastes on soil remediation, plant productivity and soil organic carbon sequestration: a review. Energies 13, 5813.

Kumari, M., Thakur, I.S., 2018. Biochemical and proteomic characterization of *Paenibacillus* sp., ISTP10 for its role in plant growth promotion and in rhizostabilization of cadmium. Bioresour. Technol. Rep. 3, 59–66.

Lal, R., 2006. Enhancing crop yields in developing countries through restoration of soil organic carbon pool in agricultural lands. Land Degrad. Dev. 17, 197–206.

Lal, R., 2008. Soil and sustainable agriculture. A review. Agron. Sustain. Dev. 28, 57–64.

Lal, K., Minhas, P.S., Shipra, C.R.K., Yadav, R.K., 2008. Extraction of cadmium and tolerance of three annual cut flowers on Cd-contaminated soils. Bioresour. Technol. 99, 1006–1011.

Larkin, R.P., 2015. Soil health and paradigms and implications for disease management. Annu. Rev. Phytopathol. 53, 199–221.

Lehtomaki, A., Huttunen, S., Rintala, J.A., 2007. Laboratory investigations on co-digestion of energy crops and crop residues with cow manure for methane production: effect of crop to manure ratio. Resour. Conserv. Recycl. 51 (3), 591–609.

Li, T., Hu, Y.J., Hao, Z.P., Li, H., Wang, Y.S., Chen, B.D., 2013a. First cloning and characterization of two functional aquaporins genes from an arbuscular mycorrhizal fungus *Glomus intraradices*. New Phytol. 197, 617–630.

Li, Y., Zou, Y.N., Wu, Q.S., 2013b. Effects of inoculation with *Diversispora spurca* on growth, root system architecture and chlorophyll contents of four citrus genotypes plants. Int. J. Agric. Biol. 15, 342–346.

Li, J., Meng, B., Chai, H., Yang, X., Song, W., Li, S., Lu, A., Zhang, T., Sun, W., 2019. Arbuscular mycorrhizal fungi alleviate drought stress in C3 (*Leymus chinensis*) and C4 grasses (*Hemarthria altissima*) via altering antioxidant enzyme activities and photosynthesis. Front. Plant Sci. 10, 499.

Li, Z., Wu, N., Meng, S., Wu, F., Liu, T., 2020. Arbuscular mycorrhizal fungi (AMF) enhance the tolerance of *Euonymus maackii* Rupr. at a moderate level of stress. PLoS One 15 (4), e0231497.

Lin, Z.Q., Haddad, S., Hong, J., Morissy, J., Banuelos, G.S., Zhang, L.Y., 2014. Use of selenium contaminated plants from phytoremediation for production of selenium-enriched edible mushrooms. In: Banuelos, G.S., Lin, Z.-Q., Yin, X.B. (Eds.), Selenium in Environment and Human Health. CRC Press, Boca Raton, FL, pp. 124–126.

Linger, P., Ostwald, A., Haensler, J., 2005. *Cannabis sativa* L., growing on heavy metal contaminated soil: growth, cadmium uptake and photosynthesis. Biol. Plant. 49 (4), 567–576.

Lubbe, A., Verpoorte, R., 2011. Cultivation of medicinal and aromatic plants for specialty industrial materials. Ind. Crops Prod. 34 (1), 785–801.

Ludvikova, M., Griga, M., 2018. Transgenic fiber crops for phytoremediation of metals and metalloids. In: Prasad, M.N.V. (Ed.), Transgenic Plant Technology for Remediation of Toxic Metals and Metalloids. Elsevier Inc, pp. 341–358.

Ma, Y., Prasad, M.N.V., Rajkumar, M., Freitas, H., 2011. Plant growth promoting rhizobacteria and endophytes accelerate phytoremediation of metalliferous soils. Biotechnol. Adv. 29, 248–258.

Maiti, S.K., Ahirwal, J., 2019. Ecological restoration of coal mine degraded lands: topsoil management, pedogenesis, carbon sequestration, and mine pit limnology. In: Pandey, V.C., Bauddh, K. (Eds.), Phytomanagement of Polluted Sites: Market Opportunities in Sustainable Phytoremediation. Elsevier, Amsterdam, pp. 83–111.

Maitra, S., Zaman, A., Mandal, T.K., Palai, J.-B., 2018. Green manures in agriculture: a review. J. Pharmacogn. Phytochem. 7 (5), 1319–1327.

Manshadi, M., Ziarati, P., Ahmadi, M., Fekri, K., 2013. Greenhouse study of cadmium and lead phytoextraction by five *Pelargonium* spices. Int. J. Farm. Allied Sci. 2 (19), 665–669.

Mardukhi, B., Rejali, F., Daei, G., Ardakani, M.R., Malakouti, M.J., Miransari, M., 2011. Arbuscular mycorrhizas enhance nutrient uptake in different wheat genotypes at high salinity levels under field and greenhouse conditions. C. R. Biol. 334, 564–571.

Marques, G., Rencoret, J., Gutierrez, A., Rio, J.C., 2010. Evaluation of the chemical composition of different non-woody plant fibers used for pulp and paper manufacturing. Open Agric. J. 4, 93–101.

Martinez, J.P., Lutts, S., Schanck, A., Bajji, M., Kinet, J.M., 2004. Is osmotic adjustment required to water stress resistance in the Mediterranean shrub *Atriplex halimus* L. Plant Physiol. 161, 1041–1051.

Mayer, J.E., Pfeiffer, W.H., Beyer, P., 2008. Biofortified crops to alleviate micronutrient malnutrition. Curr. Opin. Plant Biol. 11, 166–170.

McCarthy, J.J., Dickson, N.M., 2000. From Friibergh to Amsterdam: on the road to sustainable science. IGP Newslett. 44, 6–8.

McCutcheon, S.C., Schnoor, J.L., 2003. Phytoremediation. Transformation and Control of Contaminants. John Wiley and Sons, Inc., Publications, Hoboken, NJ, p. 987.

McGuire, S., 2015. The state of food insecurity in the world 2015: meeting the 2015 international hunger targets: taking stock of uneven progress. Rome, FAO, IFAD, and WFP. Adv. Nutr. 6 (5), 623–624.

McKendry, P., 2002. Energy production from biomass (part 1): conversion technologies. Bioresour. Technol. 83 (1), 47–54.

Meddich, A., Jaity, F., Bourzik, W., El Asli, A., Hafidi, M., 2015. Use of mycorrhizal fungi as a strategy for improving the drought tolerance in date palm (*Phoenix dactylifera*). Sci. Hortic. 192, 468–474.

Millati, R., Cahyono, R.B., Ariyanto, T., Azzahrani, I.N., Putri, R.U., Taherzadeh, M.J., 2019. Agricultural, industrial, municipal, and forest wastes: an overview. In: Taherzadeh, M.J., Bolton, K., Wong, J., Pandey, A. (Eds.), Sustainable Resource Recovery and Zero Waste Approach. Elsevier Inc., St. Louse, MO, pp. 1–22.

Mitrović, M., Pavlović, P., Lakušić, D., Stevanović, B., Djurdjevic, L., Kostić, O., Gajić, G., 2008. The potential of *Festuca rubra* and *Calamagrostis epigejos* for the revegetation on fly ash deposits. Sci. Total Environ. 72, 1090–1101.

Munns, R., 2005. Genes and salt tolerance: bringing them together. New Phytol. 167, 645–663.

Najmanova, J., Neumannova, E., Leonhardt, T., Zitka, O., Kizek, R., Macek, T., Kotrba, P., 2012. Cadmium-induced production of phytochelatins and speciation of intracellular cadmium in organs of *Linum usitatissimum seedlings*. Crops Prod. 36, 536–542.

Nichols, K.A., 2008. Indirect contribution of AM fungi and soil aggregation to plant growth and protection. In: Siddiqui, Z.A., Akhtar, M.S., Futai, K. (Eds.), Mycorrhizae: Sustainable Agriculture and Forestry. Springer Sciences, Berlin, pp. 177–194.

Nielsen, J., 2001. Metabolic engineering. Appl. Microbiol. Biotechnol. 55 (3), 263–283.

Nisanth, D., Biswas, D., 2008. Kinetics of phosphorous and potassium release from rock phosphate and waste mica enriched compost and their effect on yield and nutrient uptake by wheat (*Triticum aestivum*). Bioresour. Technol. 99, 332–3353.

Nissim, W.G., Cincinelli, A., Martinellini, T., Alvisi, L., Palm, E., Mancuso, S., Azzarelo, E., 2018. Phytoremediation of sewage sludge contaminated by trace elements and organic compounds. Environ. Res. 164, 356–366.

Nowack, B., Schulin, R., Robinson, B.H., 2006. Critical assessment of chelant-enhanced metal phytoextraction. Environ. Sci. Technol. 40, 5225–5232.

Ouziad, F., Wilde, P., Schmelzer, E., Hildebrandt, U., Bothe, H., 2006. Analysis of expression of aquaporins and Na^+/H^+ transporters in tomato colonized by arbuscular mycorrhizal fungi and affected by salt stress. Environ. Exp. Bot. 57, 177–186.

Pagano, M.C., 2014. Drought stress and mycorrhizal plant. In: Miransari, M. (Ed.), Use of Microbes for the Alleviation of Soil Stresses. Springer, New York, pp. 97–110.

Pandey, V.C., 2012. Invasive species based efficient green technology for phytoremediation of fly ash deposits. J. Geochem. Explor. 123, 13–18.

Pandey, V.C., 2013. Book review on phytoremediation and biofortification: two sides of one coin. Environ. Eng. Manag. J. 12 (4), 851–852.

Pandey, V.C., 2015. Assisted phytoremediation of fly ash dumps through naturally colonized plants. Ecol. Eng. 82, 1–5.

Pandey, V.C., 2017. Managing waste dumpsites through energy plantations. In: Bauddh, K., Singh, B., Korstad, J. (Eds.), Phytoremediation Potential of Bioenergy Plants. Springer Nature Singapore Pte. Ltd, pp. 371–386.

Pandey, V.C., 2020a. Phytomanagement of Fly Ash. Elsevier, Amsterdam, p. 334, https://doi.org/10.1016/C2018-0-01318-3.

Pandey, V.C., 2020b. Fly ash properties, multiple uses, threats, and management: an introduction. In: Phytomanagement of Fly Ash. Elsevier, Amsterdam, pp. 1–34, https://doi.org/10.1016/B978-0-12-818544-5.00001-8.

Pandey, V.V., Bajpai, O., 2019. Phytoremediation: From theory toward practice. In: Pandey, V.C., Bauddh, K. (Eds.), Phytomanagement of Polluted Sites: Market Opportunities in Sustainable Phytoremediation. Elsevier, Amsterdam, pp. 1–49.

Pandey, V.C., Bauddh, K., 2019. Phytomanagement of Polluted Sites: Market Opportunities in Sustainable Phytoremediation. Elsevier, Amsterdam, p. 602.

Pandey, V.C., Praveen, A., 2020. *Vetiveria zizanioides* (L.) Nash—more than a promising crop in phytoremediation (Authored book with contributors). In: Pandey, V.C., Singh, D.P. (Eds.), Phytoremediation Potential of Perennial Grasses. Elsevier, Amsterdam, pp. 31–62, https://doi.org/10.1016/B978-0-12-817732-7.00002-X.

Pandey, V.C., Singh, N., 2015. Aromatic plants versus arsenic hazard in soils. J. Geochem. Explor. 157, 77–80.

Pandey, V.C., Singh, V., 2019. Exploring the potential and opportunities of current tools for removal of hazardous materials from environments. In: Pandey, V.C., Bauddh, K. (Eds.), Phytomanagement of Polluted Sites: Market Opportunities in Sustainable Phytoremediation. Elsevier, Amsterdam, pp. 501–516.

Pandey, V.C., Souza-Alonso, P., 2019. Market opportunities in sustainable phytoremediation. In: Pandey, V.C., Bauddh, K. (Eds.), Phytomanagement of Polluted Sites: Market Opportunities in Sustainable Phytoremediation. Elsevier, Amsterdam, pp. 51–82.

Pandey, V.C., Singh, K., Singh, J.S., Kumar, A., Singh, B., Singh, R.P., 2012. *Jatropha curcas*: a potential biofuel plant for sustainable environmental development. Renew. Sustain. Energy Rev. 16, 2870–2883.

Pandey, V.C., Pandey, D.N., Singh, N., 2015a. Sustainable phytoremediation based on naturally colonizing and economically valuable plants. J. Clean. Prod. 86, 37–39.

Pandey, V.C., Prakash, P., Bajpai, O., Kumar, A., Sing, N., 2015b. Phytodiversity on fly ash deposits: evaluation of naturally colonized species for sustainable phytorestoration. Environ. Sci. Pollut. Res. 22 (4), 2776–2787.

Pandey, V.C., Bajpai, O., Singh, N., 2016a. Energy crops in sustainable phytoremediation. Renew. Sustain. Energy Rev. 54, 58–73.

Pandey, V.C., Bajpai, O., Sinhg, N., 2016b. Plant regeneration potential in fly ash ecosystem. Urban For. Urban Green. 15, 40–44.

Pandey, V.C., Rai, A., Korstad, J., 2019a. Aromatic crops in phytoremediation: from contaminated to waste dumpsites. In: Pandey, V.C., Bauddh, K. (Eds.), Phytomanagement of Polluted Sites: Market Opportunities in Sustainable Phytoremediation. Elsevier, Amsterdam, pp. 255–275.

Pandey, J., Verma, R.K., Singh, S., 2019b. Suitability of aromatic plants for phytoremediation of heavy metal contaminated sites: a review. Int. J. Phytoremediation 21 (5), 405–418.

Pandey, V.C., Sahu, N., Singh, D.P., 2020. Physiological profiling of invasive plant species for ecological restoration of fly ash deposits. Urban For. Urban Green. 54, 126773.

Pandey, V.C., et al., 2022. Fibre Crop-Based Phytoremediation. Elsevier, Amsterdam, ISBN: 9780128239933 (Authored Book).

Pant, R., Pandey, P., Kotocky, R., 2016. Rhizosphere mediated biodegradation of 1,4-dichlorbenzene by plant growth promoting rhizobacteria of *Jatropha curcas*. Ecol. Eng. 94, 50–56.

Pavlović, P., Mitrović, M., Djurdjevic, L., 2004. An ecophysiological study of plants growing on the fly ash deposits from the "Nikola Tesla-A" thermal power station in Serbia. Environ. Manag. 33, 654–663.

Pavlović, P., Mitrović, M., Đorđević, D., Sakan, S., Slobodnik, J., Liška, I., Csanyi, B., Jarić, S., Kostiž, O., Pavlović, D., Marinković, N., Tubić, B., Paunović, M., 2016. Assessment of the contamination of riparian soil and vegetation by trace metals—a Danube River case study. Sci. Total Environ. 540, 396–409.

Pilon-Smits, E., 2005. Phytoremediation. Annu. Rev. Plant Biol. 56, 15–39.

Porcel, R., Ruiz-Lozano, J.M., 2004. Arbuscular mycorrhiza influence on leaf water potential, solute accumulation and oxidative stress in soybean plants subjected to drought stress. J. Exp. Bot. 55, 1743–1750.

Porcel, R., Azcon, R., Ruiz-Lozano, J.M., 2004. Evaluation of the role of genes encoding Δ^1-pyrroline-5-carboxylate synthetase (P5CS) during drought stress in arbuscular mycorrhizal Glycine max and Lactuca sativa plants. Physiol. Mol. Plant Pathol. 65, 211–221.

Porcel, A., Aroca, R., Cano, C., Bago, A., Ruiz-Lozano, J.M., 2007. A gene from the arbuscular mycorrhizal fungus Glomus intraradices encoding a binding protein is up-regulated by drought stress in some mycorrhizal plants. Environ. Exp. Bot. 60, 251–256.

Pramanik, K., Mitra, S., Sarkar, A., Soren, T., 2017. Characterization of cadmium-resistant Klebsiella pneumonia MCC 3091 promoted rice seedling growth by alleviating phytotoxicity of cadmium. Environ. Sci. Pollut. Res. Int. 24 (31), 24419–24437.

Pramanik, K., Mitra, S., Sarkar, A., Maiti, T.K., 2018. Alleviation of phytotoxic effects of cadmium on rice seedlings 544 by cadmium resistant PGPR strain Enterobacter aerogenes MCC 3092. J. Hazard. Mater. 351, 317–329.

Prasad, A., Singh, A.K., Chand, S., Chanotiya, C.S., Patra, D.D., 2010. Effect of chromium and lead on yield, chemical composition of essential oil, and accumulation of heavy metals of mine species. Commun. Soil Sci. Plant Anal. 41 (18), 2170–2186.

Prasanth, L., Kattapagari, K.K., Chitturi, R.T., Baddam, V.R., Prasad, L.K., 2015. A review on role of essential trace elements in health and disease. J. NTR Univ. Health Sci. 4, 75–85.

Pratibha, Y., Krishna, S.S., 2015. Plant growth promoting Rhizobacteria: an effective tool to remediate residual organophosphate pesticide methyl parathion widely used in Indian agriculture. J. Environ. Res. Dev. 9 (4), 1138.

Praveen, A., Pandey, V.C., Marwa, N., Singh, D.P., 2019. Rhizomediation of polluted sites: harnessing plant-microbe interactions. In: Pandey, V.C., Bauddh, K. (Eds.), Phytomanagement of Polluted Sites: Market Opportunities in Sustainable Phytoremediation. Elsevier, Amsterdam, pp. 389–407.

Quiroga, G., Erice, G., Aroca, R., Chaumot, F., Ruiz-Lozano, J.M., 2017. Enhanced drought stress tolerance by the arbuscular mycorrhizal symbiosis in a drought—sensitive maize cultivar is related to a broader and differential regulation of host plant aquaporins than in a drought-tolerant cultivars. Front. Plant Sci. 8, 1056.

Rajkumar, M., Nagendran, R., Kui Jae, L., Wang, H.L., Sung Zoo, K., 2006. Influence of plant growth promoting bacteria and Cr(VI) on the growth of Indian mustard. Chemosphere 62, 741–748.

Rana, S., Pichandi, S., Parveen, S., Fangueiro, R., 2014. Natural plant fibers: production, processing, properties and their sustainability parameters. In: Muthu, S.S. (Ed.), Roadmap to Sustainable Textiles and Clothing. Springer Science + Buisness Media, Singapore, pp. 1–35.

Rayu, S., Karpouzas, D.G., Singh, B.K., 2012. Emerging technologies in bioremediation: constraints and opportunities. Biodegradation 23 (6), 917–926.

Reed, M.I.E., Glick, B.R., 2005. Growth of canola (Brassica napus) in the presence of plant growth-promoting bacteria and either copper or polycyclic aromatic hydrocarbons. Can. J. Microbiol. 51, 1061–1069.

Retamal-Morales, G., Menhert, M., Schwabe, R., Tischler, D., Zapata, C., Chavez, R., Schlomann, M., Levican, G., 2018. Detection of arsenic-binding siderophores in arsenic—tolerating Actinobacteria by a modified CAS assay. Ecotoxicol. Environ. Saf. 157, 176–181.

Rilling, M.C., 2004. Arbuscular mycorrhizae, gloaming, and soil aggregation. Can. J. Soil Sci. 84, 355–363.

Rodrigo, S., Santamaria, O., Chen, Y., 2014. Selenium speciation in malt, wort, and beer made from selenium-biofortified two-rowed barley grain. J. Agric. Food Chem. 62, 5948–5953.

Roy, M., Giri, A.K., Dutta, S., Mukherjee, P., 2015. Integrated phytobial remediation for sustainable management of arsenic in soil and water. Environ. Int. 75, 180–198.

Ruiz-Lozano, J.M., Collados, C., Barea, J.M., Azcon, R., 2001a. Arbuscular mycorrhizal symbiosis can alleviate drought induced nodule senescence in soybean plants. Plant Physiol. 82, 346–350.

Ruiz-Lozano, J.M., Collados, C., Barea, J.M., Azcon, R., 2001b. Cloning of cDNAs encoding SODs from lettuce plants which show differential regulation by arbuscular mycorrhizal symbiosis and by drought stress. J. Exp. Bot. 52, 2241–2242.

Ruiz-Lozano, J.M., del Mar, A.M., Barzana, G., Vernieri, P., Aroca, R., 2009. Exogenous ABA accentuates the differences in root hydraulic properties between mycorrhizal and non mycorrhizal maize plants through regulation of PIP aquaporins. Plant Mol. Biol. 70, 565–579.

Ruiz-Lozano, J.M., Porcel, R., Azcon, R., 2012. Regulation by arbuscular mycorrhizae of the integrated physiological response to salinity in plants: new challenges in physiological and molecular studies. J. Exp. Bot. 63, 4033–4044.

Ruiz-Sanchez, M., Armada, E., Munoz, Y., Garcia de Salamone, I.E., Aroca, R., Ruiz-Lozano, J.M., Azcon, R., 2011. Azospirillum and arbuscular mycorrhizal colonization enhance rice growth and physiological traits under well-watered and drought conditions. J. Plant Physiol. 168, 1031–1037.

Saleem, M.H., Ali, S., Seleiman, M.F., Rizwan, M., Rehman, M., Akram, N.A., Liu, L., Alotabi, M., Al-Ashkar, I., Mubushar, M., 2019a. Assessing the correlations between different traits in copper-sensitive and copper-resistant varieties of jute (*Corchorus capsularis* L.). Plants 8, 545.

Saleem, M.H., Fahad, S., Khan, S.U., Ahmar, S., Khan, M.H.U., Rehman, M., Maqbool, Z., Liu, L., 2019b. Morpho-physiological traits, gaseous exchange attributes, and phytoremediation potential of jute (*Corchorus capsularis* L.) grown in different concentrations of copper-contaminated soil. Ecotoxicol. Environ. Saf. 189, 109915.

Saleem, M.H., Ali, S., Rehman, M., Hasanuzzaman, M., Rizwan, M., Irshad, S., Shafiq, F., Iqbal, M., Alharabi, B.M., Alnusaire, T.S., Qari, S.H., 2020. Jute: a potential candidate for phytoremediation of metals—a review. Plants 9, 258.

Sampaio, C.J.S., de Souza, J.R.B., Damiao, A.O., Bahiense, T.C., Roque, M.R., 2019. Biodegradation of polycyclic aromatic hydrocarbons (PAHs) in a diesel oil-contaminated mangrove by plant growth-promoting rhizobacteria. Biotech 9, 155.

Sanchez-Romera, B., Ruiz-Lozano, J.M., Zamarreno, A.M., Garcia-Mina, J.M., Aroca, R., 2016. Arbuscular mycorrhizal symbiosis and methyl jasmonate avoid the inhibition of root hydraulic conductivity caused by drought. Mycorrhiza 26, 111–122.

Sandermann Jr., H., 1994. Higher plant metabolism of xenobiotics: the 'green liver' concept. Pharmacog. Genom. 4, 225–241.

Sangwan, N.S., Farooqi, A.H.A., Shabih, F., Sangwan, R.S., 2001. Regulation of essential oil production in plants. Plant Growth Regul. 34, 3–21.

Sarma, B.K., Yadav, S.K., Singh, D.P., 2012. Rhizobacteria mediated induced systemic tolerance in plants. Prospects for abiotic stress management. In: Maheshwari, D.K. (Ed.), Bacteria in Agrobiology: Stress Management. Springer, Berlin, pp. 225–238.

Saveyn, H., Eder, P., 2014. End-of-Waste Criteria for Biodegradable Waste Subjected to Biological Treatment (Compost & Digestate): Technological Proposals. Publications Office of the European Union.

Schröder, S., Hildebrandt, U., Bothe, H., Niehaus, K., 2001. Suppression of an elicitor-induced oxidative burst reaction in *Nicotiana tabacum* and *Medicago sativa* cell cultures by corticrocin but not by mycorradicin. Mycorrhiza 11, 101–106.

Sfilgoj-Smole, M., Stana-Kleinschek, K., Kreže, T., Strnad, S., Mandl, M., Wachter, B., 2004. Physical properties of glass fibres. Chem. Biochem. Eng. Q. 18 (1), 47–53.

Sfilgoj-Smole, M., Kreže, T., Strnad, S., Stana-Kleinschek, K., Hribernik, S., 2005. Characterisation of glass fibres. Chem. Biochem. Eng. Q 40 (20), 5349–5353.

Sfilgoj-Smole, M., Hibernik, S., Kleinscek, K.S., Kreže, T., 2013. Plant fibres for textile and technical applications. In: Stanislaw, G., Stepniewski, A. (Eds.), Advances in Agrophysical Research. IntechOpen, pp. 369–397.

Sheng, X.F., Xia, J.J., 2006. Improvement of rape (*Brassica napus*) plant growth and cadmium uptake by cadmium-resistant bacteria. Chemosphere 64, 1036–1042.

Sheng, M., Tang, M., Chan, H., Yang, B., Zhang, F., Huang, Y., 2008. Influence of arbuscular mycorrhizae on photosynthesis and water status of maize plants under salt stress. Mycorrhiza 18, 287–296.

Shu, W.S., Xia, H.P., Zhang, Z.Q., Lan, C.Y., Wong, M.H., 2002. Use of vetiver and three other grasses for revegetation of Pb/Zn mine tailings: field experiment. Int. J. Phytoremediation 4 (1), 47–57.

Singh, R., Agrawal, M., 2010. Effect of different sewage sludge applications on growth and yield of *Vigna radiata* L., field crop: metal uptake by plant. Ecol. Eng. 36, 969–972.

Singh, J.S., Pandey, V.C., Singh, D.P., 2011. Efficient soil microorganisms: a new dimension for sustainable agriculture and environmental development. Agric. Ecosyst. Environ. 140, 339–353. https://doi.org/10.1016/j.agee.2011.01.017.

Sinha, S., Mishra, R.K., Sinam, G., Mallick, S., Gupta, A.K., 2013. Comparative evaluation of metal phytoremediation potential of trees, grasses, and flowering plants from tannery-wastewater-contaminated soil in relation with physicochemical properties. Soil Sediment Contam. 22 (8), 958–983.

Smith, P., Powlson, D.S., 2007. Sustainability of soil management practices—global perspectives. In: Abbot, L.K., Mutphy, D.V. (Eds.), Soil Biological Activity—A Key to Sustainable Land Use in Agriculture. Springer, pp. 241–254.

Smykalova, I., Vrbova, M., Tejklova, E., Vetrovcova, M., Griga, M., 2010. Large scale screening of heavy metal tolerance in flax/linseed (*Linum usitatissimum* L.) tested in vitro. Ind. Crop. Prod. 32, 527–533.

Srivastava, S., Chaudhuri, M., Pandey, V.C., 2020. Endophytes—the hidden world for agriculture, ecosystem, and environmental sustainability (Edited Book). In: Pandey, V.C., Singh, V. (Eds.), Bioremediation of Pollutants. Elsevier, Amsterdam, https://doi.org/10.1016/B978-0-12-819025-8.00006-5.

Stancheva, I., Geneva, M., Markovska, Y., Tzvetkova, N., Mitova, I., Todorova, M., Petrov, P., 2014. A comparative study on plant morphology, gas exchange parameters, and antioxidant response of *Ocimum basilicum* L. and *Origanum vulgare* L. grown on industrially polluted soil. Turk. J. Biol. 38 (1), 89–102.

Suthar, S., 2010. Potential of domestic biogas digester slurry in vermitechnology. Bioresour. Technol. 101, 5419–5425.

Tang, J., Chen, J., Chen, X., 2006. Response of 12 weedy species to elevated CO_2 in low phosphorous availability soil. Ecol. Res. 21, 664–670.

Torri, S., Correa, R.S., Renella, G., Corre, A., 2014. Soil carbon sequestration resulting from biosolids application. Appl. Environ. Soil Sci. 2014, 1–9.

Truong, P., 1999. Vetiver Grass Technology for Mine Rehabilitation. Office of the Royal Development Projects Board, Bangkok.

Urbaniak, M., Wywicka, A., Tolczko, W., Serwecinska, L., Zielinski, M., 2017. The effect of sewage sludge application on soil properties and willow (Salix sp.,) cultivation. Sci. Total Environ. 586, 66–75.

Verma, S.K., Singh, K., Gupta, A.K., Pandey, V.C., Trivedi, P., Verma, R.K., Patra, D.D., 2014. Aromatic grasses for phytomanagement of coal fly ash hazard. Ecol. Eng. 73, 425–428.

Wang, W., Vinocur, B., Altman, A., 2003. Plant responses to drought, salinity and extreme temperatures: towards genetic engineering for stress tolerance. Planta 218, 1–14.

WHO (World Health Organization), 2000. World Health Report.

Wu, Q.S., Xia, R.X., 2006. Arbuscular mycorrhizal fungi influence growth, osmotic adjustment and photosynthesis of citrus under well-watered and water stress conditions. J. Plant Physiol. 163, 417–425.

Wu, Q.S., Zou, Y.N., 2009. Mycorrhizal influence on nutrient uptake of citrus exposed to drought stress. Philipp. Agric. Sci. 92, 33–38.

Wu, Y.Y., Chen, Q.J., Chen, M., Chen, J., Wang, X.C., 2005. Salt-tolerant transgenic perennial ryegrass (*Lolium perenne* L.) obtained by *Agrobacterium tumefaciens*-mediated transformation of the vacuolar Na^+/H^+ antiporter gene. Plant Sci. 169, 65–73.

Wu, C.H., Wood, T.K., Mulchandani, A., Chen, W., 2006. Engineering plant-microbe symbiosis for rhizoremediation of heavy metals. Appl. Environ. Microbiol. 72, 1129–1134.

Wu, Q.S., Xia, R.X., Zou, Y.N., Wang, G.Y., 2007. Osmotic solute responses of mycorrhizal citrus (*Poncirus trifoliata*) seedlings to drought stress. Acta Physiol. Plant. 29, 543–549.

Wu, Q.S., Xia, R.X., Zou, Y.N., 2008. Improved soil structure and citrus growth after inoculation with three arbuscular mycorrhizal fungi under drought stress. Eur. J. Soil Biol. 44, 122–128.

Wu, Q.S., He, X.H., Zou, Y.N., Liu, C.Y., Xiao, J., Li, Y., 2012. Arbuscular mycorrhizas alter root system architecture of *Citrus tangerine* through regulating metabolism of endogenous polyamines. Plant Growth Regul. 68, 27–35.

Wu, Q.-S., Srivastava, A.K., Zou, Y.-N., 2013. AMF-induced tolerance to drought stress in citrus. A review. Sci. Hortic. 164, 77–87.

Xu, H., Cooke, J.E.K., Zwiazek, J.J., 2013. Phylogenetic analysis of fungal aquaporins provides insight into their possible role in water transport of mycorrhizal associations. Botany 91, 495–504.

Yin, X., Yuan, L., Liu, Y., Lin, Z., 2012. Phytoremediation and biofortification: two sides of one coin. In: Yin, X., Yuan, L. (Eds.), Phytoremediation and Biofortification. SpringerBriefs in Molecular Sciences, Green Chemistry for Sustainability, Springer, Dordrecht, pp. 1–6.

Yoo, J.H., Lee, Y.D., Hussein, K.A., Joo, J.H., 2018. The effect of food waste compost on Chinese cabbage (*Brassica rapa* var. *glabra*) and tomato (*Solanum lycopersicum* L.) growth. Korean J. Soil Sci. Fertil. 51, 596–607.

Yooyongwech, S., Phaukinsang, N., Chaum, S., Suapaibulwatana, K., 2013. Arbuscular mycorrhiza improved growth performance in *Macadamia tetraphylla* L. grown under water deficit stress involves soluble sugar and proline accumulation. Plant Growth Regul. 69, 285–293.

Yuan, L., Zhu, Y., Lin, Z., Banuelos, G., Li, W., 2013. A novel selenocysteine-accumulating plant in selenium-mine drainage area in Enshi, China. PLoS One 8, e65615.

Zaidi, S., Usmani, S., Singh, B.R., Musarrat, J., 2006. Significance of *Bacillus subtilis* strain SJ 101 as a bioinoculant for concurrent plant growth promotion and nickel accumulation of *Brassica juncea*. Chemosphere 64, 991–997.

Zarik, L., Meddich, A., Hijri, M., Hafidi, M., Ouhammou, A., Ouhmane, L., Duponnois, R., Boumezzough, A., 2016. Use of arbuscular mycorrhizal fungi to improve the drought tolerance of *Cupressus atlantica* G. C. R. Biol. 339, 185–196.

Zhao, F.-J., McGrath, S.P., 2009. Biofortification and phytoremediation. Curr. Opin. Plant Biol. 12, 373–380.

Zheljakov, V., Jekov, D., 1996. Heavy metal content in some essential oils and plant extracts. In: Craker, L.E., Nolan, L., Shetty, K. (Eds.), Proceedings of International Symposium on Medicinal and Aromatic Plants. Acta Horticulturae, vol. 426, pp. 427–433.

Zheljakov, V.D., Craker, L.E., Xing, B., 2006. Effects of Cd, Pb, and Cu on growth and essential oil contents in dill, peppermint, and basil. Environ. Exp. Bot. 58 (1–3), 9–16.

Zhu, Y.G., Pilon-Smits, E.A.H., Zhao, F.J., Williams, P.N., Meharg, A.A., 2009. Selenium in higher plants: understanding mechanisms for biofortification and phytoremediation. Trends Plant Sci. 14, 436–442.

Zhu, Q.H., Huang, D.Y., Liu, S.L., Luo, Z.C., Rao, Z.X., Cao, X.L., 2013. Accumulation and subcellular distribution of cadmium in ramie (*Boehmeria nivea* L. Gaud.) planted on elevated soil cadmium contents. Plant Soil Environ. 59 (2), 57–61.

Ziolkowska, J.R., 2020. Biofuel technologies: an overview of feedstocks, processes, and technologies. In: Ren, J., Scipioni, A., Manzardo, A., Liang, H. (Eds.), Biofuels for More Sustainable Future. Elsevier Inc, pp. 1–19.

Adaptive phytoremediation practices for sustaining ecosystem services

Chapter outline

1 Introduction

Man-made environmental degradation has created widespread ecological, economic, and social problems around the world. Among all the pollutants, metals/metalloids have been recognized as the most widely spread soil pollutants. For example, in England and Wales they are present in more than 80% of all identified polluted sites (EA, 2009). Next in the line are organic compounds, majority of which belong to one of the following classes: polychlorinated biphenyls (PCBs), petroleum hydrocarbons (PHCs), chlorinated solvents, pesticides, dioxins, polycyclic aromatic hydrocarbons (PAHs), etc. Some of the most common contaminated soils containing the aforementioned pollutants are the abandoned industrial sites, mining sites (Roy et al., 2018b; Ahirwal and Pandey, 2021), and fly ash dumping grounds (Roy et al., 2018a; Pandey, 2020b). From an environmental perspective, the contaminated sites should be rehabilitated to the safe standard, irrespective of the costs. Till date, many countries follow excavation and off-site disposal for remediation of sites (EA, 2009). People prefer this "dig and dump" method as they are straightforward and require a short time scale. However, odor nuisance, noise, volatile emissions during the excavation process and secondary contaminations during handling, transport, and landfill are the potential risk factors. Along with the environmental concern, increasing landfill taxation makes this process nonfeasible in the long term (Cappuyns, 2013). The financial cost of remediation by the physicochemical technologies also reaches so high, it becomes impossible to treat and remediate the entire site.

Phytoremediation is the alternative sustainable solution to some of these problems (Pandey et al., 2015a; Pandey, 2020a; Pandey and Bauddh, 2018). Instead of the disruptive nature of the physicochemical methods of soil remediation that was

Adaptive Phytoremediation Practices. https://doi.org/10.1016/B978-0-12-823831-8.00008-6

previously used, phytoremediation combats pollution in an environmentally friendly and aesthetically pleasing manner. In comparison with the physicochemical processes, the cost of phytoremediation is as little as 5% of the other alternate cleanup methods (Prasad, 2003). So phytoremediation is popularly accepted as a low-cost environmental restoration technology that is viewed as a "green clean" alternative to chemical methods (Pilon-Smits, 2005). Phytoremediation technology uses plants and associated microbes either to extract, uptake, and translocate toxic compounds from soil/water to aerial tissues (phytoextraction) or contain or sequester the elements in underground parts to check their leaching from the site (phytostabilization) into the surroundings (Roy et al., 2015; Pandey and Bajpai, 2019). Phytoremediation can mitigate a wide range of organic and inorganic contaminants (Glass, 2000). Transpiration ability of plants allows them to uptake small contaminant molecules and even small ions along with water from the soil.

However, the entire credit of phytoremediation does not go to plants alone as plants achieve their target with the help of the microbes and the synergistic effect of the phyto- and bioremediation is always better than either of the individual processes. In most cases, metallophytic plants also have many heavy metal-resistant plant-growth-promoting microbes that assist the plants in metal remediation (Roychowdhury et al., 2016). In fact, plants constantly communicate with microorganisms present near the roots called rhizosphere, many of which form close, stable, and mutualistic associations with the plants. Most important examples of such symbiosis are arbuscular mycorrhizal fungi that launch a mutualism or symbiosis with land plants (Harrison, 1999), and nitrogen-fixing rhizobia that set up a symbiosis with legumes (Prell and Poole, 2006); both of these symbioses enhance phytoremediation. Recently, plant endophytic microorganisms that live inside plant tissues have been found to play a crucial role in plant health and contribute to plant protections from pathogens and pollutants (Newman and Reynolds, 2005). In a current study, engineered endophytes were found to enhance plant tolerance to toluene, and they decreased toluene transpiration into the atmosphere (Newman and Reynolds, 2005).

Based on the tolerance of the plants toward metals and metalloids, metallophytic plants can be classified as (i) accumulators (uptake metals and translocate to shoots and leaves); (ii) indicators (regulate and maintain the balance of metals within plant cell according to its concentrations in the surrounding soil and thus indicate pollutant concentration at the site); (iii) excluders (do not permit entry of metals into the root and/or their transport to the shoots) (Barrutia et al., 2010); and (iv) hyperaccumulators (specialized uptake mechanisms allow them to uptake and accumulate metals over 1% of their total dry weight) (Baker et al., 2000).

Despite the advantages, phytoremediation has its own drawbacks. It is limited to the root zone of the plant and deeply buried contaminants cannot be phytoremediated. Also, a plant cannot survive and grow in the presence of an excessively high contaminant concentration. However, various advantages overpower the limited disadvantages. The most important advantages of phytoremediation are as follows: (i) very large field sites can be remediated with a minimum cost where other remediation methods are expensive; (ii) installation and maintenance costs are lower than the other remediation methods; (iii) restoration of plantation on polluted sites helps to avoid erosion and metal leaching; (iv) contaminants can be contained within the site through special plants that phytostabilize contaminants and made them bio-unavailable; (v) important metals (Ag, Ni, Tl, and Au) can be recovered from sites with the help of certain hyperaccumulator plants that hyperaccumulate the metals from soil to their biomass and thermochemical processing of the biomass helps in recovering of the metals from ash by smelters (phytomining); (vi) biomass can be utilized to generate biofuel or simply generate heat and electricity; and (vii) sustainable land management with gradual improvement of soil fertility become feasible.

In addition to the much discussed phytoremediation of heavy metals, organic compounds are also very well managed by plants. Many reviews have shown the success of phytoremediation in cleanup of organic pollutants (Feng et al., 2017; Singh and Jain, 2003; Alkorta and Garbisu, 2001). For sustainable phytoremediation, it is advisable to select either high biomass yielding grass species (Pandey and Singh, 2020) or tree species like willows, poplars, and Jatropha (Pandey et al., 2016a). Trees like poplars that are grown in short-rotation cycles of 20 years or less (short-rotation woody crops) are suited to phytoremediation due to their favorable aspects like physiology (e.g., fast growth, elevated water usage, and extensive root system), genetics (poplars commonly hybridize and can transfer favorable traits across generations), and well-established protocols for growing them (e.g., rooted vs unrooted cuttings and planting spacing). In addition to short-rotation woody crops, biomass-producing perennial grasses like *Saccharum munja*, reed canary grass (*Phalaris arundinacea* L.), giant miscanthus (*Miscanthus* × *giganteus*), switchgrass (*Panicum virgatum* L.), and giant reed (*Arundo donax* L.) are popular for phytoremediation (Roy and Pandey, 2020). These plants are better biorefineries as various value-added products can be obtained by fractionation of their biomass/wood into their constituent components (cellulose, hemicellulose, and lignin). In addition to the commercial products, the trapped carbon, other nutrients, and solar energy present in the harvested biomass can be converted into biofuel through any one of the three valorization routes: physical, thermochemical, and biochemical.

Incineration is an example of the physical process (complete burning and obtaining heat as bioenergy), while pyrolysis, liquefaction, and gasification are examples of the thermochemical processes. Biofuels in the form of bioethanol, biogas, and biodiesel are examples of biochemical processes. Selection of plant species that provide economic benefits during the phytoextraction process is a promising alternative to overcome the long-time requirement of phytoextraction that is one of the main drawbacks of phytoremediation. Among many energy crops, *Brassica napus*, *Cynara cardunculus*, *Jatropha curcas*, *Miscanthus* spp., and *Ricinus communis*, are very popular owing to their abilities of metal resistance, metal accumulation abilities, and suitability for biofuel production (Pandey et al., 2012, 2016a; Ruiz-Olivares et al., 2013). Similarly, the practice of intercropping (commonly used in agriculture to increase produce) can be used in phytomanagement and different combinations of phytoextracting plant partners can be used in the intercropping system to stimulate remediation while providing a financial profit. In the field-scale studies, plantation of hyperaccumulators with sugar cane, maize, or ryegrass achieved high remediation efficiencies (Deng et al., 2016).

Phytomining is the commercial application of phytoextraction by which valuable metals and rare earth elements can be recovered or phytoextracted by hyperaccumulators or accumulators from the metal-laden soil/water. According to Jiang et al. (2015), phytomining can increase the financial liability of phytoextraction and eliminate the necessity of disposal of the metal-laden phytomass. According to Chaney et al. (2007), nickel recovery from phytomining showed a good profit. However, some controversies also persist on its economic feasibility. It is true that phytoremediation is indeed a slow and time-consuming process and plants either will phytoextract the metals in a slow way (phytoextraction) or contain them (phytostabilize) instead of removing completely. So, the chance is always there that metals will partially and gradually release due to changes in soil properties over time or by natural disasters like floods during the long process of phytoextraction or phytostabilization. During the valorization of the biomass from phytoremediation, metals trapped in the biomass can be released into the atmosphere if not captured properly from ashes. So, long-term monitoring and follow-up programs should be made compulsory in all phytoremediation restoration programs and ensure that the decrease in metal toxicity and improvement of soil quality remains satisfactory (Epelde et al., 2014b).

The following example shows the importance of long-lasting monitoring during the site restoration program. It was observed that birds preferred nestling on coal ash settling basins and exposed their chicks to pollutants by providing them with pollutant-exposed diet. The concentration of As, Cd, and Se in the diets of ash basins was five times more than reference food and this caused elevation of their levels in the liver, feather, and carcass with only liver Se concentrations showing a serious level. Older (> 5 days) nestlings accumulated 15% of the total body weight of As, Cd, and Se in feathers, whereas only 1% of the total body burden of Se was hidden in feathers. The concentrations of As, Cd, and Se in the feather had correlation with liver concentrations and indicated their values as nonlethal indicators of exposures (Bryan et al., 2012). Peles and Barrett (1997) assessed the uptake of metals and genetic damage in small mammals residing in a fly ash basin. So along with much needed long-term monitoring programs, researches are needed to meet the challenges and opportunities like the discovery of new analytical techniques and equipment to monitor phytoremediation process and soil conditions; novel scientific theories; budget fluctuations; execution of new laws and legislations, etc. (Epelde et al., 2014b). Because of the aforementioned, the concept of adaptive monitoring for long-term monitoring has been coined where periodic revisions are needed for the parameters selected and their proper interpretations are needed (Epelde et al., 2014b).

In this chapter, we discussed ecosystem functions and some ideas and current developments of the phytoremediation undertaken in the past. Evaluation of phytoremediation functions according to the concept of ecosystem services (ESs) gives an idea of the various environmental and social benefits phytoremediation can contribute. Discussions were made on newer approaches on adaptive phytoremediation strategies and necessities for long-time monitoring methods that together with technological advancements will contribute to better acceptability of phytoremediation as a functional, sustainable, and financially viable remediation option.

2 Adaptive phytoremediation practices—An ultimate hope for nature sustainability

Plant adaptation and phytoremediation are the two integral parts of the remediation of the contaminated lands. For successful phytoremediation, plants should be selected that could adapt themselves quickly to the gradually changing soil system. According to Cooke and Suski (2010), the integration of the ecological restoration with plant physiological knowledge would provide new dimensions on the correlation between environmental stress and plant physiological stress response and would help in the designing of restoration ecology. International guideline by the Society for Ecological Restoration has drawn some guidelines for using physiological tools to estimate the success of the restoration. Various ecophysiological models have been prepared that have proved valuable for understanding mechanisms of plant's adaptive responses toward environmental stresses that would support restoration authorities for making decisions and evaluate success of the various restoration plans undertaken (Cooke and Suski, 2010). Besides, the perception of plant biochemistry and taxonomy are key factors behind choosing the most suitable plant species for the ecorestoration of polluted sites.

Plant species that are adapted to withstand environmental stresses like high temperature, salt, drought, intense rain, and toxicity of metal(loids)/organic pollutants exhibit adoption of a wide variety of physiological responses to cope up with the aforementioned stress factors (Gajić et al., 2018). Stress factors cause oxidative stress and reactive oxygen species (ROS) generation by affecting all fundamental plant physiological machinery like photosynthesis, respiration, mineral nutrition, and water transport (Mittler, 2002). The plant achieves stress response by activating the antioxidant system with little harm to the cell integrity. The balance in between ROS generation and antioxidant biosynthesis determines success in regulating redox homeostasis (Foyer and Shigeoka, 2011).

2.1 Adaptive responses of plants toward pollutant stress

Plants have a variety of mechanisms for responding to heavy metal stresses, and various biochemical and metabolic changes of gene expression and methylation pattern have been observed under heavy metal stresses that aim to scavenge toxic heavy metals, alleviate stress, and minimize damages. Stress symptoms induce the production of stress hormones like abscisic acid, ethylene, jasmonic acid, and other antistress molecules (organic acid, phytochelatins, metallothioneins, and specific amino acids) for active chelation of metals (Dalvi and Bhalerao, 2013). Cadmium is an effective stimulator for inducing phytochelatin synthesis (Howden et al., 1995). Two special amino acids proline and histidine are also involved in inducing tolerance by chelating ion synthesis within cells and xylem sap (Rai, 2002). Chelating compounds give protection against heavy metals through sequestration and subsequent vacuole compartmentalization of the metal ions by tonoplast-associated transporters.

But the buildup of a high concentration of heavy metals in the environment triggers the generation of ROS as the aforementioned strategies become insufficient to nullify all the damaging acts. This excessive ROS causes oxidative stress that in turn activates the ROS-scavenging machinery. The ROS-scavenging machinery then removes the ROS from plant cytoplasm through activation of the enzymatic and nonenzymatic parts of the antioxidant defense system. Examples of nonenzymatic antioxidants are α-tocopherol, ascorbic acid, carotenoids, glutathione, phenolics, proline, phytochelatins (PCs), and metallothionine (MTs). Some important enzymatic antioxidants are ascorbate peroxidase (APX), catalase (CAT), glutathione peroxidase (GPX), glutathione reductase (GR), glutathione S-transferase (GST), and superoxide dismutase (SOD) (Mittler, 2002; Foyer and Noctor, 2005). ROS can also play a role as a signaling molecule in the stress-adaptive response of plants (Foyer and Noctor, 2005; Foyer and Shigeoka, 2011). Malondialdehyde, a product of membrane lipid peroxidation, functions as the buffer of ROS. It protects cells by signaling genes for the expression of antioxidant molecules that play an important role in oxidative stress response (Weber et al., 2004).

Uptake and translocation of heavy metals in plants specialized in phytoextraction are handled by different types of complexing agents and metal ion transporters. These channel proteins or H^+-coupled carrier transporter proteins situated in the root's plasma membrane uptake heavy metals and translocate them from belowground parts to aboveground parts (DalCorso et al., 2019). These transporters have been put into several transporter families, such as HMAs, MTPs, NRAMPs, and ZIPs according to sequence homologies.

For organic pollutants (e.g., polychlorinated biphenyls, polychlorinated dibenzofurans, antibiotics, herbicides, dibenzo-p-dioxins, pesticides, and bisphenol A), plants have definite uptake and transport mechanisms, but not so well studied like metals. Plants take up the organic pollutants either through the roots (soil) or sometimes through the leaves (air). Generally, plant roots can easily absorb organic compounds dissolved in soil/water. But for compounds with lower solubility, they are absorbed by passive or active uptake by roots. Higher plants, in general, possess two kinds of transport pathways for organic pollutants to exist, namely (i) short-distance transport (intracellular and intercellular transport) and (ii) long-distance transport (used in tissue transport), and the selection of a particular pathway depends on the physicochemical properties of organic pollutants and the plant itself (Zhang et al., 2017).

For other abiotic stresses, plants have well-defined stress tolerance mechanisms. They employ a whole cascade of genes in mitigating abiotic stresses like drought, salinity, and heat. The cascade begins with stress-mediated transcriptional activation of downstream genes and leads to stress tolerance of the plant and adaptation. Some relevant genes of plant stress tolerance have been discovered (Mantri et al., 2012; Leshem and Kuiper, 1996), but a lot needs to be discovered.

In the following section, discussions will be made on the different phytoremediation technologies and how the physiological features of the concerned plants allowed them to carry out the phytoremediation job sometimes alone and in most cases taking the assistance of their microbial partners.

2.2 Different phytoremediation strategies for environmental cleanup

Phytostabilization and aided phytostabilization—Plants that gather pollutants in their belowground parts are effective agents for phytostabilization or rhizostabilization. The phytostabilization process can be stimulated by adding chelators. Soil amendment is very essential for phytostabilization of heavily polluted soils. Various inorganic (phosphate fertilizers, clay, and other minerals like Mn and iron oxides) and organic additives (coal, compost, and manure) stimulate plants directly for enhanced metal sorption or indirectly by enhancing plant growth and activity by acting as plant growth promoters. In contrast to phytoextraction where contaminants mobility is increased across plant root, in phytostabilization, it is decreased. Besides, as phytostabilization does not remove the contaminants from the site there is no need for further treatments of the contaminated harvested material. The pollutants are contained within the biomass or the root zone of the soil and are not dispersed into nonpolluted sites.

For the restoration of the large contaminated area, the main purpose of remediation should be keeping the pollutants at the site and this would be best achieved by the construction of a vegetative cover. Some complex sites, for instance, mine tailings, ash dumps, etc., are characterized by poor soil structure, nutrient scarcity, high concentration of toxic metals, and pH imbalance for supporting plant growth making the establishment of a vegetative cover difficult (Burges et al., 2016). Application of organic and/or inorganic amendments are essential in metal contaminated agricultural and other sites that reduce metal bioavailability and this is called chemical stabilization. After reducing the bioavailable fraction of metals and adjusting soil pH by chemical means, highly metal-tolerant plants can be sowed as a part of initiating phytostabilization. The vegetation gradually establishes itself and makes the habitat suitable for the growth of other less tolerant plants by improving soil physical features like water-holding capacity, bulk density, and pH balance, increasing its humic content and essential minerals, reducing the bioavailability of toxic metals, etc. (Alvarenga et al., 2009). This type of phytostabilization technology that takes the help of amendments for revegetation of contaminated sites with metal-tolerant plants has been termed "aided phytostabilization" or "chemo-phytostabilization" (Alvarenga et al., 2009) and appears more promising than phytostabilization alone (Alkorta et al., 2010). Some popular organic amendments are animal slurry and compost from farmyard manure, municipal solid waste, household, and garden waste and sewage sludge, etc. (Alvarenga et al., 2009; Burges et al., 2016). Metal oxyhydrates, phosphorous compounds, aluminosilicates, natural and synthetic zeolites, lime, and fly ash are commonly used inorganic amendments (Kumpiene et al., 2009; Alkorta et al., 2010). For increasing soil pH of acidic sites like acid mine drainage sites, lime ($CaCO_3$) is ideal which, along with increasing soil pH, decreases the bioavailable metal fraction (Epelde et al., 2014a). However, as nothing excess is good, the use of amendments must be optimized before getting the desired results. An inappropriate or excessive application of organic amendments can be detrimental as it can contaminate underground and surface water bodies with antibiotics, hormones, nitrates, and toxic compounds and can decrease the bioavailability of essential nutrients. The eutrophication causes a risk to the health of aquatic animals (Burges et al., 2016; Alkorta et al., 2010).

It has been observed that in phytostabilization, rhizomicrobes play a very important role in the revegetation process by producing root exudates that cause immobilization of the metals (Roy et al., 2015; Roychowdhury et al., 2016). Bacteria and mycorrhiza actively change metal speciation, adsorb metals onto their cell walls and/or exopolysaccharide (Mukherjee et al., 2019) or through precipitation processes (Ma et al., 2011). Mycorrhiza acts like a filtration barrier and prevents the translocation of metals to the plant aboveground parts. On the contrary, bacterial endophytes manufacture siderophores like chelators that influence the translocation of metals to plant aerial parts and in this way induce their accumulation in the roots (Mastretta et al., 2009). Accumulation of metals in the root helps in the production of metal-free biofuel as the shoots are harvested only reducing the cost. The metals going back to the soil from the rooted roots would be phytostabilized again by microbes present in the roots preventing the contamination of the surroundings and retaining the metals at the sites.

Endophyte-associated phytostabilization is more pronounced for the remediation of organic pollutants. Doty et al. (2017) studied how endophyte association helped in phytoremediation of a site polluted with trichloroethylene (an industrial waste product). A strain of *Enterobacter* was inoculated in the *Populus deltoides* × *nigra* and the inoculated trees displayed better development and decreased toxicity symptoms compared to control. The inoculated trees secreted 50% more chloride ions to the rhizosphere than uninoculated trees. A remarkable reduction in the trichloroethylene and its derivatives was observed from the tree-associated groundwater plume (Doty et al., 2017). Bianconi et al. (2011) inoculated *Arthrobacter* strains with hybrid poplar clones to rhizoremediate soils contaminated with the insecticide hexachlorocyclohexane (HCH). Becerra-Castro et al. (2013) showed that the shrub *Cytisus striatus* inoculated with *Rhodococcus erythropolis* and *Sphingomonas* sp. played a role in the dissipation of the HCH.

According to recent literature, the most promising woody tree species showing good phytoremediation potential of heavy metals are the Salicaceae family (*Salix* spp., *Populus* spp.), and some other trees like *Betula pendula, Ailanthus altissima, Ginkgo biloba, Carpinus betulus, Platanus hispanica,* and *Robinia pseudoacacia* (Dadea et al., 2017). The willows and poplars are mostly applied in the largest scale phytoremediation program due to their rapid growth, quick propagation, good volume of phytomass production, deep-rooted system, and an extraordinary ability to uptake and translocate an appreciable amount of metals from the soil to the shoots (Pulford and Watson, 2003). All the woody species (poplars, eucalypts, and willows) either alone or in combination with short-rotation forestry have proven ideal candidate(s) for landscape restoration and green plantation around contaminated lands (Pulford and Watson, 2003).

In a field-scale study on polycyclic aromatic hydrocarbon (PAH) degradation and removal, Maila et al. (2005) studied the capacity of the grass species *Brachiaria serrata* and *Eleusine coracana* in the phytoremediation of PAH pollution. After 10 weeks, the naphthalene concentration became undetectable in soil cultivated with multiple native species while 96% naphthalene removal was observed in the monoplanted treatment and 63% in the barren soil. This showed the importance of consociations (multiple plant species and ecotypes) for remediation of sites having a mixed type of contaminants. In the same site, another consociation of celery (*Apium graveolens*), ryegrass (*Lolium perenne*), and white clover (*Trifolium repens*) was tested for the degradation of PAH mixtures. As expected, the consociation removed the PAHs faster than the monoculture system and control soils (Meng et al., 2011). In another work of PAH remediation, *T. repens* and *Medicago sativa* were found to possess the highest capabilities for pyrene and phenanthrene remediation abilities, respectively, while *Brassica campestris* showed low removal efficiency for PAHs. Mixed cropping (e.g., rape with *M. sativa*, white clover, or alfalfa) proved more efficient in the remediation of PAHs than monocropping (Wei and Pan, 2010). Concerning PCB or polychlorinated biphenyl degradation, Terzaghi et al. (2019) showed that *Festuca arundinacea* alone or in consociation with *Cucurbita pepo* ssp. *pepo* and *M. sativa* cultivated with compost, *Rhizobium* spp. and mycorrhizal fungi decreased 20% PCB. In another in vitro experiment, the coculture of *Petunia grandiflora* and *Gaillardia grandiflora* was shown to degrade and remove a dye mixture from the substrate within 36 h, while the single crop-based treatment took a longer time (Watharkar and Jadhav, 2014).

Phytoextraction—The phenomenon of phytoextraction, endemic to plants growing in naturally mineralized soils, was known since ancient times. Initially, this property was considered a detrimental character due to the risk of transfer of the toxic compounds from the soil into the food web. But later as the focus for bioremediation was shifted on plants, phytoextraction property received huge attention due to the natural ability of the host plants to store huge concentrations of metals in the foliage, and hyperaccumulators were given the reputation of most auspicious candidates for phytoextraction. So far, a large number of phytoextraction trials have been conducted in the last decade with the hyperaccumulators, and as of now, more than 400 plants have been recognized with hyperaccumulation ability toward different heavy metals (Boularbah et al., 2006). Examples of some popular and widely used hyperaccumulators are *Noccaea caerulescens* (former *Thlaspi caerulescens*) and *Arabidopsis halleri* for the accumulation of Zn and Cd (Baker et al., 1994), fern *Pteris vittata* for the hyperaccumulation of As (Ma et al., 2001). The tree *Pycnandra acuminata* accumulates an extraordinary 20%–25% nickel in its latex that turns blue-green color due to the formation of nickel complexes. Hyperaccumulators possess astonishing translocation power also. Hyperaccumulation evolved as an adaptive feature to cope with metalliferous environments.

Pollard et al. (2002) explained the genetic basis of hyperaccumulation. Plants suitable for phytoextraction (accumulator and hyperaccumulators) secrete acidic root exudate that contains various metal-mobilizing compounds, such as phytosiderophores, organic acids, and carboxylates that ultimately enhance the uptake process by increasing metal solubility in the soil (Yan et al., 2020; Padmavathiamma and Li, 2012). In phytoextraction, rhizospheric microbes play important role in increasing metal uptake by increasing the solubility of heavy metals for ease in uptake. Microbes secrete enzymes and chelators (siderophores and biosurfactants) into the rhizosphere zone that helps the heavy metals in the conjugate formation with the chelators increasing their uptake and translocation (Clemens et al., 2002). Clemens et al. (2002) showed that plant-growth-promoting rhizobacteria and endophytes increased the solubility of otherwise water-insoluble Cu, Ni, and Zn via secretion of organic anions or protons. Siderophores are iron-chelating compounds with a strong affinity for ferric iron (Fe^{3+}) and variable affinity for other heavy metals, such as As, Cd, Ni, and Pb (Schalk et al., 2011). It has been noticed that the plant-fungi-bacterium association formed in several tree species helped the trees in enhanced phytoremediation activity. Guarino et al. (2018) observed this association behind the enhanced phytoremediation of willow (*Salix purpurea* subsp. *lambertiana*) and hybrid poplar (*P. deltoides* × *P. nigra*). *Eucalyptus camaldulensis* illustrated enhanced efficiency in extraction and translocation of Cd after inoculation with or PGPR or arbuscular mycorrhiza (Motesharezadeh et al., 2017). In another example, *Helianthus annuus* showed the highest Ni uptake when co-inoculated with *Bacillus safensis* and *Kocuria rosea* (Mohammadzadeh et al., 2014).

Phytotransformation/phytodegradation—Phytodegradation is the degradation of organic compounds to TCA cycle intermediates or CO_2 and H_2O by enzymes of plants or plant-associated microbes (Pilon-Smits and LeDuc, 2009). Plants employ mainly three steps for transformation/degradation of organic compounds, namely (a) transformation (compounds are activated and transformed); (b) conjugation (organic compounds are conjugated with D-glucose, malonic acid, glutathione, and amino acids to make it less toxic and less mobile); (c) elimination/storage (sequestration or storage of the partially degraded transformant in organelles like vacuole, apoplast, cell walls, or elimination of the completely degraded products CO_2 and H_2O) for phytodegradation (Gajić et al., 2018). Some volatile phytodegraded products can escape directly into the atmosphere through stomata. This metabolic cascade is known as the "green liver model" because of its resemblance with the mammalian metabolism.

Rhizofiltration—Rhizofiltration is phytoextraction in water and carried out by hydrophytes. So like terrestrial plants, hydrophytes or aquatic plants can be used to remediate wastewaters. In rhizofiltration, plants take metals from polluted water and precipitate them into roots or shoots or leaves or on all parts (Pandey et al., 2014). This rhizofiltration ability of seaweeds and other hydrophytes to absorb metals and radionuclides was known for a long time, but large-scale testing was not performed due to the unavailability of cost-effective culturing, harvesting, and handling methods. The development of hydroponics systems helped in the exploitation of terrestrial plants grown hydroponically to remediate polluted wastewater. The hydroponics acquired high shoot phytomass and extensive roots to adsorb and uptake a huge amount of metals from wastewater. Three types of kinetics were found responsible for the pumping of the metals from water to roots, namely: (a) chemical and physical processes such as adsorption, chelation, and ion exchange; (b) biological processes that are dependent on plant metabolism and slowly move the metal from solutions to plant cells via intracellular uptake and translocation to the shoots and vacuolar deposition within cells; (c) root exudates mediated precipitation of the metal from the solution as insoluble compounds like phytostabilization (Pandey et al., 2014).

Phytovolatilization—Phytovolatilization is a process through which plants can remove certain metals like mercury by taking them up from polluted soils and then through transpiration release them in volatile forms into the air. As the water reaches the leaves from the root and the shoot along with the vascular system of the plant, some of the contaminants convert themselves to more volatile forms and evaporate into the atmosphere. So far, phytovolatilization has been primarily used to remove mercury. Here, the mercuric ions are transformed into less toxic elemental mercury (Meng et al., 2014). Experiments with hydroponically treated *Brassica juncea* with $HgCl_2$ showed that after uptake, phytovolatilization of Hg takes place in roots, instead of shoots due to a very low rate of metal translocation. A microbial and algal community of root was found partially responsible for such events (Moreno et al., 2008).

Selenium phytovolatilization was also found in plants growing in selenium contaminated soil where plant enzymes involved in sulfur metabolic pathways, assimilation, and volatilization converted selenium (chemical analog of sulfur) to dimethyl selenide (De Souza et al., 2000). *A. donax* with the assistance of PGPR takes up and volatilizes arsenite which is the most toxic form of carcinogenic arsenic in the soil. Arundo plant was able to remove higher than 50% As species from the soil. The plant protects itself from the toxic effect of As by DNA methylation event.

2.3 Adaptive site management and long-time monitoring

A complex and heavily polluted site with substantial technical and nontechnical challenges typically has a longer remediation time requirement, increased cost of remediation, and larger environmental footprint (such as energy use or carbon emissions) by physical and chemical treatments. On the contrary, complex sites would have a greater potential for cost savings, environmental footprint reductions, beneficial land reuse, and other societal benefits, if a proper and thoroughly planned phytomanagement program is undertaken.

The term "adaptive site management" means a flexible, comprehensive, and iterative process for the management of the remediation process. NRC (2003) coined the term "adaptive site management" referring to "a comprehensive and flexible approach for dealing with difficult-to-remediate hazardous waste sites." Long-term phytomanagement of complicated sites should begin with the design of remediation plan and should include all postplantation stages of phytoremediation process, monitoring, and appraisal of the remediation achieved. A long-term adaptive phytomanagement plan can help in the accomplishment of different sustainable developmental goals through securing human health and environment and make progress toward the site objectives. Conceptual site model (CSM) has been developed to guide in the decision making of adaptive site management about remediation potential assessment and evaluation of remediation steps as well as the state of science and practice. The CSM also guides additional site characterizations that may be needed under the long-term phytomanagement strategy. Additional site characterizations may be necessary if data show unexplainable gaps during a particular stage. The gaps may be caused due to identification of a new contamination which was not detected earlier but needs special attention or the concentration of existing contaminants are not decreasing as expected. The changes observed should be documented to determine their influence on the effectiveness of remedy, which is reflected in the periodic evaluations.

The first steps in the adaptive site management process are to identify the complexity attributes within the CSM. The next steps are to revise and refine the CSM, devise the objectives, and develop interim plans toward adaptive phytoremedial strategy. These steps are particularly relevant at sites that are undergoing an interim or final remedy or revisiting the existing remedial strategy because insufficient progress has been made toward meeting site remediation objectives. These elements can be used iteratively during the adaptive site management process. This process is eventually guided by the long-term management plan, which should be anticipated during remedy selection and implementation. Phytostabilization is the most favorable option for the remediation of heavy metal-polluted soils that need long-term monitoring programs. "Adaptive monitoring" is an intrinsic part of the long-time phytostabilization program. Soil and microbial factors are the most significant indicators for the success of restoration of soil fertility. The study should also incorporate soil microbial role as "ecological attributes" or "ecosystem functions" in order to facilitate elucidation. A "Phytostabilization Monitoring Card" should be made on the basis of ecological attributes and ecosystem services. Second part of the adaptive site management is redefining the CSM (www.archive.epa.gov). The CSM is actually an iterative tool that should be constituted and refined as the data are acquired during the review of the site history and continue throughout the remedial period. According to Site Remediation and Waste Management Program of New Jersey-Department of Environmental Protection, the purpose of a CSM is to describe relevant site characterizations and the surface and subsurface features to realize the extent of concern for the pollutants identified and the threat they pose to receivers. It is particularly important to revisit and refine the CSM periodically throughout the project life cycle. The CSM should be treated as a dynamic tool and updated as needed (e.g., by using data generated during the implementation of adaptive site management) to support remedy decisions throughout the adaptive site management process. It is the best practice to update the CSM during long-term management planning, remedy implementation, periodic evaluations of monitoring data and remedy performance data, and remedy optimization. If a remedy is not on track to meet interim objectives despite optimization, the CSM should be refined prior to revisiting the remedy. If additional site characterization is needed to fill data gaps and complete an effective remedy design, an integrated site characterization (ISC) process can improve site characterizations and maximize the effectiveness of remediation (ITRC, 2003).

3 Ecosystem services from phytoremediated polluted sites

Ecosystem services (ESs) are the advantages that humans acquire from ecosystems (Nicholson et al., 2009). As humans are degrading and depleting ecosystem resources to satisfy growing demands for food, freshwater, timber, fiber, paper, fuel, and minerals (Mertz et al., 2007; Nicholson et al., 2009), research interest in ESs is increasing rapidly and is likely to put up significantly to the sustainable management of natural resources. The dynamics of an ecosystem are affected by natural and man-made wrong activities and such changes can have direct and long-lasting damaging consequences on the spatial and temporal variations in the composition, structure, and functioning of ecosystems. For example, mining and drilling activities for obtaining fossil fuels disrupts many natural ecosystem structure and function by the removal of the topsoil, natural vegetation cover and associated flora and fauna and introduces harmful pollutants like heavy metals and acid in the nearby aquatic system (acid mine drainage). Depending on the type and scale of the mining operations, these effects can be limited to the location of the mining or, through local hydrology, can extend to nearby water bodies and groundwater reservoirs.

Similarly, fly ash landfills and ash ponds make the sites potentially unsuitable for any use and aggravate the problem of land management. However, proper and long-term phytomanagement can bring back the vegetation and restore the normal ecosystem services from the reestablished vegetation. According to Shimamoto et al. (2018), natural regeneration is best for the restoration of large ecosystems like tropical rainforest degraded by pasture. However, for excessively damaged ecosystems like mining lands human intervention and well-planned phytomanagement is essential for its restoration.

The concept of ESs was initially introduced to improve the quantitative understanding of the use and management of natural resources and quantifying postdisturbance resilience. With this goal, the UN sponsored the Millennium Ecosystem Assessment Report (2005), divided ESs into four groups, namely provisioning services, regulating services, supporting services, and cultural services (http://www.maweb.org). The results determine the success of phytorestoration strategy.

Covering large areas of fly ash and mine waste deposits and other large industrial polluted areas with vegetation cover can provide multiple ESs like stabilization of mine spoil, fly ash landfills, dispersal of airborne dust from fly ash and mining/drilling sites, decrease toxicity, bioavailability/mobility, and dispersion of toxic metals and other chemicals in the surrounding ambiance and provide the organic substances that can retain the contaminants and decrease the transfer of toxic pollutants in the food web. An exclosure was constructed on degraded communal grazing land in Aba Gerima watershed, northwestern Ethiopia (Mekuria et al., 2018). The exclosure (enclosed grazing land that prevents grazing) was connected with adjacent communal grazing land. In the exclosure, 46 plant species belonging to 32 families were recorded, whereas 18 plant species representing 13 families were found in the nearby communal grazing lands. Significant differences were noticed in woody species richness, evenness, and diversity within the exclosure and adjoining grazing land. Significant differences were found in the tree species density, aboveground woody biomass, soil properties, and basal area. The results showed degraded lands can restore native vegetation and bring back the ESs that is unavailable from degraded lands. In another instance, a global metaanalysis of ecological indicators of the ESs was conducted in restored areas and degraded areas along with reference ecosystems. Different restoration strategies were examined on the different kinds of degradation and restoration effects with time. In general, restoration actions accounted for a noticeable rise in the levels of ecological indicators of ESs including carbon pool, biodiversity, and other soil parameters in comparison with degraded sites.

Among different restoration plans, natural regeneration was found to be very fruitful. It was also noticed that reforestation with exotics instead of native species reduced the ESs of sites damaged by agricultural farming. In degraded pasturelands, the restoration of vegetation successfully preserved biodiversity, whereas in degraded agriculture areas only the carbon pool was safeguarded. These results again confirmed that with the selection of the correct plan of action, restoration can retrieve successfully much of the ESs lost by human activities (Shimamoto et al., 2018). So understanding ESs from phytoreclaimed lands is very important for its ecological and economical sustainability. Table 1 shows the specific ecosystem services gained from the different phytorestored sites. Fig. 1 shows the specific merits of phytoremediation services. Following is a short description of each category of ESs in context to services restored upon phytoremediation of sites degraded by human activities like mining or disposal of fly ash or industrial wastes.

3.1 Provisioning services

The numerous materialistic benefits that we gain every day from nature for our living constitute ecosystem provisional services. Food in the form of fruit, vegetable, fish, cattle, and other livestock is the foremost service. In addition to food, other common provisioning services, including drinking water, natural gas and oil, timber, cotton, pulp and paper, fiber, fortified crops, essential oil, medicinal plants and mushrooms, products of silviculture, and many others, form parts of our daily life. Provisioning services from natural or perennial systems made on fly ash landfills and restored mining sites have constituted significant economic benefits as shown in Table 1.

Food or agricultural activity in a newly restored fly ash/mining land is not encouraged as the pollutant load in such sites may still be high. Only after proper ecotoxicological assessment, food crops, or agriculture, may be allowed. In addition to food, obtaining timber and nontimber forest product and medicines should be encouraged after proper testing of the level of metals. Other provisional services of the plants used in phytoremediation are biochar production and raw materials for industries like biochemicals, essential oils, and paper (Pandey et al., 2016a; Schroder et al., 2008). Genetic resources, bioenergy, and nontimber forest product (NTFP)-based provisional services from phytorestored sites have been discussed in the following as the three most important provisional services of phytoremediated sites.

Nontimber Forest Product (NTFP)—The different types of NTFPs include edible and medicinal plants, cinnamon, cork, mushrooms, berries, seeds, ferns, nuts, maple syrup, moss, rubber, essential and aromatic oils, resins, and ginseng (Sheppard et al., 2020). According to the United Kingdom's Forestry Commission, NTFPs are "any biological resources found in woodlands except timber." Reforesting Scotland project defines them as "materials supplied by woodlands—except the conventional harvest of timber" (Emery et al., 2006). NTFPs or special, nonwood, minor, alternative, and secondary forest products (other terms) have been grouped into categories like floral greens, items for decorations, foods, medicinal plants, flavors and fragrances, fibers, and saps and resins. Although NTFPs are quite valuable to local villagers but have been overlooked until recently and neglected in comparison with other priority

Table 1 Ecosystem service case studies from different phytorestored sites including restored fly ash landfills and mining sites.

Ecosystem services	Phytorestored sites	Products/services from phytorestored sites	Flora/fauna	References
Provisional Services				
Phytorestored Fly Ash Disposal Sites				
Fiber	Fly ash deposits in Unchahar of Raebareli district, Uttar Pradesh, India	Roof thatch, fencing, paper, and craft (i.e., ropes, hand fans, baskets, brooms, and mats)	*S. munja* and *S. spontaneum*	Pandey et al. (2015b)
Energy	Fly ash deposit site of Utvin	Biomass and biofuel	*Miscanthus giganteus*	Lixandru et al. (2013)
	The thermal power plant of northern Indiana		*Salix* (willow)	Nienow et al. (2000)
	Metal-contaminated industrial waste disposal sites of periurban Greater Hyderabad (Bollaram, Patancheru, Bharatnagar, and Kattedan industrial areas)		*Ricinus communis*	Kiran et al. (2017)
	The greenhouse study with fly ash and other amendments		Switchgrass, eastern gamagrass, big bluestem	
	Greenhouse study with fly ash-amended soil		*Jatropha curcas*	Jamil et al. (2009)
Food	23-m-deep, 2-acre FA landfill in Endwell, New York, United States	Pollen	*Aster novae-angliae*	De Jong et al. (1977)
	Panki Thermal Power Station, Kanpur, Uttar Pradesh, India	Fruit	*Z. mauritiana*	Pandey and Mishra (2016)
Restored Mining Sites				
Food (fruits and seeds) and fishery	Mangrove forest in Bedono and Timbulsloko, Indonesia, having many abandoned mines	Seed and aquatic animals as food and livelihood	*Avicennia marina*, mud crabs (*Scylla* sp.), white shrimp (*Penaeus merguiensis*)	Damastuti and de Groot (2019)

Continued

Table 1 Continued

Ecosystem services	Phytorestored sites	Products/services from phytorestored sites	Flora/fauna	References
Fishing gear, construction material, seedling nursery, firewood	Mangrove forest in Bedono and Timbulsloko, Indonesia	Firewood and construction material were extracted from the mangroves present near seaside or privately owned ponds	*Avicennia marina* or locally called brayo	Damastuti and de Groot (2019)
Food	Flat slopes (0–5 degrees) on the iron mining site situated in the central part of Liaoning Province, China	Agriculture	Crops	Wang et al. (2017)
	Tamar estuary	Grassland is used for cattle and sheep grazing for food and wool, crops, fruits in agricultural land, salmonid fish culture in water bodies for use as food. Annual benefits of provisioning service gains included (e.g., ≈£265k for food and ≈£304k for freshwater)	Miscanthus, maize in grassland, and other agricultural products in lands restored for agriculture	Everard (2009)
Firewood, plywood	Steeper slopes on the iron mining site located in the central part of Liaoning Province, China	Forest	Timber	Wang et al. (2017)
Oil production	Mine tailings having high concentrations of Cd, Cu, Mn, Pb, and Zn in Mexico	Potential of *Ricinus communis* L. for oil production and phytoremediation of mine tailings	*Ricinus communis* L.	Olivares et al. (2012)
Other Polluted Sites				
Biomass for biofuel	Nine industrially polluted sites of the Midwestern (Illinois, Iowa, and Wisconsin) and the Southeastern United States (Alabama, Florida, and North Carolina)	Biomass values ranged from 4.4 to 15.5 Mg ha^{-1} year^{-1}	Short-rotation woody crops Poplars and their hybrids	Zalesny et al. (2019)

	Metal-contaminated industrial waste disposal sites of periurban Greater Hyderabad (Bollaram, Patancheru, Bharatnagar, and Kattedan industrial areas)	Biomass combustion for biofuel	*Ricinus communis*	Kiran et al. (2017)
Regulating Services				
Carbon sequestration	Old FA dumping site by the Panki Thermal Power Station, Kanpur, India	Carbon is trapped in the soil and plant system of fly ash deposits	*S. spontaneum, Prosopis juliflora,* and *Typha latifolia*	Pandey et al. (2016b)
	Phytoremediation sites in the Midwestern (Illinois, Iowa, Wisconsin) and Southeastern (Alabama, Florida, North Carolina) United States	A good amount of carbon is sequestered in the poplars growing on the phytoremediated sites	Short-rotation woody crops including *Populus* species and their hybrids (i.e., poplars)	Zalesny et al. (2018)
Rhizoremediation	Fly ash lagoon	Fly ash pond remediation	*Typha latifolia*	Pandey et al. (2014)
	Border of the Don river delta, the estuarine zones of small rivers entering the basin of the Sea of Azov and the northern coast of the Taganrog Bay near the city of Taganrog containing industrial enterprises, roads, and a port facing severe heavy metal pollution	Remediation and study of the suitability of the plant as a pollution bioindicator species	Cattail (*Typha australis* Schum) and other woody trees	Minkina et al. (2019)

Continued

Table 1 Continued

Ecosystem services	Phytorestored sites	Products/services from phytorestored sites	Flora/fauna	References
Phytostabilization	"Lubień" Lignite combustion site located in central Poland. The main components of combustion waste are thermally processed aluminosilicates that slowly cause leaching of sulfate, chloride, and calcium and consequent alkalization of groundwater	Stabilization of heavy metals through roots and ecorestoration of the site	Green alder (*Alnus viridis* (*Chaix*) DC. in Lam. & DC.)	Pietrzykowski et al. (2015)
	Abandoned copper mine of Lasail in the northwestern Hajar Mountains of Oman	Phytostabilization of heavy metals	Castor (*Ricinus communis* L.)	Palanivel et al. (2020)
Phytostabilization	Thermal power plant TENT-A (Serbia)	Phytostabilization of As, Cu, Mn, and Zn	*Festuca rubra*	Gajić et al. (2016)
	Thermal power plant TENT-A (Serbia)	Phytostabilization of As, B, Cu, Mo, and Se	*Dactylis glomerata*	Gajić et al. (2020)
Phytoextraction	Ash dyke of Singrauli	Heavy metal phytoextraction and Zn and Cd hyperaccumulation	*C. dactylon, I. carnea, S. munja, S. nigrum, T. angustifolia,* and *P. hysterophorus*	Kisku et al. (2018)
	Firoz Gandhi National Thermal Power Plant (FGNTPC), Unchhahar, Raibarelli (UP) Firoz Gandhi National Thermal Power Plant (FGNTPC), Unchhahar, Raibarelli (UP) Feroze Gandhi National Power Plant, UP, India	Removal of heavy metals like Cr, Cu, Zn, Mn	*Calotropis procera, Cassia tora,* and *Blumea lacera*	Gupta and Sinha (2008))
	Nikola Tesla-A power plant in Serbia	Phytoremoval of a toxic concentration of boron from ash pool and stabilization of the site for future use	*Populus* sp., *Ambrosia* sp., *Amorpha* sp., *Cirsium* sp., and *Eupatorium* sp.	Pavlović et al. (2004)

Improvement of soil quality	Fly ash disposal sites of Unchahar of Raebareli district, Uttar Pradesh, India	Improvement of microbial biomass carbon and organic C, N, and P in the naturally vegetated fly ash substrate	*Typha latifolia* L., *Cynodon dactylon* (L.) Pers., *Saccharum spontaneum* L., *Saccharum bengalense* Retz. (syn. *Saccharum munja*), *Prosopis juliflora* (Sw.) DC., *Ipomoea carnea* Jacq., and *Acacia nilotica* L.	Pandey et al. (2015b)
	Thermal power plant TENT-A (Serbia)	Improvement of the physicochemical characteristics of FA over time: increased values of hygroscopic water, silt, clay, the amount of adsorbed bases, total adsorption capacity, N, P_2O_5 and K_2O content, and the decreased levels of total and available As, B, Cu, Zn, and Mn	*Festuca rubra*	Gajić et al. (2016)
	Thermal power plant TENT-A (Serbia)	Weathering and revegetation processes increase the clay and silt fractions, decrease alkalinity and salinity, increase the cation exchangeable capacity (CEC), the content of N, P, and K, and reduce the total and available content of As, B, Cr, Cu, Mn, Ni, and Zn	*Robinia pseudoacacia*, *Amorpha fruticosa*, *Tamarix pentandra*	Kostić et al. (2018)
	Thermal power plant TENT-A (Serbia)	Woody plant species that colonized fly-ash deposits can initiate the beginning of pedogenetic processes: increased values of C, N, and phenolic compounds	*Robinia pseudoacacia* *Ailanthus altissima* *Amorpha fruticosa*	Grbović et al. (2019)

Continued

Table 1 Continued

Ecosystem services	Phytorestored sites	Products/services from phytorestored sites	Flora/fauna	References
	Asbestos deposits (Serbia)	Woody plant can improve asbestos chemical properties. High values of C, N, P_2O_5, K_2O, total phenolics, phenolic acids, and flavonoids in rhizospheric asbestos of *A. altissima* indicate changes in soil chemistry, humus formation, and initiation of pedogenesis	*Ailanthus altissima*	Grbović et al. (2020)
	Coal mine overburden dumpsites of Naya Dara colliery, Kapasara area of Eastern Coalfield Limited and adjacent forest site of Topchanchi Sanctuary of Dhanbad district, India	Plant root played a very vital role in the stabilization of dump slopes by inducing mechanical reinforcement of dump material and enhancing shear strength of dump material, which in turn increased the long-term stability of dump slopes. The factor of safety was increased from 1.2 for barren dump slope to 1.4, 1.7, 1.9, and 2.1 for the same dump slope covered with plant roots after 2, 6, 10, and 12 years of revegetation. Revegetation of mine spoil caused an increase in mineral nitrogen values by 12%, 36%, and 76%; belowground biomass values by 380, 1770, and 3750 times; nitrogen mineralization values by 0.6, 3.58, and 9.5 times; and microbial biomass values by 0.43, 2.77, and 6.07 times in 2, 6, and 12 years, respectively	Deciduous tree species *Shorea robusta* (Sal), *Butea monosperma* (Palash), *Terminalia tomentosa* (Asan), and *Dalbergia sissoo* (Shishum). On mine spoil, *Azadirachta indica* (Neem), *Dalbergia sissoo*, and *Leucaena leucocephala* (Subabool) among the tree species and *Dendrocalamus strictus* (Bas), *Cynodon dactylon* (dubo), *Saccharum spontaneum* (Kashi), and *Eragrostis tenella* among the herbaceous species	Singh et al. (2012)
Coastal protection	Bedono and Timbulsloko, Central Java, Indonesia	Decreasing the impact of wave and storm damage since the mangroves grew around their houses	Mangroves	Damastuti and de Groot (2019)

Pollution abatement	Tamar estuary	The upland, nutrient-poor zone of the Tamar catchment permitted peat formation, while lowlands were affected by drainage but restoration initiatives contributed to air and water quality improvement	Woodlands, intertidal mudflats, salt marshes and reed beds, and breeding avocets and other birdlife, otters and other mammals, butterflies, and plants including rare lichen and orchids	Everard (2009)
	Emscher River restoration in the "Ruhr Metropolitan Area" at the federal state of the North Rhine-Westphalia, western Germany	Restoration of the streams increased their self-purification potential by 38% nitrogen), 266% phosphorus, 77% carbon retention, and 39% carbon stock. Maintenance of the nursery populations and habitats also showed a clear improvement due to restoration	Native species planted	Gerner et al. (2018)
Supporting Services				
Biomass values	Long-term phytoremediation sites in the Midwestern and Southeastern United States	Biomass values ranged from 4.4 to 15.5 Mg ha^{-1} year^{-1}	Short-rotation woody crops (poplars and their hybrids)	Zalesny et al. (2019)
Biodiversity	Marginal lands of Michigan and Wisconsin, United States	Agricultural land planted with maize as monocrop showed very less biodiversity but Switchgrass showed higher biodiversity. Prairies were the most diverse ecosystem for both plant species and biomass composition than switchgrass planted land and maize planted lands. Diversity of herbivorous and predatory arthropods showed a stair-step increase from maize to the prairie that mirrored trends in plant diversity, methanotroph, bee, and breeding-bird diversity	Maize as annual monocrop, switchgrass as perennial monocrop, and prairies	Werling et al. (2014)

Continued

Table 1 Continued

Ecosystem services	Phytorestored sites	Products/services from phytorestored sites	Flora/fauna	References
Habitat services and biodiversity hotspots	Bedono and Timbulsloko, Central Java, Indonesia	Mud-crabs (*Scylla* sp.), white shrimp (*Penaeus merguiensis*) areas with high bird diversity as biodiversity hotspots. Post rehabilitation also increased the abundance of crustacean species, particularly *Scylla* sp. and *P. merguiensis*	Aquatic animals like *Scylla* sp., *Penaeus merguiensis*, different birds, crustaceans like *Scylla* sp., and mangroves like *Avicennia marina*	Damastuti and de Groot (2019)
	Tamar estuary	Tamar catchment zone	Woodlands, intertidal mudflats, salt marshes and reed beds, breeding avocets and other birdlife, otters and other mammals, butterflies, and plants including rare lichen and orchids	Everard (2009)
Pedogenesis	"Lubień" combustion fly ash disposal site located in central Poland Bełchatów thermal power station, central Poland	Fly ash amelioration (substrate quality was enhanced as a result of an increase in organic carbon, nitrogen, and phosphorus in fly ash substrate)	Alders (*Alnus* sp.) Various species of native trees and shrubs with grass mixtures in between them	Pietrzykowski et al. (2018a,b) Uzarowicz et al. (2017, 2018)
	"Nikola Tesla-A" thermal power plant in Obrenovac, Serbia Pulverized fuel ash lagoon in Lee Valley, Southern England	Soil formation was enhanced Soil formation and vegetation cover development were studied over 7–24 years in this unamended FA deposit	Spontaneously colonizing plants (more than 130) Successional change from mixed ruderal community to a woodland community dominated by birch (*Betula* spp.) and willow (*Salix* spp.)	Kostić et al. (2018) Shaw (1992)

Category	Site/Location	Findings	Species	Reference
Biodiversity conservation	Two fly ash deposits in the northern Czech Republic, Central Europe	Vanishing insects (bees and wasps) were brought back. The fly ash deposits harbored arthropod communities of high conservation values	Bees, hoverflies, moths, spiders, wasps, and ants colonized several spontaneously developed and technically reclaimed plots in two fly ash deposits	Tropek et al. (2013, 2014)
Nutrient cycling	Lubień combustion and fly ash disposal site, from Bełchatów Power Plant, located in central Poland	Fine-root biomass and the nutrients (C, N, S, P, K, Ca, Mg, Na) increased under alder plantings grown on technosols prepared from combustion wastes and extremely poor quaternary sands from sand mining	Three alder species: green, black, and gray alders	Swiatek et al. (2019)
	Coal mine spoil	Nutrient release from litter decomposition and increase in soil nutrients for uptake by other plants. *A. auriculiformis*, *E. hybrid*, and *C. equisetifolia* showed the most balanced nutrient cycling	*Acacia auriculiformis*, *Cassia siamea*, *Casuarina equisetifolia*, *Eucalyptus hybrid*, and *Grevillea pteridifolia*	Dutta and Agrawala (2001)
Photosynthesis	Coal-fired thermal power plant (NTPC, Sipat, Bilaspur)	Chlorophyll fluorescence and photosynthetic efficiency increased	*Ziziphus mauritiana*	Prajapati and Meravi (2015)
	Two fly ash lagoons, weathered for 5 and 13 years, respectively, at Serbia (Nikola Tesla-A power plant)	Certain characteristics of naturally colonized species can restore fly ash deposits and conditions can be provided for shortening the successive stages in the vegetation cover development on ash deposits after the closure of the site	*Festuca rubra* (sown) and *Calamagrostis epigejos*	Mitrović et al. (2008)
	Two fly ash lagoons of the "Nikola Tesla-A" power plant in Obrenovac, Serbia, weathered for 5 and 13 years, respectively	Thirteen years after planting salinity and toxicity reduced and *A. fruticosa* species showed the highest photosynthesis efficiency	*A. fruticosa*	Mitrović et al. (2012), Gajić et al. (2013, 2016)

Continued

Table 1 Continued

Ecosystem services	Phytorestored sites	Products/services from phytorestored sites	Flora/fauna	References
Cultural Services				
Recreational	Bauxite mine areas in southwest Australia	Recreation for bushwalkers and mountain bikers	Partially restored natural forest	Rosa et al. (2020)
Aesthetics, heritage study	Tamar Valley, South West England containing many copper mining sites	Tamar Valley forms an area of outstanding natural beauty and a site of huge historical importance	Natural landscape with hills, valleys, forests, meadows, etc.	Everard (2009)
Recreation and refreshment	Two mangrove sites in Tambaksari and Rejosari subvillages, Bedono	Beautiful mangrove scenery attracted the tourists	Mangroves and presence of a sacred tomb Tambaksari	Damastuti and de Groot (2019)
	Emscher catchment area (industrialization and mining and urbanization degraded the landscape)	Good for walking, biking, or boating Attracts researchers and environmental educators	Near-natural riverine ecosystem with dense forest vegetation in surroundings	Gerner et al. (2018)
Educational	Unchahar of Raebareli district, Uttar Pradesh, India	Provides scientific knowledge on the effect of revegetation on fly ash lagoons and fast green capping of fly ash basins	Grassland composed of mixed-grass species	Pandey et al. (2015c)

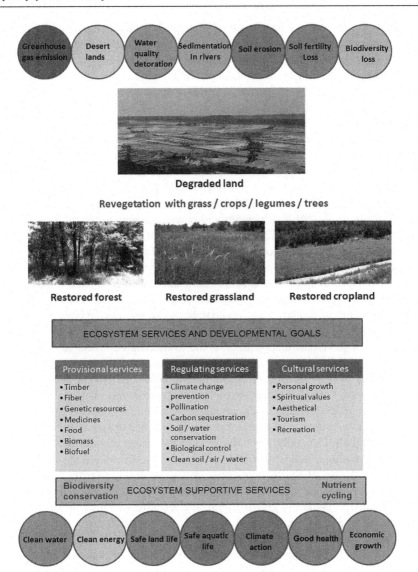

Fig. 1 Merits of phytoremediation: Degraded lands can be converted to seminatural ecosystems and deliver ecosystem services. Achievement of sustainable developmental goals by a functional ecosystem is shown.

products like timber and animal forage. Different types of NTFPs products have been obtained from forests constructed on rehabilitated mines (Damastuti and de Groot, 2019) and fly ash dumpsites (Pandey, 2020c).

Timber production—This is an age-old provisional service from the forest. This basic service from the forest ecosystem was obtained in many instances from fully or partially reclaimed forests on degraded mine lands (Wang et al., 2017).

Biomass for bioenergy—Biomass for bioenergy is controversial owing to its probable synergism or trade-offs with provisioning and regulating ESs. As fossil fuels are decreasing, people are giving more attention to renewable sources of energy. To prevent competition with crops for land, more attention is focused on the marginal lands for bioenergy crops (Meyer et al., 2015) where conditions generally limit crop productivity but allow good growth and biomass production by more stress-resistant bioenergy crops. Producing bioenergy crops on marginal lands is the alternative green way to meet energy goals while minimizing competition with food production. In the Midwestern United States, perennial grasses grown on marginal lands provided up to 25% biomass of national targets for production of cellulosic biofuel and reduced substantial amount of greenhouse gas (GHG) emission. The second-generation liquid biofuel feedstock was obtained with grasses or other perennial bioenergy feedstock (Meehan et al., 2012). Bioenergy crops were grown successfully on different FA disposal sites (Kiran et al., 2017). Zalesny et al. (2019) studied the growth and physiology of short-rotation woody crops (SRWCs) (*Populus* species and their hybrid poplars) for biomass production from prolonged (9 years) phytoremediation installations from 2012 to 2013 in the Midwestern and the Southeastern United States. Significant genotypic differences were observed and the ecosystem services provided by the contaminated poplars were comparable to ecosystem services provided by the noncontaminated poplars as bioenergy and biofuel feedstock. For example, biomass values of phytoremediation trees at the Midwestern sites ranged from 4.4 to 15.5 Mg ha^{-1} year^{-1} that was ~ 20% less in comparison with bioenergy trees.

Genetic resources—Genetic resources are important ecosystem provisional services as they are behind the phenomenon of the continuum of genes and species. Every country generally maintains a national plant genetic resource program and gene bank/seed bank to preserve the genetic resources of the native agricultural and forestry plants and animals. Generally, a positive relationship exists between species diversity and ES provisioning which is genetic diversity: a subset of biodiversity. Native breeds are species with special abilities for toleration against pests and pathogens, diseases, efficient users of nutrients, and have adapted against climate fluctuations (Mohammed and Ababa, 2019). Naturally growing native plants or revegetation on fly ash dumpsites (Pandey et al., 2015b; Kostić et al., 2018) or vegetation cover development in mining sites with native trees (Zalesny et al., 2019; Damastuti and de Groot, 2019; Wang et al., 2017) have contributed significantly to the maintenance of plant genetic resources.

3.2 Regulating services

Regulating services of the ecosystem are the services provided by the ecosystem to control natural phenomena. Regulating services include carbon sequestration, pollination, nutrient cycling, decomposition of plant and animal dead bodies/leftover, remediation of pollution, controlling soil erosion, mineral leaching, water purification, soil erosion and flood control, and mineral/nutrient cycles. As the plant cleans the air and the water, bacteria, and fungi decompose and degrade complex compounds to simpler compounds to make it available to plant, tree hold soil in place to prevent erosion, and bees pollinate flowers, all these processes together make ecosystems better, more sustainable and resilient to changes.

Managed grassland on reclaimed sites contributes to soil and water conservation practices act as filter strips for reducing runoff of surface water and also acts as buffer strips for defensing riparian zones. Reconstructed grasslands on fly ash disposal sites or mine sites treated with fly ash to reduce acidification (Roy et al., 2018b) prevent wind-mediated dispersal of fly ash particles over long distances and reduce health risks for nearby villagers (Pandey, 2015). A phytorestored site maintains all the following important regulating services of the ecosystem, and examples of them studied have been cited in Table 1.

Carbon sequestration—In general, the term carbon sequestration means long-term storage of carbon dioxide and other forms of carbon mainly in the plant body, soil, geologic formation, and the ocean to mitigate or reverse global warming. It is one of the ways to slow down the accumulation of CO_2 and other greenhouse gases, which are released by burning fossil fuels. Carbon farming is the best way for carbon sequestration that uses plants to trap CO_2. Strategic practices such as reducing tilling, planting long-rooted plants, and incorporating organic nutrients into the soil encourage the trapped carbon to move into and become trapped into the soil, further reducing GHG emissions. However, many agricultural, horticultural, forestry, and garden soils are losing more carbon than they are sequestering. The reverse movement of soil CO_2 to the atmosphere can be prevented through proper plant and soil management practice. By increasing vegetative cover, the ability of soil to sequester and store larger volumes of atmospheric carbon in a stable form can be enhanced. Soil contributes to carbon sequestration by transferring atmospheric CO_2 into plant carbon pool as soil inorganic carbon (elemental carbon and carbonate minerals like $MgCO_3$ and $CaCO_3$) and soil organic carbon (stored in plants and humic materials of soil) (Lal et al., 2015). Advantages of terrestrial C sequestration are (a) preventing climate change by decreasing atmospheric CO_2 amount, (b) improving nutrients cycling and storage, (c) restoring soil quality, (d) slowing down soil erosion, and (e) increasing biodiversity (Olson et al., 2014; Lal et al., 2015). The soil organic carbon pool is the greatest source of terrestrial carbon to 3 m depth (Olson et al., 2014). So meeting a positive soil carbon budget needs an increase in the input of aboveground (leaves and shoots; litter and understory vegetation) and belowground (roots and root exudates) biomass and application of compost and manure and decreasing soil erosion, oxidation, leaching, and mineralization (Lal et al., 2015). Revegetation of polluted and barren land like abundant fly ash ponds/landfills and open cast mine land creates a huge carbon sink and thus increases soil organic carbon (Pandey et al., 2016b). Pandey et al. (2016b) observed that CO_2 flux from vegetated sites of fly ash dump grounds was higher than from nonplanted areas. *Saccharum spontaneum* (84.29%) and *Prosopis juliflora* (92.09%) showed low CO_2 efflux rates and they liberated a minimum amount of CO_2 into the atmosphere resulting in greater CO_2 sequestration. *S. spontaneum* had the highest aboveground and belowground biomass and a very high content of organic carbon.

All these observations project that the grass species *S. spontaneum* is an ideal candidate for carbon sequestration. Research is needed to screen more grass, herbs, and trees for their suitability for carbon sequestration. In one case study, Airhwal and Maiti (2018) showed that CO_2 flux was more in forest soil (3.8 μmol CO_2 m^{-2} s^{-1}) than in one reclaimed coal mine site of 11 years old (2.4 μmol CO_2 m^{-2} s^{-1}) due to lesser biomass and soil organic carbon pool. However, as time passed, as more carbon accumulated in plant biomass, more CO_2 sequestered occurred in the revegetation planning and phytomanagement of reclaimed sites (Airhwal and Maiti, 2018).

Pollination—Pollination is an extremely important ES upon which food diversity, food security, and food prices all depend. Animal pollination like water- and wind-mediated pollination is an important ES driven by insects mainly but also by birds and mammals like bats and squirrels. According to an estimate, bees, birds, and bats are responsible for almost 35% of the world's total crop production (Spivak et al., 2011). Human activity like habitat destruction and improper pesticide misuse on animal pollinators has caused animal pollination under stress. The good news is that interest is growing among farmers of different countries to engage natural sustainable pollination services. Necessary steps taken are planting hedgerows, cheering up plant diversity, mulching, and stopping pesticides and inorganic fertilizers overuse. Also, increasing the floristic diversity of the host plants of the pollinators creates a huge pollination potential of insect pollinators. Perennial natural or phytorestored grasslands act as pollinator servicers through the diverse and generous amounts of native pollinator communities such as butterflies, bumblebees and hoverflies, and honeybees (Spivak et al., 2011). For instance, monocultures of switchgrass (*P. virgatum*) and plantings of mixed-prairie species supplied more ESs in the form of pollinator habitat, grassland bird habitat, and pest suppression than annual maize-ethanol production systems (Werling et al., 2014). Diverse floral species in the grasslands supplied not only nectar and pollen but also a peaceful habitat for ground-nesting bees. Native pollinators from grasslands also augment pollination services in nearby agricultural plantations. Commercial beekeeping services also rely on perennial forage and grazing lands for bee pasture.

Biological control—Biological control is the application of natural enemies of pests for reducing the pest population. Biological control is a natural way to keep the pest population under control and thereby reduce hazardous pesticide use. This also enhances biodiversity while ensuring pesticide-free production of food and nonfood materials (Howarth, 1991). For example, in aquaculture, Grass carp (white amur) is used that devours the weeds to keep the water clean and oxygenated (Thomas and Waage, 1996). Vegetation covers or grasslands or forests constructed on once degraded landscapes are also a great reservoir of natural pest eradicators. Biological control of weeds is an essential part of the phytomanagement practices (Roy et al., 2018a; Pandey et al., 2009).

Water quality and wastewater treatment—Water is the basic need of all types of ecosystem functioning. As vegetation is the key player in controlling water flow by binding soil, producing litter, and modifying soil structure, phytorestored sites contribute to this important ecosystem function of water conservation. Rhizofiltration is a direct wastewater treatment system and has been applied successfully in various cases (Ali et al., 2020). Water availability from the soil in an area depends not only on rainfall and water flow but also on soil structure, soil porosity, and moisture retention ability. Water storage potential of soil greatly depends on its plant cover, the amount of surface litter and litter decomposition rate, nature of soil organic compounds, and its biotic community (earthworms, bacteria, fungi, protozoa, etc.). In many instances, constructed wetlands were constructed for the remediation of wastewater (Oustriere et al., 2017).

Soil structure and maintenance of soil fertility—The purpose of ecorestoration is the maintenance of soil structure and fertility, one of the most important ESs. Important soil functions are supporting plant growth and other microbes and life forms, host biodiversity for habitats and species, nutrients and water storage, filtering, or transforming nutrients, source of various raw materials, and human activities including providing physical and cultural environment (Volchko et al., 2013). Soil pollution not only disturbs the aforementioned soil functions but have also bad effects on the soil-crop-animal-human system (Tiller, 1989). Certain soil-borne plants and microorganisms (termed as ecological engineers) always try to restore the specific soil functions and alleviate the detrimental effects of soil pollution (Mitsch and Jørgensen, 2003; Roychowdhury et al., 2016). So soil pollution can be controlled through the plantation of vegetation that also stabilizes dump slopes (Singh and Singh, 2006) and consequently maintains the ecological stability of the mining site. Successional growth and vegetation on dump slope act as hydrogeological and mechanical stabilizers where the plant root causes an increase in dump stability by intercepting rainwater and regulating evapotranspiration and pore pressure reduction (Hussain, 1995). All these regulate the hydrogeological cycle. The roots mechanically reinforce the mine dump material and increase the shear strength of the dump material. Singh et al. (2012) showed how revegetation enhanced the dump stability and fertility of the soil status in mine spoil. Here, plant and microbial biomass were the principal contributors to the ecological rejuvenation of the mine spoil.

Soil erosion—One of the important factors of vegetation cover development on degraded lands like fly ash disposal sites or sites subjected to mining and drilling is the prevention of loss of topsoil and consequent desertification. Soil erosion influences nutrient cycling and decreases soil fertility through the leaching of available nutrients. Soil erosion brings drastic modifications to the soil composition, structure, and physical-chemical properties of the soil matrix. The dust and sediments generated by mining and drilling may produce large off-site impacts. Redistribution of nutrients occurs during erosion as the topsoil migrates to other places or water bodies. It has been calculated that high rates of soil erosion influenced greater than 1.1 billion hectares of land worldwide (Berc et al., 2003) and redistributed 75 billion tons of soil (Pimente et al., 1995) containing 1.5%–5.0% C-content (Lal, 2003). In soil erosion, the soil nutrients are lost to riverine systems and eventually end up in oceans, causing major off-site impacts on the economy and wildlife. This causes the siltation of reservoirs and eutrophication of lakes (Starr et al., 2000). A vegetative cover prevents soil erosion and increases soil nutrients through the organic deposition of humus, carbon, nitrogen, solubilization of phosphorous, nitrogen fixation, etc.

3.3 Supporting services

Supporting services are essential for the production of the other three ecosystem services (provisional, regulating, and cultural services). Production of atmospheric O_2, soil formation, water and nutrient cycling, and provisioning of habitat are few important functions of the ecosystem's supporting services. Biodiversity conservation is also an important supporting service without which gaining services from ecosystems would have stopped. These processes allow Mother Earth to sustain basic life forms. Supporting services from man-made vegetation on degraded lands are comparable to those provided by natural vegetation.

Biodiversity and biodiversity conservation—Biodiversity conservation has been kept under the supporting service of the ecosystem since it supports (1) provisioning services that provide potable water, food, timber, fiber, and medicine; (2) regulating services that control our climate, pollination, disease control vectors; and (3) cultural services that influence our beliefs and traditions. It has been observed that reclaimed grasslands with native species supported a wide variety of animals and microbes (Pandey et al., 2015b; Pandey, 2020c). According to Werling et al. (2014), a careful depiction of bioenergy crops on marginal landscapes can stimulate multiple ESs in food and energy crops. Restoration of biodiversity was noticed in natural or seminatural ecosystems such as forests destroyed by mining (Damastuti and de Groot, 2019; Everard, 2009). Tropek et al. (2013, 2014) showed how vegetation development on fly ash disposal sites introduced a diverse variety of arthropod species to the sites.

Nutrient cycling—Nutrient cycling is one of the key ecosystem services that help in the sustenance of life on earth. The three main nutrient cycles are the water cycle, the carbon cycle, and the nitrogen cycle. These three cycles work together to replenish the ecosystem with the nutrients necessary to sustain life. The contribution of rhizobium and other nitrogen-fixing free-living microbes is noteworthy for their role in increasing the bioavailability of nitrogen through biological nitrogen fixation. Similarly, phosphate-solubilizing microbe contributes to increasing bioavailability of P and arbuscular mycorrhizal fungi (Gyuricza et al., 2010) are well documented for their contribution in nutrient cycling process in addition to litter decomposition and mineralization on the phytorestored sites (Dutta and Agrawala, 2001). Litter decomposition and mineralization cause the breakdown of organic materials into their monomeric forms through which nutrients become bioavailable for nutrient cycling (Ghaley et al., 2014). Landscape alteration, deforestation, and reduction of riparian forests, estuaries, and aquatic bodies cause an unbuffered flow of nutrients in between aquatic and terrestrial ecosystems.

Reforestation on deforested land, abandoned ash disposal site or mining land reestablishes stable nutrient cycling. Specific forms of biodiversity appear in the newly restored sites that are crucial for the accomplishment of the buffering mechanisms and ensure efficient use and recycling of nutrients. Swiatek et al. (2019) showed how the pool of carbon and mineral nutrients increased under alder plantings (*Alnus incana*, *Alnus viridis*, and *Alnus glutinosa*) on a nutrient-poor substrate composed of combustion wastes and sands. The establishment of nutrient cycling and decomposition are important factors of postmining forest restoration.

Microorganisms constitute a fundamental part of the nutrient cycling of N, C, and others (Littlefield et al., 2013) on severely damaged mining sites by establishing essential positive and negative interactions with plants (Jasper, 2007).

3.4 Cultural services

As the name suggests, cultural services are the nonmaterialistic benefits of an ecosystem that grants cultural development of people and show how ecosystems play a vital role in regional, national, and global cultures in the building of wisdom and spread of ideas. Cultural ESs and benefits include health, social interactions, sensory feelings, learning, cultural and symbolic importance, and identity. Like the natural forest, the lush green beauty of reconstructed grasslands is also a source of aesthetic inspiration, cultural identity, nature study, and spiritual experience. Some reconstructed forest ecosystem has created nature tourism (Damastuti and de Groot, 2019; Everard, 2009; Rosa et al., 2020; Gerner et al., 2018; Pandey, 2020c) and attracted millions of travelers worldwide. The benefit goes to both the visitors and the host and is also an income generation opportunity for sustainable conservation of the natural beauty. The reclaimed grasslands are as great as natural grasslands for outdoor activities, horse riding, bicycling, etc. According to Rosa et al. (2020), the value of reframing mine rehabilitation should incorporate cultural services in the postmining land use with the ecological goals and prove the social advantages of mine rehabilitation.

4 Opportunities and challenges in adaptive phytoremediation practices

To make phytoremediation more popular, a limited number of challenges must be identified and addressed. In recent years, many field trials were conducted to confirm the success of phytoremediation and determine the influence of various field factors on phytoremediation efficiency. For identifying new hyperaccumulator species and understanding their molecular mechanisms of stress responses, many laboratory studies were performed on soils spiked with many times higher concentrations of heavy metal(loid)s than those found at real contaminated sites (Shrestha et al., 2019; Din et al., 2020). Since 1990, a large number of phytoremediation trials including laboratory-based pot studies, small-scale, and large-scale field trials were conducted in different parts of the world (Willscher et al., 2013; Wang et al., 2019). These studies provided scientific data on the suitability and resilience of many plants under certain environmental stresses and discovered various hyperaccumulator species and practices to increase their uptake abilities. All these field trials have agreed that phytoremediation is a truly promising and cost-effective solution of pollution abatement much better than traditional physicochemical approaches.

Plant density and initial plant size (Kidd et al., 2015), cropping and harvestings like double cropping (Li et al., 2016), transplantation and double harvesting (Li et al., 2016), soil heterogeneity (Lim et al., 2017) were identified as a few success-determining factors. However, the small size of the small-scale field studies showed some variability due to their small size. A disadvantage with hyperaccumulators is that as they are adapted to grow for highly polluted soils and showed less efficiency in soils with a lesser level of contamination. Also, the dominant influential parameters could vary during outdoor field treatments, indicating that long-lasting studies that disclose annualized treatment efficiencies are desirable over shorter trials (Mertens et al., 2006). Recent studies have found less variability in larger, hectare-scale field trials than smaller studies. As an example, 2 years after conducting an agricultural trial with hyperaccumulator species *Sedum alfredii* and *P. vittata* across 11.1 ha of land contaminated with Pb (351 ppm), Cd (320 ppb), and As (37 ppm) in southwestern China (Wan et al., 2016), soluble concentrations of Pb, Cd, and As were decreased by 30.4%, 85.8%, and 55.3%, respectively. Studies were also performed to evaluate the sustainability of phytoremediation. For example, to assess the sustainability of a SRWC-based phytoremediation, a trial was made at a heavy metal(loid)-contaminated agricultural site in Belgium (Ruttens et al., 2011). It was proved that this approach is more sustainable than other remediation methods owing to its role to capture atmospheric C in plant biomass while treating soil contamination. Various organic amendments procured from different biological waste materials like compost (Touceda-González et al., 2017), manure (Galende et al., 2014), sewage sludge (Placek et al., 2016), and industrial waste (such as fly ash and red mud) have been used to increase phytoremediation efficiency, but some controversies exist regarding the harmful effects of the contaminants present in the amendments themselves (Panda et al., 2018; Kursun and Terzi, 2018) and whether they improve phytoremediation performance (Wang et al., 2019).

Unlike laboratory-based pot studies, some inconsistencies and nonreproducible results were reported in the field trials and the same plants grown in the same plots showed heterogeneous results (Li et al., 2019; Khaokaew and Landrot, 2015). Several plausible reasons have been hypothesized for this inconsistency. Foremost is the extrapolation of results from pot studies that may have made overly optimistic predictions about heavy metal uptake in the field (Vangronsveld et al., 2009). Different environmental conditions such as contamination amount, soil organic matter, pH, salinity, clay content, moisture, water content, and weather patterns that effect phytoremediation efficiency are generally constant and set at favorable levels in greenhouse/indoor studies and lead to homogenous results. But fluctuations of the same factors cause heterogeneity in field-scale studies. Second, the field heterogeneity (changes of sampling points) causes large differences making interpretation of results challenging.

Phytoremediation of urban and periurban environments by the application of ornamental (woody and flowering) plants has displayed many positive directions (Capuana, 2020). Many flowering plants contribute "multipurpose" services and play important role in the ecorestoration and aesthetic enhancement. The success of the phytomanagement depends on the selection of native species and/or genotypes matching the specific environments and pollutants (Pandey and Bajpai, 2019). Many plant species, categorized within metal accumulator groups and not hyperaccumulators, tend to be more promising for exploitation in phytoremediation as they have greater biomass, better growth potential, and more in-depth rooting. Among woody trees, Salicaceae are the most investigated species for phytoremediation purposes (Marmiroli et al., 2011; Gajić et al., 2018) that include *Salix* spp., and poplars. Due to their adaptabilities, faster growth, ease of propagations, and appreciable performances under pollutant stresses, these species have found more practical utilizations in phytoremediation practices, especially in semiurban areas. In urban localities, other woody species are more suitable, being featured by a higher ornamental value. Phytomanagement with woody species is associated with revenues from two possibilities: nutrient recovery via composting of the foliage and energy recovery via biomass productivity (Roeland et al., 2019; Pandey and Souza-Alonso, 2019). For instance, biomass from a sustainably grown tree is projected to provide 10% of the UK energy needs by 2050 and can remarkably contribute to the cut-down of GHG emissions (Roeland et al., 2019).

In contrast to woody species, herbs and grasses are characterized by greater plasticity and variability, and frequent replacements are possible. Among flowering herbaceous plants, there is a wider choice. For example, the Asteraceae family has many interesting species (Nikolić and Stevović, 2015) with sunflower representing an ideal one for the remediation of specific pollutants. With more research, several interesting species would be explored. Different plant communities interact among themselves and make phytoremediation an aesthetic experience. Finally, sincere attempts are needed to overcome the disposal problem of contaminated biomass and limiting the phytorestoration costs by the valorization of the biomass and trading other provisional services. A recent concept is to consider the phytorestored sites as nature-based solutions to achieve environmental, cultural, social, economic, and policy planning targets (Hansen and Pauleit, 2014). The integration of the nature-based concept can be viewed as a way to maximize resilience to climate change while minimizing the associated disservices (e.g., maintenance costs, set up cost). Although there is no doubt that the phytoremediated sites would reduce air pollution and greenhouse gas emission; regulate air temperature; sequester carbon; reduce storm water runoff; prevent noise pollution; and provide social, recreational, psychological, and aesthetic benefits, there is still scarce information present on the provisional ESs shown by the phytoremediated sites and how they can act as a key player toward the optimization of ESs. It is important to understand that there is a substantial potential of returning a good economic value to human societies and job prospects in urban/semiurban areas.

Like the urban forest indicators, indicators should be thought of for phytorestored forest/grasslands also that would provide quantitative assessments of ESs in vast areas and across cities. An improved understanding of ESs from phytorestored sites is essential to increase its popularity and acceptability among public entities and private companies. According to Vaughn et al. (2010), the main steps of restoring a site should be (i) evaluation of the site: a rigorous assessment of the current parameters at the restoration site is necessary for determining what kind of actions will be obligatory. Here, the reasons for ecosystem disturbance and how to stop them or reversing them should be identified (e.g., disposal of waste); (ii) setting up project goals: either to understand targets for the restoration job or searching and screening of new potential species best suited to various stresses; practitioners may visit reference target site to assess surroundings (environments and communities); (iii) removing sources of disturbance: before restoration can be successful, factors responsible for disturbance should be eliminated. For example, cessation of mining or removing toxic materials from soil or sediments is necessary to restore or revegetate the mining site; (iv) restoring processes/disturbance cycles: sometimes adopting key ecological processes such as natural flood or fire regimes can successfully restore ecosystem integrity. In these cases, native plants and animals that have evolved to tolerate or require natural disturbance regimes may come back on their own without direct action by phytomanagers; (v) rehabilitating substrates: these are the activities that can improve soil and water qualities through the addition of nutrients or other amendments for setting up of vegetation or restore the water quality and hydrological regimes; and (vi) revegetation: restoration activities involve direct revegetation of the site once conditions for vegetation set up are complete.

Careful selection of native species (woody species or grass or a mixed plantation) depending on the types of pollutants present should be analyzed before starting the vegetation plantation process from seeds/seedlings. Genetic diversity should be assured by selecting seeds or cuttings from various sources within the locality. Long-time surveillance of the restoration project is vital to understand the extent of success. It helps in the adoption of the future course of actions (e.g., harvesting biomass or periodic weed removal). Eventually given proper time, the restoration project would achieve a self-sustaining ecosystem and deliver all-natural ecosystem functions without any human intervention.

5 Summary and conclusion

Phytoremediation has proved its worth as the most promising clean green technology for the restoration of polluted lands. The advantages of phytoremediation over other physicochemical techniques are now proven. Establishment of vegetation cover by native grass, trees, shrubs, and legumes on degraded lands like mine waste site, fly ash disposal site, and other polluted site ecorestores the sites to the original natural conditions and accelerates the probability of spontaneous colonization by diverse native species. Thus, finding and selecting the most suitable native plant species is one of the most crucial factors in revegetation, permanent stability, and resilience of the ecosystem.

Phytomanagement of degraded lands like mine waste site and fly ash sites provides important ESs like (a) prevention or minimization of erosion by wind and control of flood; (b) stabilization of the steep slopes on the dyke of lagoons/ash basins; (c) reduction of toxicity and bioavailability or mobility of pollutants in the environment and protection of nearby ecosystems; (d) enhancement of soil fertility and supply of organic compounds which can bind pollutants; (e) sequestration of a large amount of CO_2 and utilization of the biomass harvested for bioenergy that in turn again would reduce dependency on fossil fuel and decrease greenhouse gas release and thus contribute to climate change prevention; (f) conservation of biodiversity of native trees, herbs, and grasses; (g) contribution to nutrient cycling; and (h) contribution to important cultural services with the clean green environment.

More field-based trials should be conducted. Local biodiversity should be explored further in search of native species that are very well adapted to the prevailing climatic conditions. Use of the heavy metal hyperaccumulator for the removal of metal from the soil is one of the most straightforward approaches for phytoremediation of metal contaminated sites, and more than hundreds of hyperaccumulator species have been picked up globally that could hyperaccumulate a variety of heavy metals. However, natural hyperaccumulators suffer from a few drawbacks like the slow process of remediation, slow growth rate, and the yield of low biomass that at the end proves to be nonprofit making for biomass-bioenergy-biofuel. Fortunately, genetic engineering approaches are coming out as a powerful tool to insert desirable characters/genes into the plants like increased heavy metal tolerance and accumulation, great biomass production, faster growth, and heightened ability to adapt quickly to existing climatic and soil conditions. Therefore, a thorough apprehension of the heavy metal uptake, translocation, accumulation, and detoxification mechanisms in plants, and characterization of genes and molecules involved will contribute to the rational design of an ideal candidate for phytoremediation via transgenic technology.

As the endophytic microbes cause degradation of organic pollutants and the rhizomicrobes cause immobilization or mobilization of metals onto their cell walls through the change of metal speciation status and other processes like precipitation the establishment of a strong microbial community can assist the plant in the revegetation process greatly. The new genomic technologies along with identifying novel genes for pollutant remediation rely more on plant-microbe interactions for remediation/degradation of complex organic wastes, and pesticides. The future direction of research should also involve plant ecophysiological responses to environmental distresses by involving biochemistry, biophysics, genetics, physiology, and "omics" tools (genomics, transcriptomics, proteomics, metabolomics, etc.) for boosting phytoremediation technologies and restoration of polluted environments.

References

Ahirwal, J., Pandey, V.C., 2021. Ecological rehabilitation of mine-degraded land for sustainable environmental development in emerging nations. Restor. Ecol. https://doi.org/10.1111/rec.13268.

Airhwal, J., Maiti, S.K., 2018. Carbon sequestration and soil CO_2 flux in reclaimed coal mine lands from India. In: Prasad, M.N.V., Favas, P.J.C., Maiti, S.K. (Eds.), Bio-Geotechnologies for Mine Site Rehabilitation. Elsevier, Amsterdam, pp. 371–392.

Ali, S., Abbas, Z., Rizwan, M., et al., 2020. Application of floating aquatic plants in phytoremediation of heavy metals polluted water: a review. Sustainability 12, 1927.

Alkorta, I., Garbisu, C., 2001. Phytoremediation of organic contaminants in soils. Bioresour. Technol. 79 (3), 273–276.

Alkorta, I., Becerril, J.M., Garbisu, C., 2010. Phytostabilization of metal contaminated soils. Rev. Environ. Health 25, 135–146.

Alvarenga, P., Goncalves, A.P., Fernandes, R.M., de Varennes, A., Vallini, G., Duarte, E., Cunha-Queda, A.C., 2009. Organic residues as immobilizing agents in aided phytostabilization: (I) effects on soil chemical characteristics. Chemosphere 74, 1292–1300.

Baker, A.J.M., McGrath, S.P., Sidoli, C.M.D., Reeves, R.D., 1994. The possibility of in situ heavy metal decontamination of polluted soils using crops of metal-accumulating plants. Resour. Conserv. Recycl. 11, 41–49.

Baker, A.J.M., McGrath, S.P., Reeves, R.D., Smith, J.A.C., 2000. Metal hyperaccumulator plants: a review of the ecology and physiology of a biological resource for phytoremediation of metal-polluted soils. In: Terry, N., Bañuelos, G. (Eds.), Phytoremediation of Contaminated Soil and Water. Lewis Publishers, Boca Raton, FL, pp. 85–107.

Barrutia, O., Garbisu, C., Hernandez-Allica, J., Garcıa-Plazaola, J.I., Becerril, J.M., 2010. Differences in EDTA-assisted metal phytoextraction between metallicolous and non-metallicolous accessions of *Rumex acetosa* L. Environ. Pollut. 158, 1710–1715.

Becerra-Castro, C., Kidd, P.S., Rodríguez-Garrido, B., Monterroso, C., Santos-Ucha, P., Prieto-Fernández, A., 2013. Phytoremediation of hexachlorocyclohexane (HCH)-contaminated soils using *Cytisus striatus* and bacterial inoculants in soils with distinct organic matter content. Environ. Pollut. 178, 202–210.

Berc, J., Lawford, R., Bruce, J., Mearns, L., et al., 2003. Conservation Implications of Climate Change: Soil Erosion and Runoff From Cropland: A Report From the Soil and Water Conservation Society. Soil and Water Conservation Society, Ankeny, IA.

Bianconi, D., De Paolis, M.R., Agnello, A.C., Lippi, D., Pietrini, F., Zacchini, M., Polcaro, C., Donati, E., Paris, P., Spina, S., Massacci, A., 2011. Field-scale rhyzoremediation of a contaminated soil with hexachlorocyclohexane (HCH) isomers: the potential of poplars for environmental restoration and economical sustainability. In: Golubev, I.A. (Ed.), Handbook of Phytoremediation. Nova Science Publishers, Inc., New York, pp. 783–794.

Boularbah, A., Schwartz, C., Bitton, G., Aboudrar, W., Ouhammou, A., Morel, J.L., 2006. Heavy metal contamination from mining sites in South Morocco: 2. Assessment of metal accumulation and toxicity in plants. Chemosphere 63, 811–817.

Bryan, A.L., Hopkins, W.A., Parikh, J.H., Jackson, B.P., Unrine, J.M., 2012. Coal fly ash basins as an attractive nuisance to birds: parental provisioning exposes nestlings to harmful trace elements. Environ. Pollut. Res. 161, 170–177.

Burges, A., Epelde, L., Benito, G., Artetxe, U., Becerril, J.M., Garbisu, C., 2016. Enhancement of ESs during endophyte-assisted aided phytostabilization of metal contaminated mine soil. Sci. Total Environ. 562, 480–492.

Cappuyns, V., 2013. Environmental impacts of soil remediation activities: quantitative and qualitative tools applied on three case studies. J. Clean. Prod. 52, 145–154.

Capuana, M., 2020. A review of the performance of woody and herbaceous ornamental plants for phytoremediation in urban areas. iForest 13, 139–151.

Chaney, R.L., Angle, J.S., Broadhurst, C.L., Peters, C.A., Tappero, R.V., Sparks, D.L., 2007. Improved understanding of hyperaccumulation yields commercial phytoextraction and phytomining technologies. J. Environ. Qual. 36, 1429–1443.

Clemens, S., Palmgren, M.G., Krämer, U., 2002. A long way ahead: understanding and engineering plant metal accumulation. Trends Plant Sci. 7, 309–315.

Cooke, S.J., Suski, C.D., 2010. Ecological restoration and physiology: an overdue integration. Bioscience 58, 957–968.

Dadea, C., Russo, A., Tagliavini, M., Mimmo, T., Zerbe, S., 2017. Tree species as tools for biomonitoring and phytoremediation in urban environments: a review with special regard to heavy metals. Arboric. Urban For. 43, 155–167.

DalCorso, G., Fasani, E., Manara, A., Visioli, G., Furini, A., 2019. Heavy metal pollutions: state of the art and innovation in phytoremediation. Int. J. Mol. Sci. 20, 3412.

Dalvi, A.A., Bhalerao, S.A., 2013. Response of plants towards heavy metal toxicity: an overview of avoidance, tolerance and uptake mechanism. Ann. Plant Sci. 2, 362–368.

Damastuti, E., de Groot, R., 2019. Participatory ecosystem service mapping to enhance community-based mangrove rehabilitation and management in Demak, Indonesia. Reg. Environ. Change 19, 65–78.

De Jong, D., Morse, R.A., Gutenmann, W.H., Lisk, D.J., 1977. Selenium in pollen gathered by bees foraging on fly ash-grown plants. Bull. Environ. Contam. Toxicol. 18 (4), 442–444.

De Souza, M.P., Lytle, C.M., Mulholland, M.M., Otte, M.L., Terry, N., 2000. Selenium assimilation and volatilization from dimethylselenoniopropionate by Indian mustard. Plant Physiol. 122, 1281–1288.

Deng, L., Li, Z., Wang, J., Liu, H., Li, N., Wu, L., Hu, P., Luo, Y., Christie, P., 2016. Long-term field phytoextraction of zinc/cadmium contaminated soil by *Sedum plumbizincicola* under different agronomic strategies. Int. J. Phytoremediation 18, 134–140.

Din, B.U., et al., 2020. Assisted phytoremediation of chromium spiked soils by *Sesbania sesban* in association with *Bacillus xiamenensis* PM14: a biochemical analysis. Plant Physiol. Biochem. 146, 249–258.

Doty, S.L., Freeman, J.L., Cohu, C.M., Burken, J.G., Firrincieli, A., Simon, A., 2017. Enhanced degradation of TCE on a superfund site using endophyte-assisted poplar tree phytoremediation. Environ. Sci. Technol. 51 (17), 10050–10058.

Dutta, R.K., Agrawala, M., 2001. Litterfall, litter decomposition and nutrient release in five exotic plant species planted on coal mine spoils. Pedobiologia 45 (4), 298–312.

Emery, M., Martin, S., Dyke, A., 2006. Wild Harvests From Scottish Woodlands: Social, Cultural And Economic Values of Contemporary Non-Timber Forest Products. F Commission, Edinburgh, Scotland.

Environment Agency (EA), 2009. Dealing with Contaminated Land in England and Wales: A Review of Progress from 2000–2007 With Part 2A of the Environmental Protection Act.

Epelde, L., Burges, A., Mijangos, I., Garbisu, C., 2014a. Microbial properties and attributes of ecological relevance for soil quality monitoring during a chemical stabilization field study. Appl. Soil Ecol. 75, 1–12.

Epelde, L., Becerril, J.M., Alkorta, I., Garbisu, C., 2014b. Adaptive long-term monitoring of soil health in metal phytostabilization: ecological attributes and ESs based on soil microbial parameters. Int. J. Phytoremediation 16, 971–981.

Everard, M., 2009. Ecosystem Services Case Studies. Environment Agency. www.environ-ment-gency.gov.uk.

Feng, N.X., Yu, J., Zhao, H.M., Cheng, Y.T., et al., 2017. Efficient phytoremediation of organic contaminants in soils using plant-endophyte partnerships. Sci. Total Environ. 583, 352–368.

Foyer, C., Noctor, G., 2005. Redox homeostasis and antioxidant signaling: a metabolic interface between stress perception and physiological responses. Plant Cell 17, 1866–1875.

Foyer, C., Shigeoka, S., 2011. Understanding oxidative stress and antioxidant functions to enhance photosynthesis. Plant Physiol. 155, 93–100.

Gajić, G., Pavlović, P., Kostić, O., Jarić, S., Đurđević, L., Pavlović, D., Mitrović, M., 2013. Ecophysiological and biochemical traits of three herbaceous plants growing on the disposed coal combustion fly ash of different weathering stage. Arch. Biol. Sci. 65 (4), 1651–1667.

Gajić, G., Đurđević, L., Kostić, O., Jarić, S., Mitrović, M., Stevanović, B., Pavlović, P., 2016. Assessment of the phytoremediation potential and an adaptive response of *Festuca rubra* L. sown on fly ash deposits: native grass has a pivotal role in ecorestoration management. Ecol. Eng. 93, 250–261.

Gajić, G., Djurdjević, L., Kostić, O., Jarić, S., Mitrović, M., Pavlović, P., 2018. Ecological potential of plants for phytoremediation and ecorestoration of fly ash deposits and mine wastes. Front. Environ. Sci. https://doi.org/10.3389/fenvs.2018.00124.

Gajić, G., Djurdjević, L., Kostić, O., Jarić, S., Stevanović, B., Mitrović, M., Pavlović, P., 2020. Phytoremediation potential, photosynthetic and antioxidant response to arsenic-induced stress of *Dactylis glomerata* L. sown on fly ash deposits. Plants 9, 657.

Galende, M.A., Becerril, J.M., Barrutia, O., Artetxe, U., Garbisu, C., Hernández, A., 2014. Field assessment of the effectiveness of organic amendments for aided phytostabilization of a Pb-Zn contaminated mine soil. J. Geochem. Explor. 145, 181–189. https://doi. org/10.1016/j.gexplo.2014.06.006.

Gerner, N.V., Nafo, I., Winking, C., Wencki, K., et al., 2018. Large-scale river restoration pays off: a case study of ecosystem service valuation for the Emscher restoration generation project. Ecosyst. Serv. 30, 327–338.

Ghaley, B.B., Porter, J.R., Sandhu, H.S., 2014. Soil-based ESs: a synthesis of nutrient cycling and carbon sequestration assessment methods. Int. J. Biodivers. Sci. Ecosyst. Serv. Manag. 10 (3), 177–186.

Glass, D.J., 2000. Phytoremediation's economic potential. In: Raskin, I., Ensley, B.D. (Eds.), Phytoremediation of Toxic Metals: Using Plants to Clean Up the Environment. John Wiley and Sons Inc., New York, pp. 15–31.

Grbović, F., Gajić, G., Branković, S., Simić, Z., Ćirić, A., Rakonjac, L., Pavlović, P., Topuzović, M., 2019. Allelopathic potential of selected woody species growing on fly-ash deposits. Arch. Biol. Sci. 71 (1), 83–94.

Grbović, F., Gajić, G., Branković, S., Simić, Z., Vuković, N., Pavlović, P., Topuzović, M., 2020. Complex effect of *Robinia pseudoacacia* L. and *Ailanthus altissima* (Mill.) Swingle growing on asbestos deposits: allelopathy and biogeochemistry. J. Serb. Chem. Soc. 85 (1), 141–153.

Guarino, C., Paura, B., Sciarrillo, R., 2018. Enhancing phytoextraction of HMs at real scale, by combining Salicaceae trees with microbial consortia. Front. Environ. Sci. 6, 137.

Gupta, A.K., Sinha, S., 2008. Decontamination and/or revegetation of fly ash dykes through naturally growing plants. J. Hazard. Mater. 153 (3), 1078–1087.

Gyuricza, V., de Boulois, H.D., Declerck, S., 2010. Effect of potassium and phosphorus on the transport of radiocesium by arbuscular mycorrhizal fungi. J. Environ. Radioact. 101, 482–487.

Hansen, R., Pauleit, S., 2014. From multifunctionality to multiple ESs? A conceptual framework for multifunctionality in green infrastructure planning for urban areas. Ambio 43, 516–529.

Harrison, M.J., 1999. Molecular and cellular aspects of the arbuscular mycorrhizal symbiosis. Annu. Rev. Plant. Physiol. Plant. Mol. Biol. 50, 361–389.

Howarth, F.G., 1991. Environmental impacts of classical biological control. Annu. Rev. Entomol. 36, 485–509.

Howden, R., Goldsbrough, P.B., Andersen, C.R., Cobbett, C.S., 1995. Cadmium-sensitive, cad1 mutants of *Arabidopsis thaliana* are phytochelatin deficient. Plant Physiol. 107, 1059–1066.

Hussain, A., 1995. Fill compaction-erosion study in reclaimed areas. Indian Min. Eng. J. 34, 19–21.

ITRC, December 2003. Technical and Regulatory Guidance for the Triad Approach: A New Paradigm for Environmental Project Management.

Jamil, S., Abhilash, P.C., Singh, N., Sharma, P.N., 2009. Jatropha curcas: a potential crop for phytoremediation of coal fly ash. J. Hazard. Mater. 172 (1), 269–275. https://doi. org/10.1016/j.jhazmat.2009.07.004.

Jasper, D.A., 2007. Beneficial soil microorganisms of the jarrah forest and their recovery in bauxite mine restoration in southwestern Australia. Restor. Ecol. 15, S74–S84.

Jiang, Y., Lei, M., Duan, L., Longhurst, P., 2015. Integrating phytoremediation with biomass valorisation and critical element recovery: a UK contaminated land perspective. Biomass Bioenergy 83, 328–339.

Khaokaew, S., Landrot, G., 2015. A field-scale study of cadmium phytoremediation in a contaminated agricultural soil at Mae Sot district, Tak Province, Thailand: (1) Determination of Cd-hyperaccumulating plants. Chemosphere 138, 883–887.

Kidd, P., et al., 2015. Agronomic practices for improving gentle remediation of trace element-contaminated soils. Int. J. Phytoremediation 17, 1005–1037.

Kiran, B., Majeti, P., Sateesh, S., 2017. *Ricinus communis* L. (castor bean) as a potential candidate for revegetating industrial waste contaminated sites in peri-urban Greater Hyderabad: remarks on seed oil. Environ. Sci. Pollut. Res. Int. 24. https://doi.org/10.1007/s11356-017-9654-5.

Kisku, G.C., Kumar, V., Sahu, P., Kumar, P., Kumar, N., 2018. Characterization of coal fly ash and use of plants growing in ash pond for phytoremediation of metals from contaminated agricultural land. Int. J. Phytoremediation 20 (4), 330–337.

Kostić, O., Jarić, S., Gajić, G., Pavlović, D., Pavlović, M., Mitrović, M., Pavlović, P., 2018. Pedological properties and ecological implications of substrates derived 3 and 11 years after the revegetation of lignite fly ash disposal sites in Serbia. Catena 163, 78–88.

Kumpiene, J., Guerri, G., Landi, L., Pietramellara, G., Nannipieri, P., Renella, G., 2009. Microbial biomass, respiration and enzyme activities after in situ aided phytostabilization of a Pb- and Cu-contaminated soil. Ecotoxicol. Environ. Saf. 72, 115–119.

Kursun, U.I., Terzi, M., 2018. Distribution of trace elements in coal and coal fly ash and their recovery with mineral processing practices: a review. J. Min. Environ. 9, 641–655.

Lal, R., 2003. Soil erosion and the global C budget. Environ. Int. 29, 437–450.

Lal, R., Negassa, W., Lorenz, K., 2015. Carbon sequestration in soil. Curr. Opin. Environ. Sustain. 15, 79–86.

Leshem, Y.Y., Kuiper, P.J.C., 1996. Is there a gas (general adaptation syndrome) response to various types of environmental stress? Biol. Plant. 38, 1.

Li, N., et al., 2016. Effects of double harvesting on heavy metal uptake by six forage species and the potential for phytoextraction in field. Pedosphere 26, 717–724.

Li, X., Wang, X., Chen, Y., Yang, X., Cui, Z., 2019. Optimization of combined phytoremediation for heavy metal contaminated mine tailings by a field-scale orthogonal experiment. Ecotoxicol. Environ. Saf. 168, 1–8.

Lim, J.E., et al., 2017. Impact of natural and calcined starfish (*Asterina pectinifera*) on the stabilization of Pb, Zn and As in contaminated agricultural soil. Environ. Geochem. Health 39, 431–441.

Littlefield, T., Barton, C., Arthur, M., Coyne, M., 2013. Factors controlling carbon distribution on reforested minelands and regenerating clearcuts in Appalachia, USA. Sci. Total Environ. 465, 240–247.

Lixandru, B., Dragomir, N., Morariu, F., Popa, M., Coman, A., Savescu, N., et al., 2013. Researches regarding the adaptation process of the species *Miscanthus giganteus* under the conditions of fly ash deposit from Utvin, Timis County. Sci. Pap. Anim. Sci. Biotechnol. 46 (1), 199–203.

Ma, L.Q., Komar, K.M., Tu, C., Zhang, W., Cai, Y., Kennelley, E.D., 2001. A fern that hyperaccumulates arsenic. Nature 409, 579.

Ma, Y., Prasad, M., Rajkumar, M., Freitas, H., 2011. Plant growth promoting rhizobacteria and endophytes accelerate phytoremediation of metalliferous soils. Biotechnol. Adv. 29, 248–258.

Maila, M., Randima, P., Cloete, T.E., 2005. Multispecies and monoculture rhizoremediation of polycyclic aromatic hydrocarbons (PAHs) from the soil. Int. J. Phytoremediation 7 (2), 87–98.

Mantri, N., Patade, V., Penna, S., Ford, R., Pang, E., 2012. Abiotic stress responses in plants: present and future. In: Ahmad, P., Prasad, M. (Eds.), Abiotic Stress Responses in Plants. Springer, New York, NY, pp. 1–19.

Marmiroli, M., Pietrini, F., Maestri, E., Zacchini, M., Marmiroli, N., Massacci, A., 2011. Growth, physiological and molecular traits in Salicaceae trees investigated for phytoremediation of heavy metals and organics. Tree Physiol. 31, 1319–1334.

Mastretta, C., Taghavi, S., Van Der Lelie, D., Mengoni, A., et al., 2009. Endophytic bacteria from seeds of *Nicotiana tabacum* can reduce cadmium phytotoxicity. Int. J. Phytoremediation 11, 251–267.

Meehan, T.D., Werling, B.P., Landis, D.A., Gratton, C., 2012. Pest-suppression potential of midwestern landscapes under contrasting bioenergy scenarios. PLoS One 7 (7), e41728.

Mekuria, W., Amare, T., Wubet, A., Feyisa, T., Yitaferu, B., 2018. Restoration of degraded landscapes for ESs in North-Western Ethiopia. Heliyon 4, e00764.

Meng, L., Qiao, M., Arp, H., 2011. Phytoremediation efficiency of a PAH-contaminated industrial soil using ryegrass, white clover, and celery as mono- and mixed cultures. J. Soil. Sediment. 11 (3), 482–490.

Meng, M., Li, B., Shao, J.J., Wang, T., He, B., Shi, J.B., Ye, Z.H., Jiang, G.B., 2014. Accumulation of total mercury and methylmercury in rice plants collected from different mining areas in China. Environ. Pollut. 184, 179–186.

Mertens, J., Vervaeke, P., Meers, E., Tack, F.M.G., 2006. Seasonal changes of metals in willow (Salix sp.) stands for phytoremediation on dredged sediment. Environ. Sci. Technol. 40, 1962–1968.

Mertz, O., Ravnborg, H.M., Lövei, G.L., Nielsen, I., Konijnendijk, C.C., 2007. ESs and biodiversity in developing countries. Biodivers. Conserv. 16, 2729–2737.

Meyer, M.A., Chand, T., Priess, J.A., 2015. Comparing bioenergy production sites in the Southeastern US regarding ES supply and demand. PLoS One 10 (3), e0116336.

Minkina, T.M., Fedorenko, G.M., Nevidomskaya, D.G., et al., 2019. Bioindication of soil pollution in the delta of the Don River and the coast of the Taganrog Bay with heavy metals based on anatomical, morphological and biogeochemical studies of macrophyte (*Typha australis* Schum. & Thonn). Environ. Geochem. Health. https://doi.org/10.1007/s10653-019-00379-3.

Mitrović, M., Pavlović, P., Lakusić, D., Stevanović, B., Djurdjevic, L., Kostić, O., et al., 2008. The potential of *Festuca rubra* and *Calamagrostis epigejos* for the revegetation on fly ash deposits. Sci. Total Environ. 72, 1090–1101.

Mitrović, M., Jaric, S., Kostić, O., Gajić, G., Karadzić, B., Lola Djurdjević, L., et al., 2012. Photosynthetic efficiency of four woody species growing on fly ash deposits of a Serbian 'Nicola Tesla-A' thermoelectric plant. Pol. J. Environ. Stud. 21 (5), 1339–1347.

Mitsch, W.J., Jørgensen, S.E., 2003. Bioremediation restoration of contaminated soils. In: Mitsch, W.J., Jørgensen, S.E. (Eds.), Ecological Engineering and Ecosystem Restoration, second ed. John Wiley and Sons, Inc., New York, pp. 263–286.

Mittler, R., 2002. Oxidative stress, antioxidants and stress tolerance. Trends Plant Sci. 7, 405–410.

Mohammadzadeh, A., Tavakoli, M., Chaichi, M.R., Motesharezadeh, B., 2014. Effects of nickel and PGPBs on growth indices and phytoremediation capability of sunflower (*Helianthus annuus* L.). Arch. Agron. Soil Sci. 60 (12), 1765–1778.

Mohammed, J., Ababa, A., 2019. The role of genetic diversity to enhance ES. Am. J. Biol. Environ. Stat. 5 (3), 46–51.

Moreno, F.N., Anderson, C.W.N., Stewart, R.B., Robinson, B.H., 2008. Phytofiltration of mercury-contaminated water: volatilisation and plant-accumulation aspects. Environ. Exp. Bot. 62, 78–85.

Motesharezadeh, B., Kamalpoor, S., Alikhani, H.A., Zarei, M., Azimi, S., 2017. Investigating the effects of plant growth promoting bacteria and *Glomus mosseae* on cadmium phytoremediation by *Eucalyptus camaldulensis* L. Pollution 3, 575–588.

Mukherjee, P., Mitra, A., Roy, M., 2019. Halomonas rhizobacteria of *Avicennia marina* of Indian Sundarbans promote rice growth under saline and heavy metal stresses through exopolysaccharide production. Front. Microbiol. https://doi.org/10.3389/fmicb.2019.01207.

National Research Council (NRC), 2003. Environmental Cleanup at Navy Facilities: Adaptive Site Management. The National Academies Press, Washington, DC, https://doi.org/10.17226/10599.

Newman, L.A., Reynolds, C.M., 2005. Bacteria and phytoremediation: new uses for endophytic bacteria in plants. Trends Biotechnol. 23, 6–8.

Nicholson, E., Mace, G.M., Armsworth, P.R., Atkinson, G., et al., 2009. Priority research areas for ESs in a changing world. J. Appl. Ecol. 46, 1139–1144.

Nienow, S., McNamara, K.T., Gillespie, A.R., 2000. Assessing plantation biomass for co-firing with coal in northern Indiana: a linear programming approach. Biomass Bioenergy 18 (2), 125–135.

Nikolić, M., Stevović, S., 2015. Family Asteraceae as a sustainable planning tool in phytoremediation and its relevance in urban areas. Urban For. Urban Green. 14, 782–789.

Olivares, A.R., Carrillo, R., Gonzalez, C., et al., 2012. Potential of castor bean (*Ricinus communis* L.) for phytoremediation of mine tailings and oil production. J. Environ. Manag. 114, 316–323.

Olson, K.R., Al-Kaisi, M.M., Lal, R., Lowery, B., 2014. Experimental consideration, treatments and methods in determining soil organic carbon sequestration rates. Soil Sci. Soc. Am. J. 78, 348–360.

Oustriere, N., Marchand, L., Roulet, E., Mench, M., 2017. Rhizofiltration of a Bordeaux mixture effluent in pilot-scale constructed wetland using *Arundo donax* L. coupled with potential Cu-ecocatalyst production. Ecol. Eng. 105, 295–305.

Padmavathiamma, P.K., Li, L.Y., 2012. Rhizosphere influence and seasonal impact on phytostabilisation of metals—a field study. Water Air Soil Pollut. 223, 107–124.

Palanivel, T.M., Pracejus, B., Victor, R., 2020. Phytoremediation potential of castor (*Ricinus communis* L.) in the soils of the abandoned copper mine in Northern Oman: implications for arid regions. Environ. Sci. Pollut. Res. 27, 17359–17369.

Panda, D., Panda, D., Padhan, B., Biswas, M., 2018. Growth and physiological response of lemongrass (*Cymbopogon citratus* (D.C.) Stapf.) under different levels of fly ash-amended soil. Int. J. Phytoremediation 20, 538–544.

Pandey, V.C., 2015. Assisted phytoremediation of fly ash dumps through naturally colonized plants. Ecol. Eng. 82, 1–5. https://doi.org/10.1016/j.ecoleng.2015.04.002.

Pandey, V.C., 2020a. Phytomanagement of Fly Ash. Elsevier, Amsterdam, p. 334, https://doi.org/10.1016/C2018-0-01318-3.

Pandey, V.C., 2020b. Fly ash properties, multiple uses, threats, and management: an introduction. In: Phytomanagement of Fly Ash. Elsevier, Amsterdam, pp. 1–34, https://doi.org/10.1016/B978-0-12-818544-5.00001-8.

Pandey, V.C., 2020c. Fly ash ecosystem services. In: Phytomanagement of Fly Ash. Elsevier, Amsterdam, pp. 257–288, https://doi.org/10.1016/B978-0-12-818544-5.00009-2.

Pandey, V.C., Bajpai, O., 2019. Phytoremediation: from theory toward practice. In: Pandey, V.C., Bauddh, K. (Eds.), Phytomanagement of Polluted Sites: Market Opportunities in Sustainable Phytoremediation. Elsevier, Amsterdam, pp. 1–49.

Pandey, V.C., Bauddh, K., 2018. Phytomanagement of Polluted Sites: Market Opportunities in Sustainable Phytoremediation. Elsevier, Amsterdam, p. 602.

Pandey, V.C., Mishra, T., 2016. Assessment of *Ziziphus mauritiana* grown on fly ash dumps: prospects for phytoremediation but concerns with the use of edible fruit. Int. J. Phytoremediation 20 (12), 1250–1256.

Pandey, V.C., Singh, D.P., 2020. Phytoremediation Potential of Perennial Grasses. Elsevier, Amsterdam, ISBN: 9780128177327, p. 371, https://doi.org/10.1016/C2017-0-00586-4.

Pandey, V.C., Souza-Alonso, P., 2019. Market opportunities in sustainable phytoremediation. In: Pandey, V.C., Bauddh, K. (Eds.), Phytomanagement of Polluted Sites: Market Opportunities in Sustainable Phytoremediation. Elsevier, Amsterdam, pp. 51–82.

Pandey, V.C., Abhilash, P.C., Singh, N., 2009. The Indian perspective of utilizing fly ash in phytoremediation, phytomanagement and biomass production. J. Environ. Manag. 90 (10), 2943–2958.

Pandey, V.C., Singh, K., Singh, J.S., Kumar, A., Singh, B., Singh, R.P., 2012. *Jatropha curcas*: a potential biofuel plant for sustainable environmental development. Renew. Sustain. Energy Rev. 16, 2870–2883.

Pandey, V.C., Singh, N., Singh, R.P., Singh, D.P., 2014. Rhizoremediation potential of spontaneously grown *Typha latifolia* on fly ash basins: study from the field. Ecol. Eng. 71, 722–727.

Pandey, V.C., Pandey, D.N., Singh, N., 2015a. Sustainable phytoremediation based on naturally colonizing and economically valuable plants. J. Clean. Prod. 86, 37–39.

Pandey, V.C., Prakash, P., Bajpai, O., Kumar, A., Singh, N., 2015b. Phytodiversity on fly ash deposits: evaluation of naturally colonized species for sustainable phytorestoration. Environ. Sci. Pollut. Res. 22, 2776–2787.

Pandey, V.C., Bajpai, O., Pandey, D.N., Singh, N., 2015c. *Saccharum spontaneum*: an underutilized tall grass for revegetation and restoration programs. Genet. Resour. Crop. Evol. 62 (3), 443–450. https://doi.org/10.1007/s10722-014-0208-0.

Pandey, V.C., Bajpai, O., Singh, N., 2016a. Energy crops in sustainable phytoremediation. Renew. Sustain. Energy Rev. 54, 58–73.

Pandey, V.C., Sahu, N., Behera, S.K., Singh, N., 2016b. Carbon sequestration in fly ash dumps: comparative assessment of three plant association. Ecol. Eng. 95, 198–205.

Pavlović, P., Mitrović, M., Djurdjević, L., 2004. An ecophysiological study of plants growing on the fly ash deposits from the "Nikola Tesla-A" thermal power station in Serbia. Environ. Manag. 33 (5), 654–663.

Peles, J.D., Barrett, G.W., 1997. Assessment of metal uptake and genetic damage in small mammals inhabiting a fly ash basin. Bull. Environ. Contam. Toxicol. 59, 279–284.

Pietrzykowski, M., Krzaklewski, W., Woś, B., 2015. Preliminary assessment of growth and survival of green alder (*Alnus viridis*), a potential biological stabilizer on fly ash disposal sites. J. For. Res. 26 (1), 131–136.

Pietrzykowski, M., Woś, B., Pająk, M., Wanic, T., Krzaklewski, W., Chodak, M., 2018a. The impact of alders (Alnus spp.) on the physico-chemical properties of technosols on a lignite combustion waste disposal site. Ecol. Eng. 120, 180–186.

Pietrzykowski, M., Woś, B., Pająk, M., Wanic, T., Krzaklewski, W., Chodak, M., 2018b. Reclamation of a lignite combustion waste disposal site with alders (Alnus sp.): assessment of tree growth and nutrient status within 10 years of the experiment. Environ. Sci. Pollut. Res. Int. 25 (17), 17091–17099.

Pilon-Smits, E., 2005. Phytoremediation. Annu. Rev. Plant Biol. 56, 15–39.

Pilon-Smits, E., LeDuc, D.L., 2009. Phytoremediation of selenium using transgenic plants. Curr. Opin. Biotechnol. 20, 207–212.

Pimente, D., Harvey, C., Resosudarmo, P., Sinclair, K., et al., 1995. Environmental and economic costs of soil erosion and conservation benefits. Science 267, 1117–1123.

Placek, A., Grobelak, A., Kacprzak, M., 2016. Improving the phytoremediation of heavy metals contaminated soil by use of sewage sludge. Int. J. Phytoremediation 18 (6), 605–618. https://doi.org/10.1080/15226514.2015.1086308.

Pollard, A.J., Powell, K.D., Harper, F.A., Smith, J.A.C., 2002. The genetic basis of metal hyperaccumulation in plants. Crit. Rev. Plant Sci. 21 (6), 539–566.

Prajapati, S.K., Meravi, N., 2015. Effect of fly ash on the photosynthetic parameters of *Zizyphus mauritiana*. Adv. For. Lett. 4, 1–5. https://doi.org/10.14355/afl.2015.04.001 1.

Prasad, M.N.V., 2003. Phytoremediation of metal-polluted ecosystems: hype for commercialization. Russ. J. Plant Physiol. 50, 686–700.

Prell, J., Poole, P., 2006. Metabolic changes of rhizobia in legume nodules. Trends Microbiol. 14, 161–168.

Pulford, I.D., Watson, C., 2003. Phytoremediation of heavy metal-contaminated land by trees—a review. Environ. Int. 29, 529–540.

Rai, V.K., 2002. Role of amino acids in plant responses to stress. Biol. Plant. 45, 481–487.

Roeland, S., Moretti, M., Amorim, J.H., Branquinho, C., et al., 2019. Towards an integrative approach to evaluate the environmental ESs provided by urban forest. J. For. Res. 30 (6), 1981–1996.

Rosa, J.C.S., Geneletti, D., Morrison-Saunders, A., Sánchez, L.E., Hughes, M., 2020. To what extent can mine rehabilitation restore recreational use of forest land? Learning from 50 years of practice in Southwest Australia. Land Use Policy 90, 104290.

Roy, M., Pandey, V.C., 2020. Role of microbes in grass-based phytoremediation. In: Pandey, V.C., Singh, D.P. (Eds.), Phytoremediation Potential of Perennial Grasses, first ed. Elsevier Press, pp. 303–336.

Roy, M., Giri, A.K., Dutta, S., Mukherjee, P., 2015. Integrated phytobial remediation for sustainable management of arsenic in soil and water. Environ. Int. 75, 180–198.

Roy, M., Roychowdhury, R., Mukherjee, P., 2018a. Remediation of fly ash dumpsites through bioenergy crop plantation and generation: a review. Pedosphere 28 (4), 561–580.

Roy, M., Roychowdhury, R., Mukherjee, P., Roy, A., et al., 2018b. Phytoreclamation of abandoned acid mine drainage site after treatment with fly ash. In: Akinyemi, S., Gitari, M. (Eds.), Coal Fly Ash Beneficiation—Treatment of Acid Mine Drainage with Coal Fly Ash. IntechOpen, Croatia.

Roychowdhury, R., Mukherjee, P., Roy, M., 2016. Identification of chromium resistant bacteria from dry fly ash sample of Mejia MTPS thermal power plant, West Bengal, India. Bull. Environ. Contam. Toxicol. 96 (2), 210–216.

Ruiz-Olivares, A., Carrillo-Gonzalez, R., Gonzalez-Chavez, M.C., Soto Hernandez, R.M., 2013. Potential of castor bean (*Ricinus communis* L.) for phytoremediation of mine tailings and oil production. J. Environ. Manag. 114, 316–323.

Ruttens, A., et al., 2011. Short rotation coppice culture of willows and poplars as energy crops on metal contaminated agricultural soils. Int. J. Phytoremediation 13, 194–207.

Schalk, I.J., Hannauer, M., Braud, A., 2011. New roles for bacterial siderophores in metal transport and tolerance. Environ. Microbiol. 13, 2844–2854.

Schroder, P., Herzig, R., Bojinov, B., Ruttens, A., Nehnevajova, E., et al., 2008. Bioenergy to save the world. Producing novel energy plants for growth on abandoned land. Environ. Sci. Pollut. Res. Int. 15, 196–204.

Shaw, P.J.A., 1992. A preliminary study of successional changes in vegetation and soil development on unamended fly ash (PFA) in southern England. J. Appl. Ecol. 29, 728–736.

Sheppard, J.P., Chamberlain, J., Agúndez, D., et al., 2020. Sustainable forest management beyond the timber-oriented status quo: transitioning to co-production of timber and nonwood forest products—a global perspective. Curr. For. Rep. 6, 26–40.

Shimamoto, C.Y., Padial, A.A., da Rosa, C.M., Marques, M.C.M., 2018. Restoration of ESs in tropical forests: a global meta-analysis. PLoS One 13 (12), e0208523.

Shrestha, P., Bellitürk, K., Görres, J.H., 2019. Phytoremediation of heavy metal-contaminated soil by switchgrass: a comparative study utilizing different composts and coir fiber on pollution remediation, plant productivity, and nutrient leaching. Int. J. Environ. Res. Public Health 16, 1261.

Singh, O.V., Jain, R.K., 2003. Phytoremediation of toxic aromatic pollutants from soil. Appl. Microbiol. Biotechnol. 63, 128–135.

Singh, A.N., Singh, J.S., 2006. Experiments on ecological restoration of coalmine spoil using native trees in a dry tropical environment, India: a synthesis. New For. 31, 25–39.

Singh, R.S., Tripathi, N., Chaulya, S.K., 2012. Ecological study of revegetated coal mine spoil of an Indian dry tropical ecosystem along an age gradient. Biodegradation 23, 837–849.

Spivak, M., Eric, M., Mace, V., Euliss, N.H., 2011. The plight of the bees. Environ. Sci. Technol. 45 (1), 34–38.

Starr, G.C., Lal, R., Malone, R., Hothem, D., Owens, L., Kimble, J., 2000. Modeling soil C transported by water erosion processes. Land Degrad. Dev. 11, 83–91.

Swiatek, B., Woś, B., Chodak, M., Maiti, S.K., Jozefowska, A., Pietrzykowski, M., 2019. Fine root biomass and the associated C and nutrient pool under the alder (Alnus spp.) plantings on reclaimed technosols. Geoderma 337, 1021–1027.

Terzaghi, E., Vergani, L., Mapelli, F., Borin, S., 2019. Rhizoremediation of weathered PCBs in a heavily contaminated agricultural soil: results of a biostimulation trial in semi field conditions. Sci. Total Environ. 686, 484–496.

Thomas, M., Waage, J., 1996. Integration of Biological Control and Host-Plant Resistance Breeding: A Scientific and Literature Review. CTA Publishers, Wageningen. 99 pp.

Tiller, K.G., 1989. Heavy metals in soils and their environmental significance. In: Stewart, B.A. (Ed.), Advances in Soil Science. vol. 9. Springer-Verlag, New York.

Touceda-González, M., et al., 2017. Aided phytostabilisation reduces metal toxicity, improves soil fertility and enhances microbial activity in Cu-rich mine tailings. J. Environ. Manag. 186, 301–313.

Tropek, R., Cerna, I., Straka, J., Cizek, O., Konvicka, M., 2013. Is coal combustion the last chance for vanishing insects of inland drift sand dunes in Europe? Biol. Conserv. 162, 6064.

Tropek, R., Cerna, I., Straka, J., Kadlec, T., Pech, P., Tichanek, F., et al., 2014. Restoration management of fly ash deposits crucially influence their conservation potential for terrestrial arthropods. Ecol. Eng. 73, 45–52.

Uzarowicz, Ł., Zagórski, Z., Mendak, E., Bartmiński, P., Szara, E., et al., 2017. Technogenic soils (Technosols) developed from fly ash and bottom ash from thermal power stations combusting bituminous coal and lignite. Part I. Properties, classification, and indicators of early pedogenesis. Catena 157, 75–89.

Uzarowicz, L., Kwasowski, W., Śpiewak, O., Switoniak, M., 2018. Indicators of pedogenesis of technosols developed in an ash settling pond at the Belchatów thermal power station (central Poland). Soil Sci. Annu. 69 (1), 49–59.

Vangronsveld, J., et al., 2009. Phytoremediation of contaminated soils and groundwater: lessons from the field. Environ. Sci. Pollut. Res. 16, 765–794.

Vaughn, K.J., Porensky, L.M., Wilkerson, M.L., Balachowski, J., Peffer, E., Riginos, C., Young, T.P., 2010. Restoration ecology. Nat. Educ. Knowl. 3 (10), 66.

Volchko, Y., Norman, J., Bergknut, M., Rosen, L., Soderqvist, T., 2013. Incorporating the soil function concept into sustainability appraisal of remediation alternatives. J. Environ. Manag. 129, 367–376.

Wan, X., Lei, M., Chen, T., 2016. Cost–benefit calculation of phytoremediation technology for heavy-metal contaminated soil. Sci. Total Environ. 563–564, 796–802.

Wang, J., Zhao, F., Yang, J., Li, X., 2017. Mining site reclamation planning based on land suitability analysis and ecosystem services evaluation: a case study in Liaoning Province, China. Sustainability 9, 890.

Wang, L., et al., 2019. Field trials of phytomining and phytoremediation: a critical review of influencing factors and effects of additives. Crit. Rev. Environ. Sci. Technol. https://doi.or g/10.1080/10643389.2019.1705724.

Watharkar, A.D., Jadhav, J.P., 2014. Detoxification and decolorization of a simulated textile dye mixture by phytoremediation using *Petunia grandiflora* and *Gaillardia grandiflora*: a plant-plant consortial strategy. Ecotoxicol. Environ. Saf. 103, 1–8.

Weber, H., Chetelat, A., Reymond, P., Farmer, E.E., 2004. Selective and powerful stress gene expression in Arabidopsis in response to malondialdehyde. Plant J. 37, 877–888.

Wei, S.Q., Pan, S.W., 2010. Phytoremediation for soils contaminated by phenanthrene and pyrene with multiple plant species. J. Soil. Sediment. 10 (5), 886–894.

Werling, B.P., Dickson, T.L., Isaacs, R., Gaines, H., 2014. Perennial grasslands enhance biodiversity and multiple ESs in bioenergy landscapes. PNAS 111 (4), 1652–1657.

Willscher, S., et al., 2013. Field scale phytoremediation experiments on a heavy metal and uranium contaminated site, and further utilization of the plant residues. Hydrometallurgy 131–132, 46–53.

Yan, A., Wang, Y., Tan, S.N., Yusof, M.L.M., Ghosh, S., Chen, Z., 2020. Phytoremediation: a promising approach for revegetation of heavy metal-polluted land. Front. Plant Sci. 11, 359.

Zalesny Jr., R.S., Headlee, W.L., Soolanayakanahally, R.Y., Richardson, J., 2018. Short rotation woody crop production systems for ecosystem services and phytotechnologies. Printed Edition of the Special Issue Published in Forests by MDPI. https://www.fs.fed.us/nrs/pubs/ jrnl/2019/nrs_2019_zalesny_005.pdf.

Zalesny, R.S., Headlee, W.L., Gopalakrishnan, G., Bauer, E.O., et al., 2019. Ecosystem services of poplar at long-term phytoremediation sites in the Midwest and Southeast, United States. Wiley Interdiscip. Rev. Energy Environ. https://doi.org/10.1002/wene.349.

Zhang, C., Feng, Y., Liu, Y.W., Chang, H.Q., Li, Z.J., Xue, J.M., 2017. Uptake and translocation of organic pollutants in plants: a review. J. Integr. Agric. 16 (8), 1659–1668.

Designer plants for climate-resilient phytoremediation

Chapter outline

1 Overview

The meltdown of glaciers, abnormal rainfalls in many places, frequent hurricanes, and unpredictable weather events are giving warning signals of continuously rising earth's temperature and global warming. In 2015, the Paris Agreement was approved and adopted by 195 nations to defend food security and finishing poverty and hunger and reducing the impacts of climate change phenomena (COP21, 2015; IPCC, 2007, 2012). Paris Climate Change committee concluded that if the earth's temperature increase is restricted to less than 2°C, adaptation support would be necessary by the developing countries (IPCC, 2007, 2012). Climate change has already changed global weather patterns, and in the near future, weather of an area may be entirely different from its past weather pattern. In the context of the global 2°C goal, all countries promised for strengthening climate resilience, enhancement of adaptive capacity, and decreasing vulnerability to climate change, and contribute to sustainable growth and development. Nevertheless, issues such as climate variabilities and vulnerabilities in the form of extreme events are not only suspected to play havoc on crop quality and yield affecting future food production (Rehmani et al., 2014); serious concerns are coming on ecosystem functioning as the natural resources (land, water, and biodiversity) would be shrinking (Evenson and Gollin, 2003). Phytoremediation is currently the most well-accepted technique that plays an indispensable part in proper ecosystem functioning. It is an in situ technology that works along with natural attenuation and is lesser energy demanding than traditional remediation methods, and

Adaptive Phytoremediation Practices. https://doi.org/10.1016/B978-0-12-823831-8.00007-4

it most importantly produces biomass, the most important resource for biofuel (Yan et al., 2020). Unfortunately, till now, the resilience of remediation to climate change so far has been very poorly explored (Al-Tabbaa et al., 2007; Simon, 2015). The Environmental Protection Agency of United States (USEPA) in 2014 carried out a screening analysis for assessing the vulnerability of remediation of Superfund sites in view of climate change, and concluded that clean-up projects are not protected from climatic vulnerability (US EPA, 2011, 2014). A study was performed in the perchloro-ethylene (PCE)-contaminated site of the San Francisco Bay area to see whether reme-diation was susceptible to mounting sea-level rise (O'Connor et al., 2019). Life cycle assessment (LCA) (predicts primary cum secondary effects caused by the site to the residents/biota of the site) was used to understand both primary (impacts caused by the pollution) and secondary impacts (impacts caused by the phytoremediation program itself) caused by the system in response to changes in sea level and hydroclimatic con-ditions. In the phytoremediation project involving native eucalyptus trees, the biomass was utilized as saw timber to render a net environmental benefit. The study proposed that with an increase in the sea level, the hydraulic gradient of the polluted site would decrease in the presence of a fixed upgradient of groundwater head. At the same time, a reduced hydraulic gradient together with natural degradation of PCE and allied prod-ucts would result in a slower movement of the plume. Overall, the study came to the conclusion that the success of the remediation system was dependent on the sea-level increase and that the amount of sea-level rise and the local hydrogeological system would be the ultimate impact determining factors (O'Connor et al., 2019).

Hydroclimatic changes normally influence phytoremediation performances. So, scientists are now trying to construct transgenic plants that can withstand the climatic variability and provide us environmental protection at the altered climatic conditions too. Identification of genes that render robustness and adaptive capacity to the plants to survive in contaminated areas is the most important task to understand evolution-ary processes involved and in due course will improve our understanding of the phy-toremediation process under adverse climatic conditions. A genome-wide-scan study was undertaken to identify locus under divergent selection pressure in between four *Arabidopsis halleri* populations propagating on both metals polluted or unpolluted lo-cations (Meyer et al., 2009). Proof for selection came from amplified fragment length polymorphism (AFLP) study, which found some AFLP markers in each sampled pop-ulation. The four loci deviated from neutrality from both metallophytic and nonmetal-lophytic groups and represented as good candidate genes for acclimatization to metal contamination. As some candidates also differed between the two metallicolous popu-lations, that separate loci may be involved in assigning adaptive ability in the different metallicolous populations (Meyer et al., 2009). Newer varieties are needed with higher phytoremediation efficiencies and higher adaptabilities to take over the older ones. The Crop Science Society of America proposed that research investments are needed to understand the molecular and physiological basis of adaptation to heat, cold, water deficiency, salinity, and biotic stresses coming from climate change. Genetic engineer-ing has a great role to play to add new climate-resilient features in candidate plants for phytoremediation. Keeping pace with the enormous technological improvements made in plant genomics and its application for the development of new traits in food crops, more investment is required to develop climate-resilient varieties of accumulators, hyperaccumulators, and bioenergy crops for phytoremediation in the near future.

CRISPER system of bacteria discovered in 2012 has emerged as a multipurpose game-changing platform that can precisely handle genomic modification for the insertion of a new trait, promote gene regulation, and take part in protein engineering (Jinek et al., 2012; Cong et al., 2013). Among the CRISPR-Cas systems, *Streptococcus pyogenes*-derived type II CRISPR-SpCas9 system has appeared like versatile genome-editing tools (Hsu et al., 2014), and many plants have been converted into climate-resilient forms through this technology (see review by Nadakuduti and Enciso-Rodríguez, 2020; Hsu et al., 2014).

The present paper has discussed advanced gene-editing technologies along with the drawbacks and prospects. This chapter has highlighted various transgenic plants created through conventional genetic engineering and advanced gene-editing technology of CRISPER that would play a leading role in phytoremediation in altered climatic conditions.

2 Biotechnological strategies for generating climate-resilient phytoremediation

Plant species that can grow on heavy metal-contaminated soils are called metallophytes. Accumulators are the metallophytes that can accumulate metal in various plant tissues in a concentration less than the surrounding soil. In contrary to accumulators, in hyperaccumulators metal concentration in their various tissues exceeds the concentration present in the surrounding soil. Although hyperaccumulators have better phytoextraction capacity, accumulators have better growth rates and better biomass-producing abilities, which is an essential quality to be an ideal bioenergy crop. Given the growing interest in creating climate-resilient varieties, various accumulator plants with enhanced phytoextraction, phytostabilization, and phytofiltration abilities have been created (Pauwels et al., 2008). In the following section, two main strategies have been discussed through which desired traits can be introduced into the target plant.

2.1 Omics-based breeding approach for introducing adaptive traits

Before going into the gene-editing part, scientists look for the total variety of genes (gene pool) available in the population and select the desired ones. This is a part of population genetics whose aim is to identify the loci giving adaptive advantages in the different varieties available in the natural population. Pseudometallophytes which can grow and survive on both metal-contaminated and metal-noncontaminated soils are of great significance as they are the ideal models for studying the local adaptation of the plants in reaction to metals and to understand the role of the relevant genes responsible for the adaptation. Positive selection always operates on genetic differences, and the genes contributing to local adaptations are selected for further gene-editing purposes. So, scanning the whole genome for polymorphism at certain loci on the representative candidates from diverse populations is done to differentiate locus-specific effects from genome-wide effects (Luikart et al., 2003). Genome scans have been found very useful in relating adaptation to environmental gradients (Jump et al., 2006). It has also linked adaptation to respective habitats (Namroud et al., 2008) and natural selection during speciation (Wilding et al., 2001; Minder and Widmer, 2008). All the genomic scan investigations have found evidence that supports the hypothesis that divergent selection operates on some marker loci. Developments in genotyping technology have encouraged genomic scans on diverse types of plants. The studies supported that certain genomic regions are susceptible to selection pressure. To acquire new information into the genetic make-up of adaptation of plants to heavy metal contamination, Meyer et al. (2009) performed one genotyping or genome scan study on a much-studied and well-characterized model pseudometallophyte species, *A. halleri*. The plant was characterized for the study of Zn tolerance in contaminated and noncontaminated sites (Pauwels et al., 2005, 2006). Similarly, natural variation can be used to scan other abiotic stress tolerance genes like heat, cold, salinity, etc. (Rus et al., 2006). For example, Zhang et al. (2017a) showed natural variations of *Oryza sativa* in response to low-temperature adaptation.

Different "omics" approaches, namely, epigenomics, genomics, metabolomics, proteomics, and transcriptomics, have enhanced the breeding programs of the staple food crops, phytoremediator, and bioenergy crops (Parry and Hawkesford, 2012; Kumar et al., 2019) for making them climate-smart. Knowledge obtained from proteomics, metabolomics, and genomics studies of plants has enriched our understanding of the strategies to manage abiotic stresses (Behr et al., 2015). Genomic sequences of model phytoremediators, *A. halleri* (Zn and Cd hyperaccumulator), *Brassica juncea* (multiple heavy metal accumulator), *Hirschfeldia incana* (mustard variety that uptakes Pb), *Pteris vittata* (As hyperaccumulator), *Noccaea caerulescens* (former name *Thlaspi caerulescens*; hyperaccumulator of the Cd, Ni, and Zn), and several other species, including many bioenergy crops, have been fully or partially exposed. Modification of their genome may assist the identification and characterization of critical genetic elements during phytoremediation processes, like phytostabilization, phytoextraction, rhizofiltration, phytovolatilization, and phytodegradation.

Along with genomics, epigenomics should also be studied as plants frequently change their gene expression without changing gene sequence through DNA methylation or posttranslational modification of histones (Kakutani, 2002). Epigenetic variations can be reversible (transient and inheritable expression) or irreversible (permanent and heritable). Epigenetic variations induced by environmental stressors have been described to play a vital role in augmenting phenotypic plasticity in response to climate change in many plants (Tsaftaris and Polidoros, 1999). While epigenetic variation was investigated in few food crops, no reports are available to date on its application on the betterment of phytoremediation ability of plants.

2.2 CRISPR/Cas9-based genome editing

With the fast advancement of next-generation sequencing technologies, exposure of genomic information of an increasing number of plant species is coming in an almost exponential fashion. The genome editing services that have almost reached a milestone are now providing the chance to edit genes with great precision. The basic tactic of all types of genome editing processes (e.g., zinc finger nuclease or ZFN; meganuclease; transcription activators like effector or TALENs, and CRISPER-Cas9) involves a sequence-specific nuclease that makes a DNA double-strand break (DSB) at the desired point. Afterward, the donor-dependent homology-directed repair (HDR) pathway or the error-prone nonhomologous end-joining (NHEJ) pathway repairs the strand breakage. While the error-prone NHEJ makes the deletion or addition of nucleotides resulting in gene knockouts, the HDR pathway makes accurate base replacement in the guidance of donor DNA template (Chen and Gao, 2014). So, an effective genome-editing test requires the delivery of the appropriately designed target sequences and the genome-editing cargo loads DNA, RNA, or ribonucleoproteins (RNPs) into proper vectors. Once inside the plant cell, the cargo does the required editing duty. Next, the transformed calli with the desired gene of interest or the desired gene knockout is selected for the desired mutants, which finally gives rise to the transgenic plants. CRISPR/Cas system is better than other recent gene-editing systems (ZFNs and TALENs) in terms of the simplicity of the system, efficiency, success rate, cost affordability, and finally by its capability to target multiple genes (Cong et al., 2013).

The discovery of CRISPR/Cas9 in 2012 by Niklas Elmehed (honored with "The Nobel Prize in Chemistry, 2020") was a grand biotechnological innovation. CRISPR protein is always present in most bacteria and archaea that provides them adaptive immunity against phage virus. CRISPER in bacteria is composed of the Cas nuclease and 2 distinctive RNA constituents. One of them is a programmable crRNA or CRISPR RNA, and the other one is a fixed tracrRNA or trans-activating crRNA. Bacteria use Cas nuclease to mutilate the foreign invading phage DNA into tiny wreckages, which are then incorporated into the CRISPR array. Consequently, the CRISPR array is transcribed to generate CRISPR RNA and the complementary trans-activating CRISPER RNA, which together form a dsRNA structure that further employs Cas for cleavage (Datsenko et al., 2012). On the viral DNA, a short sequence called protospacer-adjacent motif (PAM) lies in line with the crRNA-targeted sequence and guides the CRISPR-Cas complex to bind the target DNA. PAM sequence is responsible for the affinity between Cas and the target DNA, and it discriminates nonself-target sequences from nontarget sequences. The discovery of this PAM sequence was one of the critical factors that paved the way for CRISPR to become the best genome-editing technology so far discovered. Another critical finding was the observation that the acquired spacer sequences are highly identical to one another at PAMs and that this sequence is very critical for the CRISPR tool to work. Lastly, the demonstration that a CRISPR system from one bacterium can be easily transferred to other bacterial strains made the system flexible and versatile (Mei et al., 2016). Till now, three types of CRISPR-Cas systems (I, II, and III) have been recognized. The type II system isolated from *S. pyogenes* requires only a single Cas protein: Cas9. The Cas9 protein functions in both RNA-guided DNA detection depending on single guide RNA (sgRNA) formed by crRNA and tracrRNA fusion and cleavage (Cong et al., 2013). As they depend on DNA-RNA hybridization by sequence-specific nucleases, the CRISPR-Cas systems are targeted to present DSBs at the desired sites (Zhu et al., 2020) for genetic engineering purpose. After its first use in plants in 2013 (Shan et al., 2013; Nekrasov et al., 2013), CRISPR-Cas was continuously used and further improved for genome editing in different crop species. In addition to its ability to induce precise nucleotide alterations, CRISPR-Cas is equipped to do much more than just causing changes at specific loci and the full potential of CRISPER technology for genome modification can be found in some current review articles (Mei et al., 2016; Adli, 2018).

Due to these distinguishing characters, CRISPR/Cas9 has been rapidly exploited in plants (Lowder et al., 2015, 2018), and up to date, many crops like apple, barley, camelina, cotton, cucumber, flax, grapes, grapefruit, lettuce, rice, maize, oranges, potato, rapeseed, soybean, sorghum, tomato, wheat, and watermelon have been transformed by this technique (Gao, 2018). Another significant advantage of CRISPR over other genome-editing tools is their capacity for multiplexing, i.e., the editing of multiple target sites in a synchronized way (Cong et al., 2013). For multiplexing, researchers used Gibson Assembly kit of Golden Gate cloning system and pulled together multiple sgRNAs into a single Cas9/sgRNA expression vector where multiple sgRNAs were transcribed by distinct promoters (Silva and Patron, 2017). Multiple sgRNAs are also exploited to modify a single gene in crops where transformation or editing efficiencies are low. The tRNA processing system was also hired by the multiplex editing service of CRISPR/Cpf1 (Ding et al., 2018).

In the last 6–7 years, more than 10 diverse types of CRISPR-Cas proteins have been repurposed for bioengineering (examples include spCas9, FnCas9, AaCas9, NmCas9, St1Cas9, St3Cas9, AsCpF1, LbCpf1, and Cas 13). Among these, most noteworthy are Cpf1 proteins (a class II type V endonuclease) from *Acidaminococcus* sp. (AsCpf1), *Lachnospiraceae* bacterium (LbCpf1) (Zetsche et al., 2015), and *Prevotella* and *Francisella* (CpF1) (Zaidi et al., 2017; Ma et al., 2016). Cpf1 is a dual nuclease that processes its CRISPR RNA and cleaves its target DNA (Zetsche et al., 2015). In contrary to the native Cas9 that requires two short RNAs, only one sgRNA is required by Cpf1. It also differs from Cas9 in the cutting site of the target DNA. As it cuts DNA at 3′ downstream of the PAM sequence in a staggering fashion, 5′ overhang is generated instead of the blunt end. Taking advantage of this advanced feature of CpF1, Wang et al. (2017a,b) coaxed CRISPR/Cpf1 with a short DR-guide and validated the likelihood of multiplex gene editing with Cpf1. The expression of a gene or genes of interest can be upregulated or downregulated by fusing transcription factors with dCas9 (dead or catalytically inactive Cas9), and thus, transcription by RNA polymerase can be repressed or enhanced (Miglani, 2017). The repression known as Crispr (interference) and activation known as CRISPRa (activation) were able to regulate gene expression over a 1000-fold range and were very successful in creating desired genetically modified plants (Lowder et al., 2018).

3 Designing and developing climate-resilient plants for adaptive phytoremediation involving OMICS approach and CRISPER-Cas9

Exploration of CRISPR system for plant genetic engineering for climate-resilient phytoremediation is less explored and auspicious endeavor to try. Trait improvement through conventional breeding approaches has long turn-around times due to the complexities associated with high ploidy numbers, small occurrences of homologous recombination, etc., making site-specific mutagenesis hard (Estrela and Cate, 2016). So, there is a growing need for advanced gene editing through CRISPER for increased plant growth, biomass, improved metal tolerance and metal accumulation, and increased disease and climate resistances. In the last decade, plentiful of experiments reported cloning of the desired gene from plant and bacterial sources to the target plants, rendered desired effects for phytoremediation. In the same line, CRISPR-Cas9 systems are now being exploited to modify the genome of plants popular for their contribution toward phytoremediation processes (Miglani, 2017; Fan et al., 2015; Aken, 2008). According to Noman et al. (2016) there are several advantages (pros) and disadvantages (cons) of CRISPR/Cas9 systems:

Advantages/pros of CRSIPR Cas9 system

- off-target effects of this system can be minimized by selecting unique crRNA sequence
- guide RNA can be produced by in vitro transcription, which is a cost-effective procedure
- multiple guide RNA can be tested, and it does not require protein engineering steps
- at the same time, it is possible for several genes to be edited
- low mutation rate
- high target recognition rate
- fast and easily designed procedure in laboratory for new target site
- creation of large-scale libraries

Disadvantages/cons of CRSIPR-Cas9 system

- PAM sites may lead to the undesired cleavage of DNA sequences causing unexpected mutations
- insufficient Cas9 codon optimization leading to the inefficient translation of Cas9 proteins
- improper vectors can stop system proceedings
- some gene family members can cause false editing of target sequence

3.1 Omics approach (cloning and expression)-mediated transgenic plant generation

Plant resistance to heavy metals and metalloids relies on the uptake followed by translocation and accumulation of the heavy metals in specialized storage organelles (e.g., vacuoles), storage tissues, and trichomes. Different chelators like siderophores, phenolics, and organic acids synthesized by plants help in the uptake process. Some important class of proteins associated with heavy metal transport and metal-ion homeostasis in plants are the cation diffusion facilitator family proteins (CDF), heavy metal (CPx-type) ATPases, zinc-iron permease family proteins (ZIP), and natural resistance-associated macrophage family of proteins (Nramp), etc. (Yang et al., 2005a,b; Williams et al., 2000). After transport, accumulation and storage are accomplished by the intracellular complex formation with peptide ligands like phytochelatins (PCs) and metallothioneins (MTs) (Yang et al., 2005b). Within the cytosol, plants employ certain cellular high-affinity ligands that chelate the heavy metals as a means of detoxification (Hall, 2002).

Poplar (*Populus* species) and willow (*Salix* species) have been widely used in various phytoremediation projects owing to their high-biomass productions, high growth rates, easy propagations, vegetative reproductions, well-developed dense and deep root systems, quick adaptabilities to environmental changes, and marked tolerances to organic and inorganic pollutants (Baldantoni et al., 2014). Maize (*Zea mays*), mustard (*B. juncea*), tobacco (*Nicotiana tabacum*), sunflower (*Helianthus annuus*), barley (*Hordeum vulgare* L.) etc. are some other popular fast-growing, high-biomass-yielding grass species with outstanding metal sequestration capabilities and are widely used in phytoremediation, phytomining, and bioenergy generation programs (Anderson et al., 2005; Ali et al., 2013; Meers et al., 2010; Pandey et al., 2016).

Metal uptake can be enhanced by increasing uptake and intracellular binding sites and enhancing the specificity of the uptake system. Table 1 shows some climate-smart plants created through the omics approach. *Populus angustifolia* (poplar), *N. tabacum* (tobacco), *H. vulgare* (barley), *H. incana* (mustard), *B. juncea* (Indian mustard), *Arabidopsis* etc. were genetically engineered to overexpress genes like glutamylcysteine synthetase, nicotianamine synthase 1, metallothionein-encoding genes, and phytochelatin synthase genes and thus served superior heavy metal sequestration abilities in comparison with their wild-type plants (Fulekar et al., 2008).

For enhancing multiple metal tolerance in *Arabidopsis* and tobacco plants, nicotianamine synthase 1 gene from barley was overexpressed in them, proving that nicotinamine plays a crucial character in metal detoxification and this can be a potent tool for use in phytoremediation (Pianelli et al., 2005). The plants showed an increased uptake and tolerance for metals like Cd, Cu, Mn, Mg, Ni, Fe, and Zn (Kim et al., 2005). In contrast to nicotinamines (ubiquitous metal-chelating molecule in higher plants), cysteine-rich metallothioneins are a group of small, well-conserved, metal-binding proteins responsible for binding most of the heavy metals. They also provide oxidative stress protection and buffering against toxic heavy metals. Various plants like *Arabidopsis*, poplar, and tobacco showed an increased metal accumulation after overexpressing metallothionein genes (*MTA1*, *MT1*, and *MT2*) (Turchi et al., 2012; Lv et al., 2013; Xia et al., 2012). The overexpression of the metallothionein gene *MT2b* and up-/downregulation of genes involved in abscisic acid (ABA) biosynthesis and catalysis increased the lead (Pb) tolerance ability of mustard (*H. incana*) (Auguy et al., 2016). Gasic and Korban (2007) developed transgenic Indian mustard (*B. juncea*) plants that overexpressed *Arabidopsis* phytochelatin synthase (*AtPCS1*) gene and displayed improved resistance for As and Cd. LeDuc et al. (2006) worked on selenium phytoremediation and produced selenium-resistant transgenic *Arabidopsis* and Indian mustard plants by incorporating selenocysteine methyltransferase (*SMT*) gene from the Se hyperaccumulator, *Astragalus bisulcatus* (LeDuc et al., 2004). The resulting transgenic plants successfully displayed an enhanced Se accumulation from selenite but not from selenate. Double transgenic plants were then produced to enhance Se accumulation from selenate (and improve selenate phytoremediation) by overexpressing ATP sulfurylase gene in addition to SMT gene. The transgene called *APSxSMT* caused four to nine times increased Se accumulation from selenate (LeDuc et al., 2006).

However, in some cases, desired results were not achieved. For example, overexpression of a plasma membrane protein (NtCBP4) in transgenic tobacco plants caused increased lead accumulation, but unfortunately, the plant became significantly

Table 1 Cloning and expression of abiotic stress response genes in transgenic plants.

Transgenic plant	Transgene	Promoters used	Effect of the modification	Reference
Tobacco	Ion exclusion (transporters) Na^+/H^+ antiporter (SOS1)	Constitutive	Altered root and shoot accumulation of Na^+ and K^+; improved biomass production; improved germination	Yue et al. (2012)
Tobacco	Ion exclusion (transporters) Na^+/H^+ antiporter (SbSOS1)	Constitutive promoter	Improved salinity tolerance; improved shoot and root accumulation of Na^+ and K^+; high biomass production; and improved germination	Yadav et al. (2012)
Rice	Ion exclusion (transporters) Na^+/H^+ antiporter (SOS2)	Constitutive promoter	Improved salinity tolerance; improved shoot accumulation of Na^+ and K^+; high biomass production; and improved germination	Zhao et al. (2006)
Rice	Na^+ transporter (HKT subfamily 1) (*AtHKT1;1*)	Cell type specific	Improved salinity tolerance; improved shoot accumulation of Na^+ and K^+; high biomass production; improved germination	Plett et al. (2010a)
Barley	Na^+/K^+ transporter (*HKT* subfamily 2)	Constitutive	Improved salinity tolerance; improved shoot accumulation of Na^+ and K^+; high biomass production; improved germination	Mian et al. (2011)
Rice	Na^+ ATPase *PpENA1*	Constitutive	Better salinity tolerance; improved shoot accumulation of Na^+ and K^+; improved biomass production and germination	Jacobs et al. (2011)
Rice	Na^+ ATPase *PpENA1*	Cell type specific	Improved salinity tolerance; improved shoot accumulation of Na^+ and K^+; enhanced biomass production and germination	Plett et al. (2010b)
Buckwheat Cotton Tomato Poplar Kiwifruit Wheat *Brassica,* Poplar	*AtNHX1*	Constitutive	Altered Na^+, K^+, and Cl^- accumulation; Improved biomass production; Reduced leaf senescence	Chen et al. (2008), Xue et al. (2004), Qiao et al. (2011), Rajagopal et al. (2007)

Continued

Table 1 Cloning and expression of abiotic stress response genes in transgenic plants—cont'd

Transgenic plant	Transgene	Promoters used	Effect of the modification	Reference
Rice (*O. sativa*)	*PpENA1*	Cell type specific	Improved shoot and root biomass production; Altered Na$^+$ and K$^+$ accumulation; Increased proline content	Plett et al. (2010b)
Tobacco (*N. tabacum*)	Mitogen-activated protein kinase (*GbMPK2*)	Constitutive	Altered Na$^+$, K$^+$, and Cl$^-$ accumulation; Improved biomass production; Reduced leaf senescence	Zhang et al. (2011a)
Alfalfa	Trehalose-6-phosphate phosphatase *ScTPS2*	Stress inducible	Increased compatible solute accumulation; Improved plant survival; Increased growth; reduced wilting; Maintenance of photosynthetic efficiency	Suárez et al. (2009)
Tomato	Calcineurin-B like interacting protein kinases (*CIPK*) (*SlCIPK24*)	constitutive	Altered Na$^+$, K$^+$, and Cl$^-$ accumulation	Huertas et al. (2012)
Wheat	Transcription factors *GhDREB*	Constitutive	Improved biomass production; reduced leaf senescence; improved drought and salt tolerance Improved germination; improved biomass; improved chlorophyll retention; improved resistance to freezing, salt, and drought	Gao et al. (2009)
Tobacco	Mannitol-1-phosphate dehydrogenase (*mt1D*)	Constitutive	Increased compatible solute accumulation; Improved plant survival; increased growth; reduced wilting; Maintenance of photosynthetic efficiency	Karakas et al. (1997)
Tobacco	*DREB1A* (*rd29A*) DRE-binding protein	Stress-inducible promoter	Improved drought and low-temperature stress tolerance	Kasuga et al. (2004)
Tomato	*Transcription factor CBF1* (CRT/DRE binding factor 1)	Stress-inducible promoter	Higher superoxide dismutase; higher nonphotochemical quenching; lower malondialdehyde content; enhanced cold tolerance	Zhang et al. (2011b)
Arabidopsis	RNA chaperon (*AtCSP3*) *Cold shock protein*	Stress-inducible promoter	Enhanced freezing tolerance	Kim et al. (2009)

Organism	Gene	Promoter/Condition	Effect	Reference
Arabidopsis; tobacco white poplar	Nicotianamine synthase gene of metallothionein A1 gene of *Pisum sativum*	Constitutive	Increased tolerance to nickel and other heavy metals	Kim et al. (2005), Pianelli et al. (2005) Turchi et al. (2012)
Arabidopsis	Metallothioneins BcMT1 and BcMT2 from *Brassica campestris*	Constitutive	Enhanced tolerance to Cu and Cd; decreased production of ROS	Lv et al. (2013)
Tobacco	*Elsholtzia haichowensis* metallothionein 1 (EhMT1)	Constitutive	Enhanced accumulation and Cu tolerance; decreased hydrogen peroxide production	Xia et al. (2012)
Brassica juncea	ATP sulfurylase and selenocysteine methyltransferase	Constitutive	Enhanced selenium phytoremediation	LeDuc et al. (2006)
Arabidopsis thaliana and tobacco	Bacterial Mercury Membrane Transport Protein MerC	Cauliflower mosaic virus 35S promoter	Mercury hyperaccumulation	Sasaki et al. (2006)
Arabidopsis thaliana and rice	Napthalene dioxygenase from Pseudomonas	Constitutive	Improved phenanthrene phytoremediation	Peng et al. (2014)
Barley	citrate transporter genes SbMATE and FRD3	High Al^{3+}	Enhancing the aluminum tolerance	Zhou et al. (2014)
Rice	DRO1	Drought	Enhances deep rooting for better water stress tolerance	Uga et al. (2013)

sensitive to the metal (Arazi et al., 1999). Similarly, the expression of the *MerC* gene caused twofold higher mercury accumulation in transgenic *Arabidopsis* and tobacco in comparison with their wild-type counterparts, but the transgenic plants became hypersensitive to mercury (Sasaki et al., 2006).

Among other examples of transgenic plants made tolerant to heavy metals are tobacco plants with foreign gene *CAX-2* (vacuolar transporters) inserted from *Arabidopsis thaliana* for increased tolerance to Cu, Mn, and Cd (Hirschi et al., 2000); *Arabidopsis* made tolerant to Cu, Al, and Na with gene glutathione-S-transferase from tobacco (Ezaki et al., 2000); ferritin gene cloned from soybean into tobacco (Goto et al., 1998) and rice (Goto et al., 1999) for increasing iron accumulation; transgenic *Arabidopsis* co-expressing *SRSIp/ArsC* and *ACT 2p/γ-ECS* together showing extreme high tolerance to As than transgenic plants expressing γ-*ECS* or *ArsC* alone (Dhankher et al., 2002). New metabolic pathways were engineered into *Arabidopsis* plants by cloning *Escherichia coli ars C* and γ-*ECS* genes for the transport of oxyanion arsenate to leaf and shoot, reduce to arsenite and detoxify it as thiol peptide complexes (Eapen and D'Souza, 2005).

Along with metallophytes that remediate heavy metals from soil/water, many plants are equipped with machinery for detoxifying different types of organic xenobiotics like polychlorinated biphenyls (PCBs), phthalates, polycyclic aromatic hydrocarbons (PAHs), different pesticides, herbicides (e.g., atrazine, triazine), explosives like Royal Demolition Explosive (RDX), and 2,4,6-trinitro-toluene (TNT), etc. Plant enzymes responsible for the initial detoxification of organic pollutants to lesser toxic forms are P450 monoxygenases and carboxylesterases. After the initial detoxification, the partially transformed pollutants are conjugated to different biomolecules like glutathione (GSH), glucose, amino acids etc., and the conjugates are more soluble, polar, and less toxic. For example, glutathione S-transferase (GST) causes the detoxification of herbicides by catalyzing conjugate formation between the herbicides with GSH (Lamoureux et al., 1970). The final phase of detoxification is compartmentalization and storage of the soluble conjugates either in the vacuoles within the cell or apoplast or in the cell wall matrix. ATP-dependent cellular transport or the membrane pump transfers the glutathione S-conjugate formed by GST to the vacuole or the apoplast. Rylott et al. (2011) performed genetic engineering of *Arabidopsis* plant for RDX detoxification. Successful cloning and expression of bacterial XplA and XplB genes in plants enabled them to remove and detoxify RDX through the activation of the cytochrome P450-reductase complex.

Endophytes and rhizomicrobes, which live in the different tissues and roots of plants, respectively, greatly help the host plants in the degradation process of aromatic compounds and help the plant survive in the organically polluted soil. Peng et al. (2014) successfully cloned microbial naphthalene dioxygenase genes into transgenic *Arabidopsis* and rice plants which thereafter tolerated and metabolized naphthalene and phenanthrene. Overexpression of the biphenyl dioxygenase gene by transgenic alfalfa plants caused a substantial improvement of their tolerance and degradation ability of polychlorinated biphenyls (PCBs) and 2,4-dichlorophenol (2,4-DCP) (Wang et al., 2015). Rylott et al. (2015) reviewed the detoxification and biodegradation of organic pollutants by transgenic plants.

3.2 CRISPER-Cas-mediated transgenic plant generation

For enhancing phytoremediation potential, CRISPER-Cas system targeted genes responsible for biosynthesis of metal-ligands (e.g., phytochelatins and metallothioneins), metal transport proteins (nine superfamilies, namely, HMA, OPT, COPT, NRAMP, CDF/MTP, YSL, ZIP, PDR, and MRP), root exudates (particularly LMWOA and siderophores) and plant growth hormones (AUXs, GAs, CKs, SAs, ABA) (Basharat et al., 2018). gRNA-Cas9 of CRISPER system simplified the targeting of multiple sequences, and this simplified editing of multiple traits simultaneously. Golden-Braid cloning system enabled the conjugation of previously prepared DNA elements with multigene constructs (Liu and Stewart, 2015; Xing et al., 2014).

The data arising from plant genetic engineering can be harnessed for boosting future CRISPR research for the improvement of phytoremediation ability of plants toward detoxification of organic contaminants and/or heavy metals. The common platforms of the various approaches that can be tried to insert superior properties in the plants during phytoremediation are the direct transfection of Cas9 along with the gRNAs into the plant protoplasts, T-DNA-delivery of gRNA-Cas9, use of Golden-Braid or Gibson assembly for modular cloning or cloning-free strategy, and regeneration of the whole plant from the single cells. In plants, *Agrobacterium*-mediated delivery and transformation is most commonly used (Feng et al., 2013). Few binary vectors were also created to deliver the CRISPR/Cas9 system into plant genomes (Li et al., 2013).

CRISPR/Cas9 technology was applied for the generation of herbicide-resistant transgenics. Many herbicide-resistant transgenic rice varieties were created by diverse ways like inactivation of DNA ligase 4, a component of NHEJ repair pathway (Endo et al., 2016), NHEJ-mediated intron targeting (Li et al., 2016a,b), using two sgRNAs targeting the repair template (Sun et al., 2016), and through the use of chimeric single-guide RNAs (cgRNAs) bearing both target site and repair template sequences. Herbicide-resistant maize and soybean were developed by co-transforming CRISPR/Cas9 and donor DNAs by particle bombardment methods (Svitashev et al., 2015; Butler et al., 2016). In cassava, tolerance toward the herbicide glyphosate was created by a promoter swap and dual amino acid substitution method targeting EPSPS locus gene (required for the biosynthesis of vitamins, some plant hormones, and aromatic amino acids, and many secondary plant metabolites) (Hummel et al., 2018). CRISPR/Cas9-mediated gene insertion and replacement methods introduced herbicide resistance, drought resistance, and other abiotic stress resistance properties in maize (Shi et al., 2017), rice (Miao et al., 2013), orange (Jia and Wang, 2014), tomato (Pramanik et al., 2021; Bari et al., 2021), and in many other plants. Zhang et al. (2014) tried to understand the patterns, specificity, and heritability of CRISPER-mediated target gene mutations. He targeted 11 genes of 2 rice subspecies and found that in T0 generation the target genes were edited in nearly half of the transformed embryogenic cells before their first cell division, and in T1 generation, all the gene mutations were successfully carried to the next generation (T1) without any reversion/mutation. Additionally, there was no proof of large-scale off-targeting (Zhang et al., 2014).

A toolkit was developed that facilitated a transient but stable expression of the CRISPR/Cas9 in a range of plant species, specialized in high-efficiency generation of multiple gene mutations (Xing et al., 2014). The CRISPR/Cas9 binary vector was developed for multiplex genome editing depending on the pGreen or pCAMBI backbone and a gRNA module vector set. Except, *Bsa*I restriction enzyme that generated final constructs harboring maize codons optimized for Cas9 and one or more gRNAs, no restriction enzymes were used. The toolkit was experimentally applied to transgenic *Arabidopsis* lines, maize protoplasts, and transgenic maize lines with high efficiency and specificity. Additionally, the toolkit generated transgenic seedlings of the T1 generation from targeted mutations of three *Arabidopsis* genes (Xing et al., 2014).

Recent CRISPER system was also used for identifying many genes used in phytoremediation of metals by gene-knockout study, and experiments are going for the generation of more accumulators, hyperaccumulators, and bioenergy crops with enhanced phytoremediation ability. Some recent reviews have given the details on *Z. mays* (Agarwal et al., 2018; Shukla et al., 2009) and *Populus* genome editing (Fan et al., 2015) through CRISPR/Cas9.

4 Application of CRISPER to improve plant growth-promoting microbes and other features of plant growth-promoting microbes

Due to the rising tension over climate change, we should try to exploit the beneficial soil microbes, plant-microbe interaction, and the plant microbiome for improved soil fertility and plant health in climatically adverse conditions. A plant is always mutually accompanied by an extremely rich, highly varied, and well-orchestrated microbial community called the phytomicrobiome. It is composed of root-associated soil rhizomicrobes, plant tissue-associated endophytes, and plant surface or phyllosphere-associated microbes. The phytomicrobiome is very important for overall growth and vigor of the host plant. It assists the plant in nutrient uptake, nitrogen fixation, and biotic (pathogen and pest control) and abiotic stress reduction (heavy metal stress, salinity stress etc.) and produces indole acetic acid (IAA) and other plant hormones required for plant development. Signal compounds like recently discovered recently lipo-chitooligosaccharide (LCO) and thuricin 17 have been found to have important roles in the plant-microbe interaction (Smith et al., 2015). The phytomicrobiome and the signal compounds interchanged between microbes and plants are critical factors for crop harvests, especially in the presence of abiotic/biotic stresses (Mueller and Sachs, 2015) and stresses linked with climate change (drought, flooding, too high/low temperatures, salinity etc.) (Smith and Zhou, 2014; Kashyap et al., 2017). A significant part of the phytomicrobiome is occupied by the fungi, including plant growth-promoting fungi, endophytic, ectomycorrhizal, and most importantly arbuscular mycorrhizal (AM) fungi. AM fungi are vital part of the soil microbial community and enjoy mutualistic relations with the roots of > 80% of all terrestrial plants (Öpik et al., 2010). AM fungi plays an active role in the alleviation of salt stress by accumulating compatible osmolytes like proline, glutamate, glycine-betaine, trehalose, and other soluble sugars; improves host nutrition by solubilizing phosphorous;

maintains higher K^+/Na^+ ratios in plant tissues; and protects the plant from soil pathogens (Porcel et al., 2012). Arbuscular mycorrhizal fungi are indispensable part of metal phytoremediation as they alleviate heavy metal stress by accumulating the metal within itself and filtering out excess toxic heavy metals and protect the plants from metal toxicity (Karimi et al., 2011). AM symbiosis also causes an increased production of enzymatic and nonenzymatic antioxidants that take part in plant protection under severe drought condition. Plants made sufficient increase in relative apoplastic water flow in the presence of AM fungi in the root than plants where AM fungi was absent, and this enhancement of water flow was obvious in both watered and without water conditions (Barzana et al., 2012).

The enzyme 1-aminocyclopropane-1-carboxylate deaminase (ACCD) is an important stress-relieving enzyme produced by certain plant growth-promoting bacteria (PGPR) present either as rhizobacteria in the root zone or as tissue endophytes. Treatment of the plants with such bacteria made them more resistant to stress as a consequence of the presence of ACC deaminase, which decreased the production of stress hormone ethylene (Singh et al., 2015).

Our understanding of this very complex system of phytomicrobiome needs to be explored in more depth as manipulation of the phytomicrobiome and the plant-microbe interaction can be exploited for better crop yield and prevent yield loss in situation of climate change (Mabood and Smith, 2005; Mabood et al., 2006).

In view of the environmental impacts of pesticides and widespread fertilizer misuse, PGPR and the signal molecules operating in plant-microbe interaction are offering environmentally safe alternate plans for increasing crop yields with less use of fertilizer and pesticide and will contribute to the development of more sustainable climate change-resilient agricultural systems (Glick, 2010).

As bacterial genome can be modified more easily than plant, climate-resilient characters can be introduced to the plant indirectly by modulating the genome of the selected microbe.

CRISPER technology has also been harnessed to alter the genome of important plant growth-promoting microbes (PGPRs) (Novo et al., 2018; Compant et al., 2010).

It is expected that climate change-induced altered environmental conditions are likely to affect PGPRs by inducing changes in root exudation and plant physiology. Raised carbon dioxide in the atmosphere may lead to heightened carbon sharing in the root zone, thus affecting the composition of the root exudates. Other climate-related adjustments might include alterations in the availability of nutrients, different C/N ratio, synthesis of chemoattractants or signal compounds, pH, and osmotic balance, etc. (Kandeler et al., 2006; Haase et al., 2008; Chinnaswamy et al., 2018) and might substantially impact the colonization, density, and diversity of the soil microbes as well as PGPR microbes (Drigo et al., 2008; Boivin et al., 2016). Altered climatic conditions may directly change the phytomicrobiome or indirectly through reformed plant physiology. But how this change would affect the plant growth and ultimately ecosystem functioning is still not well understood.

Research has started to identify genes that directly or indirectly influence the PGPR-plant interaction (Lyu et al., 2020). Among these, most noteworthy are genes accountable for the synthesis of phytohormones like auxin (*ipd/ppd*), siderophores, phosphate solubilizing (*pqq* genes), and the nitrogenase complex (*nif* genes), Acc deamination (*acd* genes), and hydrogen cyanide synthesis (*hcn* genes) (Bruto et al., 2014). Equipping the microbes of the phytomicrobiome with these abiotic and other biotic control genes could render the plant able to meet its nutritional requirements (nitrogen, iron, and phosphorous), phytohormones production, and render the plant powerful enough to keep away the plant pathogens (Novo et al., 2018). Considering the significant role played by the AM fungi, they should also be enriched with other genes that would make them more climatically resistant. Equipping the plant with more climatically resilient microbes to tolerate environmental stresses of temperature fluctuations, drought resistance, salinity tolerance etc. would render the plant capable of facing and surviving a possible climate change. Overall, bacterial and plants' CRISPR-derived updates can upgrade phytoremediation to the next height, permitting an effective recovery of contaminated soil and water.

5 Adaptive and climate-resilient phytoremediation practices

Although much progress has not been achieved for new transgenic metallophytes, much progress is underway for developing the biotic and abiotic stress-tolerant plants. Few authors (Massel et al., 2021; Jaganathan et al., 2018; Ahmed et al., 2019) have reviewed the various applications of CRISPER for the improvement of climate-resilient properties in cereals and crops and have shown the progress made on GE plants for increasing their various stress tolerances.

5.1 Genome editing for cold tolerance

Most significant among them are the CBF genes (C-repeat binding factor) which impart cold adaptations and signal transductions (Zhou et al., 2011). Along with CBF group of genes, many cold-inducible genes have been identified in plants that can tolerate cold stress (Thomashow, 1999; Sanghera et al., 2011). Most of these genes operate under the regulation of the C-repeat binding factor/dehydration-responsive element binding (CBF/DREB1) transcription factors. When these stress response genes are overexpressed in plant tissues, plants can respond better toward salt stress and drought stress. CBF/DREB1 transcriptional factor particularly binds with the dehydration-responsive element (DRE)/C-repeat (CRT) and regulates the expression of many stress-inducible genes in *Arabidopsis*. Three CBF/DREB1 genes (CBF3/DREB1a, CBF1/DREB1b, and CBF2/DREB1c) owing to the AP2/DREBP family of DNA-binding proteins have been ascertained in *Arabidopsis* (Kasuga et al., 1999, 2004). Cloning and overexpression of CBF1: CRT/DRE binding protein into *Arabidopsis* made them tolerant to chilling without any side effects on the growth and vigor of plants (Jaglo-Ottosen et al., 1998). In the last decade, a massive attempt was undertaken to isolate and characterize COR or cold-responsive genes. Many COR genes were isolated from *Arabidopsis* that were thought to provide defense against dehydration (Hajela et al., 1990).

A substantial number of the up-regulated COR genes were found to be involved in metabolism causing extensive physiological changes during cold acclimatization (Fowler and Thomashow, 2002), a thorough study of which may give leads in understanding and improving freezing tolerance. Pennycooke et al. (2003) suggested that engineering raffinose metabolism provides one way of enhancing the freezing tolerance of plants. Cytosolic malate dehydrogenase gene is another gene that participates in cold tolerance (Yao et al., 2011). Overexpression of a cytosolic malate dehydrogenase gene (*MdcyMDH*) improved the tolerance of the transgenic apple plants to salt and chill conditions by delivering increased reductive redox potential and salicylic acid production (Wang et al., 2016b). Cloning and expression of the CBF1 cDNA into *Solanum lycopersicum* (tomato) under the *CaMV35S* promoter improved cold, drought, and salinity tolerance but introduced some growth retardation effects displaying dwarf phenotype, reduction in fruit set, reduction in seed number etc. (Hsieh et al., 2002). Introduction of bacterial cold shock proteins (CSPs) like *CspA* from *E. coli* and *CspB* from *Bacillus subtilis* endorsed low-temperature adjustments in some species (Castiglioni et al., 2008). CRISPER technology has identified annexin gene *OsAnn3* to be responsible for cold tolerance in rice (Shen et al., 2017).

Recently, plant sphingolipids have been found very suitable for generating cold tolerance in crops. In one study, a single modification of membrane lipids in the otherwise cold-hardy plant *A. thaliana* was found to convert the plant into extremely cold-sensitive plant. This variation, a desaturation of the long-chain fatty acids of sphingolipids, is moderately unusual in nature and is found only in very few cold-tolerant plant species. Few studies have pointed sphingolipid desaturation as a major factor behind coldtolerance in few freezing-tolerant plants. Many scientists postulate that structural alternation of sphingolipid can be a key factor for improving cold tolerance to cold-sensitive plants. More study on this is underway, and it is expected that transgenic plants can be created soon that would be able to survive freezing (Huby et al., 2020).

5.2 Genome editing for salt tolerance

Biotech salt-tolerant crops have been developed for sugarcane, rice, barley, wheat, tomato, and soybean (Tammisola, 2010). Na^+ transport within a plant is regulated by the high-affinity potassium transporter (HKT) gene family (Hauser and Horie, 2010; Munns and Tester, 2008) and the salt overly sensitive (SOS) pathway (Kudla et al., 2010). Tempering the expression of these genes can alter sodium levels in the aboveground parts of the plant (Jesus et al., 2015). A marker-assisted selection (MAS) was successfully used in wheat for inserting novel HKT alleles to increase its salinity tolerance (Munns et al., 2012). Up to date, three main mechanisms have been recorded that are responsible for salt tolerance in plant shoots. The mechanisms are sodium ion build-up in the vacuoles, biosynthesis and accumulation of other compatible nontoxic solutes, and activities of enzymes catalyzing detoxification of ROS, generation of which is common in plants under abiotic stress. So, for the generation of salt-tolerant transgenic plants, various researchers took various means like increasing the abundance of vacuolar Na^+/H^+ antiporters (NHX), increasing the activity of vacuolar H^+ pyrophosphatases, overexpression of enzymes responsible for the biosynthesis of osmotic balance maintaining solutes like glycine-betaine and proline, and enhanced activity of ROS-detoxifying enzymes (Barragan et al., 2012; Ferjani et al., 2011).

5.3   Genome editing for heat tolerance

High temperature is a well-known factor responsible for reducing the crop yield. So, because of global warming, there is an urgent surge to construct heat-resistant transgenic crops and other plants. Although breeding for heat-tolerant crops has given few varieties, the long-time requirement needed for breeding can be bypassed by identifying genes responsible for heat tolerance and then modulating them through CRISPER or other gene-editing tools. Few reviews (e.g., Driedonks et al., 2016) explained the effects of heat stress on plant physiology and crop productivity and discussed the strategies for the breeding of cultivars that can withstand high temperatures. Generation of the thermotolerant crops is quite demanding but puzzling task as plant heat sensitivity changes throughout different developmental and reproductive stages and high temperature triggers a cascade of events, modulating the expression of a plethora of genes. Annotation of the gene expression profile (Lavania et al., 2015) was exploited to identify the pathway of events involved in temperature acclimation to heat stress. Most plants can reprogram their signal transduction pathways by altering the transcription factors and proteins associated with metabolism in response to heat stress. However, most plants can do so within a certain limit, and above the threshold point, plants die. A considerable similarity was noticed among the different heat stress response genes of the different plants residing in different climatic regimes. Moreover, there is a conserved induction of cascade of genes programming for enzymes that control the membrane fluidity if a cell goes through sudden heat exposure. For example, overexpression of the enzyme glycerol-3-phosphate acyltransferase caused a quick revival after heat exposure in comparison with nonengineered tobacco plants (Yan et al., 2008). Stress causes the formation of harmful reactive oxygen species (ROS) from various aerobic metabolic pathways and the ROS-free radicles start damaging plant molecules and tissues. To prevent ROS-aided cell damage and maintain constant redox homeostasis, ROS scavenging machinery is switched on and production of mRNA and proteins that scavenge ROS is increased (Chou et al., 2012). Another significant plant response against high temperature is the generation of heat shock proteins (HSPs) (Wang et al., 2004). HSPs act as molecular chaperones, prevent protein misfolding, and eliminate heat-caused protein aggregations. Heat shock transcription factors (HSFs) regulate the expression of HSPs. Overexpressing HSPs or HSFs have resulted in increased thermotolerance in *Arabidopsis*, rice, tobacco, and tomato (reviewed in Grover et al., 2013). DREBIA gene was cloned from *A. thaliana* and expressed in chrysanthemum where the heat-responsive HSP70 proteins were highly expressed (Hong et al., 2009).

5.4   Genome editing for drought tolerance

Climate change would cause drought in many parts of the world. Equipping the phytoremediator plants with genes that help to survive the long periods of drought is very urgent. Drought tolerance is known to be a complicated and quantitative polygenic attribute. Certain physiological features like proline and abscisic acid content, water potential, root and leaf traits, antioxidation, osmotic fine-tuning, and osmoprotection permit plants to adjust easily with the low water stress condition (Luo, 2010). Transgenic plants containing water stress-resistant genes have already been constructed for *Arabidopsis*, groundnut, maize, rice, potato, papaya, sugarcane, tobacco, tomato, and wheat (Joshi et al., 2016). Regulatory genes coding for dehydration-responsive factors, NAC transcription factor, element binding factors, zinc finger proteins, and structural genes coding for key enzymes expressing redox proteins, osmolyte biosynthesis (proline, glycine/betaine, trehalose, mannitol) proteins, and stress-induced LEA (late embryogenesis abundant) proteins have been used. The synthesis of hydrophilic proteins is a key mechanism to manage water shortage. Overexpression of LEA proteins is one of the primary lines of defense to avoid intercellular loss of water during drought (Veeranagamallaiaha et al., 2011).

Various phytochemicals like ROS, ABA, calcium, and particularly phytohormones play as signaling molecules and cross-talk among themselves during transmission of dehydration stress signals (Hu and Xiong, 2014). Dehydration stress induces the expression of various dehydration-responsive genes encoding ion transporters, calcineurin interacting protein kinases (CIPKs), calcineurin B-like interacting protein kinase (CIPK), calcium-dependent protein kinases (CDPKs), mitogen-activated protein kinases (MAPKs), sucrose non-fermenting protein (SNF1)-related kinase 2 (SnRK2), etc. (Fang and Xiong, 2015). A wide range of transcription factors, including the AREB/ABFs, AP2/EREBF, NAC, MYB, and zinc-finger transcription factors, was identified to induce dehydration tolerance in plants (Joshi et al., 2016). Through gene-knockout study, CRISPR/Cas9 system identified some genes like *SINPR1* gene in rice, *NPR1* gene in tomato, and *OsSAPK2* gene in rice that conferred drought and other abiotic stress tolerances (Quilis et al., 2008; Li et al., 2019; Lou et al., 2017). Roca-Paixão et al. (2019) performed an epigenetic modification of *Arabidopsis* to improve drought stress tolerance. The mutants enjoyed better survival under drought stress.

5.5 Genome editing for increasing yields

CRISPR/Cas9 system presents an effective tool for improving yield-related traits knocking out negative regulators, to affect grain number (*OsGn1a*), grain size (*OsGS3*), grain weight (*TaGW2, OsGW5, OsGLW2, Ta, GASR7*), panicle size (*OsDEP1, TaDEP1*), and tiller number (*OsAAP3*) (Liu et al., 2017a,b; Lu et al., 2018). In *O. sativa*, simultaneous knockout grain weight-related genes (*GW2, GW5, TGW6*) lead to the increased grain weight (Xu et al., 2016) (Table 2). Furthermore, CRISPR-Cas9-mediated editing of gene for cytokinin homeostasis can increase cereal yield, i.e., in *O. sativa* editing gene *LOGL5* which encodes a cytokinin-activation enzyme (Wang et al., 2020) or in *Triticum* sp. knocking out the gene *CKX* that encodes cytokinin oxidase/dehydrogenase, enzyme that catalyzes cytokinin degradation (Zhang et al., 2012). Thus, the yield of phytoremediators (i.e., oilseed-based biofuel crops, essential oil-producing aromatic crops, etc.) can be increased by using CRISPR/Cas9 system.

Table 2   Designer plants in agriculture through CRISPR/Cas9 system.

Plants	Target traits	Target gene	Reference
Oryza sativa	Cold tolerance	*OsCOLD1*	Ma et al. (2015)
	Drought tolerance	*OsMODD, OsNAC14, OsDERF1, OsPMS3, OsEPSPS, OsMSH1, OsMYB5*	Tang et al. (2016), Shim et al. (2018)
	Heat tolerance	*OsTT1*	Li et al. (2015)
	Salt tolerance	*OsSIT1*	Li et al. (2014)
		OsRR2	Zhang et al. (2019a,b)
		TaHKT1;5	Dubcovsky et al. (1996)
		OsRAV2	Duan et al. (2016)
	Growth, yield, nutrition	*Gn1a, DEP1, GS3, IPA1, GS3, GW2, GW5, TGW6, PYL, DEP1, BADH2, OsGSTU, OsAnP, OsMRP15, OsWaxy, LOGL5*	Ma et al. (2015), Li et al. (2016a,b), Gaoneng et al. (2017), Miao et al. (2018), Zhang et al. (2019a,b), Wang et al. (2016a,b,c)
	Resistance to disease	*OsERF922, OsSWEET13, OsSWEET11, OsSWEET14, eIF4G*	Jiang et al. (2013), Zhou et al. (2015), Macovei et al. (2018)
Triticum aestivum	Low temperature	*VRN1, CBFs*	Dhillon et al. (2012), Stockinger et al. (2007), Knox et al. (2010), Francia et al. (2007)
	Salt tolerance	TaHKT1;5	Dubcovsky et al. (1996)
	Growth, yield, nutrition	*TaGW2-A1, -B1, -D1; TaVIT2*	Liang et al. (2016), Connorton et al. (2017)
	Resistance to disease	*TaEDR1, TaMLO, TaLpx*-1, *EDR*1	Wang et al. (2014), Wang et al. (2018a,b)
Zea mays	Drought tolerance	*ARGOS*	Shi et al. (2017)
	Growth, yield, nutrition	*ZmTMS5, ZmIPK1A, ZmIPK, ZmMRP4, PSY1, ZmAgo18a, ZmAgo18b, a1, a4*	Liang et al. (2014), Li et al. (2017), Char et al. (2017)
Glycine max	Growth, yield, nutrition	*01gDDM1, 11gDDM1, Glyma04g36150, miR1509, miR1514, Glyma06g18790, GFP; GmFT2a; GmFEI2, GmSHR; GmPDS11, GmPDS18*	Cai et al. (2015), Jacobs et al. (2015), Du et al. (2016), Cai et al. (2017)
	Resistance to disease	*Rps4, Rps6*	Fang and Tyler (2016)

Table 2 Designer plants in agriculture through CRISPR/Cas9 system—cont'd

Plants	Target traits	Target gene	Reference
Gossypium hirsutum	Drought tolerance	GhPIN1-3	Dass et al. (2017), He et al. (2017)
	Salt tolerance	*GhCLA1, GhVP*	Zhao et al. (2016)
	Growth, yield, nutrition	*GhARG, GhMYB25-like A, GhMYB25-like D*	Wang et al. (2017a,b), Li et al. (2017)
	Resistance to disease	*CLCuMuB, RepGhARG*	Iqbal et al. (2016)
Solanum lycopersicum	Drought tolerance	*SlMAPK3*	Wang et al. (2017a,b)
	Heat tolerance	*BZR1*	Yin et al. (2018)
	Chilling tolerance	*CBF1*	Li et al. (2018a,b)
	Growth, yield, Nutrition	*DELLA, SlCIV1, SlCLV2, SlCLV3, SlRRA3a, CLC-WUS, FW2.2, SP5GSI-AG07, SlSHR, SlPDS, SlPIF4, ANT2, SGR1, LCY-E, LCY-B1, LCY-B2, Blc*	Symington and Gauter (2011), Ron et al. (2014), Xu et al. (2015), Pan et al. (2016), Čermak et al. (2015), Rodriguez-Leal et al. (2017), Soyk et al. (2017), Li et al. (2018a,b)
	Parthenocarpy	*SlAGL6*	Klap et al. (2017)
	Resistance to disease	*CP, Rep, DCL2, DMR6, MLO1, MLO7, ANT1, PMR, MAPK3, CRTISO, PSY1, TYLCV, JAZ2*	Ali et al. (2015), Čermak et al. (2015), Nekrasov et al. (2017), Dahan-Meir et al. (2018)
Solanum tuberosum	Growth, yield, nutrition	*StIAA2, GBSS, StGBSS, St16DOX*	Wang et al. (2018a,b), Nakayasu et al. (2018)
Cucumis sativus	Resistance to disease	*eIF4E*	Chandrasekaran et al. (2016)
Lactuca sativa	Resistance to disease	*NCED4*	Bertier et al. (2018)
Citrus sp.	Resistance to disease	*CIPDS*	Jia et al. (2017)
		CsLOB1	Peng et al. (2017)
Vitis vinifera	Resistance to disease	*MLO-7, VvWRKY52*	Malnoy et al. (2016), Wang et al. (2018a,b)
Musa sp.	Cold, salt tolerance	*MaSWEET-4d, MaSWEET-1b, MaSWEET-4b; MaAPS1, MaAPL3*	Miao et al. (2017a,b)
Carica papaya	Cold, heat, drought tolerance	*CpDreb2, CpRap2.4b*	Figueroa-Yañez et al. (2016), Arroyo-Herrera et al. (2016)
Nicotiana benthamiana	Resistance to disease	*BeYDV*	Baltes et al. (2015)

Continued

Table 2   Designer plants in agriculture through CRISPR/Cas9 system—cont'd

Plants	Target traits	Target gene	Reference
Manihot esculenta	Cold, drought, salt tolerance	*MeKUPs, MeMAPKKK*	Ye et al. (2017), Ou et al. (2018)
Saccharum officinarum	Cold, drought tolerance	ScNsLTP	Chen et al. (2017)
Panicum virgatum	Cold tolerance, Tillering	*Osa-miR393a*	Liu et al. (2017a,b)

5.6   Genome editing for improving quality/nutrition

Quality traits of crops, such as starch content, nutritional value, and fragrance, are important for agricultural production. Grain of *O. sativa* or *Z. mays* with low amylose content is better for eating and cooking, and it is generated by knocking out *Waxy* via CRISPR/Cas9 system by the disruption of gene *GBSS1* (granule-bound starch synthase 1), which is crucial for amylose biosynthesis (Gao et al., 2020; Xu et al., 2021) (Table 2). However, high-amylose foods are beneficial for health, and if the CRISPR/Cas9 system is employed, it is possible to produce targeted mutagenesis (the starch branching enzyme genes *SBEI* and *SBEIIb*) (Sun et al., 2017). Gluten protein in cereal grains can cause celiac disease, and CRISPR/Cas9 editing offers knockout of the most conserved domain of the α-*gliadin gene* family members which create the low-gluten plants (Sanchez-Leon et al., 2018) (Table 2). CRISPR/Cas9 system editing is also used for the breeding of high-quality crops that are enriched with high oleic acid oil, carotenoids, lycopene, and γ-aminobutyric acid content (GABA), or they have a reduced content of toxic steroidal glycoalkaloids, phytic acids, etc. (Jaganathan et al., 2018; Islam, 2019; Chen et al., 2019). Thus, the quality of phytoremediation-mediated biofortification can be improved by using CRISPR/Cas9 system.

5.7   Genome editing for disease stress resistance

CRISPR/Cas9 gene-knockout plant species can increase resistance to disease stress to fungal, bacterial, and viral diseases (Islam, 2019). The disease stress resistance of *O. sativa* to bacterial light caused by *Xanthomonas oryzae* pv. *oryzae* can be achieved by the deletion of the *OsSWEET11*, *OsSWEET13*, and *OsSWEET14* promoters using CRISPR/Cas9 system (Jiang et al., 2013; Zhou et al., 2015) (Table 2). Resistance to fungal powdery mildew in *Triticum* sp. can be achieved by knocking out of six *TaMLO* alleles (Wang et al., 2014) and in *S. lycopersicum* by knockout of *MLO* gene in *S. lycopersicum* via CRISPR/Cas9 system (Nekrasov et al., 2017) (Table 2). In *Triticum* sp., fungi *Blumeria graminis* f.sp. *tritici* causes powdery mildew and CRISPR/Cas9 system can be used for editing *EDR1* gene (enhance disease resistance 1), which encodes MAPK kinase that inhibits defense response and mutation of that gene plants enhances resistance to this fungi (Zhang et al., 2017b), whereas in *S. lycopersicum* mutation of three mildew-resistance locus O (MLO) by CRISPR/Cas9 system can create plant resistance to fungi (Wang et al., 2014). Plant immune system against viruses is enhanced by using CRISPR/Cas9 system: TYLCV, tomato yellow leaf curl DNA virus to engineer *Nicotiana benthamiana* and *S. lycopersicum* (Ali et al., 2015; Tashkandi et al., 2018); beet severe curly top (BSCTV) geminivirus to engineer *N. benthamiana* (Ji et al., 2015). Therefore, the disease stress resistance of phytoremediators can be achieved by using CRISPR/Cas9 system.

6 Designer plants in agriculture and phytoremediation through CRISPR/Cas9 system

6.1 Designer plants in agriculture via CRISPR/Cas9 system

6.1.1 Genome editing in monocots

O. sativa (rice), *Triticum aestivum* (wheat), and *Z. mays* (maize) are the main food crops grown worldwide. CRISPR/Cas9 technology can be used to target a number of traits in plant genome as response to abiotic and biotic stress and improvement of nutritional quality (Jaganathan et al., 2018) (Fig. 1, Table 2).

CRISPR/Cas9 genome editing system has been demonstrated in *O. sativa* for:

(a) resistance to abiotic stress with targeted genes for editing: *OsCOLD1* for cold tolerance (Ma et al., 2015); *OsMODD, OsNAC14, OsDERF1, OsPMS3, OsEPSPS, OsMSH1, OsMYB5* for drought tolerance (Tang et al., 2016; Shim et al., 2018); *OsTT1* for thermotolerance (Li et al., 2015); *OsSIT1*, TaHKT1;5, OsRAV2 for salt tolerance (Dubcovsky et al., 1996; Li et al., 2014; Duan et al., 2016); flooding Sub1A, SK1, SK2 (Xu et al., 2006; Fukao et al., 2006; Hattori et al., 2009).

(b) resistance to diseases with specific genes for editing: *OsERF922* for blast resistance (Wang et al., 2016a,b,c); *OsSWEET* for bacterial blight resistance (Zhou et al., 2015; Jiang et al., 2013); *eIF4G* for tungro spherical virus (Macovei et al., 2018).

(c) increased quality of yield with specific genes for editing: Os*Gn1a, OsDEP1, OsGS3, IPA1, GW2, GW5, TGW6, PYL* (Li et al., 2016a,b; Miao et al., 2018; Chen et al., 2020);

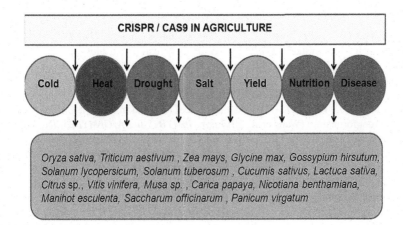

Fig. 1 CRISPR/Cas9 system in agriculture.

(d) increased nutrition with specific genes for editing: *BADH2* for fragrant rice variety (Gaoneng et al., 2017); *OsGSTU, OsAnP, OsMRP15* for decreased anthocyanin content and *OsWaxy* for the reduced synthesis of amylase (Ma et al., 2015).

CRISPR/Cas9 genome editing system has been demonstrated in *T. aestivum* for:

(a) resistance to abiotic stress with targeted genes for editing: VRN1 and CBFs for low-temperature tolerance (Dhillon et al. (2012); Stockinger et al., 2007; Knox et al., 2010; Francia et al., 2007); TaHKT1;5 for salinity stress (Dubcovsky et al., 1996); resistance to diseases with specific genes for editing: *TaEDR1, TaMLO,* and *TaLpx-1* for powdery mildew resistance (Wang et al., 2014, 2018a,b).
(b) increased quality of yield with specific genes for editing: α-*Glutein-encoding gene* for low gluten that leads to the reduced celiac disease in humans; *TaGW2-A1, -B1, -D1* for grain weight (Liang et al., 2016); *TaVIT2* for Fe content (Connorton et al., 2017).

CRISPR/Cas9 genome editing system has been demonstrated in *Z. mays* for:

(a) resistance to abiotic stress with targeted genes for editing: overexpression of *ARGOS* (*AUXIN REGULATED GENE INVOLVED IN ORGAN SIZE*) gene family enhances the drought tolerance (Shi et al., 2017).
(b) high yielding with *ZmTMS5* (*thermosensitive genic male-sterile 5*) (Li et al., 2017).
(c) nutritional quality with targeted genes for editing: knockout of genes involved in phytic acid synthesis (*ZmIPK1A, ZmIPK, ZmMRP4*) (Liang et al., 2014) and PSY1 gene for carotenoid synthesis (Zhai et al., 2016), and anthocyanin biosynthesis (*ZmAgo18a, ZmAgo18b, a1, a4*) (Char et al., 2017).

6.1.2   Genome editing in dicots

A number of dicots are important food crops, such as vegetables, fruits, and plants with seed oil reserves, and they are good candidates for testing CRISPR/Cas9 editing system (Jaganathan et al., 2018). Therefore, CRISPR/Cas9 genome editing system has been demonstrated in tolerance to abiotic stress, improved quality of yield, plant growth and nutrient, as well as in resistance to pathogen with specific genes for editing (Fig. 1, Table 2):

(a) *Glycine max* (soybean)—*01gDDM1, 11gDDM1, Glyma04g36150, miR1509, miR1514, Glyma06g18790, GFP* (quality); *GmFT2a* (vegetative growth), *GmFEI2, GmSHR* (improved hairy root system), *GmPDS11, GmPDS18* (carotenoid biosynthesis) (Cai et al., 2015, 2017; Jacobs et al., 2015; Du et al., 2016); *Rps4, Rps6* (virus resistance) (Fang and Tyler, 2016).
(b) *Gossypium hirsutum* (cotton)—GhRDL1, GhPIN1-3 (drought tolerance) (Dass et al., 2017; He et al., 2017); *GhCLA1, GhVP* (salt tolerance) (Zhao et al., 2016); *GhARG* (improved lateral root formation) (Wang et al., 2017a,b); *GhMYB25-like A, GhMYB25-like D* (fiber development) (Li et al., 2017); *CLCuMuB, RepGhARG* (leaf curl disease) (Iqbal et al., 2016).
(c) *S. lycopersicum* (tomato)—*SlMAPK3* (drought tolerance) (Wang et al., 2017a,b); *BZR1* (heat tolerance) (Yin et al., 2018); *CBF1* (chilling tolerance) (Li et al., 2018a,b); *DELLA, SlCIV1, SlCLV2, SlCLV3, SlRRA3a, CLC-WUS, FW2.2, SP5GSI-AG07* (increased yield and growth) (Symigton and Gauter, 2011; Xu et al., 2015; Rodriguez-Leal et al., 2017; Soyk et al., 2017); *SlSHR* (root development) (Ron et al., 2014); *SlPDS, SlPIF4* (carotenoid biosynthesis) (Pan et al., 2016); *ANT2* (anthocyanin content) (Čermak et al., 2015); *SGR1,*

LCY-E, LCY-B1, LCY-B2, Blc (lycopene content) (Li et al., 2018a,b); *CP, Rep, DCL2, DMR6, MLO1, MLO7, ANT1, PMR, MAPK3, CRTISO, PSY1, TYLCV, JAZ2* (powdery mildew, resistance to the geminivirus and bacteria) (Čermak et al., 2015; Nekrasov et al., 2017; Dahan-Meir et al., 2018; Ali et al., 2015).

(d) *Solanum tuberosum* (potato)—*StIAA2* (fruit development); *GBSS, StGBSS* (starch quality) (Wang et al., 2019); *St16DOX* (reduce steroidal glycoalkaloides) (Nakayasu et al., 2018).

(e) *Cucumis sativus* (cucumber)—*eIF4E* (virus resistance) (Chandrasekaran et al., 2016).

(f) *Lactuca sativa* (lettuce)—*9-cis-EPOXYCAROTENOID DIOXYGENASE4* (*NCED4*) (increased seed germination at high temperatures) (Bertier et al., 2018).

(g) *Citrus* sp. (citrus)—*CIPDS* (resistance to citrus cancer by *Xanthomonas citri* subsp. *citri* (Xcc)) (Jia et al., 2017); *CsLOB1* (citrus canker resistance) (Peng et al., 2017).

(h) *Vitis vinifera* (grape)—*MLO-7* (resistance to powdery mildew) (Malnoy et al., 2016); *VvWRKY52* (resistance to fungal infection by *Botrytis cinerea*) (Wang et al., 2018a,b).

(i) *Musa* sp. (banana)—*MaSWEET-4d, MaSWEET-1b, MaSWEET-4b; MaAPS1, MaAPL3* for cold and salt tolerance (Miao et al., 2017a,b).

(j) *N. benthamiana* (tobacco)—*BeYDV* (resistance to bean yellow dwarf virus) (Baltes et al., 2015)

(k) *Manihot esculenta* (cassava)—*MeKUPs, MeMAPKKK* (cold, drought, salt tolerance) (Ye et al., 2017; Ou et al., 2018)

(l) *Saccharum officinarum* (sugarcane)—*ScNsLTP* (cold and drought tolerance) (Chen et al., 2017)

(m) *Panicum virgatum* (switchgrass)—*Osa-miR393a* (cold tolerance and tillering) (Liu et al., 2017a,b).

6.2   Designer plants in phytoremediation through CRISPR/Cas9 system

CRISPR/Cas9 genome editing technology helps to engineer the genome of plant species that can be used in phytoremediation of polluted sites that are enriched by metal(loid)s and organic pollutants (Basharat et al., 2018) (Fig. 2). Furthermore, designing plant growth-promoting rhizobacteria (PGPR) through CRISPR/Cas9 system technology and modification of genes that are responsible for biosynthesis of siderophores, nitrogenase complex, or phytohormones may benefit the plants to overcome stress caused by pollution and enable Fe sequestration, N fixation, and P solubilization (Mosa et al., 2016; Chinnaswamy et al., 2018; Thode et al., 2018). CRISPR/Cas9 system can be used for the modification of genes responsible for plants heavy metal uptake (metal transporters, such as ZIP family, CDF, HMA, etc.), their accumulation (metallothioneins, phytochelatins), or biosynthesis of root exudates (siderophores) (Saxena et al., 2020). CRISPR/Cas9 system can be used to modify the plant genome for multiple characteristics simultaneously, such as increased growth rate and yield, plant biomass (root system), resistance to environmental stress, pathogens, and tolerance to heavy metals (Bortesi and Fischer, 2015).

Hyperaccumulators whose genomes are sequenced can be used successfully in CRISPR/Cas9 technology as model plants with high phytoremedial potential: *A. halleri* for Zn and Cd; *N. caerulescens* for Ni, Zn, Cd; *Pteris vittata* for As (Basharat et al., 2018). Tang et al. (2017) reported new hybrid lines of *O. sativa* with low Cd

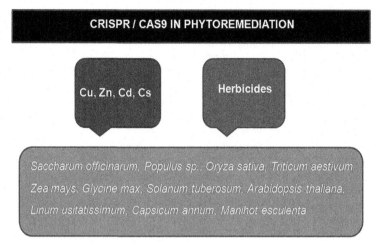

Fig. 2 CRISPR/Cas9 system in phytoremediation.

concentrations in grains (0.05 mg/kg) by knockout of *OsNramp5* transporter using CRISPR/Cas9 system compared to the wild type (0.33–2.90 mg/kg) (Table 3). In *Phoenix dactylifera* (Date palm), CRISPR/Cas9 system can be used for Cd and Cr with targeted genes for editing, such as *Pdpcs* and *Pdmt* (Chaâbene et al., 2018), whereas in *S. officinarum*, ScAPX6 gene can be used for Cu tolerance (Liu et al., 2017a,b) (Table 3). Knockout of *OsHAK1* gene, i.e., inactivation of the K transporter in *O. sativa* by CRISPR/Cas9 system, ^{137}Cs uptake and transport were decreased in plants grown in Fukushima soil (Nieves-Cordones et al., 2017) (Table 3). In addition, CRISPR/Cas9 system can be exploited to edit genomic sequence in energy crops, such as *Populus tomentosa* (Fan et al., 2015), *Sorghum bicolor* (Jiang et al., 2013), and *Z. mays* (Liang et al., 2014). Therefore, CRISPR/Cas9 system can be used in *Populus* sp. for an enhanced accumulation of Cu, Zn, and Cd through overexpression of MTA1, MT1, and MT2 (metallothionein-encoding genes) (Turchi et al., 2012) (Table 3). According to Swaminathan et al. (2010), genome of *S. bicolor* can be used in the assembly of gene-space sequence of *Miscanthus giganteus*, i.e., the coding fraction of the *M. giganteus* genome is largely identical to that of other grasses (Andropogoneae). Furthermore, 12 abundant families of repeat sequences and their derived RNAs composed of 24 nucleotides were found abundant in the rhizomes, leaves, and inflorescences of *M. giganteus* (Swaminathan et al., 2010).

CRISPR/Cas9 technology is mostly used in phytoremediation for herbicide-resistant plants, and the most common targeting genes are *ALS* (acetolactate synthase), *EPSPS* (5-enolpyruvylshikimate-3-phosphate synthase), *CESA3* (cellulose synthase A catalytic subunit 3), and *SF3B1* (splicing factor 3B subunit 1) (Han and Kim, 2019) (Fig. 2, Table 3).

ALS is a key enzyme that catalyzes the biosynthesis of essential amino acids, such as valine, leucine, and isoleucine (Garcia et al., 2017). It was noted that point mutations in the ALS gene can lead to the substitution of amino acids creating the plant herbicide tolerance (Powles and Yu, 2010). Therefore, the precise gene editing of *ALS*

Table 3 Designer plants in phytoremediation through CRISPR/Cas9 system.

Plants	Pollutant/ herbicide	Target gene	Reference
Saccharum officinarum	Cu	*ScAPX6*	Liu et al. (2017a,b)
Populus sp.	Cu, Zn, Cd	*MTA1, MT1, MT2*	Turchi et al. (2012)
Oryza sativa	Cd	*OsNramp5*	Tang et al. (2017)
	^{137}Cs	*OsHAK1*	Nieves-Cordones et al. (2017)
	Imazamox, Bispyribac sodium, Imidazolinone	*ALS*	Sun et al. (2016), Shimatani et al. (2017), Wang et al. (2020)
	Glyphosphate	*EPSPS*	Li et al. (2016a,b)
	SGR GEX1A	*SF3B1*	Butt et al. (2019)
	Haloxyfop	*ACCase (C2186R, W2125C, P1927F)*	Li et al. (2018a,b)
	Benzobicyclon	*HIS1*	Komatsu et al. (2020)
Triticum aestivum	Nicosulfuron, Mesosulfuron, Imazapic	*ALS*	Zhang et al. (2019a,b)
	Quizalofop	*ACCase (A1992V)*	Zhang et al. (2019a,b)
Zea mays, Glycine max	Chlorosulfuron	*ALS*	Li et al. (2015), Svitashev et al. (2015)
Solanum tuberosum	Imidazolinone	*ALS*	Butler et al. (2016)
Arabidopsis thaliana	C17, CBI (isoxaben, flupoxam, dichlobenil, indaziflam)	*CESAs*	Somerville (2006)
	2,4-D herbicide	PDR9	Ito and Gray (2006)
Linum usitatissimum	Glyphosphate	*EPSPS*	Sauer et al. (2016)
Capsicum annum	Glyphosphate	*EPSPS*	Ortega et al. (2018)
Manihot esculenta	Glyphosphate	*EPSPS*	Hummel et al. (2018)

is done by CRISPR/Cas9 system for the herbicide resistance found in different plant species:

- *O. sativa* (Imazamox, Bispyribac sodium, Imidazolinone) (Sun et al., 2016; Shimatani et al., 2017; Wang et al., 2020)
- *Z. mays* and *G. max* (Chlorosulfuron) (Li et al., 2015; Svitashev et al., 2015)
- *S. tuberosum* (Imidazolinone) (Butler et al., 2016)
- *T. aestivum* (Nicosulfuron, Mesosulfuron, Imazapic) (Zhang et al., 2019a,b).

EPSPS is a key enzyme on the shikimate pathway, which is important for the synthesis of aromatic amino acids and catalyzes the transfer of enolpyruvyl moiety from PEP (phosphoenol pyruvate) to skimimate-3-phosphate creating EPSP and Pi (Schonbrunn et al., 2001). Active compound of many herbicides is glyphosate, and in mutant herbicide-resistant plant, simultaneous mutation of threonine and proline, threonine to isoleucine, and proline to serine or alanine mutation provides changes for resistance to this herbicide (Han and Kim, 2019). The gene editing of *EPSPS* by CRISPR/Cas9 system for the glyphosate resistance is found in different plant species:

- *O. sativa* (Li et al., 2016a,b)
- *Linum usitatissimum* (Sauer et al., 2016)
- *Capsicum annum* (Ortega et al., 2018)
- *M. esculenta* (Hummel et al., 2018)

Herbicides, such as C17 (5-(4-chlorophenyl)-7-(2-methoxyphenyl)-1,5,6,7-tetrahydro-[1,2,4]triazolo[1,5-a] pyrimidine), and cellulose biosynthesis inhibitors (CBI) (isoxaben, flupoxam, dichlobenil, indaziflam) operate in different modes of action and inhibit the growth of dicotyledons crops, i.e., inhibit cellulose synthase catalytic subunits (*CESAs*) which catalyze the conversion of UDP-glucose to cellulose (Kimura et al., 1999). Generally, C17 action leads to weakened cell wall, reduced content of cellulose, and decreased elongation of plant hypocotyl, whereas C17-tolerant mutants had single-nucleotide missense mutations at *CESAs* (*CESA1* and *CESA3*) (Hu et al., 2016). According to Somerville (2006), *A. thaliana* have 10 *CESA* proteins and at least 3 of them (*CESA1*, *CESA3*, and *CESA6*) takes part in primary cell wall formation (Lei et al., 2012). Hu et al. (2019) showed that a C17-resistant *CESA3* allele can be used as transformation selection marker by using CRISPR/Cas9 system, and it is possible to engineer C17 tolerance in isoxaben-resistant line resulting in double herbicide-resistant plants, i.e., there is the possibility to transfer CBI multiresistance to any crop species (Table 3).

CRISPR/Cas9 system can be used for resistance to splicing inhibitors, such as pladienolide B and herboxidiene in *O. sativa* edited splicing factor 3B subunit 1 gene (*SF3B1*) and created SGR GEX1A-resistant plants (Butt et al., 2019) (Table 3).

Acetyl coenzyme A carboxylase (ACCase) is an important enzyme in the biosynthesis of lipids catalyzing the carboxylation of acetyl-CoA to malonyl-CoA (Yu et al., 2013) and presenting herbicide target. CRISPR/Cas9 editing ACCase genes from dicotyledonous plants can be used for grass weed control. ACCase inhibiting herbicides (aryloxyphenoxypropionate, cyclohexanedione, and phenylpyrazoline) can be used in dicotyledonous plants for grass weed control (*Lolium, Alopecurus, Avena*) (Kaundum, 2014).

CRISPR/Cas9 system can edit the PDR9 (pleiotropic drug resistance 9) that is ABC transporter in *Arabidopsis*, and it is important for efflux of 2,4-D herbicide (2,4-dichlorphenoxyacetic acid) outside of plant cells, i.e., designated mutant *pdr9-1* in which only single amino acid change showed the tolerance to 2,4-D herbicide (Ito and Gray, 2006). Furthermore, haloxyfop resistance in *O. sativa* is achieved through CRISPR/Cas9 editing, i.e., amino acid substitutions in ACCase (C2186R, W2125C, P1927F) (Li et al., 2018a,b), whereas quizalofop resistance in *T. aestivum* is created by introducing A1992V substitution in ACCase (Zhang et al., 2019a,b).

The production of feed *O. sativa* varieties depends on the control and suppression of volunteer rice infestation, and through CRISPR/Cas9 system, it was possible to design strains sensitive to herbicide BBC (Benzobicyclon), which belongs to bTHs (beta-triketone herbicides) (Komatsu et al., 2020). Generally, the *O. sativa HIS1* gene (*HPPD INHIBITOR SENSITIVE 1*) confers resistance to BBC and bTHs herbicides, and it is important for creating BBC-sensitive strain through knockout of the *HIS1* gene (Komatsu et al., 2020) (Table 3). According to Maeda et al. (2019), BBC-sensitive cultivar absorbed BBC-OH hydrolysate, which can act as an inhibitor in the roots and leaves of *O. sativa*, but in the transformed *HIS1* line (through nucleotide substitution), BBC-OH was not detected in the leaves.

7 Challenges and opportunities

Transgenic plants are yet to receive universal acceptance. The objections to GM technologies are centered around either on human health issue or on their adverse effects on biodiversity (Fischer et al., 2015). So far, after the introduction of GM foods in 1994, no evidence has come up against the safety of GM crops on health or ecosystem but the technology was successful in cutting down the use of pesticide and insecticide and contributed to environmental protection (NASEM, 2016, 2017). For sustainable growth of humankind and protecting the environment from the altered climatic conditions, scientists should look for climate-resilient development, which means that developmental activities should come up with existing climate variability and be able to adapt to future climate changes with stable development allowing damages to be minimum. The goal of climate-resilient development should be keeping the consequences of climate impacts minimum without hindering development goals.

Because of the climate impacts being felt today, it is high time to identify the climatic stressors and link them to phytoremediation to address climate variability and development planning. As science cannot predict climate change and the impacts are uncertain and will remain so, development activities should adaptively employ multiple approaches to manage risk. Adaptation is essentially a slow and continuous process and as new information about climate stressors turn up should be integrated into responses. United States Agency for International Development (USAID) in 2007 issued a manual on adaptation to climate variability to help development practitioners to determine whether a new or existing project is susceptible to climate stressors and the different ways to recognize and act together to reduce climatic susceptibility (USAID, 2007).

CRISPR-Cas tool of plants holds incredible potential toward ensuring food security and fighting climate change induced biotic and abiotic stresses, and all these lead to climate resilience sustainability. Despite the usefulness of CRISPR-Cas technology established extensively, many challenges remain for widespread applications of CRISPER-Cas technology in plant biotechnology. The rigid and thick plant cell wall of plants poses one of the biggest challenges of CRISPER as delivery of the gene-editing reagents through this thick extracellular matrix of complex polysaccharides is tough.

CRISPR-Cas reagents are generally transported into plant cells either by plasmid-based methods or by non-DNA-based methods for either stable or for transitory transformation. As transient methods are temporarily expressed and are not incorporated in chromosomes, they are preferred over stable transformations. A low number of nontarget mutations and nonheritable DNA integration due to the transient expression results in reduced regulatory load. Transient expression methods through plasmid vehicles have found fewer applications as most plant species don't accept these plasmid constructs. The alternate method of DNA-free delivery where the CRISPR-Cas complex is transported straight into plant cells without any DNA vector is more popular. Preassembled ribonucleoproteins (RNPs) are one of the DNA-free delivery systems where protoplast transfection (Woo et al., 2015) or particle bombardment (Liang et al., 2017) is used. Protoplasts are created by removing the cell walls, and the absence of cell walls facilitates the delivery of CRISPR reagents. On the other hand, regeneration of the whole plant from the protoplast is another technical challenge and unfortunately not demonstrated in most species (Eeckhaut et al., 2013). Particle bombardment is another DNA-free delivery option that permits reagents to crack the cell wall, even it escalates the chance of irreversible cell damage and has a low transformation efficiency. The editing efficiency of both protoplast transfection and particle bombardment through gene gun is < 10% (Svitashev et al., 2016). As germline transformation is not possible in many plants, somatic embryogenesis is the only viable option that is laborious, time-consuming, and arduous specifically in monocots. Moreover, within a species, different genotypes behave differently to tissue culture. Ellison et al. (2020) successfully transported the CRISPR-Cas reagents through a RNA virus. The editing efficiency was of 90%–100% in infected somatic tissue and caused heritable mutations in multiple genes simultaneously. However, as RNA virus has a limited host range, this was not applicable in other plants.

Another challenge of fruitful gene insertion through HDR is the need for simultaneous delivery of donor templates and the CRISPR-Cas complex. Success stories with HDR in plants are quite low. Another potential option is the delivery of the reagents in form of nanomaterials as only nanoparticles can easily breach all barriers and are accessible to remote locations, inaccessible by others (Demirer et al., 2021). Nanomaterials can give targeted delivery and the released cargoes are also protected from degradation. An additional advantage is that nanomaterials can work in a species-independent manner. Although transporting DNA and proteins directly into plant cells using nanomaterials has been quite successful in the past, nanomaterial-mediated CRISPR-Cas genome editing in plants has not progressed much due to the noble physicochemical qualities of CRISPR reagents and the high delivery precisions required to achieve successful genome editing in plants.

Regulatory laws on transgenics and societal acceptance may generate further obstacles to introduce the technology or technology-generated products in the open market of many countries.

8   Conclusion

CRISPR-Cas-mediated plant genome editing and subsequent traditional transformation and regeneration processes have shown great achievements in many plant species. Despite the success that CRISPR has shown for genome engineering in some plants, certain uncertainties remain in other plants, including factors like choice of the target site, selection of gRNA, Cas9/Cpf1 action, probability of off-target mutations, and choice of delivery systems. Challenge remains to overcome the delivery-related barriers in certain species, and new technologies should be innovated to bring them under this approach. Nevertheless, with the current speed of CRISPER research, significant breakthroughs can come at any time that would contribute significantly toward the enrichment of climate-resilient phytoremediation. In summary, CRISPR-mediated genome engineering embraces huge potential to cause a paradigm shift in phytoremediation research.

References

Adli, M., 2018. The CRISPR tool kit for genome editing and beyond. Nat. Commun. 9, 1911.

Agarwal, A., Yadava, P., Kumar, K., Singh, I., Kaul, T., Pattanayak, A., Agrawal, P.K., 2018. Insights into maize genome editing via CRISPR/Cas9. Physiol. Mol. Biol. Plants 24, 175–183.

Ahmed, S., Zhang, Y., Abdullah, M., et al., 2019. Current status, challenges, and future prospects of plant genome editing in China. Plant Biotechnol. Rep. 13, 459–472.

Aken, B.V., 2008. Transgenic plants for phytoremediation: helping nature to clean up environmental pollution. Trends Biotechnol. 26 (5), 225–227.

Al-Tabbaa, A., Harbottle, M., Evans, C., 2007. Robust sustainable technical solutions. In: Dixon, T., Raco, M., Catney, P., Lerner, D.N. (Eds.), Sustainable Brownfield Regeneration. Blackwell Publishing, Oxford.

Ali, H., Khan, E., Sajad, M.A., 2013. Phytoremediation of heavy metals—concepts and applications. Chemosphere 91, 869–881.

Ali, Z., Abulfaraj, A., Idris, A., Ali, S., Tashkandi, M., Mahfouz, M., 2015. CRISPR/Cas9 mediated viral interference in plants. Genome Biol. 16, 238.

Anderson, C., Moreno, F., Meech, J., 2005. A field demonstration of gold phytoextraction technology. Miner. Eng. 18, 385–392.

Arazi, T., Sunkar, R., Kaplan, B., Fromm, H., 1999. A tobacco plasma membrane calmodulin-binding transporter confers Ni2 + tolerance and Pb2 + hypersensitivity in transgenic plants. Plant J. 20, 171–182.

Arroyo-Herrera, A., Figueroa-Yanez, L., Castano, E., Santamari'a, J., Pereira-Santana, A., Espadas-Alcocer, J., et al., 2016. A novel Dreb2-type gene from *Carica papaya* confers tolerance under abiotic stress. Plant Cell Tissue Organ Cult. 125 (1), 119–133.

Auguy, F., Fahr, M., Moulin, P., El Mzibri, M., Smouni, A., Filali-Maltouf, A., Béna, G., Doumas, P., 2016. Transcriptome changes in *Hirschfeldia incana* in response to lead exposure. Front. Plant Sci. 6, 1–13.

Baldantoni, D., Cicatelli, A., Bellino, A., Castiglione, S., 2014. Different behaviours in phytoremediation capacity of two heavy metal tolerant poplar clones in relation to iron and other trace elements. J. Environ. Manag. 146, 94–99.

Baltes, N.J., Hummel, A.W., Konecna, E., Cegan, R., Bruns, A.N., Bisaro, D.M., et al., 2015. Conferring resistance to geminiviruses with the CRISPR–Cas prokaryotic immune system. Nat. Plants 1, 15145.

Bari, V.K., Nassar, J.A., Aly, R., 2021. CRISPR/Cas9 mediated mutagenesis of MORE AXILLARY GROWTH 1 in tomato confers resistance to root parasitic weed *Phelipanche aegyptiaca*. Sci. Rep. 11 (1), 3905.

Barragan, V., Leidi, E.O., Andrés, Z., Rubio, L., De Luca, A., Fernandez, J.A., Cubero, B., Pardo, J.M., 2012. Ion exchangers NHX1 and NHX2 mediate active potassium uptake into vacuoles to regulate cell turgor and stomatal function in Arabidopsis. Plant Cell 24, 1127–1142.

Barzana, G., Aroca, R., Paz, J.A., Chaumont, F., Martinez-Ballesta, M.C., 2012. Arbuscular mycorrhizal symbiosis increases relative apoplastic water flow in roots of the host plant under both well-watered and drought stress conditions. Ann. Bot. 109, 1009–1017.

Basharat, Z., Novo, L.A.B., Yasmin, A., 2018. Genome editing weds CRISPR: what is in it for phytoremediation? Plan. Theory 7, 51.

Behr, M., Legay, S., Hausman, J.F., Guerriero, G., 2015. Analysis of cell wall-related genes in organs of *Medicago sativa* L. under different abiotic stresses. Int. J. Mol. Sci. 16, 16104–16124.

Bertier, L.D., Ron, M., Huo, H., Bradford, K.J., Britt, A.B., Michelmore, R.W., 2018. High-resolution analysis of the efficiency, heritability, and editing outcomes of CRISPR-Cas9-induced modifications of NCED4 in lettuce (*Lactuca sativa*). G3 (Betheseda) 8 (5), 1513–1521.

Boivin, S., Fonouni-Farde, C., Frugier, F., 2016. How auxin and cytokinin phytohormones modulate root microbe interactions. Front. Plant Sci. 7, 1–12.

Bortesi, L., Fischer, R., 2015. The CRISPR/Cas9 system for plant genome editing and beyond. Biotechnol. Adv. 33, 41–52.

Bruto, M., Prigent-Combaret, C., Muller, D., et al., 2014. Analysis of genes contributing to plant-beneficial functions in plant growth-promoting rhizobacteria and related Proteobacteria. Sci. Rep. 4, 6261.

Butler, N.M., Baltes, N.J., Voytas, D.F., Douches, D.S., 2016. Geminivirus-mediated genome editing in potato (*Solanum tuberosum* L.) using sequence-specific nucleases. Front. Plant Sci. 7, 1045.

Butt, H., Eid, A., Momin, A.A., Bazin, J., Crespi, M., Arold, S.T., Mahfouz, M.M., 2019. CRISPR directed evolution of the spliceosome for resistance to splicing inhibitors. Genome Biol. 20, 73.

Cai, Y.P., Chen, L., Liu, X.J., Sun, S., Wu, C.X., Jiang, B.J., 2015. CRISPR/Cas9-mediated genome editing in soybean hairy roots. PLoS One 10, 0136064.

Cai, Y., Chen, L., Liu, X., Guo, C., Sun, S., Wu, C., Jiang, B., Han, T., Hou, W., 2017. CRISPR/Cas9-mediated targeted mutagenesis of GmFT2a delays flowering time in soybean. Plant Biotechnol. J. 16 (1), 176–185.

Castiglioni, P., Warner, D., Bensen, R.J., Anstrom, D.C., Harrison, J., et al., 2008. Bacterial RNA chaperones confer abiotic stress tolerance in plants and improved grain yield in maize under water-limited conditions. Plant Physiol. 147, 446–455.

Čermak, T., Baltes, N.J., Čegan, R., Zhang, Y., Voytas, D.F., 2015. High-frequency, precise modification of the tomato genome. Genome Biol. 16, 232.

Chaâbene, Z., Rorat, A., Hakim, I., Bernard, F., Douglas, G.C., Elleuch, A., et al., 2018. Insight into the expression variation of metal-responsive genes in the seedling of date palm (*Phoenix dactylifera*). Chemosphere 197, 123–134.

Chandrasekaran, J., Brumin, M., Wolf, D., Leibman, D., Klap, C., Pearlsman, M., et al., 2016. Development of broad virus resistance in non-transgenic cucumber using CRISPR/Cas9 technology. Mol. Plant Pathol. 17, 1140–1153.

Char, S.N., Neelakandan, A.K., Nahamoun, H., Frame, B., Maim, M., Spalding, M.H., Becraft, P.W., Meyers, B.C., Walbot, V., Wang, K., Yang, B., 2017. An agrobacterium-delivered CRISPR/Cas9 system for high frequency targeted mutagenesis in maize. Plant Biotechnol. J. 15, 257–268.

Chen, K., Gao, C., 2014. Targeted genome modification technologies and their applications in crop improvements. Plant Cell Rep. 33, 575–583.

Chen, L.H., Zhang, B., Xu, Z.Q., 2008. Salt tolerance conferred by overexpression of Arabidopsis vacuolar Na+/H+ antiporter gene AtNHX1 in common buckwheat (*Fagopyrum esculentum*). Transgenic Res. 17, 121–132.

Chen, Y., Ma, J., Zhang, X., et al., 2017. A novel non-specific lipid transfer protein gene from sugarcane (NsLTPs), obviously responded to abiotic stresses and signaling molecules of SA and MeJA. Sugar Technol. 19 (1), 17.

Chen, K., Wang, Y., Zhang, R., Zhang, H., Gao, C., 2019. CRISPR/Cas genome editing and precision plant breeding in agriculture. Annu. Rev. Plant Biol. 70, 28.1–28.30.

Chen, Y., Hu, A., Xue, P., Wen, X., Cao, Y., Wang, B., Zhang, Y., Shah, L., Cheng, S., Cao, L., Zhang, Y., 2020. Effects of GS3 and GL3.1 for grain size editing by CRISPR/Cas9 in rice. Rice Sci. 27 (5), 405–413.

Chinnaswamy, A., Coba de la Peña, T., Stoll, A., de la Peña Rojo, D., Bravo, J., Rincón, A., Lucas, M.M., Pueyo, J.J., 2018. A nodule endophytic *Bacillus megaterium* strain isolated from *Medicago polymorpha* enhances growth, promotes nodulation by *Ensifer medicae* and alleviates salt stress in alfalfa plants. Ann. Appl. Biol. 172, 295–308.

Chou, T.S., Chao, Y.Y., Kao, C.H., 2012. Involvement of hydrogen peroxide in heat shock- and cadmium-induced expression of ascorbate peroxidase and glutathione reductase in leaves of rice seedlings. J. Plant. Physiol. 169, 478–486.

Compant, S., Heijden, M., Sessitsch, A., 2010. Climate change effects on beneficial plant microorganism interactions. FEMS Microbiol. Ecol. 73, 197–214.

Cong, L., Ran, F.A., Cox, D., Lin, S., Barretto, R., Habib, N., et al., 2013. Multiplex genome engineering using CRISPR/Cas systems. Science 339, 819–823.

Connorton, J.M., Jones, E.R., Rodriguez-Ramiro, I., Fairweather-Tait, S., Uauy, C., Balk, J., 2017. Wheat vacuolar, iron, transporter TaVIT2 transporter Fe and Mn and is effective for biofortfication. Plant Physiol. 174, 2434–2444.

COP21, 2015. Paris Agreement Under the United Nations Framework on Climate Change. Paris Climate Agreement. 12 December.

Dahan-Meir, T., Filler-Hayut, S., Melamed-Bessudo, C., Bocobza, S., Czosnek, H., Aharoni, A., Levy, A., 2018. Efficient in planta gene targeting in tomato using geminiviral replicons and the CRISPR/Cas9 system. Plant J. 95, 5–16.

Dass, A., Abdin, M.Z., Reddy, V.S., Leelavathi, S., 2017. Isolation and characterization of the dehydration stress-inducible GhRDL1 promoter from the cultivated upland cotton (*Gossypium hirsutum*). J. Plant Biochem. Biotechnol. 26 (1), 113–119.

Datsenko, K.A., Pougach, K., Tikhonov, A., Wanner, B.L., Severinov, K., Semenova, E., 2012. Molecular memory of prior infections activates the CRISPR/Cas adaptive bacterial immunity system. Nat. Commun. 3, 945.

Demirer, G.S., Silva, T.N., Jackson, C.T., Thomas, J.B., Ehrhardt, D.W., et al., 2021. Nanotechnology to advance CRISPR–Cas genetic engineering of plants. Nat. Nanotechnol. 16, 243–250.

Dhankher, O.P., Li, Y., Rosen, B.P., Shi, J., Salt, D., Senecoff, J.F., Sashti, N.A., Meagher, R.B., 2002. Engineering tolerance and hyperaccumulation of arsenic in plants by combining arsenate reductase and g-glutamylcysteine synthetase expression. Nat. Biotechnol. 20, 1140–1145.

Dhillon, T., Pearce, S.P., Stockinger, E.J., Distelfeld, A., Li, C., Knox, A.K., et al., 2012. Regulation of freezing tolerance and flowering in temperate cereals: the VRN-1 connection. Plant Physiol. 153 (4), 1846–1858.

Ding, D., Chen, K., Chen, Y., Li, H., Xie, K., 2018. Engineering introns to express RNA guides for Cas9- and Cpf1-mediated multiplex genome editing. Mol. Plant 11, 542–552.

Driedonks, N., Rieu, I., Vriezen, W.H., 2016. Breeding for plant heat tolerance at vegetative and reproductive stages. Plant Reprod. 29, 67–79.

Drigo, B., Kowalchuk, G.A., van Veen, J.A., 2008. Climate change goes underground: effects of elevated atmospheric CO2 on microbial community structure and activities in the rhizosphere. Biol. Fertil. Soils 44, 667–679.

Du, H., Zheng, X., Zhao, M., Cui, X., Wang, Q., Yang, H., Cheng, H., Yu, D., 2016. Efficient targeted mutagenesis in soybean by TALENs and CRISPR/Cas9. J. Biotechnol. 217, 90–97.

Duan, Y.B., Li, J., Qin, R.Y., Xu, R.F., Li, H., Yang, Y.C., Ma, H., Li, L., Wei, P.C., Yang, J.B., 2016. Identification of a regulatory element responsible for salt induction of rice OsRAV2 through ex situ and in situ promoter analysis. Plant Mol. Biol. 90 (1–2), 49–62.

Dubcovsky, J., María, G.S., Epstein, E., Luo, M.C., Dvořák, J., 1996. Mapping of the K+/Na+ discrimination locus Kna1 in wheat. Theor. Appl. Genet. 92 (3–4), 448–454.

Eapen, S., D'Souza, S.F., 2005. Prospects of genetic engineering of plants for phytoremediation of toxic metals. Biotechnol. Adv. 23, 97–114.

Eeckhaut, T., Lakshmanan, P.S., Deryckere, D., Van Bockstaele, E., Van Huylenbroeck, J., 2013. Progress in plant protoplast research. Planta 238, 991–1003.

Ellison, E.E., et al., 2020. Multiplexed heritable gene editing using RNA viruses and mobile single guide RNAs. Nat. Plants 6, 620–624.

Endo, M., Mikami, M., Toki, S., 2016. Biallelic gene targeting in rice. Plant Physiol. 170, 667–677.

Estrela, R., Cate, J.H.D., 2016. Energy biotechnology in the CRISPR-Cas9 era. Curr. Opin. Biotechnol. 38, 79–84.

Evenson, R.E., Gollin, D., 2003. Assessing the impact of the green revolution, 1960 to 2000. Science 300, 758–762.

Ezaki, B., Gardner, R.C., Ezaki, Y., Matsumoto, H., 2000. Expression of aluminium induced genes in transgenic Arabidopsis plants can ameliorate aluminium stress and/or oxidative stress. Plant Physiol. 122, 657–665.

Fan, D., Liu, T., Li, C., Jiao, B., Li, S., Hou, Y., Luo, K., 2015. Efficient CRISPR/Cas9-mediated targeted mutagenesis in Populus in the first generation. Sci. Rep. 5, 12217.

Fang, Y., Tyler, B.M., 2016. Efficient disruption and replacement of an effector gene in the oomycete *Phytophthora sojae* using CRISPR/Cas9. Mol. Plant Pathol. 17, 127–139.

Fang, Y., Xiong, L., 2015. General mechanisms of drought response and their application in drought resistance improvement in plants. Cell. Mol. Life Sci. 72, 673–689.

Feng, Z., Zhang, B., Ding, W., Liu, X., Yang, D.L., et al., 2013. Efficient genome editing in plants using a CRISPR/Cas system. Cell Res. 23, 1229–1232.

Ferjani, A., Segami, S., Horiguchi, G., Muto, Y., Maeshima, M., Tsukaya, H., 2011. Keep an eye on PPi: the vacuolar-type H+-pyrophosphatase regulates postgerminative development in Arabidopsis. Plant Cell 23, 2895–2908.

Figueroa-Yañez, L., Pereira-Santana, A., Arroyo-Herrera, A., Rodriguez-Corona, U., et al., 2016. RAP2.4a is transported through the phloem to regulate cold and heat tolerance in papaya tree (*Carica papaya* cv. Maradol): implications for protection against abiotic stress. PLoS One 11 (10), 0165030.

Fischer, K., Ekener-Petersen, E., Rydhmer, L., Björnberg, K.E., 2015. Social impacts of GM crops in agriculture: a systematic literature review. Sustainability 7 (7), 8598–8620.

Fowler, S., Thomashow, M., 2002. Arabidopsis transcriptome profiling indicates that multiple regulatory pathways are activated during cold acclimation in addition to the CBF cold response pathway. Plant Cell 14, 1675–1690.

Francia, E., Barabaschi, D., Tondelli, A., Laidò, G., Rizza, F., et al., 2007. Fine mapping of a HvCBF gene cluster at the frost resistance locus Fr-H2 in barley. Theor. Appl. Genet. 115 (8), 1083–1091.

Fukao, T., Xu, K., Ronald, P.C., Bailey-Serres, J., 2006. A variable cluster of ethylene response factor–like genes regulates metabolic and developmental acclimation responses to submergence in rice. Plant Cell 18, 2021–2034.

Fulekar, M.H., Singh, A., Bhaduri, A.M., 2008. Genetic engineering strategies for enhancing phytoremediation of heavy metals. Afr. J. Biotechnol. 8 (4), 529–535.

Gao, C., 2018. The future of CRISPR technologies in agriculture. Nat. Rev. Mol. Cell Biol. 19, 275–276.

Gao, S.Q., Chen, M., Xia, L.Q., Xiu, H.J., Xu, Z.S., Li, L.C., Zhao, C.P., Cheng, X.G., Ma, Y.Z., 2009. A cotton (Gossypium hirsutum) DRE-binding transcription factor gene, GhDREB, confers enhanced tolerance to drought, high salt, and freezing stresses in transgenic wheat. Plant Cell Rep. 28, 301–311.

Gao, H., Gadlage, M.J., Lafitte, H.R., Lenderts, B., Yang, M., Schroder, M., Farrell, J., Snopek, K., Peterson, D., Feigenbutz, L., Jones, S., St Clair, G., Rahe, M., Sanyour-Doyel, N., Peng, C., Wang, L., Yong, J.K., Beatty, M., Dahlke, B., Hazerbroek, J., Greene, T.W., Cigan, A.M., Chilcoat, N.D., Meeley, R.B., 2020. Superior field performance of waxy corn engineered using CRISPR-Cas9. Nat. Biotechnol. 38, 579–581.

Gaoneng, S., Lihong, X., Guiai, J., Xiangjin, W., Zhonghua, S., Shaoging, T., Peisong, H.U., 2017. CRISPR/Cas9-mediated editing of the fragrant gene Badh2 in rice. Rice Sci. 6 (2), 216.

Garcia, M.D., Nouwens, A., Lonhienne, T.G., Guddat, L.W., 2017. Comprehensive understanding of acetohydroxyacid synthase inhibition by different herbicide families. Proc. Natl. Acad. Sci. U. S. A. 114, E1091–E1100.

Gasic, K., Korban, S.S., 2007. Transgenic Indian mustard (Brassica juncea) plants expressing an Arabidopsis phytochelatin synthase (AtPCS1) exhibit enhanced As and Cd tolerance. Plant Mol. Biol. 64 (4), 361–369.

Glick, B.R., 2010. Using soil bacteria to facilitate phytoremediation. Biotechnol. Adv. 28, 367–374.

Goto, F., Yoshihara, T., Saiki, H., 1998. Iron accumulation in tobacco plants expressing soybean ferritin gene. Transgenic Res. 7, 173–180.

Goto, F., Yoshihara, T., Shigemoto, N., Toki, S., Takaiwa, F., 1999. Iron accumulation in rice seed by soybean ferritin gene. Nat. Biotechnol. 17, 282–286.

Grover, A., Mittal, D., Negi, M., Lavania, D., 2013. Generating high temperature tolerant transgenic plants: achievements and challenges. Plant Sci. 205–206, 38–47.

Haase, S., Philippot, L., Neumann, G., Marhan, S., Kandeler, E., 2008. Local response of bacterial densities and enzyme activities to elevated atmospheric CO_2 and different N supply in the rhizosphere of Phaseolus vulgaris L. Soil Biol. Biochem. 40, 1225–1234.

Hajela, R.K., Norvath, D.P., Gilmour, S.J., Thomashow, M.F., 1990. Molecular cloning and expression of cor (cold-regulated) genes in Arabidopsis thaliana. Plant Physiol. 93, 1246–1252.

Hall, J.L., 2002. Cellular mechanisms for heavy metal detoxification and tolerance. J. Exp. Bot. 53 (366), 1–11.

Han, Y.-J., Kim, J.-L., 2019. Application of CRISPR/Cas9-mediated gene editing for the development of herbicide-resistant plants. Plant Biotechnol. Rep. 13, 447–457.

Hattori, Y., Nagai, K., Furukawa, S., Song, X.J., Kawano, R., et al., 2009. The ethylene response factors SNORKEL1 and SNORKEL2 allow rice to adapt to deep water. Nature 460 (7258), 1026–1030.

Hauser, F., Horie, T., 2010. A conserved primary salt tolerance mechanism mediated by HKT transporters: a mechanism for sodium exclusion and maintenance of high $K+/Na+$ ratio in leaves during salinity stress. Plant Cell Environ. 33, 552–565.

He, P., Zhao, P., Wang, L., Zhang, Y., Wang, X., Xiao, H., et al., 2017. The PIN gene family in cotton (*Gossypium hirsutum*): genome-wide identification and gene expression analyses during root development and abiotic stress responses. BMC Genomics 18, 507.

Hirschi, K.D., Korenkov, V.D., Wilganowski, N.L., Wagner, G.J., 2000. Expression of Arabidopsis CAX2 in tobacco altered metal accumulation and increased manganese tolerance. Plant Physiol. 124, 125–133.

Hong, B., Ma, C., Yang, Y., Wang, T., Yamaguchi-Shinozaki, K., Gao, J., 2009. Over-expression of AtDREB1A in chrysanthemum enhances tolerance to heat stress. Plant Mol. Biol. 70 (3), 231–240.

Hsieh, T.H., Lee, J.T., Yang, P.T., Chiu, L.H., Charng, Y., Wang, Y.C., Chan, M.T., 2002. Heterology expression of the Arabidopsis C-repeat/dehydration response element binding factor 1 gene confers elevated tolerance to chilling and oxidative stresses in transgenic tomato. Plant Physiol. 129, 1086–1094.

Hsu, P.D., Lander, E.S., Zhang, F., 2014. Development and applications of CRISPR-Cas9 for genome engineering. Cell 57, 1262–1278.

Hu, H., Xiong, L., 2014. Genetic engineering and breeding of drought-resistant crops. Annu. Rev. Plant Biol. 65, 715–741.

Hu, Z., Vanderhaeghen, R., Cools, T., Wang, Y., De Clercq, I., Leroux, O., Nguyen, L., Belt, K., Millar, A.H., Audenaert, D., Hilson, P., Small, I., Mouille, G., Vernhettes, S., Van Breusegem, F., Whelan, J., Hofte, H., De Veylder, L., 2016. Mitochondrial defects confer tolerance against cellulose deficiency. Plant Cell 28, 2276–2290.

Hu, Z.B., Zhang, T., Rombaut, D., Decaestecker, W., Xiang, A.M., D'Haeyer, S., Hofer, R., Vercauterern, I., Karimi, M., Jacobs, T., De Veylder, L., 2019. Genome editing-based engineering of CESA3 dual cellulose-inhibitor resistant plants. Plant Physiol. 180, 827–836.

Huby, E., Napier, J.A., Baillieul, F., Michaelson, L.V., Dhondt-Cordelier, S., 2020. Sphingolipids: towards an integrated view of metabolism during the plant stress response. New Phytol. 225, 659–670.

Huertas, R., Olias, R., Eljakaoui, Z., Galvez, F.J., Li, J., et al., 2012. Overexpression of SlSOS2 (SlCIPK24) confers salt tolerance to transgenic tomato. Plant Cell Environ. 35, 1467–1482.

Hummel, A.W., Chauhan, R.D., Cermak, T., Mutka, A.M., Vijayaraghaavan, A., Boyher, A., Starker, C.G., Bart, R., Voytas, D.F., Taylor, N.J., 2018. Allele exchange at the EPSPS locus confers glyphosate tolerance in cassava. Plant Biotechnol. J. 16, 1275–1282.

IPCC, 2007. Climate change 2007: impacts, adaptation and vulnerability. In: Parry, M.L., Canziani, O.F., Palutikof, J.P., van der Linden, P.J., Hanson, C.E. (Eds.), Contribution of Working Group II to the Fourth Assessment Report of the Intergovernmental Panel on Climate Change. Cambridge University Press, Cambridge. 976 pp.

IPCC, 2012. Managing the risks of extreme events and disasters to advance climate change adaptation. In: Field, C.B., Barros, V., Stocker, T.V., Qin, D., Dokken, D.J., Ebi, K.L., Mastrandrea, M.D., Mach, K.J., Plattner, G.-K., Allen, S.K., Tignor, M., Midgley, P.M. (Eds.), A Special Report of Working Groups I and II of the Intergovernmental Panel on Climate Change. 582 pp.

Iqbal, Z., Sattar, M.N., Shafiq, M., 2016. CRISPR/Cas9: a tool to circumscribe cotton leaf curl disease. Front. Plant Sci. 7, 475.

Islam, T., 2019. CRISPR-Cas technology in modifying food crops. CAB Rev. 14 (50), 1–16. https://doi.org/10.1079/PAVSNNR201914050.

Ito, H., Gray, W.M., 2006. A gain-of-function mutation in the *Arabidopsis* pleitropic drug resistance transporter PDR9 confers resistance to auxinic herbicides. Plant Physiol. 142, 63–74.

Jacobs, A., Ford, K., Kretschmer, J., Tester, M., 2011. Rice plants expressing the moss sodium pumping ATPase PpENA1 maintain greater biomass production under salt stress. Plant Biotechnol. J. 9, 838–847.

Jacobs, T.B., LaFayette, P.R., Schmitz, R.J., Parott, W.A., 2015. Targeted genome modifications in soybean with CRISPR/Cas9. BMC Biotechnol. 15, 1–10.

Jaganathan, D., Ramasamy, K., Sellamuthu, G., Jayabalan, S., Venkataraman, G., 2018. CRISPR for crop improvement: an update review. Front. Plant Sci. 9, 985.

Jaglo-Ottosen, K.R., Gilmour, S.J., Zarka, D.G., Schabenberger, O., Thomashow, M.F., 1998. Arabidopsis CBF1 overexpression induces COR genes and enhances freezing tolerance. Science 280, 104–106.

Jesus, J.M., Danko, A.S., Fiúza, A., Borges, M.T., 2015. Phytoremediation of salt-affected soils: a review of processes, applicability, and the impact of climate change. Environ. Sci. Pollut. Res. Int. 22 (9), 6511–6525.

Ji, X., Zhang, H., Zhang, Y., Wang, Y., Gao, C., 2015. Establishing a CRISPR-Cas-like immune system conferring DNA virus resistance in plants. Nat. Plants 1, 15144.

Jia, H., Wang, N., 2014. Targeted genome editing of sweet orange using Cas9/sgRNA. PLoS One 9, e93806.

Jia, H., Zhang, Y., Orbovic, V., Xu, J., White, F.F., Jones, J.B., Ang, N., 2017. Genome editing of the disease susceptibility gene CsLOB1 in citrus confers resistance to citrus canker. Plant Biotechnol. J. 15, 817–823.

Jiang, W., Zhou, H., Bi, H., Fromm, M., Yang, B., Weeks, D.P., 2013. Demonstration of CRISPR/Cas9/sgRNA-mediated targeted gene modification in Arabidopsis, tobacco, sorghum and rice. Nucleic Acids Res. 41, 188.

Jinek, M., Chylinski, K., Fonfara, I., Hauer, M., Doudna, J.A., Charpentier, E.A., 2012. Programmable dual-RNA-guided DNA endonuclease in adaptive bacterial immunity. Science 337, 816–821.

Joshi, R., Wani, S.H., Singh, B., Bohra, A., Dar, Z.A., Lone, A.A., Pareek, A., Singla-Pareek, S.L., 2016. Transcription factors and plants response to drought stress: current understanding and future directions. Front. Plant Sci. 7, 1029.

Jump, A.S., Hunt, J.M., Martinez-Izquierdo, J.A., Penuelas, J., 2006. Natural selection and climate change: temperature-linked spatial and temporal trends in gene frequency in Fagus sylvatica. Mol. Ecol. 15, 3469–3480.

Kakutani, T., 2002. Epi-alleles in plants: inheritance of epigenetic information over generations. Plant Cell Physiol. 43, 1106–1111.

Kandeler, E., Mosier, A.R., Morgan, J.A., Milchunas, D.G., King, J.Y., Rudolph, S., Tscherko, D., 2006. Response of soil microbial biomass and enzyme activities to the transient elevation of carbon dioxide in a semi-arid grassland. Soil Biol. Biochem. 38, 2448–2460.

Karakas, B., Ozias-Akins, P., Stushnoff, C., Suefferheld, M., Rieger, M., 1997. Salinity and drought tolerance of mannitol-accumulating transgenic tobacco. Plant Cell Environ. 20, 609–616.

Karimi, A., Khodaverdiloo, H., Sepehri, M., Sadaghiani, M.R., 2011. Arbuscular mycorrhizal fungi and heavy metal contaminated soils. Afr. J. Microbiol. Res. 5 (13), 1571–1576.

Kashyap, P.L., Rai, P., Srivastava, A.K., Kumar, S., 2017. Trichoderma for climate resilient agriculture. World J. Microbiol. Biotechnol. 33, 155.

Kasuga, M., Liu, Q., Miura, S., Yamaguchi-Shinozaki, K., Shinozaki, K., 1999. Improving plant drought, salt, and freezing tolerance by gene transfer of a single stress-inducible transcription factor. Nat. Biotechnol. 17, 287–291.

Kasuga, M., Miura, S., Shinozaki, K., Yamaguchi-Shinozaki, K.A., 2004. A combination of the Arabidopsis DREB1A gene and stress-inducible rd29A promoter improved drought and low-temperature stress tolerance in tobacco by gene transfer. Plant Cell Physiol. 45, 346–350.

Kaundum, S.S., 2014. Resistance to acetyl-CoA carboxylase-inhibiting herbicides. Pest Manag. Sci. 70, 1405–1417.

Kim, S., Takahashi, M., Higuchi, K., Tsunoda, K., Nakanishi, H., Yoshimura, E., Mori, S., Nishizawa, N.K., 2005. Increased nicotianamine biosynthesis confers enhanced tolerance of high levels of metals, in particular nickel, to plants. Plant Cell Physiol. 46, 1809–1818.

Kim, M.H., Sasaki, K., Imai, R., 2009. Cold shock domain protein 3 regulates freezing tolerance in *Arabidopsis thaliana*. J. Biol. Chem. 284, 23454–23460.

Kimura, S., Laosinchai, W., Itoh, T., Cui, X., Linder, C.R., Brown, R.M., 1999. Immunogold labeling of rosette terminal cellulose-synthesizing complexes in the vascular plant *Vigna angularis*. Plant Cell 11, 2075–2086.

Klap, C., Yeshayahou, E., Bolger, A.M., Arazi, T., Gupta, S.K., Shabtai, S., et al., 2017. Tomato facultative parthenocarpy results from SlAGAMOUS-LIKE 6 loss of function. Plant Biotechnol. J. 15, 634–647.

Knox, A.K., Dhillon, T., Cheng, H., Tondelli, A., Pecchioni, N., Stockinger, E.J., 2010. CBF gene copy number variation at frost resistance-2 is associated with levels of freezing tolerance in temperate-climate cereals. Theor. Appl. Genet. 121 (1), 21–35.

Komatsu, A., Ohtake, M., Shimatani, Z., Nishida, K., 2020. Production of herbicide-sensitive strain to prevent volunteer rice infestation using a CRISPR-Cas9 cytidine deaminase fusion. Front. Plant Sci. 11, 925.

Kudla, J., Batistic, O., Hashimoto, K., 2010. Calcium signals: the lead currency of plant information processing. Plant Cell 22, 541–563.

Kumar, J., Choudhary, A.K., Gupta, D.S., Kumar, S., 2019. Towards exploitation of adaptive traits for climate-resilient smart pulses. Int. J. Mol. Sci. 20, 2971.

Lamoureux, G.L., Shimabukuro, R.H., Swanson, H.R., Frear, D.S., 1970. Metabolism of 2-chloro-4-ethylamino-6-isopropylamino-s-triazine (atrazine) in excised sorghum leaf sections. J. Agric. Food Chem. 18, 81–86.

Lavania, D., Dhingra, A., Siddiqui, M.H., Grover, A., 2015. Current status of the production of high temperature tolerant transgenic crops for cultivation in warmer climates. Plant Physiol. Biochem. 86, 100–108.

LeDuc, D.L., Tarun, A.S., Montes-Bayon, M., Meija, J., Malit, M.F., Wu, C.P., et al., 2004. Overexpression of selenocysteine methyltransferase in Arabidopsis and Indian mustard increases selenium tolerance and accumulation. Plant Physiol. 135 (1), 377–383.

LeDuc, D.L., AbdelSamie, M., Móntes-Bayon, M., Wu, C.P., Reisinger, S.J., Terry, N., 2006. Overexpressing both ATP sulfurylase and selenocysteine methyltransferase enhances selenium phytoremediation traits in Indian mustard. Environ. Pollut. 144, 70–76.

Lei, L., Li, S., Gu, Y., 2012. Cellulose synthase complexes: composition and regulation. Front. Plant Sci. 3, 75.

Li, J.F., Norville, J.E., Aach, J., McCormack, M., Zhang, D., Bush, J., Church, G.M., Sheen, J., 2013. Multiplex and homologous recombination-mediated genome editing in Arabidopsis and *Nicotiana benthamiana* using guide RNA and Cas9. Nat. Biotechnol. 31, 688–691.

Li, C.H., Wang, G., Zhao, J.L., Zhang, L.Q., Ai, L.F., Han, Y.F., et al., 2014. The receptor-like kinase SIT1 mediates salt sensitivity by activating MAPK3/6 and regulating ethylene homeostasis in rice. Plant Cell 26 (6), 2538–2553.

Li, Z., Liu, Z.B., Xing, A., Moon, B.P., Koellhoffer, J.P., Huang, L., Ward, R.T., Clifton, E., Falco, S.C., Cigan, A.M., 2015. Cas9-guide RNA directed genome editing in soybean. Plant Physiol. 169, 960–970.

Li, J., Meng, X.B., Zong, Y., Chen, K.L., Zhang, H.W., Liu, J.X., Li, J.Y., Gao, C.X., 2016a. Gene replacements and insertions in rice by intron targeting using CRISPR-Cas9. Nat. Plants 2, 16139.

Li, M., Li, X., Zhou, Z., Wu, P., Fang, M., Pan, X., Lin, Q., Luo, W., Wu, G., Li, H., 2016b. Reassessment of the four yield-related genes Gn1a, DEP1, GS3, and IPA1 in rice using a CRISPR/Cas9 system. Front. Plant Sci. 7, 377.

Li, J., Zhang, H., Si, X., Tian, Y., Chen, K., Liu, J., Chen, H., Gao, C., 2017. Generation of thermosensitive male-sterile maize by targeted knockout of the ZmTMS5 gene. J. Genet. Genomics 44 (9), 465–468.

Li, C., Zong, Y., Wang, Y., Jin, S., Zhang, D., Song, Q., Zhang, R., Gao, C., 2018a. Expanded base editing in rice and wheat using a Cas9-adenosine deaminase fusion. Genome Biol. 19, 59.

Li, Y., Yang, X., Yu, Y., Si, X., Zhai, X., Zhang, H., 2018b. Domestification of wild tomato is accelerated by genome editing. Nat. Biotechnol. 36, 1160–1163.

Li, R., Liu, C., Zhao, R., Wang, L., Chen, L., Yu, W., Zhang, S., Shen, J., Shen, L., 2019. CRISPR/Cas9-mediated SlNPR1 mutagenesis reduces tomato plant drought tolerance. BMC Plant Biol. 19, 38.

Liang, Z., Zhang, K., Chen, K., Gao, C., 2014. Targeted mutagenesis in Zea mays using TALENs and CRISPR/Cas system. J. Genet. Genomics 41, 63–68.

Liang, X.H., Shen, W., Sun, H., Migawa, M.T., Vickers, T.A., Crooke, S.T., 2016. Translation efficiency of mRNA is increased by antisense oligonucleotides targeting upstream open reading frames. Nat. Biotechnol. 34, 875–880.

Liang, Z., et al., 2017. Efficient DNA-free genome editing of bread wheat using CRISPR/Cas9 ribonucleoprotein complexes. Nat. Commun. 8, 14261.

Liu, W., Stewart, C.N., 2015. Plant synthetic biology. Trends Plant Sci. 20, 309–317.

Liu, F., Huang, N., Wang, L., Ling, H., Sun, T., Ahmad, W., et al., 2017a. A novel L-ascorbate peroxidase 6 gene, ScAPX6, plays an important role in the regulation of response to biotic and abiotic stresses in sugarcane. Front. Plant Sci. 8, 2262.

Liu, J., Chen, J., Zheng, X., Wu, F., Lin, Q., Heng, Y., Tian, P., Cheng, Z., Yu, X., Zhou, K., Zhang, X., Guo, X., Wang, J., Wang, H., Wan, J., 2017b. GW5 acts in the brassinosteroid signaling pathway to regulate grain width and weight in rice. Nat. Plants 3, 17043.

Lou, D., Wang, H., Liang, G., Yu, D., 2017. OsSAPK2 confers abscisic acid sensitivity and tolerance to drought stress in rice. Front. Plant Sci. 8, 993.

Lowder, L.G., Zhang, D., Baltes, N.J., Paul, J.W., Tang, X., et al., 2015. A CRISPR/Cas9 toolbox for multiplexed plant genome editing and transcriptional regulation. Plant Physiol. 169, 971–985.

Lowder, L.G., Zhou, J., Zhang, Y., Malzahn, A., Zhong, Z., Hsieh, T.F., Voytas, D.F., Zhang, Y., Qi, Y., 2018. Robust transcriptional activation in plants using multiplexed CRISPR-Act2.0 and mTALE-act systems. Mol. Plant 11, 245–256.

Lu, K., Wu, B., Wang, J., Zhu, W., Nie, H., Qia, J., Huang, W., Fang, Z., 2018. Blocking amino acid transporter OsAPP3 improves grain yield by promoting outgrowth buds and increasing tiller number in rice. Plant Biotechnol. J. 16, 1710–1722.

Luikart, G., England, P.R., Tallmon, D., Jordan, S., Taberlet, P., 2003. The power and promise of population genomics: from genotyping to genome typing. Nat. Rev. Genet. 4, 981–994.

Luo, L.J., 2010. Breeding for water-saving and drought-resistance rice (WDR) in China. J. Exp. Bot. 61, 3509–3517.

Lv, Y., Deng, X., Quan, L., Xia, Y., Shen, Z., 2013. Metallothioneins BcMT1 and BcMT2 from *Brassica campestris* enhance tolerance to cadmium and copper and decrease production of reactive oxygen species in *Arabidopsis thaliana*. Plant Soil 367, 507–519.

Lyu, D., Backer, R., Subramanian, S., Smith, D.L., 2020. Phytomicrobiome coordination signals hold potential for climate change-resilient agriculture. Front. Plant Sci. 11, 634.

Ma, Y., Dai, X., Xu, Y., Luo, W., Zheng, X., Zeng, D., Pan, Y., Lin, X., Liu, H., Zhang, D., Xiao, J., Guo, X., Xu, S., Niu, Y., Jin, J., Zhang, H., Xu, X., Li, L., Wang, W., Qian, Q., Ge, S., Chong, K., 2015. COLD1 confers chilling tolerance in rice. Cell 160, 1209–1221.

Ma, X., Zhu, Q., Chen, Y., Liu, Y.G., 2016. CRISPR/Cas9 platforms for genome editing in plants: developments and applications. Mol. Plant 9, 961–974.

Mabood, F., Smith, D.L., 2005. Pre − incubation of *Bradyrhizobium japonicum* with jasmonates accelerates nodulation and nitrogen fixation in soybean (*Glycine max*) at optimal and suboptimal root zone temperatures. Physiol. Planar. 125, 311–323.

Mabood, F., Zhou, X., Lee, K.-D., Smith, D.L., 2006. Methyl jasmonate, alone or in combination with genistein, and *Bradyrhizobium japonicum* increases soybean (*Glycine max* L.) plant dry matter production and grain yield under short season conditions. Field Crops Res. 95, 412–419.

Macovei, A., Sevilla, N.R., Cantos, C., Jonson, G.B., Slamet-Loedin, I., Čermak, T., Voytas, D.F., Choi, I.-C., Chadha-Mohanty, P., 2018. Novel alleles of rice eIF4G generated by CRISPR/Cas9-targeted mutagenesis confer resistance to rice tungro spherical virus. Plant Biotechnol. J. 16, 1918–1927.

Maeda, H., Murata, K., Sakuma, N., Takei, S., Yamazaki, A., Karim, M., Kawata, M., Hirose, S., Kwagashi-Kobayashi, M., Tanugchi, Y., Suzuki, S., Sekino, K., Ohshima, M., Kato, H., Yoshida, H., Tozawa, Y., 2019. A rice gene that confers broad-spectrum resistance to β-triketone herbicides. Science 365, 393–396.

Malnoy, M., Viola, R., Jung, M.H., Koo, O.J., Kim, J.S., Velasco, R., Kanchiswamy, C.N., 2016. DNA-free genetically edited grapevine and apple protoplast using CRISPR/Cas9 ribonucleoproteins. Front. Plant Sci. 7, 1904.

Massel, K., Lam, Y., Wong, A.C.S., et al., 2021. Hotter, drier, CRISPR: the latest edit on climate change. Theor. Appl. Genet. https://doi.org/10.1007/s00122-020-03764-0.

Meers, E., Van Slycken, S., Adriaensen, K., Ruttens, A., Vangronsveld, J., Du Laing, G., Witters, N., Thewys, T., Tack, F.M.G., 2010. The use of bio-energy crops (*Zea mays*) for 'phytoattenuation' of heavy metals on moderately contaminated soils: a field experiment. Chemosphere 78, 35–41.

Mei, Y., Wang, Y., Chen, H., Sun, Z.S., Ju, X.D., 2016. Recent progress in CRISPR/Cas9 technology. J. Genet. Genomics 43 (2), 63–75.

Meyer, C.L., Vitalis, R., Saumitou-Laprade, P., Castric, V., 2009. Genomic pattern of adaptive divergence in *Arabidopsis halleri*, a model species for tolerance to heavy metal. Mol. Ecol. 18, 2050–2062. Blackwell Publishing Ltd.

Mian, A., Oomen, R.J.F.J., Isayenkov, S., Sentenac, H., Maathuis, F.J.M., Véry, A.A., 2011. Over-expression of an Na+- and K+-permeable HKT transporter in barley improves salt tolerance. Plant J. 68, 468–479.

Miao, J., Guo, D., Zhang, J., Huang, Q., Qin, G., Zhang, X., Wan, J., Gu, H., Qu, L.J., 2013. Targeted mutagenesis in rice using CRISPR-Cas system. Cell Res. 23, 1233–1236.

Miao, H., Sun, P., Liu, Q., Miao, Y., Liu, J., Zhang, K., et al., 2017a. Genome-wide analyses of SWEET family proteins reveal involvement in fruit development and abiotic/biotic stress responses in banana. Sci. Rep. 7, 3536.

Miao, H., Sun, P., Liu, Q., Miao, Y., Liu, J., Xu, B., et al., 2017b. The AGPase family proteins in banana: genome wide identification, phylogeny, and expression analyses reveal their involvement in the development, ripening, and abiotic/biotic stress responses. Int. J. Mol. Sci. 18 (8), 1581.

Miao, C., Xiao, L., Hua, K., Zou, C., Zhao, Y., Bressen, R.A., Zhu, J.-K., 2018. Mutations in a subfamily of abscisic acid receptor genes promote rice growth and productivity. PNAS 115, 6058–6063.

Miglani, G.S., 2017. Genome editing in crop improvement: present scenario and future prospects. J. Crop Improv. 31, 453–559.

Minder, A.M., Widmer, A., 2008. A population genomic analysis of species boundaries: neutral processes, adaptive divergence and introgression between two hybridizing plant species. Mol. Ecol. 17, 1552–1563.

Mosa, K.A., Saadoun, I., Kumar, K., Helmy, M., Dhanker, O.P., 2016. Potential biotechnological strategies for the cleanup of heavy metals and metalloids. Front. Plant Sci. 7, 303.

Mueller, U.G., Sachs, J.L., 2015. Engineering microbiomes to improve plant and animal health. Trends Microbiol. 23, 606–617.

Munns, R., Tester, M., 2008. Mechanisms of salinity tolerance. Annu. Rev. Plant Biol. 59, 651–668.

Munns, R., James, R.A., Xu, B., Athman, A., Conn, S.J., Jordans, C., Byrt, C.S., Hare, R.A., et al., 2012. Wheat grain yieldon saline soils is improved by an ancestral Na + transportergene. Nat. Biotechnol. 30, 360–364.

Nadakuduti, S.S., Enciso-Rodríguez, F., 2020. Advances in genome editing with CRISPR systems and transformation technologies for plant DNA manipulation. Front. Plant Sci. 11, 637159.

Nakayasu, M., Akiyama, R., Lee, H.J., Osakabe, K., Osakabe, Y., 2018. Generation of α-solanine-free hairy roots of potato by CRISPR/Cas9 mediated genome editing of the St16DOX gene. Plant Physiol. Biochem. 131, 70–77.

Namroud, M.C., Beaulieu, J., Juge, N., Laroche, J., Bousquet, J., 2008. Scanning the genome for gene single nucleotide polymorphisms involved in adaptive population differentiation in white spruce. Mol. Ecol. 17, 3599–3613.

NASEM (National Academies of Sciences, Engineering and Medicine), 2016. Genetically Engineered Crops: Experiences and Prospects. The National Academies Press, Washington, DC, https://doi.org/10.17226/23395.

National Academies of Sciences, Engineering, and Medicine, 2017. Preparing for Future Products of Biotechnology. The National Academies Press, Washington, DC, https://doi.org/10.17226/24605.

Nekrasov, V., Staskawicz, B., Weigel, D., Jones, J.D., Kamoun, S., 2013. Targeted mutagenesis in the model plant *Nicotiana benthamiana* using Cas9 RNA-guided endonuclease. Nat. Biotechnol. 31, 691–693.

Nekrasov, V., Wang, C., Win, J., Lanz, C., Weigel, D., Kamoun, S., 2017. Rapid generation of a transgene-free powdery mildew resistant tomato by genome deletion. Sci. Rep. 7, 482.

Nieves-Cordones, M., Mohamed, S., Tanoi, K., Kobayashi, N., Takagi, K., Vernet, A., 2017. Production of low-Cs$^+$ rice plants by inactivation of the K$^+$ transporter OsHAK1 with the CRISPR-Cas system. Plant J. 92, 43–56.

Noman, A., Aqeel, M., He, S., 2016. CRISPR-Cas9: tool for qualitative and quantitative plant genome editing. Front. Plant Sci. 7, 1740.

Novo, L.A.B., Castro, P.M.L., Alvarenga, P., da Silva, E.F., 2018. Plant growth–promoting rhizobacteria-assisted phytoremediation of mine soils. In: Bio-Geotechnologies for Mine Site Rehabilitation. vol. 1. Elsevier, New York, NY, ISBN: 0471394351, pp. 281–295.

O'Connor, D., Zhengb, X., Houa, D., Shena, Z., Lia, G., et al., 2019. Phytoremediation: climate change resilience and sustainability assessment at a coastal brownfield redevelopment. Environ. Int. 130, 104945.

Öpik, M., Vanatoa, A., Vanatoa, E., Moora, M., Davison, J., Kalwij, J.M., Reier, U., Zobel, M., 2010. The online database MaarjAM reveals global and ecosystemic distribution patterns in arbuscular mycorrhizal fungi (Glomeromycota). New Phytol. 188, 223–241.

Ortega, J.L., Rajapakse, W., Bagga, S., Apodaca, K., Lucero, Y., Sengupta-Gopalan, C., 2018. An intragenic approach to confer glyphosphate resistance in chile (*Capsicum annuum*) by introducing an in vitro mutagenized Chile EPSPS gene encoding for a glyophosphate resistant EPSPS protein. PLoS One 13, e0194666.

Ou, W., Mao, X., Huang, C., Tie, W., Yan, Y., Ding, Z., et al., 2018. Genome-wide identification and expression analysis of the KUP family under abiotic stress in cassava (*Manihot esculenta* Crantz). Front. Physiol. 9, 17.

Pan, C., Ye, L., Qin, L., Liu, X., He, Y., Wang, J., Chen, L., Lu, G., 2016. CRISPR/Cas9-mediated efficient and heritable targeted mutagenesis in tomato plants in the first and later generations. Sci. Rep. 6, 24765.

Pandey, V.C., Bajpai, O., Singh, N., 2016. Energy crops in sustainable phytoremediation. Renew. Sust. Energ. Rev. 54, 58–73.

Parry, M.A., Hawkesford, M.J., 2012. An integrated approach to crop genetic improvement. J. Integr. Plant Biol. 54, 250–259.

Pauwels, M., Saumitou-Laprade, P., Holl, C., Petit, D., Bonin, I., 2005. Multiple origin of metallicolous populations of the pseudometallophyte *Arabidopsis halleri* (Brassicaceae) in central Europe: the cpDNA testimony. Mol. Ecol. 14, 4403–4414.

Pauwels, M., Frérot, H., Bonin, I., Saumitou-Laprade, P., 2006. A broadscale analysis of population differentiation for Zn tolerance in an emerging model species for tolerance study: *Arabidopsis halleri* (Brassicaceae). J. Evol. Biol. 19, 1838–1850.

Pauwels, M., Willems, G., Roosens, N.C.J., Frérot, H., Saumitou-Laprade, P., 2008. Merging methods in molecular and ecological genetics to study the adaptation of plants to anthropogenic metal-polluted sites: implication for phytoremediation. Mol. Ecol. 17, 108–119.

Peng, R.H., Fu, X.Y., Zhao, W., Tian, Y.S., Zhu, B., Han, H.J., Xu, J., Yao, Q.H., 2014. Phytoremediation of phenanthrene by transgenic plants transformed with a naphthalene dioxygenase system from pseudomonas. Environ. Sci. Technol. 48, 12824–12832.

Peng, A., Chen, S., Lei, T., Xu, L., He, Y., Wu, L., et al., 2017. Engineering canker resistant plants through CRISPR/Cas9-targeted editing of the susceptibility gene CsLOB1 promoter in citrus. Plant Biotechnol. J. 10, 1011–1013.

Pennycooke, J.C., Jones, M.L., Stushnoff, C., 2003. Down-regulating α-galactosidase enhances freezing tolerance in transgenic petunia. Plant Physiol. 133, 901–909.

Pianelli, K., Mari, S., Marquès, L., Lebrun, M., Czernic, P., 2005. Nicotianamine over-accumulation confers resistance to nickel in *Arabidopsis thaliana*. Transgenic Res. 14, 739–748.

Plett, D., Safwat, G., Møller, I., Gilliham, M., Roy, S.J., Shirley, N.J., Jacobs, A., Johnson, A.A.T., Tester, M., 2010a. Improved salinity tolerance of rice through cell type-specific expression of AtHKT1;1. PLoS One 5, e12571.

Plett, D., Johnson, A., Jacobs, A., Tester, M., 2010b. Cell type-specific expression of sodium transporters improves salinity tolerance of rice. GM Crops 1, 273–275.

Porcel, R., Aroca, R., Ruiz-Lozano, J.M., 2012. Salinity stress alleviation using arbuscular mycorrhizal fungi. A review. Agron. Sustain. Dev. 32, 181–200.

Powles, S., Yu, Q., 2010. Evolution in action: plants resistant to herbicides. Annu. Rev. Plant Biol. 61, 317–347.

Pramanik, D., Shelake, R.M., Park, J., Kim, M.J., Hwang, I., Park, Y., Kim, J.Y., 2021. CRISPR/Cas9-mediated generation of pathogen-resistant tomato against tomato yellow leaf curl virus and powdery mildew. Int. J. Mol. Sci. 22 (4), 1878.

Qiao, G., Zhuo, R., Liu, M., Jiang, J., Li, H., Qiu, W., Pan, L., Lin, S., Zhang, X., Sun, Z., 2011. Over-expression of the Arabidopsis Na+/H+ antiporter gene in *Populus deltoides* CL × *P. euramericana* CL "NL895" enhances its salt tolerance. Acta Physiol. Plant. 33, 691–696.

Quilis, J., Peñas, G., Messeguer, J., Brugidou, C., Segundo, B.S., 2008. The Arabidopsis AtNPR1 inversely modulates defense responses against fungal, bacterial, or viral pathogens while conferring hypersensitivity to abiotic stresses in transgenic rice. Mol. Plant-Microbe Interact. 21, 1215–1231.

Rajagopal, D., Agarwal, P., Tyagi, W., Singla-Pareek, S., Reddy, M.K., Sopory, S.K., 2007. *Pennisetum glaucum* Na+/H+ antiporter confers high level of salinity tolerance in transgenic *Brassica juncea*. Mol. Breed. 19, 137–151.

Rehmani, M.I.A., Wei, G., Hussain, N., Li, G., Ding, C., Liu, Z., Wang, S., Ding, Y., 2014. Yield and quality responses of two indica rice hybrids to post-anthesis asymmetric day and night open-field warming in lower reaches of Yangtze River delta. Field Crops Res. 256, 231–241.

Roca-Paixão, J.F., Gillet, F.X., Ribeiro, T.P., Bournaud, C., Lourenco-Tessutti, T., Noriega, D.D., et al., 2019. Improved drought stress tolerance in Arabidopsis by CRISPR/dCas9 fusion with a histone acetyl transferase. Sci. Rep. 9, 8080.

Rodriguez-Leal, D., Lemmon, Z.H., Man, J., Bartlett, M.E., Lippman, Z.B., 2017. Engineering quantitative trait variation for crop improvement by genome editing. Cell 171, 470–480.

Ron, M., Kajala, K., Pauluzzi, G., Wang, D., Reynoso, M.A., Zumstein, K., Garcha, J., Winte, S., Masson, H., Inagaki, S., Federici, F., Sinha, N., Deal, R.B., Bailey-Serres, J., Brady, S.M., 2014. Hairy root transformation using *Agrobacterium rhizogenes* as a tool for exploring cell type-specific gene expression and function using tomato as a model. Plant Physiol. 166, 455–469.

Rus, I., Baxter, B., Muthukumar, J., Gustin, B., Lahner, E., Yakubova, D.E., 2006. Salt natural variants of AtHKT1 enhance Na+ accumulation in two wild populations of Arabidopsis. PLoS Genet. 2, 1964–1973.

Rylott, E.L., Lorenz, A., Bruce, N.C., 2011. Biodegradation and biotransformation of explosives. Curr. Opin. Biotechnol. 22, 434–440.

Rylott, E.L., Johnston, E.J., Bruce, N.C., 2015. Harnessing microbial gene pools to remediate persistent organic pollutants using genetically modified plants—a viable technology? J. Exp. Bot. 66, 6519–6533.

Sanchez-Leon, S., Gil-Humanes, J., Ozuna, C.V., Gimenez, M.J., Sousa, C., Voytas, D.F., Barro, F., 2018. Low-gluten, nontransgenic wheat engineered with CRISPR/Cas9. Plant Biotechnol. J. 16 (4), 902–910.

Sanghera, G.S., Wani, S.H., Hussain, W., Singh, N.B., 2011. Engineering cold stress tolerance in crop plants. Curr. Genomics 12 (1), 30–43.

Sasaki, Y., Hayakawa, T., Inoue, C., Miyazaki, A., Silver, S., Kusano, T., 2006. Generation of mercury-hyperaccumulating plants through transgenic expression of the bacterial mercury membrane transport protein MerC. Transgenic Res. 15, 615–625.

Sauer, N.J., Narvaez-Vasquez, J., Mozoruk, J., Miller, R.B., Warburg, Z.J., Woodward, M.J., Mihiret, Y.A., Lincoln, T.A., Segami, R.E., Sanders, S.L., Walker, K.A., Beetham, P.R., Scopke, C.R., Gocal, G.F., 2016. Oligonucleotide-mediated genome editing provides precision and function to engineered nucleases and antibiotics in plants. Plant Physiol. 170, 1917–1928.

Saxena, P., Singh, N.K., Harish, Singh, A.K., Pandey, S., Thanki, A., Yadav, T.C., 2020. Recent advances in phytoremediation using genome engineering CRISPR-Cas9 technology. In: Pandey, V.C., Singh, V. (Eds.), Bioremediation of Pollutants From Genetic Engineering to Genome Engineering. Elsevier, pp. 125–141.

Schonbrunn, E., Eschenburg, S., Shuttleworth, W.A., Schloss, J.V., Amrhein, N., Evans, J.N., Kabsch, W., 2001. Interaction of the herbicide glyphosate with its target enzyme 5-enolpyruvylshikimate 3-phosphate synthase in atomic detail. Proc. Natl. Acad. Sci. U. S. A. 98, 1376–1380.

Shan, Q., et al., 2013. Targeted genome modification of crop plants using a CRISPR-Cas system. Nat. Biotechnol. 31, 686–688.

Shen, C., Que, Z., Xia, Y., et al., 2017. Knock out of the annexin gene OsAnn3 via CRISPR/Cas9-mediated genome editing decreased cold tolerance in rice. J. Plant Biol. 60, 539–547.

Shi, J., Gao, H., Wang, H., Lafitte, H.R., Archibald, R.L., Yang, M., et al., 2017. ARGOS8 variants generated by CRISPR-Cas9 improve maize grain yield under field drought stress conditions. Plant Biotechnol. J. 15, 207–216.

Shim, J.S., Oh, N., Chung, P.J., Kim, Y.S., Choi, Y.D., Kim, J.K., 2018. Overexpression of osnac14 improves drought tolerance in rice. Front. Plant Sci. 9, 310.

Shimatani, Z., Kashojiya, S., Takayama, M., Terada, R., Miura, K., Ezura, H., Nishida, K., Ariizumi, T., Kondo, A., 2017. Targeted base editing in rice and tomato using a CRISPR-Cas9 cytidine deaminase fusion. Nat. Biotechnol. 35, 441–443.

Shukla, V.K., Doyon, Y., Miller, J.C., DeKelver, R.C., Moehle, E.A., Worden, S.E., et al., 2009. Precise genome modification in the crop species Zea mays using zinc-finger nucleases. Nature 459, 437–441.

Silva, N.V., Patron, N.J., 2017. CRISPR-based tools for plant genome engineering. Emerg. Top. Life Sci. 1, 135–149.

Simon, J.A., 2015. Editor's perspective—the effects of climate change adaptation planning on remediation programs. Remediat. J. 25, 1–7.

Singh, R.P., Shelke, G.M., Kumar, A., Jha, P.N., 2015. Biochemistry and genetics of ACC deaminase: a weapon to "stress ethylene" produced in plants. Front. Microbiol. 6, 937.

Smith, D., Zhou, X., 2014. An effective integrated research approach to study climate change in Canada Preface. Can. J. Plant Sci. 94, 995–1008.

Smith, D.L., Praslickova, D., Ilangumaran, G., 2015. Inter-organismal signalling and management of the phytomicrobiome. Front. Plant Sci. 6, 722.

Somerville, C., 2006. Cellulose synthesis in higher plants. Annu. Rev. Cell Dev. Biol. 22, 53–78.

Soyk, S., Lemmon, Z.H., Oved, M., Fishers, J., Liberatore, K.L., Park, S.J., Goren, A., Jiang, K., Ramos, A., van der Knaap, E., Van Eck, J., Zamir, D., Eshed, Y., Lippman, Z.B., 2017. Bypassing negative epistasis on yield on tomato imposed by a domestication gene. Cell 169, 1142–1155.

Stockinger, E.J., Skinner, J.S., Gardner, K.G., Francia, E., Pecchioni, N., 2007. Expression levels of barley Cbf genes at the frost resistance-H2 locus are dependent upon alleles at Fr-H1 and Fr-H2. Plant J. 51 (2), 308–321.

Suárez, R., Calderón, C., Iturriaga, G., 2009. Enhanced tolerance to multiple abiotic stresses in transgenic Alfalfa accumulating trehalose. Crop Sci. 49, 1791–1799.

Sun, Y., Zhang, X., Wu, C., He, Y., Ma, Y., Hou, H., Guo, X., Du, W., Zhao, Y., Xia, L., 2016. Engineering herbicide-resistant rice plants through CRISPR/Cas9-mediated homologous recombination of acetolactate synthase. Mol. Plant 9, 628–631.

Sun, Y., Jiao, G., Liu, Z., Zhang, X., Li, J., Guo, X., Du, W., Du, J., Francis, F., Zhao, Y., Xia, L., 2017. Generation of high–amylose rice through CRISPR/Cas9-mediated targeted mutagenesis of starch branching enzymes. Front. Plant Sci. 8, 298.

Svitashev, S., Young, J.K., Schwartz, C., Gao, H., Falco, S.C., Cigan, A.M., 2015. Targeted mutagenesis, precise gene editing, and site-specific gene insertion in maize using Cas9 and guide RNA. Plant Physiol. 169, 931–945.

Svitashev, S., Schwartz, C., Lenderts, B., Young, J.K., Cigan, A.M., 2016. Genome editing in maize directed by CRISPR–Cas9 ribonucleoprotein complexes. Nat. Commun. 7, 13274.

Swaminathan, K., Alabady, M.S., Varala, K., De Paoli, E., Ho, I., Rokhsar, D.S., Arumuganathan, A.K., Minh, R., Green, P.J., Meyers, B.C., Moose, S.P., Hudson, M.E., 2010. Genomic and small RNA sequencing of *Miscanthus* x *giganteus* shows the utility of sorghum as a reference genome sequence for Andropogoneae grasses. Genome Biol. 11, R12.

Symigton, L.S., Gauter, J., 2011. Double-strand break end resection and repair pathway choice. Annu. Rev. Genet. 45, 241–271.

Tammisola, J., 2010. Towards much more efficient biofuel crops—can sugarcane pave the way? GM Crops 1 (4), 181–198.

Tang, N., Ma, S., Zong, W., Yang, N., Lv, Y., Yan, C., Guo, Z., et al., 2016. MODD mediates deactivation and degradation of OsbZIP46 to negatively regulate ABA signaling and drought resistance in rice. Plant Cell 28 (9), 2161–2177.

Tang, L., Mao, B., Li, Y., Lv, Q., Zhang, L., Chen, C., et al., 2017. Knockout of OsNramp5 using the CRISPR/Cas9 system produces low Cd-accumulating indica rice without compromising yield. Sci. Rep. 7, 14438.

Tashkandi, M., Ali, Z., Aljedani, F., Shami, A., Mahfouz, M.M., 2018. Engineering resistance against tomato yellow leaf curl virus via the CRISPR/Cas9 system in tomato. Plant Signal. Behav. 13 (10), e1525996.

Thode, S.K., Rojek, E., Kozlowski, M., Ahmad, R., Haugen, P., 2018. Distribution of siderophores gene systems on a Vibrionaceae phylogeny: database searches, phylogenetics analyses and evolutionary perspectives. PLoS One 13, 1–20.

Thomashow, M.F., 1999. Plant cold acclimation: freezing tolerance genes and regulatory mechanisms. Annu. Rev. Plant Physiol. Plant Mol. Biol. 50, 571–599.

Tsaftaris, A.S., Polidoros, A.N., 1999. DNA methylation and plant breeding. Plant Breed. Rev. 18, 87–176.

Turchi, A., Tamantini, I., Camussi, A.M., Racchi, M.L., 2012. Expression of a metallothionein A1 gene of *Pisum sativum* in white poplar enhances tolerance and accumulation of zinc and copper. Plant Sci. 183, 50–56.

Uga, Y., Sugimoto, K., Ogawa, S., Rane, J., Ishitani, M., et al., 2013. Control of root system architecture by DEEPER ROOTING 1 increases rice yield under drought conditions. Nat. Genet. 45 (9), 1097–1102.

US EPA, 2011. Agriculture and Food Supply: Climate Change, Health and Environmental Effects. April 14, 2011.

US EPA, 2014. U.S. Environmental Protection Agency Climate Change Adaptation Plan. June 2014 https://www.epa.gov/sites/production/files/2015-08/documents/adaptationplans2014_508.pdf.

USAID, 2007. Adapting to Climate Variability and Change: A Guidance Manual for Development Planning. https://pdf.usaid.gov/pdf_docs/PNADJ990.pdf. August 2007.

Veeranagamallaiaha, G., Prasanthi, J., Reddy, K.E., Pandurangaiahb, M., Babub, O.S., Sudhakar, C., 2011. Group 1 and 2 LEA protein expression correlates with a decrease in water stress induced protein aggregation in horsegram during germination and seedling growth. J. Plant Physiol. 168, 671–677.

Wang, W., Vinocur, B., Shoseyov, O., Altman, A., 2004. Role of plant heat-shock proteins and molecular chaperones in the abiotic stress response. Trends Plant Sci. 9, 244–252.

Wang, Y., Cheng, X., Shan, Q., Zhang, Y., Liu, J., Gao, C., Qiu, J.-L., 2014. Simultaneous editing of three homoeoalleles in hexaploid bread wheat confers heritable resistance to powdery mildew. Nat. Biotechnol. 32, 947–951.

Wang, Y., Ren, H., Pan, H., Liu, J., Zhang, L., 2015. Enhanced tolerance and remediation to mixed contaminates of PCBs and 2,4-DCP by transgenic alfalfa plants expressing the 2,3-dihydroxybiphenyl-1,2-dioxygenase. J. Hazard. Mater. 286, 269–275.

Wang, F., Wang, C., Liu, P., Lei, C., Hao, W., Gao, Y., et al., 2016a. Enhanced rice blast resistance by CRISPR/Cas9-targeted mutagenesis of the ERF transcription factor gene OsERF922. PLoS One 11, e0154027.

Wang, H., Wang, H., Shao, H., Tang, X., 2016b. Recent advances in utilizing transcription factors to improve plant abiotic stress tolerance by transgenic technology. Front. Plant Sci. 7, 67.

Wang, Q.J., Sun, H., Dong, Q.L., Sun, T.Y., Jin, Z.X., Hao, Y.J., Yao, Y.X., 2016c. The enhancement of tolerance to salt and cold stresses by modifying the redox state and salicylic acid content via the cytosolic malate dehydrogenase gene in transgenic apple plants. Plant Biotechnol. J. 14 (10), 1986–1997.

Wang, L., Chen, L., Li, R., Zhao, R., Yang, M., Sheng, J., et al., 2017a. Reduced drought tolerance by CRISPR/Cas9-mediated SlMAPK3 mutagenesis in tomato plants. J. Agric. Food Chem. 65, 8674–8682.

Wang, M., Mao, Y., Lu, Y., Tao, X., Zhu, J.K., 2017b. Multiplex gene editing in rice using the CRISPR-Cpf1 system. Mol. Plant 10, 1011–1013.

Wang, W., Pan, Q., He, F., Akhunova, A., Chao, S., Trick, H., Akhunov, E., 2018a. Transgenerational CRISPR-Cas9 activity facilitates multiplex gene editing in allopolyploid wheat. CRISPR J. 1, 65–74.

Wang, X., Tu, M., Wang, D., Liu, J., Li, Y., Li, Z., Wang, Y., Wang, X., 2018b. CRISPR/Cas9-mediated efficient targeted mutagenesis in grape in the first generation. Plant Biotechnol. J. 16, 844–855.

Wang, H., Wu, Y., Zhang, Y., Yang, J., Fan, W., Zhang, H., Zhao, S., Yuan, L., Zhang, P., 2019. CRISPR/Cas9-based mutagenesis of starch biosynthetic genes in sweet potato (*Ipomoea batatas*) for the improvement of starch quality. Int. J. Mol. Sci. 20 (19), 4702.

Wang, C., Wang, G., Gao, Y., Lu, G., Habben, J.E., Mao, G., Chen, G., Wang, J., Yang, F., Zhao, X., Mo, H., Qu, P., Liu, J., Greene, T.W., 2020. A cytokinin-activation enzyme-like gene improves grain yield under various field conditions in rice. Plant Mol. Biol. 102, 373–388.

Wilding, C.S., Butlin, R.K., Grahame, J., 2001. Differential gene exchange between parapatric morphs of *Littorina saxatilis* detected using AFLP markers. J. Evol. Biol. 14, 611–619.

Williams, L.E., Pittman, J.K., Hall, J.L., 2000. Emerging mechanisms for heavy metal transport in plants. Biochim. Biophys. Acta 1465, 104–126.

Woo, J.W., et al., 2015. DNA-free genome editing in plants with preassembled CRISPR-Cas9 ribonucleoproteins. Nat. Biotechnol. 33, 1162–1164.

Xia, Y., Qi, Y., Yuan, Y., Wang, G., Cui, J., Chen, Y., Zhang, H., Shen, Z., 2012. Overexpression of *Elsholtzia haichowensis* metallothionein 1 (EhMT1) in tobacco plants enhances copper tolerance and accumulation in root cytoplasm and decreases hydrogen peroxide production. J. Hazard. Mater. 233–234, 65–71.

Xing, H.L., Dong, L., Wang, Z.P., et al., 2014. A CRISPR/Cas9 toolkit for multiplex genome editing in plants. BMC Plant Biol. 14, 327.

Xu, K., Xu, X., Fukao, T., Canlas, P., Maghirang-Rodriguez, R., Heuer, S., et al., 2006. Sub1A is an ethylene response-factor-like gene that confers submergence tolerance to rice. Nature 442 (7103), 705–708.

Xu, C., Liberatore, K.L., MacAlister, C.A., Huang, Z., Chu, Y.H., Jiang, K., Brooks, C., Ogawa-Onishi, M., Xiong, G., Pauly, M., Van Eck, J., Matsubayashi, Y., van der Knaap, E., Lippman, Z.B., 2015. A cascade of arabinosyltransferases controls shoot meristem size in tomato. Nat. Genet. 47, 784–792.

Xu, R., Yang, Y., Qin, R., Li, H., Qui, C., Li, L., Wei, P.-C., Yang, J., 2016. Rapid improvement of grain weight via highly efficient CRISPR/Cas9-mediated multiplex genome editing in rice. J. Genet. Genomics 43, 529–532.

Xu, Y., Lin, Q., Li, X., Wang, F., Chen, Z., Wang, J., Li, W., Fan, F., Tao, Y., Jiang, Y., Wei, X., Zhang, E., Zhu, Q.-H., Bu, Q., Yang, J., Gao, C., 2021. Fine-tuning the amylase content of rice by precise base editing of Wx gene. Plant Biotechnol. J. 19, 11–13.

Xue, Z.Y., Zhi, D.Y., Xue, G.P., Zhang, H., Zhao, Y.X., Xia, G.M., 2004. Enhanced salt tolerance of transgenic wheat (*Triticum aestivum* L.) expressing a vacuolar Na+/H+ antiporter gene with improved grain yields in saline soils in the field and a reduced level of leaf Na+. Plant Sci. 167, 849–859.

Yadav, N., Shukla, P., Jha, A., Agarwal, P., Jha, B., 2012. The SbSOS1 gene from the extreme halophyte *Salicornia brachiata* enhances Na+ loading in xylem and confers salt tolerance in transgenic tobacco. BMC Plant Biol. 12, 188.

Yan, K., Chen, N., Qu, Y.Y., Dong, X.C., Meng, Q.W., Zhao, S.J., 2008. Overexpression of sweet pepper glycerol-3-phosphate acyltransferase gene enhanced thermotolerance of photosynthetic apparatus in transgenic tobacco. J. Integr. Plant Biol. 50, 613–621.

Yan, A., Wang, Y., Tan, S.N., Yusof, M.L.M., et al., 2020. Phytoremediation: a promising approach for revegetation of heavy metal-polluted land. Front. Plant Sci. 11, 359.

Yang, X., Jin, X.F., Feng, Y., Islam, E., 2005a. Molecular mechanisms and genetic bases of heavy metal tolerance/hyperaccumulation in plants. J. Integr. Plant Biol. 47 (9), 1025–1035.

Yang, X., Feng, Y., He, Z., Stoffella, P., 2005b. Molecular mechanisms of heavy metal hyperaccumulation and phytoremediation. J. Trace Elem. Med. Biol. 18, 339–353.

Yao, Y.X., Dong, Q.L., Zhai, H., You, C.X., Hao, Y.J., 2011. The functions of an apple cytosolic malate dehydrogenase gene cytosolic malate dehydrogenase gene in growth and tolerance to cold and salt stresses. Plant Physiol. Biochem. 49, 257–264.

Ye, Y., Li, P., Xu, T., Zeng, L., Cheng, D., Yang, M., Luo, J., Lian, X., 2017. Ospt4 contributes to arsenate uptake and transport in rice. Front. Plant Sci. 8, 2197.

Yin, Y., Qin, K., Song, X., Zhang, O., Zhou, Q., Xia, X., Yu, J., 2018. BZR1 transcription factor regulates heat stress tolerance through FERONIA receptor-like kinase-mediated reactive oxygen species signaling in tomato. Plant Cell Physiol. 59, 2239–2254.

Yu, Q., Ahmad-Hamdani, M.S., Han, H., Christoffers, M.J., Powles, S.B., 2013. Herbicide resistance–endowing ACCase gene mutation in hexaploid wild oar (*Avena fatua*): insights into resistance evolution in a hexaploid species. Heredity 110, 220–231.

Yue, Y., Zhang, M., Zhang, J., Duan, L., Li, Z., 2012. SOS1 gene overexpression increased salt tolerance in transgenic tobacco by maintaining a higher K+/Na+ ratio. J. Plant Physiol. 169, 255–261.

Zaidi, S.S.A., Mahfouz, M.M., Mansoor, S., 2017. CRISPR-Cpf1: a new tool for plant genome editing. Trends Plant Sci. 22, 550–553.

Zetsche, B., et al., 2015. Cpf1 is a single RNA-guided endonuclease of a class 2 CRISPR-Cas system. Cell 163, 759–771.

Zhai, S., Li, G., Sun, Y., et al., 2016. Genetic analysis of phytoene synthase 1 (Psy1) gene function and regulation in common wheat. BMC Plant Biol. 16, 228. https://doi.org/10.1186/s12870-016-0916-z.

Zhang, L., Xi, D., Li, S., Gao, Z., Zhao, S., Shi, J., Wu, C., Guo, X., 2011a. A cotton group C MAP kinase gene, GhMPK2, positively regulates salt and drought tolerance in tobacco. Plant Mol. Biol. 77, 17–31.

Zhang, Y.J., Yang, J.S., Guo, S.J., Meng, J.J., Zhang, Y.L., Wan, S.B., He, Q.W., Li, X.G., 2011b. Over-expression of the Arabidopsis CBF1 gene improves resistance of tomato leaves to low temperature under low irradiance. Plant. Biol. 13 (2), 362–367.

Zhang, L., Zhao, Y.-.L., Gao, L.-.F., Zhao, G.-.Y., Zhou, R.-.H., Zhang, B.-.S., Jia, J.-.Z., 2012. *TaCKX6-D1*, the ortholog of rice *OsCKX2*, is associated with grain weight in hexaploid wheat. New Phytol. 195, 574–584.

Zhang, H., Zhang, J., Wei, P., Zhang, P., Gou, F., et al., 2014. The CRISPR/Cas9 system produces specific and homozygous targeted gene editing in rice in one generation. Plant Biotechnol. J. 12, 797–807.

Zhang, Z., Li, J., Pan, Y., Li, J., Zhou, L., Shi, H., Zeng, Y., et al., 2017a. Natural variation in CTB4a enhances rice adaptation to cold habitats. Nat. Commun. 8, 14788.

Zhang, Y., Bai, Y., Wu, G., Zou, S., Chen, Y., Gao, C., Tang, D., 2017b. Simultaneous modification of three homoeologs of TaEDR1 by genome editing enhances powdery mildew resistance in wheat. Plant J. 91 (4), 714–724.

Zhang, A., Liu, Y., Wang, F., Li, T., Chen, Z., Kong, D., Bi, J., Zhang, F., Luo, X., Wang, J., Tang, J., Yu, X., Liu, G., Lup, L., 2019a. Enhanced rice salinity tolerance via CRISPR/Cas9-targeted mutagenesis of the OsRR22 gene. Mol. Plant Breed. 39, 47.

Zhang, R., Liu, J., Chai, Z., Chen, S., Bai, Y., Zong, Y., Chen, K., Li, J., Jiang, L., Gao, C., 2019b. Generation of herbicide tolerance traits and a new selectable marker in wheat using base editing. Nat. Plants 5, 480–485.

Zhao, F., Guo, S., Zhang, H., Zhao, Y., 2006. Expression of yeast SOD2 in transgenic rice results in increased salt tolerance. Plant Sci. 170, 216–224.

Zhao, X., Lu, X., Yin, Z., Wang, D., Wang, J., Fan, W., Wang, S., Zhang, T., Ye, W., 2016. Genome-wide identification and structural analysis of pyrophosphatase gene family in cotton. Crop Sci. 56 (4), 1831–1840.

Zhou, M.Q., Shen, C., Wu, L.H., Tang, K.X., Lin, J., 2011. CBF-dependent signaling pathway: a key responder to low temperature stress in plants. Crit. Rev. Biotechnol. 31 (2), 186–192.

Zhou, G., Pereira, J.F., Delhaize, E., Zhou, M., Magalhaes, J.V., Ryan, P.R., 2014. Enhancing the aluminium tolerance of barley by expressing the citrate transporter genes SbMATE and FRD3. J. Exp. Bot. 65 (9), 2381–2390.

Zhou, J., Peng, Z., Long, J., Sosso, D., Liu, B., Eom, J.S., Huang, S., Liu, S., Vera Cruz, C., Frommer, W.B., White, F.F., Yang, B., 2015. Gene targeting by the TAL effector PthXo2 reveals cryptic resistance gene for bacterial blight of rice. Plant J. 82, 632–643.

Zhu, H., Li, C., Gao, C., 2020. Applications of CRISPR–Cas in agriculture and plant biotechnology. Nat. Rev. Mol. Cell Biol. 21, 661–677.

Making biomass from phytoremediation fruitful: Future goal of phytoremediation

7

Chapter outline

1 Introduction

The term "biomass" generally means biological mass mainly of plant origin. Trees, grasses, agricultural crops, forest residues, or other biological materials like animal waste or municipal household waste sludge can be considered as biomass. Among different types of biomass, heavy metal-contaminated biomass (HMCB) has been pulling attraction of researchers across the world because of the potential ability of HMCB to address energy crisis and environmental restoration together. Different types of metallophytic plants (certain plants able to resist and grow in the presence of heavy metals) able to phytoaccumulate heavy metals (HMs) are first used to phytoremediate the polluted soils. The whole plants or the above-ground parts are then repeatedly harvested, and the harvested biomass is called HMCB. There are several potential opportunities for utilizing phytoremediated biomass and well described by Pandey and Souza-Alonso (2019). The biomass should be able to generate electric power, heat, industrially important chemical, liquid, or solid fuel, animal feed etc. In view of the recent energy crisis, bioenergy production from biomass is getting increasing concern. The growing attention over climate change is another determinant showing environmental advantage of biomass application as fuel source. The crude oil spill disaster of April 2010 in the deep sea of Gulf of Mexico has highlighted the risks of overexploitation of fossil fuel. Last decade witnessed many scientists investigating the biochemical

Adaptive Phytoremediation Practices. https://doi.org/10.1016/B978-0-12-823831-8.00001-3

and thermochemical pathways for the conversion of the biomass into biofuel as a source of bioenergy. Although the fine composition and physiochemical properties of biomass vary according to source, in general it is a complicated mixture of organic compounds like carbohydrates (cellulose, hemicellulose, and starch), lignin, proteins, and fats. Carbohydrates and lignins are the primary components of the biomass from plant/agricultural origin, and they vary according to plant/crop type. Biofuel is derived from biomass through physical, biological, chemical, and/or a combination of processes. Bioethanol, biogas, biodiesel etc. representing different types of biofuel are produced via enzymatic/microbial fermentations with or without using physical and chemical pretreatment steps (Carucci et al., 2005). Thermochemical processes, such as combustion or partial incineration, thermal liquefaction, torrefaction, pyrolysis, and gasification, are most commonly used for the conversion of the biomass into bio-oil, syn-gas, biochar, and others (Catoire et al., 2008). Advancements in thermochemical processes can address present energy crisis and become a major contributor of the practical and sustainable energy solution.

This chapter has discussed pros and cons of different thermochemical ways (pyrolysis, gasification, combustion, and liquefaction) of bioenergy recovery from HMCB obtained from phytoremediation project of contaminated land/water (Tanger et al., 2013). Strengths have been given on biomass recovery from woody plants grown on marginal lands (FA dump sites, mining regions etc.). This not only reduces complexities associated with bioenergy production from agricultural crops (Zhang et al., 2010) but also solves the dilemma of food versus fuel debate as diverting agricultural land for biofuel production (current demand of the time) would compromise with growing demand of food for an ever-increasing population. The need for the execution of the biomass valorization and metal-free biofuel production for sustainable phytoextraction has been critically viewed. Literature has been cited on the most efficient phytoremediation approaches, different possibilities of the cost-effective production of metal-free biofuel from HMCB using diverse thermochemical routes, effect of various parameters (feedstock type, class of metals present in HMCB, operating conditions, reactor types, effect of pre- and posttreatment) on the fate of HMs during biomass valorization, and the quality of end products and by-products. Lastly, based on the relevant and empirical findings and techno-economic assessment (TEA) study of the various types of thermochemical conversions for biofuel production, this chapter has found pyrolysis as the most promising, appropriate, and reliable thermochemical technique for the large-scale bioenergy recovery from HMCB.

2 Phytoremediation and generation of heavy metal-contaminated biomass (HMCB)

Heavy metals, polycyclic aromatic hydrocarbons, pesticides, solvents, explosives, etc. are the widespread pollutants that affect human health in several ways. However, in this chapter, discussions were limited to HMs only. Among the HMs, some (Cu, Co, Fe, Mo, Mn, Ni, and Zn) have biological functions and others (As, Ag, Cd, Hg, Pb, and Sb) have no biological functions. They are exposed into the environment through industrial effluents, metal smelters, excessive use of fertilizers and pesticides for agricultural cultivation, etc. (Davis et al., 2001). In most cases, HMs are present in the soils in trace quantity as free metal ions, organically bound metals, soluble metal complexes with different ligands, exchangeable metal ions, precipitated or inorganic compounds such as oxides, hydroxides, carbonates, and silicates. Heavy metals are nonbiodegradable, are hazardous in nature, and change their valence states according to pH of the media, and various HMs can cause different extents of damage to various organs of our body. The mechanisms of toxicity of the HMs have been reviewed in detail in various types of literature (DalCorso et al., 2013). In plant, HM toxicity may cause overproduction of reactive oxygen species (ROS) that causes peroxidation of many crucial elements of the cell including DNA. To prevent this damage, HM-tolerant plant developed an efficient defense strategy that comprised a set of enzymatic and nonenzymatic antioxidants. Some examples of enzymatic antioxidants are ascorbate peroxidase, catalase, glutathione reductase, glutathione peroxidase, superoxide dismutase, and glutathione-s-transferase, which are capable of converting the superoxide radicals into hydrogen peroxide and subsequently to water and oxygen molecules. Examples of low-molecular-weight, nonenzymatic antioxidants are ascorbic acid, carotenoids, glutathione, proline, polyphenols, vitamin C, vitamin E, etc. that directly scavenge those ROS (Xu et al., 2009) and prevent the cells from oxidative stress. These HM-tolerant plants not only tolerate and grow in the presence of the HMs but also remediate the soil by various ways. This is called phytoremediation, and the different types of remediation methods are described in the following sections. It is important to mention here that since the 1990s, phytotechnologies have emerged as alternatives to conventional remediation techniques. This biological system of environmental remediation is more beneficial than the conventional physicochemical methods like soil washing, stripping, excavation and reburial, or capping and chemical treatments (Assink, 1988).

Different types of phytoremediation techniques are phytoextraction (root-mediated uptake of pollutants, transport, and storage in plant tissues), rhizofiltration (phytoextraction in hydrophytes or plants cultured in hydroponic cultures or constructed wetlands), phytostabilization (containment or immobilization of contaminants in the soil by substances released by roots or root-associated rhizomicrobes), phytodegradation and rhizodegradation (degradation of organic compounds by plants or endospheric microbes), and phytovolatilization/phytoevaporation (volatilization and release of the absorbed contaminants from plants as volatile compounds) (Roy et al., 2015; Pandey and Bajpai, 2019).

Phytoextraction or phytoaccumulation is the uptake and concomitant translocation of HMs present in the soil into the plant roots and then into above-ground components of the plants. Some plants, called hyperaccumulators, are able to absorb and accumulate very large quantity of HMs in their different parts (1%–5% of total body weight) (Maestri et al., 2010). Hyperaccumulation is defined as the process of uptake, translocation, and accumulation of the heavy metals at concentrations higher than the surroundings in different parts of the plant (Maestri et al., 2010). So far, more than 340 plant species of hyperaccumulators have been recognized, which can store certain amount of specific heavy metals. The detoxification efficiency of HMs in the concerned plant is dictated by antioxidant capacity of the plant. Antioxidant genes for cysteine, O-acetylserine, and reduced glutathione were found in overexpressed states in certain hyperaccumulator plants (Anjum et al., 2016). Most hyperaccumulators are herbaceous, and very few (less than 20%) are ligneous. Their reduced root system cannot reach sufficient remediation depths. In contrast to most bioenergy crops, they are not good biomass generating also. As hyperaccumulators also cannot uptake insoluble HM, to further increase phytoextraction of insoluble HMs, chelators like EDTA and EDDS can be added (induced phytoextraction). But the amount of chelators should be controlled to avoid overleaching of the excess solubilized HMs.

The success-determining factor of phytoextraction is definitely the adoption of the plant, which should have a good tolerance to high concentrations of HMs and would be able to accumulate HM to a substantial degree and translocate the metals from roots to above-ground parts. An ideal plant for phytoextraction must generate abundant biomass and should have a high cut top for ease in harvesting. It would not be a desirable phytoremediation trait if a plant accumulates the HMs in the roots and has low translocation ability. This increases the risk of return of the HMs to the soil after death of the plant. Uprooting an entire plant with intact root increases the cost. So plants with higher translocation ability are better suited for phytoextraction purpose as the collection of the above-ground parts would be enough. However, if clean biofuel generation is the main objective, then selecting plant that phytoextracts and accumulates metals in the roots may be desired as the biofuel production from heavily HM-contaminated shoot biomass would be undesired. Efficiency of phytoextraction is quantified by the accumulation of metals per unit of biomass. So biomass-producing plants like crop plants (e.g., *Brassica juncea* (L.) Czern.) able to accumulate metals to some extent are called accumulators and serve as alternative to hyperaccumulators (Tomović et al., 2013).

Rhizofiltration is another method of phytoremediation approach where plant roots absorb the contaminants from water, accumulate, and precipitate them through the plant root system into/onto the roots and above-ground shoots (Verma et al., 2006). In rhizofiltration, the plants are grown hydroponically and transplanted back into the HM-polluted water to absorb and accumulate the HMs in their body parts. Sometimes the root exudate lowers the rhizosphere pH, which decreases metal solubility. As a result, the HMs are precipitated onto the root surfaces. Once the roots are saturated with the precipitated metal contaminants, the roots or the whole plants are harvested for disposal (Padmavathiamma and Li, 2007). This method is mainly used to remove HMs such as cadmium, lead, zinc, and uranium isotopes from contaminated groundwater (Ghosh and Singh, 2005). Specific wetland plants have good rhizofiltration capacity, and biomass production from them should be investigated (Ghosh, 2010). But the contaminated biomass disposal-related problem hinders the commercial application of rhizofiltration just like phytoextraction.

The phytostabilization is another very good approach of phytoremediation that uses abilities of certain plant root exudates for the reduction of the solubility and concurrent bioavailability of the toxic HMs and prevents their migration in the surroundings (Cheraghi et al., 2009). Root exudate plays a very important role in two ways: It precipitates the ions of the toxic HMs as insoluble salts (e.g., lead in the form of lead phosphate) and/or reduces the harmful ion concentration (such as CrO_4^{2-} and

CrO_7^{2-} to Cr^{3+}) by altering the soil oxidation–reduction (redox) potential. In contrary to phytoextraction, model plants for phytostabilization should have a low degree of metal translocation from roots to shoots (Wong and Bradshaw, 2002). However, like phytoextraction, they should have an extensive root system, and they should be also good biomass generating.

The most promising phytotechnology that can be utilized over vast contaminated lands is the aided phytostabilization that uses plants along with other amendments (chemical or biological or both) to enhance the retention of contaminants in soils through root uptake or reduction or chelation (Mench and Bes, 2009). Short rotation coppice (SRC) and other rapidly growing trees like willows and poplars are most suitable for phytostabilization due to their wide-spreading root system and high biomass-producing abilities that save the surface soil from erosion (Pulford and Watson, 2003) with minimum food chain transfer of the contaminants (Bert et al., 2017). They are proven bioenergy trees ideal for sustainable phytoremediation with ample economic opportunities (Chalot et al., 2012). China et al. (2014) found phytostabilization more suitable than phytoextraction in the copper tailing of Mosaboni mine that was nutrient-poor, was acidic in nature, and had an elevated concentration of HMs. Addition of chicken manure and soil caused an increased biomass production and a decreased accumulation of metals (except Pb) in both shoots and roots of the lemon grass. Grass *Phalaris arundinacea, Pennisetum purpureum*, and *Phragmites* species can be used in phytostabilization along with bioenergy production (Pandey et al., 2020a,b; Pandey and Maiti, 2020).

Phytoevaporation or phytovolatilization is less used phytoremediation associated with the phenomena of uptake, assimilation, and transpirational release of the pollutants by the plant into the air either in intact or in transformed form. This process is suitable for both soil and water contaminated with mercury, arsenic, selenium, and their volatile derivatives (Erdei et al., 2005). For example, plants take up selenium in the form of soluble ions SeO_4^{2-} and SeO_3^{2-} and secrete them in the form of dimethyl selenide (de Souza et al., 1999). Advantage of phytoevaporation is that it does not generate any waste, but the serious disadvantage comes from the fact that the volatile forms are more toxic than their parent forms and dangerous for human life and the environment (Cargnelutti et al., 2006). So phytovolatilization has less practical application.

Phytoremediation efficiency of plants is calculated by some parameters of the plants like BAC (biological absorption coefficient), BCF (bioconcentration factor), TF (translocation factor) etc. While BAC is the ratio of HM concentrations in plants and soil, BCF is their ratio in root and soil and TF is their ratio in shoot and root. Plant with higher BAC (> 1) is good for phytoextraction, whereas a plant with low TF (< 1) and high BCF (> 1) is ideal for phytostabilization (Tangahu et al., 2011).

Postharvest biomass treatment involves compaction and composting (Blaylock and Huang, 2000). But the leachate generated by composted biomass should be collected and treated. Instead of burning and disposing the dried biomass in landfill as ash, it should be turned into profit-making business by either phytomining of the HMs (Keller et al., 2005) or energy generation through biochemical or thermochemical biomass-to-bioenergy conversion ways.

Some credit of phytoremediation should also go to the invisible microbial community of the plant. They are present within the plant itself called endophytic microbes and in the root and rhizosphere soil called rhizomicrobes. Certain members of this plant endobiome or rhizobiome are called plant growth-promoting rhizomicrobes (PGPRs). They not only stimulate root proliferation and promote plant growth directly, but also help the plant in phytoremediation of metals (Mishra et al., 2017; Mukherjee et al., 2017). They may participate in metal uptake and accumulation for phytoextraction or metal phytostabilization around plant root or degradation of the organic pollutants (phytodegradation) or methylate HM compounds producing volatile derivatives (phytovolatilization) (Summers and Silver, 1978). Abou-Shanab et al. (2003a,b) showed that the addition of *Alyssum murale*, *Microbacterium liquefaciens*, *Microbacterium arabinogalactanolyticum*, and *Sphingomonas macrogoltabidus* in serpentine soil greatly increased nickel uptake as a result of soil pH reduction compared to control samples that were not inoculated. Endospheric microbes present in all plant compartments greatly enhance the process of hydrocarbon degradation. Benefits of the synergy of plant and microorganisms are of immense importance as they perform multiple ecosystem functions such as enhanced phytoremediation, enhanced plant growth, and improved soil quality, and also contribute to high microbial biodiversity.

For an effective phytoremediation of HM-contaminated soil, plants should have the following ideal properties: (i) an efficient metal uptake and good translocation capability as harvest of the above-ground parts are easy, (ii) high metal tolerance and accumulation ability without showing toxic symptoms, and (iii) an extensive well-developed root system and huge shoot biomass. Although properties of hyperaccumulator species match the first two points, their applications for large-scale phytoremediation purposes are limited due to their slow growth rate and poor biomass (Rascio and Navari-Izzo, 2011). Due to the barriers of using the natural hyperaccumulators, transgenic alternatives have been thought of where the metal-accumulating ability of a hyperaccumulator can be transferred to a higher biomass-producing bioenergy plant (Brewer et al., 1999). Recombinant DNA technology has allowed researchers to transfer genes from one organism to another irrespective of their evolutionary relationship or sexual compatibility. With the help of advanced next-generation sequencing and modern bioinformatic algorithms, researchers are trying to screen potential high biomass crop species and apply them for phytoremediation in the field. Research is going on the engineering of biomass-generating plants with genes from model metallophytic species and identification of potential genes from nonmetallophytes. Based on the information acquired on the molecular determinants of the hyperaccumulators' metal accumulation and tolerance ability, transgenic research has focused attention on the creation of novel transgenic hyperaccumulator species without the drawbacks of natural hyperaccumulators (Reeves et al., 2017). As real field condition may contain multiple HMs and other toxic elements, one has to look into the interactive effects of metals on the growth, hyperaccumulation/accumulation, and metal resistance of plants, which may be significantly different from laboratory conditions.

Whatever the technique of phytoremediation, the phytoremediated biomass should be valorized for making the phytoremediation self-sustaining and attractive to farmers. Even though some valorization techniques have been scrutinized (Bert et al., 2017;

Chalot et al., 2012; Delplanque et al., 2013), phytomanagement is still limited to a few practical field-based experiments (Cundy et al., 2015). Recently, phytotechnologies have shifted to sustainable phytomanagement whose main objective is valorization or economic utilization of the plant biomass grown in polluted lands (Evangelou and Deram, 2014; Pandey and Bauddh, 2018; Pandey and Bajpai, 2019). Since arable lands are limited and should be used only for growing food crops to meet increasing food demand of ever-increasing population, phytomanagement generally implies growing bioenergy crop in noncultivable lands. Most of the bioenergy crops are suitable for growth in these agriculturally unsuitable lands, and they grow and convert the land suitable for food crops cultivation. Repeated harvest should be made until the pollutant loads go down below safe limits. After taking up the soil HMs, every part of the plant owns them but distributes them differentially to handle accordingly. In some plants, the highest concentrations of HMs are found in the below-ground parts and in others the highest concentrations are found in the above-ground parts like leaves, stems, and flowers (Jiang et al., 2015). This is decided by the translocation ability of plants. For phytoextraction, high translocation ability is desired, but for phytostabilization, low translocation ability is desired. Several studies verified that HMs accumulated in the actively growing tissues of sprouts, leaf (Chakraborty et al., 2013), and bark (Unterbrunner et al., 2007). Pulford and Watson (2003) found that the Zn and Cd concentrations were present maximum in the foliage. Knowledge about the distribution of HMs in selected plants is important before planning remediation of contaminated lands to prevent re-exposure of the environment with the HMs contained in the plant biomass during processing. Plants with a higher translocation efficiency (i.e., store more HM in the below-ground parts) are better than plants with lower translocation efficiency from phytoremediation perspective. This raises some issues on food chain transfer risk of HMs but as harvesting of the root is not needed, the whole phytomanagement plan becomes cost-effective. This HM-laden biomass needs to be disposed of carefully. One of the primary concerns about the use of phytomanagement is the postharvest disposal of HM-rich biomass. Phytoextraction and rhizofiltration technologies are generally applied to remove metals from contaminated soil and waste water, respectively, and revive the soil and water for agricultural purposes. This is generally carried out by repeated cropping and harvesting of the plants in the polluted soil/water, until the HM concentration drops down to acceptable level. The accumulation of large amounts of harvested biomass is hazardous and should be processed properly. Therefore, HMCB is the most promising candidate of bioenergy production with the following aims: (i) producing clean metal-free biofuel, (ii) contributing global energy demands in the renewable sector, (iii) mitigating soil and water pollution via sustainable phytoextraction, and (iv) resolving the controversy of food crisis versus biofuel production.

3 The fate of contaminated biomass from phytoremediation

The main disadvantage of phytoremediation of HM-polluted sites and utilization of the metal-laden biomass is the fate of the hazardous HMs. Scholz and Ellerbrock (2002) showed that metals found in biomass of bioenergy crops which were fertilized and harvested regularly for a number of years varied in the range of Cd 0.3–2.2; Cu 2.6–22.6; Pb < 1 to ~ 4; and Zn 15–135 mg kg^{-1}.

After harvest, to reduce transportation and disposal cost, contaminated biomass is usually sun-dried first to reduce wet weight and then further reduced by composting followed by compacting (Kumar et al., 1995). In densification (compaction), mechanical compressor is applied to increase the mass density of the solid-harvested waste (Mani et al., 2003). The compacted biomass is treated with microbes that aerobically decompose it to form a stable product called compost, which when added to soil can increase plant available nutrients (Epstein et al., 1996). Both compaction and composting should be done with utmost care to keep the compost and the leachate free from metals, and therefore, proper disposal is necessary (Ghosh and Singh, 2005).

This plant biomass after harvesting possesses a significant calorific value. To utilize this, recent concept is "integrated phytoremediation," which aims to valorize metal-rich biomass and thus coupling land reclamation with energy production (Fig. 1). The energy entrapped within plant biomass is usually released by one of the three methods: combustion, gasification, and pyrolysis.

In combustion or incineration, the biomass rapidly reacts with oxygen forming CO_2, heat, and water vapor, which can be used to produce energy. It causes a 90% volume reduction of the biomass and the exhaust fume captures the volatized metal and metalloid elements such as Hg, As, and Se whose escape into the environment must be prevented (Elekes et al., 2010).

In gasification, the plant biomass is converted to a mixture of CO and H_2 gas called as syngas by treatment with substoichiometric quantity of O_2 at 700–800°C. The syngas is used to produce electricity. Charcoal and ash are the two by-products from the gasification reaction. During gasification, different polynuclear aromatic compounds such as methyl naphthalene, methyl acenaphthylene, and methyl toluene and heavy metals and gaseous compounds form which condense on fly ash particles during cooling (Vervaeke et al., 2006). So ash should be collected and disposed properly and should not be used as fertilizer (Huber et al., 2006). Surprisingly, not only HMCB, but problem persists even with burning of unpolluted wood and straw as considerable amount of cadmium forms in the fly ash and Cd has detrimental health effect (Scholz and Ellerbrock, 2002).

Unlike incineration and gasification which use O_2 as the oxidizing agent, pyrolysis is anaerobic thermal decomposition of biomass conducted at ~ 500°C. At this temperature, biomass undergoes a rapid volume reduction (20%–35%), and a mixture of gas, aerosols, char (or ash). and a liquid fraction called pyrolysis oil is produced. The pyrolysis oil can be further utilized as fuel (Bridgwater, 2012). The percentage of gas, liquid, or solid differs according to reaction parameters. Flash pyrolysis is conducted at 650°C, and slow pyrolysis is conducted at 450–500°C. Flash pyrolysis gives rise to a higher quantity of liquid products, known as bio-oil (Huber et al., 2006).

Two main problems encountered in the thermochemical processing of HMCB are (i) relatively high humidity of the harvested plant biomass and (ii) volatilization of certain HMs (Cd, Pb, Zn, Se) at the high temperatures. Water content of biomass is reduced by sun drying, which not only maximizes its calorific power but also makes the process economically feasible. For heavy metal volatilization, the ash part should

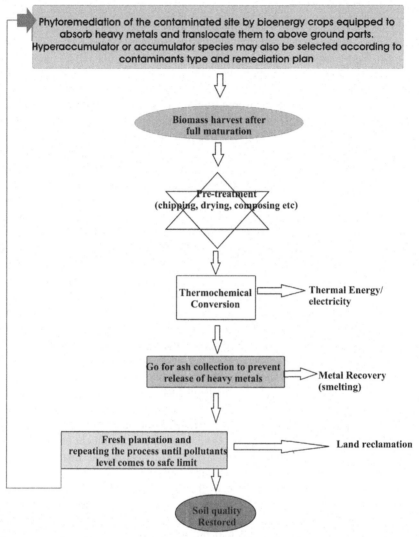

Fig. 1 Valorization of heavy metal-contaminated biomass and thus coupling land reclamation with energy production.

be collected and disposed or treated properly, ensuring that HMs don't return back to environment (Nzihou and Stanmore, 2013). In the laboratory-scale experiment, it has been noticed that after pyrolysis, > 98% of the metal is retained in the char (Koppolu et al., 2003).

Study on the fate of HMs during thermal processing of biomass is the focus of many studies, and research is trying to develop new processes to recover metals from ash/char using hydro-metallurgical or electrochemical approaches (Machado et al., 2010). In an application on biomass gasification and investigation on the fate of HMs, a small-scale fixed-bed downdraft gasifier was installed to convert biomass to electricity and heat. In that gasification experiment, 1 kg of wood was found to produce 40 g of ashes, 9 MJ of heat, and 1.2 kWh of electricity. Among the total 40 g ash, 31 g was bottom ash, 2 g was filter ash, and 7 g was cyclone ash. In addition, 18 g of gasifier bed ash was also obtained. Concentration of Cd, Pb, and Zn in the fly ash was 7–100 times higher than in the bottom ash. 30%–40% Cd, Cr, Ni, and Pb was recovered. Among the recovered fraction, > 60% was recovered from finer filter and cyclone ash. Some HMs (Cr, Cu, and Ni) were mainly observed in the bottom ashes and a little bit in the bed ashes. The bottom ash formed the highest fraction. Among all the HMs, only cadmium and zinc exceeded the Flemish threshold points by a small margin to be used as a fertilizer. Jiang et al. (2016) studied fates of the heavy metals in different gasification temperatures. He noticed that Cd, As, Pb, and Zn tended to transform into gaseous states at ~ 1000°C, while Ni, Cu, Mn, and Co converted to gaseous forms within 1000–1200°C. Other metals like Cr, Al, Fe, and Mg stayed back in the solid form even at > 1200°C.

Lu et al. (2012) explored the suitability of the combustion process for the disposal of HM-loaded hyperaccumulator plants obtained through phytoextraction process. Efficient fabric filters were designed to capture highly volatile HMs (Cd, Pb, Zn) from the ash fraction below the filter. This fulfilled the existing thresholds sanctioned by European Union directive for large combustion plants (Chalot et al., 2012). It was observed that in the absence of fabric filter, a high concentration of Cd and Zn released. Modern sophisticated waste-to-energy incinerators should be equipped with advanced pollution control devices like cyclone devices, electrostatic precipitators, and filter fly devices to capture considerable amounts of HMs (Chalot et al., 2012).

Another research reported the wonderful capability of natural and modified limestone for the capture of HMs during wood sawdust combustion in a combustor. The natural limestone was found to capture Cu and Pb at temperatures > 700°C, but not for Cr and Zn. Modification of the limestone with sulfite enhanced its ability to absorb all studied HMs (i.e., Cu, Pb, Zn, and Cr) (Zheng et al., 2017). Treatment of the limestone with sulfite removed impurities from inner pores of limestone and increased its pore sizes and specific surface area. This boosted physical and chemical adsorption abilities of limestone at 700°C and 900°C, respectively (Zheng et al., 2017).

There is a controversy on whether the combustion of heavy metal-contaminated biomass (HMCB) produced during phytoextraction process should be considered as a waste, or as a legal commercial fuel for clean bioenergy recovery (Delplanque et al., 2013). In some previous studies, a majority of contaminated Willow or Salix species were recommended for direct combustion (Delplanque et al., 2013; Mayer et al., 2012), but enough data are not available on the effect of pretreatment techniques on the combustion of HMCBs or the effect of the type of HMCBs on HMs distribution etc. Many scientists have voted for the use of bioenergy plants, which have low translocation efficiency of HMs and accumulate the HMs predominantly in the root and not shoot/leaf. Plants like *Miscanthus* × *giganteus* are well known for effective phytoextraction of the HMs and storing them in the roots. So the combustion of the aboveground biomass, i.e., stem and leaves, is relatively safe due to the low accumulation of HMs. However, whatever be the technique and source of the wood source, it is always advisable to use the industrial-scale boilers provided with efficient filtration systems to capture volatile HMs (Delplanque et al., 2013). HMs and fine particles are accumulated primarily in the fly ash and flue gases, respectively (Boman et al., 2006). Horizontal tube furnace is better than entrained tube furnace for causing higher volatilization rate of HMs (Lu et al., 2012). Most scientists agree that among thermochemical conversion processes, pyrolysis is the best choice for clean bioenergy production (Cornelissen et al., 2008). On the other hand, few scientists (Keller et al., 2005; Kovacs and Szemmelveisz, 2017) are against bioenergy production from HMCB and consider combustion as the best method of HMCB disposal.

In an investigation on the HM removal efficiency from end products by different bioreactors, it was found that the emission percentage of HMs (Cd, Pb, Zn) in the flue gas remained below the standard (Guo and Zhong, 2018; Guo et al., 2017). Heavy metal (Cd, Cu, and Pb)-enriched biomass of the plant *Sedum plumbizincicola* was co-combusted with sewage sludge and coal at different temperatures in a large-scale fluidized bed and a tube furnace (Guo and Zhong, 2018). Analysis of the fate of HMs found co-combustion as a suitable option for the safe disposal of *S. plumbizincicola* and sludge.

Guo et al. (2017) found that high temperature helped in the volatilization of the HMs and the order of volatilization was Pb > Cd > Zn. He also found that the co-combustion in the presence of air is beneficial to the disposal of HMs. Lu et al. (2012) reported that in the incineration of HMCBs, the ash content generated was almost three times less than that by pyrolysis. But the oxidizing environment of combustion was unfavorable for the volatilization of HMs to the gaseous phase. Kuppens et al. (2015) claimed that pyrolysis of HMCB through the implementation of a novel dynamic techno-economic assessment would prove best option in terms of energy and heat recovery in comparison with gasification and combustion.

An integrated phytoremediation with bioenergy production project was undertaken in the Campine region in the northeast Belgium and the southwest Netherlands (Gomes, 2012). Smelter activities for a long time in the region have made the site contaminated with multiple HMs causing HM-contaminated food and financial losses (Thewys and Kuppens, 2008; Thewys et al., 2010). Cultivation of several energy crops like maize, rape seed, wheat, and short-rotation coppice was tried here. The phytoextraction abilities of the plants, the fate of HMs in the plants, and biomass valorization were explored by different thermochemical conversion techniques (anaerobic digestion, incineration, gasification, and plant oil production). For example, anaerobic digestion and biogas formation from *Zea mays* resulted in 33,000–46,000 kWh of renewable electrical and thermal energy per hectare per year. Generation of similar amount of nonrenewable thermal and electrical energy would have contributed about 21×103 kg ha^{-1} year^{-1} of CO_2 in a coal-fed power plant (Meers et al., 2010). Another positive outcome was that the metal concentrations in grains were also quite low according to European animal feed criteria (Meers et al., 2010).

4 Bioeconomy via products recovery from contaminated biomass

Building a bio-based economy through contaminated biomass produced from phytoremediation of polluted sites is a novel idea toward sustainable development. So bioprocesses and bioengineering should be boosted for building this bio-based economy.

Thermochemical energy conversion is the best method for valorization of the biomass as the HM-enriched waste cannot be used in fodder and fertilizer.

Bio-based economy is future vision of sustainable use of renewable energy resources. Multidisciplinary study involving science, technology, and management is needed as it is one of the most rapidly advancing segments of the world's economy due to the enormous societal benefits it offers. Bio-based economy concept is vast and includes all areas of utilizations of biological resources, including plants, animals, and microorganisms. Bio-energy sector is the greatest beneficiary of bio-based economy. Up to now, coal, oil, and natural gas (carbon-based fossil fuel) constituted the backbone of energy and chemical sectors. But now it seems possible that in the near future, bio-based fuel would totally replace fossil fuel. Production of these bio-based products (biofuels, bioenergy, and bio-based chemical) from waste materials is increasing day-by-day reflecting their significant role in the environment. In the near future, fossil fuel would be depleted and biofuel can become one and the only energy source. At present, it can save nature from ongoing climatic destruction due to increased fossil fuel-mediated greenhouse gas (GHG) emission. Scientists from all over the world are trying to convert all types of biological waste materials (agricultural waste, forestry residue, solid municipal waste) into advanced environmentally friendly biofuel and other industrially important products as by-products.

Most important environmental benefit of biofuel is that it generates significantly lower GHG emissions (especially CO_2) in comparison with petroleum and diesel. Advancement of biofuel technology and popularization of biofuel among carbon-intensive industries like mining, transportation, etc. would further bring down carbon footprint. This green bio-based economy has the potential to offer novel opportunity to address dual challenges of climate change and resource scarcity.

5 Technologies for contaminated biomass conversion into bioenergy

Renewable biofuels have full potential to displace nonrenewable fossil fuels, which contribute to global warming by GHG emission. Biochemical and thermochemical processes are the two technologies for transformation of biomass to biofuel. Biomass feed stocks preparation is the first step for starting the biomass conversion process. Crushing, oil extraction, fermentation, anaerobic digestion, and transesterification are the main steps of biomass feed stock preparation, which are further refined in the thermochemical conversion plant. Then, energy yield of the crop is determined from the dry weight and heating values of the biomass. Generally, the energy content of dried biomass (ash-free) lies in the range 17–21 MJ/kg for most bioenergy plants (McKendry, 2002). Therefore, biomass yield is positively linked with energy yield. Fig. 2 shows the overview of biofuel generation from biomass. Biodiesel and bioethanol are the two most popular renewable fuels that are produced through biochemical transformation of biomass.

Another interesting fact is that bioethanol and biodiesel produced from bioenergy plants grown on nonagricultural land or marginal land (not suitable for food crop

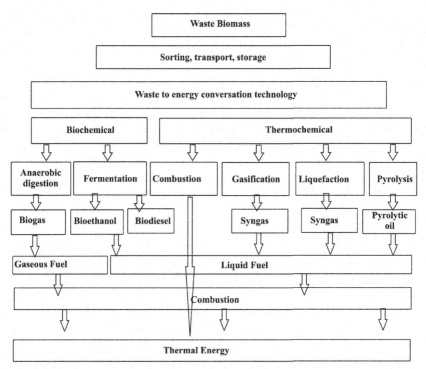

Fig. 2 Potential thermochemical technologies for valorization of biomass.

cultivation) cannot replace fossil-based transport fuel. But their partnerships with bio-diesel from microalgae have the ability to completely replace petroleum-based transport fuels without interfering with the production of agricultural crops. Even the most high-yielding oil crops (e.g., sugarcane or oil palm) are not comparable to microalgae's ability to generate such copious volume of biodiesel in a short time and with minimum investment (Chisti, 2008). Microalgae can be grown in waste water and so adds another dimension of sustainable phytoremediation of water bodies with good economic return. So along with biofuel production from HMCB, fuel production from microalgae grown on contaminated waste water would be discussed in the next section.

5.1 Biochemical processes

Conversion of biomass to bioethanol is a biochemical process having high efficiency and selectivity. The biochemical steps for second-generation bioethanol production are pretreatment of lignocellulosic material, enzymatic hydrolysis, microbe-mediated sugar fermentation, and finally distillation of the produced bioethanol and dehydration. The first step, i.e., biomass pretreatment, needs biological, chemical, and physical (heat) catalysts. In the next step, biocatalysts or enzymes are used to speed up the hydrolysis of oligosaccharide bonds of polysaccharides, and then different kinds of fermentative yeast or bacteria are used for the fermentation of mixed sugars. Some biomass produce an increased amount of gaseous or liquid bioethanol than others.

The proportion of cellulose and lignin in the source decides the final biomass energy yield. As cellulose has better biodegradability than lignin, biomass enriched in cellulose gives a better bioethanol yield than lignin-rich biomass. Because of this, switchgrass, which has as high as 90% of biodegradable cellulose/hemicellulose and lesser amount of lignin (5%–20%), gives up to 280 L ton^{-1} of ethanol instead of usual 205 L ton^{-1} of ethanol produced by wood-based biomass, which has an increased amount of lignin (up to 30%) (McKendry, 2002; Tao et al., 2014). Ammonia pretreatment can be useful for increased yield as ammonia removes the lignin and causes decrystallization of cellulose (Kim and Lee, 2005). As biomass feedstock cost determines the biodiesel production cost (75%), therefore the selection of proper feedstock is a salient feature of sustainable biodiesel production. A detail discussion on biodiesel, biogas, and bioethanol has been made in the next section.

5.2   Thermochemical processes

Thermochemical processes of biomass conversions are lucrative and possess many benefits over biochemical processes in terms of better productivity, availability of a wide range of feedstocks to choose, ability to completely utilize the feedstocks leading to multiple products, better control, and independency of climatic conditions (Verma et al., 2014). In the following section, short discussions have been made on the different thermochemical processes of pyrolysis, liquification, gasification, and combustion. Table 1 compares some of the applied thermochemical processes for the conversion/disposal of HMCBs.

5.2.1   Pyrolysis

Pyrolysis is the anaerobic thermal decomposition of biomass at 350–700°C into solid (bio-char), liquid (bio-oil), and gaseous products (Goyal et al., 2008). A combination of heat and mass transfers take place during pyrolysis reactions (Bridgwater, 2012). Different types of pretreatments enhance end product quality and quantity (He et al., 2020; Wang et al., 2020).

Pyrolysis gases are mainly composed of CO_2, CO, H_2, CH_4, C_2H_4, and C_2H_6 and trace quantities of water vapor and high molecular weight hydrocarbons (Goyal et al., 2008). The pyrolytic gas can be directly combusted to generate heat or can be converted into liquid synfuel. The pyrolytic liquid oil is a mixture of oxygenated aromatic and aliphatic compounds (Meier and Faix, 1999). It has two parts: a pyroligneous acid portion and a tar fraction. The tar part has aldehydes, aromatic compounds, carbohydrates, native resins, phenols, condensation products of the reactants, and other derivatives. The pyroligneous acid part has approximately 50% acetic acid (CH_3OH), acetone (C_3H_6O), phenols, and water. On the other hand, bio-char is mainly composed of elemental C, H, and inorganic K, N, P, Si, etc. (Goyal et al., 2008). Bio-char has certain properties that make bio-char an attractive soil amendment and a good adsorbent of metals. The porous nature, affinity for metal adsorption, and ability to slowly release nutrients make bio-char a cheap but good-quality fertilizer. The high longevity of bio-char (100–1000 years in soil) allows it to store large amount of CO_2.

Table 1 Thermochemical conversion processes of some of the heavy metal-loaded feedstocks and the fate of the heavy metals.

Biomass Feedstock	HM in feedstock	Thermo-chemical conversion	Reactor	Temperature (°C)	Pretreatments	HM content in end product/biofuel	References
Pyrolysis							
Avicennia marina	Cd, As, Cr, Ni, and Pb	Ferric salts catalyzed the pyrolysis process	Quartz tube furnace	300–700	Ferric salts	More HMs concentrated in bio-char and low in gas	He et al. (2020)
Hydrocotyle verticillata species, an aquatic plant	Pb	Pyrolysis	Quartz tube furnace filled with N_2 gas	350 and 450	Phosphoric acid (PA)	More HMs concentrated in bio-char and low in gas	Wang et al. (2020)
Broussonetia papyrifera	Cu, Cd	Slow pyrolysis	Laboratory-scale tube furnace	450	Crushed and preheated	Low HM in bio-oil	Han et al. (2018)
Willow	Cd	Fast pyrolysis	Full-scale reactor	350–650	–	NR	Kuppens et al. (2015)
Willow	Cu, Cd, Pb, Zn	Flash pyrolysis	Laboratory-scale, semicontinuous stirred fluidized bed reactor	350–550	Crushed and preheated	Very low	Stals et al. (2010)
Willow	Cu, Cd, Pb, Zn	Fast pyrolysis	Quartz horizontal tube reactor	350	Air-dried and crushed	No HM in bio-oil	Lievens et al. (2009)
Willow and *Thlaspi caerulescens*	Cd, Zn	Slow pyrolysis	Laboratory-scale tube furnace	25–900	–	NR	Keller et al. (2005)
Sedum plumbizincicola	Cd, Pb, Zn	Slow and fast pyrolysis	Horizontal quartz tube (laboratory scale)	450–750	–	High HM in bio-oil	Zhong et al. (2016)
Switchgrass	Cu, Fe	Fast pyrolysis	Fluidized bed reactor	450–500	Crushed and preheated	NR	Mullen and Boateng (2013)

Continued

Table 1 Continued

Biomass Feedstock	HM in feedstock	Thermo-chemical conversion	Reactor	Temperature (°C)	Pretreatments	HM content in end product/biofuel	References
Switchgrass and timothy grass	Pb	Fast pyrolysis	Micropyrolyzer	500	Crushed, preheated, and hydrolyzed using mild acid and enzymes	NR	Balsamo et al. (2015)
Birch wood and sunflower	Cu, Cd, Pb, Zn	Fast pyrolysis	Quartz horizontal tube reactor	400–600	–	No HM in bio-oil	Lievens et al. (2008)
Agricultural residues and manure	Cd, Cu, Zn, Pb, Cr, Ni, Co, Mn, Fe, A	Fast pyrolysis	Fluidized bed reactor	400–700	–	Almost no	Leijenhorst et al. (2016)
Mixed feedstock (pine and spruce)	Al, Zn, Fe,	Fast pyrolysis	Pilot-scale pyrolyzer	750	Crushed and preheated	Low HM in bio-oil	Wiinikka et al. (2015)
Fir sawdust	Cu	Fast pyrolysis	(Batch) vertical drop fixed bed reactor	450–600	Crushed and preheated	Almost no	Liu et al. (2012)
Gasification							
Willow	Cd, Zn, Cu, Pb, Cr, Ni, Co, Al, Fe, Mg, Mn, Sn	Gasification with CO_2	Laboratory-scale fixed bed gasifier	450–950	–	High HM in syngas	Said et al. (2017)
Willow and *Thlaspi caerulescens*	Cd, Zn	Gasification	Laboratory-scale tube furnace	25–900	Crushed and preheated	NR	Keller et al. (2005)

Feedstock	Heavy metals	Process	Reactor	Temperature	Pretreatment	Result	Reference
Rice straw, wheat straw, corn stover, switchgrass, Miscanthus, Jose tall wheatgrass, and Douglas fir wood	Cd, Cu, Zn, Pb, Cr, Ni	Gasification	Furnace	950	Crushed, preheated, and undergone leaching pretreatment	NR	Yu et al. (2014)
Leucaena leucocephala and Saccharum officinarum	Cd, Zn, Pb, Cu, Ni, Fe, Al, Co, Ti, Cr, Mo, V, Mn	Gasification with steam	Bench-scale fluidized bed gasifier reactor	800	Dried and crushed	Low HM in syngas	Cui et al. (2013)
Mixed feedstock (flax, oak and beech)	Cd, Zn, Cu, Pb, Ni, Fe, Mn, Ti	Gasification with steam	Atmospheric fluidized bed gasifier	855	–	High HM in syngas	Šyc et al. (2011)
Miscanthus giganteus, S. hermaphrodita, S. pectinata, P. virgatum (bioenergy crops)	Cd, Pb, Zn	Gasification with nitrogen	Laboratory-scale tube furnace	1500	Crushed	NR	Werle et al. (2016)
Commercially wood pellets	Cu, Pb, Cr, Ni, Co, As, Hg, Fe, Al, Mn	Gasification	Pilot-scale (fixed bed) autothermal downdraft system	900–1200	–	NR	Tafur-Marinos et al. (2014)
Straw char and glycol	As, Cd, Pb, Cr, Hg, Ni, V	Gasification with enriched air	Pilot-scale atmospheric entrained flow gasifier REGA	1700	–	negligible	Pudasainee et al. (2014)

Continued

Table 1 Continued

Biomass Feedstock	HM in feedstock	Thermo-chemical conversion	Reactor	Temperature (°C)	Pretreatments	HM content in end product/biofuel	References
Combustion							
Willow (*Salix viminalis* and commercial *S. tora*)	Cd, Zn, Cu, Cr, Co, Ni, Mn	Combustion	Industrial-scale boiler	900–1000	Crushed and then digested by aqua regia and HNO₃ solution (65%).	HM present in ash	Delplanque et al. (2013)
Willow (*Salix caprea*)	Cd, Zn, Cu, Pb	Combustion	Fluidized bed reactor	850	–	HM present in ash	Šyc et al. (2012)
Willow (Salix leaves)	Cd, Zn	Combustion	Quartz glass tube reactor	25–900	–	HM present in ash	Keller et al. (2005)
Biomass char (willow)	Cd, Zn, Cu, Fe	Combustion	TGA-MS/DSC	25–900	Crushed and predried	HM present in ash	Mayer et al. (2012)
S. plumbizincicola and *Sedum alfredii*	Cd, Zn, Cu, Pb	Combustion	Entrained flow tube furnace and horizontal tube furnace	650–950	Prewashed by distilled water, then crushed, and dried at 85°C	HM present in ash	Lu et al. (2012)
Poplar	Cd, Zn, Cu, Pb, Cr, Ni	Combustion	Industrial-scale boiler	509–940	Crushed and air-dried	HM present in ash	Chalot et al. (2012)
Willow (*Salix viminalis* and commercial *S. tora*)	Cd, Zn, Cu, Cr, Co, Ni, Mn	Combustion	Industrial-scale boiler	900–1000	Crushed and then digested by aqua regia and HNO₃ solution (65%)	HM present in ash	Delplanque et al. (2013)

Depending on the operating temperature and reaction time, pyrolysis can be classified into three types: slow pyrolysis, fast pyrolysis, and ultra-fast/flash pyrolysis. Conventional pyrolysis or slow pyrolysis is also called as torrefaction (Vander Stelt et al., 2011) and is performed at 200–300°C. In earlier days, it was used to produce charcoal. In slow pyrolysis, oil produced is minimal, whereas in fast and ultra-fast pyrolysis, the oil and gas productions are maximized. The low-temperature pyrolysis or torrefaction is actually thermal upgrading of biomass into a more homogeneous product, which is made more concentrated and more energy-dense product. This product called torrefied pellet has much similarity with coal, and like coal or charcoal, this can be used for domestic heating, cooking, co-firing power generation, and gasification (Agar and Wihersaari, 2012). Fast pyrolysis is more commonly used today in research and industry as it offers a quick thermal decomposition of the carbonaceous biomass without the presence of O_2 at a moderate heating rate of ~ 50°C/s. The products from fast pyrolysis are liquid condensates (30%–60%), gas mixtures (CO, CO_2, H_2, CH_4, and low molecular weight hydrocarbons) (15%–35%), and bio-char (10%–15%). In ultra-fast or flash pyrolysis, thermal decomposition happens extreme rapidly with a high heating rate of 100–10,000°C/s and with a short residence time. Flash pyrolysis produces gases (60%–80%), liquid condensate (~ 10%–20%), and bio-char (10%–15%) (Demirbaş, 2000). Stals et al. (2010) showed that pyrolysis temperature was an important factor during the transfer of heavy metals from metal contaminated biomass of Willow into volatile pyrolysis products. Here, minimal transfer of Zn and Cd into the pyrolysis oil was detected at 623 K. Other than temperature, entrained flow was another prime factor that significantly affected the movement of target HMs to volatile pyrolysis products. Use of hot-gas filter can reduce the amount of Cd and Zn in pyrolysis oil, without affecting the yields of oil or char (Stals et al., 2010).

Commonly used pyrolyzers where pyrolysis is performed are fluidized-bed pyrolyzers or rotating cone and Auger pyrolyzer. Solar pyrolysis is a special type of pyrolysis and is relevant for pyrolysis of HMCB where carbonaceous feedstock (biomass or carbon-containing biowaste) is used as chemical reactants and solar energy high-temperature provider. In comparison with conventional pyrolysis, solar pyrolysis has lower carbon footprint and produces products with higher caloric value as solar energy is chemically stored in the form of solar fuels (Zeng et al., 2017). Zeng et al. (2020) conducted solar pyrolysis of HMCB from willow at different temperatures (600–1600°C) and studied the effects of temperature and HMs contamination on bio-char properties. According to the study, aromatic char was formed in an ordered manner with increasing pyrolysis temperature. Increased temperature caused an increase in char carbon content from 70.0% to 88.4% but oxygen and hydrogen contents decreased. Heavy metals (Cu or Ni) caused a significant declinement of char hydrogen and oxygen contents compared with raw willow char. It may be due to the fact that copper or nickel can promote depolymerization of cellulose and hemicellulose (C–H and C–O bonds cleavage). They also have a noticeable catalytic activity that may be responsible for tar cracking and reforming into H_2 and CO, which could further decrease hydrogen and oxygen contents in char (Said et al., 2018).

Pyrolysis offers many environmental and economic advantages as low energy-rich biomass can be converted into high energy-rich liquid, which can be used in numerous economic ways. In contrast to other thermochemical methods, the comparative lower pyrolysis temperature averts HMs from getting volatilized, while the worthy pyrolysis oil is made. Pyrolysis is the preferred option due to less transfer of heavy metals to products. Considering all the facts, pyrolysis is considered best thermochemical conversion due to the less transfer of HMs to the products (Dastyar et al., 2019).

5.2.2 Liquefaction

There are two classes of liquefaction techniques, namely, hydrothermal liquefaction and solvent liquefaction (He et al., 2017). In hydrothermal liquefaction, under high pressure of 5–20 MPa, wet biomass slurries are heated at 200–500°C to generate a mixture of products. Major part of the liquefied mixture is bio-oil (15%–75%) and gaseous products (10%–20%) and a minor amount of solids (Elliott et al., 2015). Liquefied products or the bio-oil containing high amount of nutrients and organics can be used as biofuel or chemical industries feedstock (Chiaramonti et al., 2017). Hydrothermal liquefaction allows the separation of the HMs from the bio-oil, and the final solid residues can be reused as soil fertilizer. The solvent liquefaction process causes direct dissolution of low molecular weight molecules and hemicelluloses in the feedstock into micelle-like materials, and the celluloses and lignose polymers are broken down into smaller molecules. Finally, physiochemical processes of dehydration, dehydrogenation, decarboxylation, and deoxygenation degrade the micelle-like materials into smaller compounds generating the bio-oil and gaseous products (Mukherjee et al., 2016). In the solvent-based liquefaction, the nature of solvent is the vital variable that determines the process rate, quality, and quantity of products. Qian et al. (2018) studied the prospect of subcritical hydrothermal liquefaction reaction to recycle biomass of Zn-contaminated *S. plumbizincicola* to valorized products, such as bio-oil, carboxylic acids, hydrochar, and other chemicals. Yang (2010) developed a chemical extraction technique for HM separation from crude bio-oil of *S. plumbizincicola* contaminated with Cu, Pb, and Zn through hydrothermal upgrading. Effect of pressure (18.00–25.40 MPa), temperature (270–421°C), PS (1.00–4.75 mm), and time (10–120 s) on the yield of bio-oil and separation efficiency was studied. Temperature and pressure were found to have positive impacts on the yield of the bio-oil and gaseous products as the feedstocks were continuously decomposed. Almost 100% efficiency was obtained with a bio-oil yield of > 63% at 370°C and pressure of 22.10 MPa. But the efficiency of HM removal remained the same despite the increase in pressure and temperature. It was concluded that the high-temperature liquefaction is a realistic opportunity for bio-oil extraction from HM-polluted hyperaccumulators. Guo et al. (2017) also certified solvent liquefaction approach as an ideal means to dispose HMCBs. They also urged for further treatment of the liquefied products for the production of polyurethane compounds.

5.2.3 Gasification

Partial oxidation of the biomass with stoichiometry quantity of O_2 at high temperature of 800–900°C leads to a combustible gaseous fuel through gasification. The gaseous fuel mixture known as producer gas or syngas (synthesis gas H_2/CO) consists of CO, CO_2, H_2, CH_4, N_2, H_2O, and small unwanted char particles, ash and tars (oxygenated hydrocarbons) in the air, oxygen, or steam (Goyal et al., 2008; McKendry, 2002). The syngas can be used to produce methanol, dimethyl ether through Fischer-Tropsch process, or burned directly to generate steam and heat and put in the internal combustion engines to give rise to bio-electricity and thus contributes to CO_2 mitigation. However, the gas has little calorific value of 1000–1200 kcal/Nm^3. About 1 kg of dried biomass can give rise to almost 2.5–3.0 Nm^3 of syngas through gasification. One of the main problems of gasification is the occurrence of tar (occasionally containing HMs), which leads to serious troubles such as blockage of pipes and filters, corrosions, and engine failures (Pudasainee et al., 2014).

5.2.4 Combustion

For HMCBs disposal, combustion route is recognized by many authors as the one of the most environmentally safe technique than other technologies like composting, pyrolysis etc. (Kovacs and Szemmelveisz, 2017; Sas-Nowosielska et al., 2004). Combustion of HMCB leads to emissions of undesirable CO, NO_x, fly ash, solid, and gaseous metal compounds. Even combustion of unpolluted biomass from uncontaminated sites has also shown heavy metal emission (Nzihou and Stanmore, 2013). Combustion temperature can volatilize HMs (Karimanal and Hall, 1996), and they exit from the combustion chamber either as solid particles or as gaseous form. After combustion of HMCB is over, the heavy metals exit from the combustion chamber in the form of solid particles in the combustor (bed ash) and fly ash or in gaseous form in the flue gas (exhausted gas/flue gas). The combustion parameters and furnace type determine the partitioning of the HMs in various forms. The ash should not be disposed as the leached HMs can create environmental damage at waste yards (Christensen et al., 2001). The different reactors for combustion are fluid bed, circulating fluid bed, entrained flow, and stoker grate.

In addition to the above-mentioned four thermochemical conversion processes that are most commonly used, many new innovations are coming. One of them is carbonization, in which charcoal is produced slowly by the partial oxidation of woody feedstocks (Bailis, 2009). Another one is hydrothermal approach through which biomass is decomposed into solid, liquid, and gaseous intermediates in an aqueous environment, at moderate temperatures of 200–600°C and high pressure at 5–40 MPa (Peterson et al., 2008).

6 Biodiesel, biogas, and bioethanol recovery from contaminated biomass

Worldwide, biodiesel and biogas are recognized as key sectors of renewable energy. While biodiesel is known as transportation fuel, biogas is known for electricity and heat-generating gas. Biogas is also used for the production of pure biomethane by sophisticated purification technique and delivered to natural gas grid to be used as transportation fuel. Source of biogas is mainly waste materials (agricultural waste, landfills, animal manure, and sludge from wastewater treatment plant), and raw material for biodiesel is generally rapeseed or other oil-yielding crops that are used as food. This factor raises the controversy of "food or fuel," which is pushing scientists to pay attention on nonfood feedstock for biodiesel and bioethanol production. In both of the cases, HMCB act as a feedstock. After fermentation or anaerobic digestion of the treated biomass into bioethanol or biogas, the digestate can be send back to the same soil as fertilizer. This method cannot remediate the site but can confine the HMs at a particular site, and at the same time, the site acts as a renewable source of energy.

6.1 Biodiesel

The main difference of biodiesel from petroleum diesel is that the biodiesel possesses higher octane number and almost 10 O_2 in various fatty acids and is devoid of any aromatic compounds, making it more environmentally safe. It not only releases low quantity of hydrocarbons and CO, but also has much reduced impact on GHG emission as the carbon atoms in biodiesel originate from atmospheric CO_2 that was fixed by photosynthesis. Therefore, CO_2 emission by the combustion of biodiesel falls within the Earth's natural carbon cycle. For example, GHG emission of biodiesel (category B100) is 4.5 times less than gasoline and 3 times beneath than petroleum diesel (Christopher et al., 2014). Biodiesel production history from vegetable feedstock dates back to 1853 (Demirbas, 2007). The first oil crisis during second industrial revolution in the 20th century again brought vegetable oil in the limelight as a source of biodiesel as an alternate to the petroleum fuel (Demirbas, 2007). The vegetable oil is renewable, low cost, easily available, and biodegradable with a high heating value and low sulfur and aromatic contents. To overcome the extremely viscous nature of the vegetable oil, which is 10–20 times more viscous than petroleum diesel, catalyst-mediated transesterification of the vegetable oil with an alcohol was performed (Demirbas, 2009). Choice of biodiesel feedstock is made according to the availability of the vegetable at the particular geographical location. For instance, sunflower and rapeseed oil are mostly used in Europe, soybean oil in the United States, canola oil in Canada, and palm oil in the tropical countries (Cao et al., 2008). They are regarded as the first-generation biodiesel feedstock. As all these oil-producing plants have been used as food, it raised the dispute of "food versus fuel," which leads way to produce second-generation biodiesel from nonedible sources like jatropha, karanja, and pongamia, as well as microalgae and other microorganisms. Second-generation biodiesel cost is also cheaper than the first-generation biodiesel. The generalized processing steps of biodiesel production from biodiesel-producing oil seeds like canola, camelina, jatropha seeds, rapeseed, mustard, safflower, soybeans, sunflower etc. include cleaning, preconditioning by heating, flake preparation, cooking the flakes, pressing the flakes, extracting the oil by mixing the flakes with hexane, and finally purification of the crude oil by degumming, neutralization (removes the phospholipids and free fatty acids), and bleaching at high temperature (removes color pigments and odoriferous compounds). The oil then undergoes one of the following methods for conversion to proper biodiesel: (i) direct blending, (ii) emulsification, (iii) pyrolysis, and (*iv*) transesterification (Abbaszaadeh et al., 2012). Biodiesel obtained by pyrolysis and microemulsion methods is an incomplete combustion product. Transesterification (or alcoholysis) is currently regarded as the best choice for biodiesel production from oil and vegetable fat (Borges and Díaz, 2012). Transesterification reaction involves the reaction of triglycerides of vegetable oil/animal fat with alcohol in the presence of a catalyst (a strong acid/base). Transesterification reaction requires a catalyst as without catalyst, triglycerides of oil/fat and alcohol wouldn't come to contact with each other as they are not miscible with each other. The catalyst here facilitates the intimacy between the reactants and accordingly increases the reaction rates and yield of biodiesel (Tan et al., 2009). Catalyst can be homogeneous (acts in the same liquid phase as the reactants) or heterogeneous (acts in the different phase from the reactants, i.e., solid). Both an acid and a base can act as catalyst (Borges and Díaz, 2012). Base-catalyzed reaction is selected for oils with lower amount of free fatty acids, whereas acid-catalyzed reaction is more ideal for oils with higher free fatty acid content (Schuchardt et al., 1998). In transesterification reactions, the existing ester bonds between fatty acids and OH groups of glycerol break down, and new ester bond is formed between free fatty acids (FFA) with alcohol molecules present in the transesterification mixture. End product of the reaction is a mixture composed of alkyl esters, free fatty acids, and glycerol (Ma and Hanna, 1999). Higher alcohols like propanol, butanol etc. can also be used but the lower-degree alcohols are mostly used. If methanol is used, the fatty acid methyl esters (FAMEs) form and fatty acid ethyl esters (FAEEs) form if ethanol is used. Both FAMEs and FAEEs are used in biodiesel engines. Process efficiency of transesterification is dependent on alcohol-to-oil ratio, temperature, type of catalyst, water, and free fatty acid content etc. (Ali et al., 1995).

Phytoremediated biomass from contaminated sites is another attractive feedstock of second-generation biodiesel. Azad (2017) studied the prospective of the nonedible mandarin seed oil as a potential biodiesel. The mandarin seeds were dried and crushed, and crude oil was extracted with *n*-hexane solvent, and then transesterification was performed after determining the acid value (mg KOH/g) of the crude oil. Consequences of changes in various reaction parameters like temperature, ratio of methanol to oil, and concentration of catalyst required were studied on the yield of biodiesel conversion. After successful oil-to-diesel conversion, the physiochemical properties of the fuel were quantified following the relevant ASTM standards. The results obtained were correlated with standard biodiesel ASTM D6751 and ultra-low-sulfur diesel (ULSD). About 49.23% (by weight) oil obtained was successfully converted to biodiesel with 96.82% conversion efficiency with the transesterification reaction designed. According to GC-MS analysis, the biodiesel was dominated by palmitic acid 26.80%, oleic acid 21.43% (both *cis* and *trans*), stearic acid 4.93%, linoleic acid 4.07%, and other fatty acids with less than 1% each.

6.2 Biogas

Like other types of biofuel, biogas is an important source of renewable energy and its production-and-use cycle generates no net carbon dioxide. Biogas is primarily a mixture of methane and carbon dioxide gases (Ma and Hanna, 1999). Sometimes the presence of hydrogen sulfide adds a bad odor to it. Sometimes it is formed naturally in marshy areas when the gases hydrogen, carbon monoxide, and methane generated are oxidized. Biogas is used as fuel or for cooking. The gas is also used to run a gas engine where the energy in the gas is converted into electricity and heat to run the engine. Alternatively, the gas can be compressed by the removal of CO_2 in the same manner by which compressed natural gas, or CNG is produced and used to run small transport vehicles. It was estimated in United Kingdom that biogas has the potential to replace around 17% of vehicle fuel. Another application of biogas is biogas that can be used for industrial production of biomethane (Fantozzi and Buratti, 2009).

For industrial production, closed anaerobic fermenter or bioreactor is used. In the anaerobic fermenter, the biomass is treated and digested anaerobically by microbes. Before the actual fermentation, pretreatment is sometimes carried out, which makes the biomass amenable to microbial attack. The pretreatment disrupts the closely packed tight nature of the biomass and helps the microbes to invade it. Various types of physical agents (irradiation crushing and grinding), chemical agents (acids, alkali, wet oxidation etc.), or biological agents (fungi, bacteria, or enzymes), or a combination of them are used to disrupt the compactness of the biomass. The four consecutive steps of anaerobic digestion for biogas formation from biomass are hydrolysis, acidogenesis, acetogenesis, and methanogenesis. The first step that involves hydrolysis of the complex organic biomass into simpler compounds is the rate-limiting step. In the second step of acidogenesis, the hydrolyzed molecules are further converted into smaller molecules like acetic acid, hydrogen, and carbon dioxide (Chen et al., 2008). In the third step, a complex consortium of acetogenic and methanogenic bacteria and archaea (Bundhoo et al., 2016) take part and ferment the simpler compounds and organics into organic acids and hydrogen. In the last step, methane and carbon dioxide are produced from organic acids and hydrogen by methanogenic bacteria/archaea. The characteristics of these two main groups of biogas-producing bacteria, i.e., the acid-formers and the biomethane-forming microorganisms, are quite different with reference to their growth rate, physiology, nutritional requirements, and sensitivity to environmental factors (Chen et al., 2008). Keeping proper balance between these two groups of microorganisms is crucial for bioreactor stability. Another reason of sudden reactor failure is the generation of inhibitory substances, which are normally present in substantial concentrations in wastewaters and sludges. Solid-state anaerobic digestion is also used that works in the same principle as solid-state fermentation. The solid-state anaerobic digester had several advantages, including requirement of less agitation, less energy for operation, and smaller reactor with fewer moving parts.

With the use of this technology, the disposal problem of huge amounts of liquid effluents and floating and stratification of fibers were no more. Moreover, the finished digestate formed compost that could be disposed of easily or can be reused as a fertilizer. Bundhoo et al. (2016) performed solid-state anaerobic digestion for biogas generation from palm oil biomass. As the pretreated biomass was composed of fibrous matter dominantly, it was easily embarked by gases and floated to the reactor's surface forming a mat-like scum layer.

Shiratori et al. (2017) developed one innovative technology for utilizing biogas for stable power generation through solid oxide fuel cell (SOFC). This SOFC power generation technology together with fermentation technology could generate energy from local biowaste like sludge, bagasse, and molasses. Laboratory-scale methane fermentation experiment conducted with a mixture of local biomass feedstock generated good amount of biogas (> 400 L kgVS^{-1}), and H$_2$S concentration was under 50 ppm. The biogas without any posttreatment processing was delivered to an SOFC unit, and a stable generation of power was recorded by the application of the paper structured catalyst technology (Shiratori et al., 2017).

Various workers studied the biogas generation from HMCBs, and accordingly, heavy metals were found to influence the total procedure and affected the end product generation. Biochemical reactions that take place during anaerobic digestion processes of HMCB are negatively influenced by HMs. Heavy metals like copper, cadmium, chromium, lead, nickel, and zinc have been reported to be inhibitory by many authors. Inhibitory effects depended on the concentrations of the HMs, speciation forms and solubility, and amount and distribution of the biomass in the chamber of the anaerobic digestion process. Studies have shown that cadmium, chromium, and nickel exert their toxic effects by disturbance of the enzyme composition and activity by binding to the different groups like thiol on protein molecules, or they replace the natural metal cofactor of the enzyme (Mudhoo and Kumar, 2013).

6.3 Bioethanol

Bioethanol is a produced by glucose fermentation of polysaccharide/sugar-rich fruit and vegetables like beet, cassava, sugarcane, maize, etc. First-generation bioethanol is prepared from sugar-rich biomass or starch-rich biomass. But it raised the debate of food vs fuel for agricultural arable land, urging to move to alternative sources from nonfood crop origin. Lignocellulosic biomass (rich in both starch and sugar), including HMCB, was found ideal for the second-generation bioethanol. Now other industrial byproducts like whey or crude glycerol, as feedstock, have been tried successfully (Robak and Balcerek, 2018).

As the energy content of ethanol is 34% (per unit volume) less than gasoline, it is commonly blended with gasoline before using it as transport fuel because if used alone consumption of ethanol would be almost double than gasoline. It has been observed that ethanol-amalgamated fuels such as E85 (85% ethanol and 15% gasoline) can bring down the net emission of GHG by 37.1%. In 2016, the use of ethanol-blended gasoline reduced 43.5 million metric tons GHG emissions in transportation, which is comparable to removing of 9.3 million vehicles for the period of 1 year (Robak and Balcerek,

2018). Brazil and the United States were the first bioethanol-producing countries. In United States alone, in 1998, 6.4 billion liters of bioethanol was manufactured from corn starch (Berg, 1999). Currently, more than 200 biorefineries are present in the United States, who together produces ~ 60.64 billion L of ethanol per year (Robak and Balcerek, 2018). Currently, countries from European Union manufacture > 2 billion L of bioethanol annually (Robak and Balcerek, 2018).

The steps of bioethanol formation include pretreatment, enzymatic hydrolysis, fermentation, distillation, and dehydration. As a result of the pretreatment, surface area of the biomass increases, which aids for enzymatic saccharification. It also reduces the amount of inhibitors present. Next step is the most important enzyme hydrolysis step that releases fermentable sugars. In the next step of microbial fermentation, sugars are converted into ethanol. Before carrying out fermentation, detoxification is needed to remove harmful inhibitors like furan derivatives, phenol components, and weak organic acids. Research has been done on the utilization of heavy metal-contaminated biomass for bioethanol production. Switchgrass and Miscanthus are potential assets for phytoremediation and bioethanol production (Pandey et al., 2016; Praveen and Pandey, 2020; Patel and Pandey, 2020). Ko et al. (2017) produced bioethanol from Napier grass biomass grown on HM-contaminated soil by the enzymatic hydrolysis method. The harvested HMCB biomass was pretreated using the steam explosion conditions with 1.5% H_2SO_2 at 180°C for 10 min and subjected to enzymatic hydrolysis. Fermentation efficiencies were 90%, 77%, and 77% for Zn-, Cr-, and Cd-contaminated soils, respectively, and in all cases, the ethanol yields were higher than the biomass obtained from unpolluted soil. Concentrations of ethanol produced after fermentation was 8.69–12.68 g L^{-1}, 13.03–15.50 g L^{-1}, and 18.48–19.31 g L^{-1} for Cd-, Cr-, and Zn-contaminated biomasses, respectively. So HMs exerted a positive effect on bacterial fermentation. However, too severe heavy metal pollution lowered fermentation efficiency. Thus, according to Ko et al. (2017), the prospect of Napier grass for bioethanol production and phytoremediation is quite promising. In another experiment, Dhiman et al. (2016) observed bioethanol production through saccharification of Canola biomass. Canola was grown on various metal-amended soils. It successfully phytoextracted ~ 95% of the zinc and accumulated it in the root the biomass. About 74.1% and 74.4% saccharification yields were obtained with nickel- and copper-contaminated biomasses, respectively. Under similar parameters, commercial cellulase gave a saccharification yield of 73.4% (Dhiman et al., 2016). Vintila et al. (2016) worked on the concept that after processing the biomass for biofuel production, the digestate can be returned as a fertilizer to the same polluted soil where the plant was grown. The purpose was to increase the soil nutrients and keep the HMs confined or recycled back in the polluted soil. The advantage of the concept is that no waste will be produced to dispose of and would pose no risk to surroundings. With this idea, they produced sorghum biomass on the HM soil. The authors wanted to assess the biorefinery production capability of sorghum biomass acquired from crops grown on HM-contaminated soil with emphasis on metal distribution. First bagasse (dry pulpy fibrous residue remaining after juice or liquid extraction from biomass) was obtained. Processing of the lignocellulosic bagasse generated ethanol and anaerobic digestion of fermentation residues produced biogas and organic fertilizer. The productivity of juice extracted after milling and pressing sorghum biomass was 0.16–0.27 g. Cadmium was not detected in the juice extracted from sorghum biomass produced on HM-polluted soil, and the concentrations of Pb and Zn were under safe levels. The finding was significant as it allowed sorghum to use as a crop on polluted soil and the extracted juice can be readily fermented to ethanol. The study gave additional information on the fate of different HMs during each step of biofuel production. After the pretreatment with NaOH/steam, the metals were found concentrated only in the solid sorghum bagasse but a minor part slipped to the liquid phase. Except Cu, Zn, and part of Pb which were extracted into the distillate, concentrations of rest of the metals increased in the solid phase of the hydrolysis/fermentation broth because of the solubilization of the main fraction of the organic solids. Lead mainly appeared in the first distillation fraction, and in the subsequent distillation steps, the concentration decreased. In the case of Zn and Cd, they could not be extracted from the broth by distillation and appeared in the vinasse or distillation residue. Thus, in ethanol production from sorghum HMCB, a major part of the HMs remained in the solid part with a small part in the distillation residual fraction and negligible Pb in the distilled ethanol. The solid residue part with the HM content was used for biogas production via anaerobic digestion, and the digestate was brought back to the same polluted field as fertilizer. Majority of the HM originally present in the soil was recycled back to its original place without getting exposed to the environment and thus allowed sustainable production of bioethanol. More research should be undertaken for the improvement of the pretreatment steps, enzymatic hydrolysis, fermentation stages, and increasing the fermentation efficiency so that all pentose and hexose sugars released during the previous steps are fermented into ethanol. The fluctuations in biomass composition and generation of inhibitors in presaccharification treatment, ethanol accumulation, end-product inhibition, and osmotic and oxidative stress are the main technical barriers, which should be overcome to make bioethanol more popular for use along with gasoline. More stress should be given to increase the cost-effectiveness of the environmentally safe bioethanol production from HMCB and optimization of the parameters for successful transition of the production scale from the laboratory to the industry.

6.4 Microalgal biofuel

Microalgae represent another recent promising source of biofuel (Chisti, 2008). They are found in moist habitats like surfaces of mud or water bodies. Minimum amount of wastewater, sunlight, or artificial light can produce algal bloom. Through photosynthetic respiration, they change CO_2 to O_2 and give rise to cellular energy and embed the energy in the form of cellular biomolecules of protein, lipid, and sugar (Hu et al., 2008). As carbohydrate, protein, and lipid constitute the main backbone of divergent biofuel and biogas, microalgal biomass enriched with the carbohydrate, protein, and lipid make them an ideal candidate for biofuel production. Advantages of selecting microalgae for biofuel include easy harvesting of energy-rich body of microalgae, inexpensive culture approaches, good capacity of CO_2 fixation, speedy growth rate throughout the year with little requirement of sunlight/artificial light, addition of oxygen to the environment, no clash with human or animal food chains, and their ability to grow on in any type of aqueous media such as freshwater, brackish water, waste water, and highly polluted water. Important microalgal species for biofuel production are *Chlamydomonas reinhardtii*, *Chlorella* sp., *Chlorella vulgaris*, *Chlorococcum* sp., *Botryococcus braunii*, *Ostreococcus tauri*, *Phaeodactylum tricornutum*, *Nannochloropsis* sp., *Symbiodinium* sp., *Phytoplanktons*, *Cyanobacterial mats*, *Saccharina japonica*, *Spirulina platensis*, and *Spirogyra* sp. etc. Microalgae have a short life cycle, and this short harvest time with dense biomass leads to higher productivity of the desired biofuel. Numerous types of biofuels (biodiesel, bioethanol, bio-hydrogen, bio-oil, biomethane, and others) were drawn out from microalgae (Brennan and Owende, 2010). Brennan and Owende (2010) reviewed all aspects of biofuel production from microalgae. Currently, scientists are conducting more research toward the sophistication of biofuel technologies from microalgae. Implications of nano-additives like the addition of nano-droplets, nano-fibers, nano-sheets, nano-tubes, nano-particles, and other nano-structures on biofuel production by microalgae are investigated by several scientists. The nano-additive application was found to boost biofuel production by microalgae (Hossain et al., 2019). However, more research is needed to understand the most suitable nano-additive for the biofuel. Other than advanced research on nano-additive research on algal biofuel, insights are needed to emphasize on policy drafting, understanding socioeconomic impact, careful analysis on pros and cons for the overall system to attract the government and nongovernment fuel industries. Microalgal biofuel is the most recent addition on renewable energy research, and so it known as third-generation biofuel. This microalgal fuel can accelerate further reduction of GHG emission. Biomass from microalgae can produce more oil (100 times more) than soybean or other crops grown on land can produce. It can supply 5000–20,000 gal oil per year per acre (7–30 times higher than crops).

7 Conversion of biomass to bioelectricity

Another utilization of biomass is bioelectricity production from renewable feed stocks. Agricultural biomass and forest residual biomass as a source for bioelectricity generation is getting popularity. The immediate raw material for bioelectricity is the lignocellulose feedstock whose combustion gives rise to bioelectricity. Lignocellulose is obtained through biomass of agricultural products (corn stover, straw sugarcane bagasse etc.) and forestry (saw mill and paper mill discards). Farine et al. (2012) showed the prospect of bioelectricity generation from the combustion of lignocellulosic biomass obtained from forestry and agricultural biomass and thus reduced greenhouse gas emission. In Australia, ongoing forest and agronomics contributed 15% of the country's total electricity generation and brought a great impact on GHG emissions (Farine et al., 2012). For United States, White et al. (2013) examined the prospect of integrating the forest and agriculture sectors and found that a synergistic effect on the production of renewable bioelectricity using simulated standards. According to this model, the joint sectors of agriculture and forest have the ability to deliver 10%–20% of the future electricity generation where bioenergy crops will supply the majority of the biomass feedstock. The model also estimated that with this trend about 27 million ton of CO_2 emission can be reduced in 15 years. In China also, bioelectricity generation from agricultural crops was studied. Here, a target of 30,000 GW electricity capacities by 2020 was set up (Clare et al., 2016). Jiang et al. (2017) reported another case study in Amsterdam, the Netherlands, on the bioelectricity generation from waste stream, and a model was set up to project future electricity making. These analyses played an important role in helping the policy-making to a sustainable electricity system and also gave stress on locally available biomass. Bioelectricity can be produced through thermochemical conversion method too including combustion, where biomass forms heat, water and carbon dioxide. Biomass is transformed into char and volatiles with the production of light and heat (volatile gases react with O_2 to produce heat). The heat is used to produce steam that is fed into the steam turbine to produce bioelectricity. The efficiency of electricity generation through steam turbine can be increased by the construction of different types of turbine blades or changing the operation modes of the steam turbine (Brown, 2011). Biomass gasification is equally suitable for bioelectricity generation. Biomass is converted into syngas and some other products like char, chlorides, tar, and sulfides through gasification. Here, the syngas is used for electricity production and gives better yield than combustion with lower quantities of air pollutants (Punia et al., 2017). Gasification can be applied for rural electricity using local biomass wastes. The synergism of the combustion boilers with gasifiers significantly enhances the electricity generation efficiency (~ 35%) (Brown, 2011). Another promising renewable technology for bioelectricity generation is the microbial fuel cell (MFC) technology (Moqsud et al., 2013). It is a bio-electrochemical bioengineered device that can give rise to a steady electric current when chemical energy released by microbes during oxidation of the complex carbon of biomass is converted to electrical energy. The electrons released during microbial metabolism are captured and connected through a circuit within an assembly called MFC. MFC has an abiotic cathode and a biotic anode separated by a proton exchange membrane. Bamboo charcoal (an environment-friendly material) is example of a good anode material, and carbon fiber is a good cathode material. Microbes oxidize the organic fuel (any carbon-rich waste biomaterial) at the anode and give rise to protons that penetrate through the membrane to the cathode. Then electrons after crossing the anode enter in an external circuit that generates a current. MFCs are into two general types: mediated and unmediated. In the mediated type, a chemical compound passes electrons from the bacterial cell to the anode. Unmediated MFCs discovered in the 1970s applied electrogenic bacteria that were able to use cytochromes (electrochemically active redox proteins) situated on the outer membranes of the bacterial cells to transfer electrons to the anode directly under anaerobic condition (Chatzikonstantinou et al., 2018). Chatzikonstantinou et al. (2018) showed that food residue biomass acted as a very good substrate in comparison with other types of biomass and the energy released during the hydrolysis of food was thought to be behind the increase. Akman et al. (2013) observed that $Ti-TiO_2$ electrode can serve better than platinum electrode. MFC has shown great prospect as a green and sustainable process of energy generation, and more research is needed to make this new technology acceptable for commercial biofuel companies.

8   Other valorization of heavy metal-contaminated biomass

8.1   Essential oils recovery from contaminated biomass of aromatic plants

Aromatic grasses, producers of essential oils, possess a feature of potential candidate for phytoremediation. As essential oil is plant secondary metabolite and plants enhance their secondary metabolite production under stress, it has been found that plants grown under stress increase the production of essential oil (Kumar and Patra, 2012). In addition to the financial contribution of the aromatic grasses toward the farmers for essential oil production, their unpalatable taste, perennial growth, cost-effectiveness for cultivation, minimum water requirement, tolerance against abiotic stresses (i.e., salinity, pH variation, drought, heavy metals toxicity) make themselves ideal for using in phytomanagement (Pandey and Singh, 2017; Verma et al., 2015). Presence of silica phytoliths and strong smell make the aromatic grasses toxic for being consumed by herbivores, and this reduces risk of food chain transfer of accumulated HM (Pandey and Singh, 2015; Verma et al., 2014). In addition, as hydro-distillation extracts only the essential oil from the aromatic grass, the essential oil is also safe to use to in perfume/medicinal industry (Verma et al., 2014, 2015). The essential oils have important industrial values having a wide range of uses in cosmetics, soaps, perfumes, aromatherapy, insect repellents, and medicines (Quintans-Júnior et al., 2008). Examples of important aromatic grasses are *Lavandula angustifolia*, *Lavandula officinalis*, *Ocimum basilicum*, *Matricaria recutita*, *Mentha piperita*, *Mentha arvensis*, *Thymus vulgaris*, and *Salvia officinalis*. They are used for remediation of heavy metal-contaminated soil without affecting their essential oils quality and yield (Pandey et al., 2019b). For example, the yields of essential oils from *Cymbopogon citratus*, *Cymbopogon martini*, *Cymbopogon flexuosus*, *Cymbopogon winterianus*, and *Vetiveria zizanioides* grown in HM-contaminated areas were higher than yields found when the plants were cultivated under normal unpolluted soil. However, slight changes in the biochemical composition and little HM presence were observed with plants grown under HM stress (Pandey et al., 2019a,b, 2020a,b,c; Lal et al., 2008, 2013). However, no effect on essential oil production in the roots of the *V. zizanioides* was observed due to compartmentalizations and HM-chelating mechanisms of the plant (Andra et al., 2010; Anjum et al., 2015; Pandey and Praveen, 2020). Some other essential oil-producing grasses are *V. zizanioides* (root), *C. citrates* (leave), *C. winterianus* (leave), *C. flexuosus* (leave), and *C. martini* (leaves and inflorescence). In all the cases, the HMs mainly were aggregated in below-ground parts and negligible HMs were translocated into the above-ground parts, indicating their phytostabilization abilities (Danh et al., 2009). Hence, the essential oils from the leaves can be safely extracted and used without post-treatment modification for HM removal.

8.2   Dye recovery from plants grown on contaminated lands

Dye-producing plants or dyeing plants are another important group of plants that give good industrial returns but themselves do not require special cares for growth or cultivation. Even they adapt to grow well comfortably in nonfarming lands such as on sands and wetlands as well as contaminated lands. Dyeing plants can resist abiotic stresses, climate changes, and soil nutrient scarcity. The dye products derived from the plants have a variety of applications. Some lichens and fruits, flowers, seeds, roots, bark, leaves, and whole body of certain plants produce dyes. Even different parts of the same plants show different colors and shades. Therefore, dye recovery from plants grown on contaminated sites can be used for dyeing textiles, leather, and wood. *Lawsonia inermis* is a dye-producing plant and has been assessed for phytoremediation potential on iron ore tailing (Chaturvedi et al., 2015). Dyeing plants, namely, *Calophyllum inophyllum* L., *Schleichera oleosa* (lour.) Oken and *Bixa orellana* L., were also evaluated for their phytoremediation potential on iron ore tailings (Chaturvedi et al., 2012). Therefore, dye-yielding plants should be screened for heavy metals' phytoremediation purposes with economic returns (in dye form). Other dyeing plants such as *Curcuma longa*, *B. orellana*, *Indigofera tinctoria*, and *Camellia sinensis* should be tested for potential uses in phytoremediation programs.

8.3 Bioenergy from farm waste via pyrolysis

As much as 90% of biomass can be generated from farm as farm wastes (Oasmaa et al., 2010). Biological processing of farm waste is challenging due to stricter regulations, economic feasibilities, and technical limitations. Once the technical and economic challenges are overcome, innovative ideas can be thought of to create various chances to recover profitable products from huge quantity of farm waste biomass. Anaerobic digestion would be energetically beneficial but would be dependent on climatic factors, variations in waste composition, and microbial partners. Bad odor, nutrient leaching, and low net energy inputs are some drawbacks of farm waste composting which researchers are trying to solve. On the other hand, pyrolysis of farm wastes can convert waste into wealth with lesser carbon footprint. The stored carbon fixed by the plant during photosynthesis is used as food, animal feed, fuel, and other essential commodities. Per unit mass of animal feed composed of about six to seven times plant biomass is consumed by animals in the farm houses and generates remarkable volume of animal manure as farm wastes (Badger and Fransham, 2006). The animal manure can be separated into $\geq$ 60% w/w solids and a liquid stream via chemical or mechanical process. The liquid stream can be treated by biological process to remove the toxins and pathogens and then added to the soil. This maintains water balance and adds up essential minerals and nutrients, particularly P and N. The solid biomass part of the manure could be transferred into a drying unit. The drying unit operated by pyrolytic gas further reduces moisture of the solid manure to $\leq$ 10% w/w. The bio-oil, by-product of pyrolysis of biomass from manure, could be used either for heating or for electricity generation for running the pyrolytic unit or sold to a bio-oil refinery network for purification and production of value-added products. The bio-char can be added to the soil to increase nutrient level and enhance other physicochemical characteristics of soil (porosity, density, and pH) (Abdullah and Wu, 2009). In addition, a high self-life of 100–1000 years of bio-char would capture a significant portion of atmospheric CO_2 in soil. This solution would retain a remarkable portion of the present GHG for a long term until a more integrated and comprehensive approach is designed. To validate this scheme, pyrolytic formation of farm wastes (e.g., plant residues, animal manure) is being subjected to life cycle analysis and techno-economic study. So far, only handful of researches was undertaken on pyrolytic transformation of farm wastes to combustible gases, bio-oil, biochar etc. (Cantrell et al., 2008). With increasing fossil fuel crisis, many countries have initiated/planned subsidies and working on biomass to energy options. In present days, although many business enterprises are based on forest biomass/residue, farm residue/wastes also should be regarded as an alternative biomass source. With the increasing consumption of farm goods, farm waste generation would increase, the transformation of which would lead to carbon sequestration, energy production, and other industrially valuable products generation and help to minimize the net use of fossil fuels.

9 Techno-economic assessment (TEA)

The major techno-economic barriers for the commercialization of the technology are (i) the long-time period needed for complete phytoremediation of the site and (ii) the cost burden at the experimental stage for the selection of optimum thermochemical process for valorization. So to make phytoextraction of HM-contaminated soil financially feasible and profit generating for farmers, options should be there for either bioenergy production or metal mining from the HM-laden biomass. A techno-economic assessment (TEA) is based on the quantification of biomass and assessed on new technologies applied for environmental remediation. TEA is essential to analyze the valorization potential of recent techniques to give proper idea of potential ambiguities to decision makers (Ma, 2011; Man et al., 2014; Thewys et al., 2010).

Some of the technical barriers toward TEA of thermochemical processes including pyrolysis are (i) unavailability of cost-benefit data as these thermochemical treatments are quite recent findings, (ii) occurrences of significant variations in the cost data analysis of pyrolysis and others by different plants, (iii) no large datasets are available, and (iv) certain uncertainty is present for the capital cost of processes for which insufficient information is available.

Kuppens et al. (2015) prepared a TEA and cost recovery model on thermochemical conversions of Cd-containing willow tree. It showed a positive net present value (NPV) indicating the technology's profitability. According to this study, willow was selected because of its well-known efficiency for Cd removal from Cd-contaminated soil. According to the results and interpretations, among different thermochemical conversion routes used, fast pyrolysis gave higher profitability than combustion and gasification. It was also revealed that the profit was highly dependent on the operation scale, oil yield, and subsidies of the policy support.

The financial risk can be decreased by (i) operation scale-up using complementing feedstock(s) (Cornelissen et al., 2009) and (ii) valorization of the bio-char by its consecutive transformation to activated carbon and direct use as soil nutrient (Kuppens et al., 2015), etc. The companies can also decrease the operational risk by adopting co-pyrolysis of miscellaneous suitable biomasses and increasing the scale of operation and outputs maximization and upgrading. Although current investigations are emphasizing fast pyrolysis as the most promising thermochemical conversion method of HMCB for the generation of HM free bio-oil, it should be remembered that many potential pre- and posttreatments processes should be carried out for the production of clean, green energy with minimized HMs transfer to products. So more research should be conducted on fast pyrolysis of HMCB to upgrade it from its current infancy stage to fully established one.

Further opportunities lie in the following areas: (i) More research is needed on potential sorbents like zeolites, char, fly ash, activated carbon, etc. (ii) More study should be conducted on demineralization and leaching behavior of the above sorbents due to their abilities to remove or immobilize dissolved HMs from HMCBs. These steps would create more chances to improve quantity and quality of bio-oil with enhanced HM recovery. (iii) Implementation of suitable posttreatment strategies is necessary that will reduce the risk of HMs dispensed from the reactor or reduce the amount of HMs in the biofuel produced. Accurate determination of the amount of the HMs and other pollutants before re-utilization of products/by-products should be practiced and existing national or international environmental regulations should be followed. (iv) The companies can increase profit by co-pyrolysis of various biomass (Abnisa et al., 2014), increasing the scale of operation (Cornelissen et al., 2009), outputs optimization/upgrading, etc. Further assessment regarding technical uncertainties of fast pyrolysis method of HMCBs is required through experimental researches. Quality of bio-oil and bio-char produced at various temperatures and the moisture conditions from HMCB should be checked to know the relation between changes in heating values and profitability.

Lamers et al. (2015) showed the importance of preparing biomass depots for decentralized biomass processing facilities to analyze feedstock quantity, quality, and cost necessary to develop the future bio-economy. They further made three different depot configurations, including specifications of conventional pelleting to advanced pretreatment technologies for explaining technical differences and economic performances. According to them, the integration of this technology should be aggressively pursued as the advantages of incorporating depots into the overall biomass feedstock supply chain will go beyond depot processing costs.

10 Summary and conclusion

In summary, phytoremediation and biofuel production can be tied up to meet growing energy demands and at the same time restore environment. Among all the heavy metals commonly present in contaminated phytoextracted biomass, the relatively nonvolatile ones (e.g., Cr, Ni, V) are contained in bottom ash and the more volatile metals (e.g., As, Cd, Hg, Pb) go into the FA, particularly into the smallest particles. Installation of an efficient particle removal system can contain the HMs and reduce their liberation directly into the atmosphere to a large extent. Through adaptive phytoremediation, a considerable part of the nutrients and minerals present in FA can be brought back to the soil. Among the valorization technologies for biomass production for biofuel, it can be said that thermochemical conversion processes have many advantages over "biological conversion processes" as there is no need of various pretreatment steps for lignocelluloses biomass making it more economical. Among the thermochemical processes, combustion and gasification are more favorable commercially for electricity/heat production due to the simplicity of operation. This popularity was partially due to the less information present about their role on sustainable development. Recently, it has been found that both the processes have limited utilizations as heat and energy-generating processes in comparison with pyrolysis. Pyrolysis is slowly gaining more popularity over combustion and gasification as it can contribute better to the sustainable development by generating more useful by-products like bio-char, bio-oil etc., which have miscellaneous uses as liquid fuels, energy, and chemical feedstock. The bio-char enhances soil properties and bio-oil has applications as heating and electricity generation without net energy requirement. Thus, the products of pyrolysis are cleaner, free from HM, and provide additional environmental benefits like bioremediation, carbon sequestration, and soil enhancement. However, the selection of the thermochemical process technology should depend on the end product desired and availability of the feedstock. Research is going to fulfill the drawbacks of the current technologies and enhance the economics and efficiency of the existing production technologies. Further researches and techno-economical assessments on potential HMCB conversion processes or pretreatment technologies are indispensable that would show distinct net energy generations and give cleaner carbon neutral biofuel.

References

Abbaszaadeh, A., Ghobadian, B., Omidkhah, M.R., Najafi, G., 2012. Current biodiesel production technologies: a comparative review. Energy Convers. Manag. 63, 138–148.

Abdullah, H., Wu, H., 2009. Biochar as a fuel: 1. Properties and grindability of biochars produced from the pyrolysis of mallee wood under slow-heating conditions. Energy Fuel 23 (8), 4174–4181.

Abnisa, F., Wan, D., Wan, M.A., 2014. A review on co-pyrolysis of biomass: an optional technique to obtain a high-grade pyrolysis oil. Energy Convers. Manag. 87, 71–85.

Abou-Shanab, R.A., Angle, J.S., Delorme, T.A., Chaney, R.L., van Berkum, P., Moawad, H., Ghanem, K., Ghozlan, H.A., 2003a. Rhizobacterial effects on nickel extraction from soil and uptake by *Alyssum murale*. New Phytol. 158 (1), 219–224.

Abou-Shanab, R.A., Delorme, T.A., Angle, J.S., Chaney, R.L., Ghanem, K., Moawad, H., 2003b. Phenotypic characterization of microbes in the rhizosphere of *Alyssum murale*. Int. J. Phytoremediation 5, 367–379.

Agar, D., Wihersaari, M., 2012. Bio-coal, torrefied lignocellulosic resources. Key properties for its use in co-firing with fossil coal-their status. Biomass Bioenergy 44, 107–111.

Akman, D., Cirik, K., Ozdemir, S., Ozkaya, B., Cinar, O., 2013. Bioelectricity generation in continuously-fed microbial fuel cell: effects of anode electrode material and hydraulic retention time. Bioresour. Technol. 149, 459–464.

Ali, Y., Hanna, M.A., Cuppett, S.L., 1995. Fuel properties of tallow and soybean oil esters. J. Am. Oil Chem. Soc. 72 (12), 1557–1564.

Andra, S.S., Datta, R., Sarkar, D., Makris, K.C., Mullens, C.P., Sahi, S.V., et al., 2010. Synthesis of phytochelatins in vetiver grass upon lead exposure in the presence of phosphorus. Plant Soil 326, 171–185.

Anjum, N.A., Hasanuzzaman, M., Hossain, M.A., et al., 2015. Jacks of metal/metalloid chelation trade in plants—an overview. Front. Plant. Sci. 6, 192.

Anjum, N.A., Sharma, P., Gill, S.S., Hasanuzzaman, M., Khan, E.A., Kachhap, K., et al., 2016. Catalase and ascorbate peroxidase-representative H_2O_2-detoxifying heme enzymes in plants. Environ. Sci. Pollut. Res. 23 (19), 19002–19029.

Assink, J.W., 1988. Physico-chemical treatment methods for soil remediation. In: Wolf, K., Van Den Brink, W.J., Colon, F.J. (Eds.), Contaminated Soil '88. Springer, Dordrecht, https://doi.org/10.1007/978-94-009-2807-7_138.

Azad, A.K., 2017. Biodiesel from mandarin seed oil: a surprising source of alternative fuel. Energies 10, 1689.

Badger, P.C., Fransham, P., 2006. Use of mobile fast pyrolysis plants to densify biomass and reduce biomass handling costs—a preliminary assessment. Biomass Bioenergy 30 (4), 321–325.

Bailis, R., 2009. Modeling climate change mitigation from alternative methods of charcoal production in Kenya. Biomass Bioenergy 33, 1491–1502.

Balsamo, R.A., Kelly, W.J., Satrio, J.A., et al., 2015. Utilization of grasses for potential biofuel production and phytoremediation of heavy metal contaminated soils. Int. J. Phytoremediation 17, 448–455.

Berg, C., 1999. World Ethanol Production and Trade to 2000 and beyond. The Online Distillery Network for Distilleries and Fuel Ethanol Plants Worldwide, Winchester, VA. Available from: http://www.distill.com/berg/.

Bert, V., Allemon, J., Sajet, P., Dieu, S., Papin, A., Collet, S., Gaucher, R., Chalot, M., Michiels, B., Raventos, C., 2017. Torrefaction and pyrolysis of metal-enriched poplars from phytotechnologies: effect of temperature and biomass chlorine content on metal distribution in end-products and valorization options. Biomass Bioenergy 96, 1–11.

Blaylock, M., Huang, J., 2000. Phytoextraction of metals. In: Raskin, I., Ensley, B.D. (Eds.), Phytoremediation of Toxic Metals: Using Plants to Clean up the Environment. J. Wiley & Sons.

Boman, C., Öhman, M., Nordin, A., 2006. Trace element enrichment and behavior in wood pellet production and combustion processes. Energy Fuels 20, 993–1000.

Borges, M.E., Díaz, L., 2012. Recent developments on heterogeneous catalysts for biodiesel production by oil esterification and transesterification reactions: a review. Renew. Sustain. Energy Rev. 16 (5), 2839–2849.

Brennan, L., Owende, P., 2010. Biofuels from microalgae—a review of technologies for production, processing, and extractions of biofuels and co-products. Renew. Sust. Energ. Rev. 14, 557–577.

Brewer, E., Saunders, J., Angle, J., Chaney, R., Mcintosh, M., 1999. Somatic hybridization between the zinc accumulator *Thlaspi caerulescens* and *Brassica napus*. Theor. Appl. Genet. 99, 761–771.

Bridgwater, A.V., 2012. Review of fast pyrolysis of biomass and product upgrading. Biomass Bioenergy 38, 68–94.

Brown, R.C., 2011. Thermochemical Processing of Biomass: Conversion into Fuels, Chemicals and Power. John Wiley & Sons, Ltd, Chichester.

Bundhoo, Z.M.A., Mauthoor, S., Mohee, R., 2016. Potential of biogas production from biomass and waste materials in the Small Island Developing State of Mauritius. Renew. Sustain. Energy Rev. 56, 1087–1100.

Cantrell, K.B., Ducey, T., Ro, K.S., Hunt, P.G., 2008. Livestock waste-to-bioenergy generation opportunities. Bioresour. Technol. 99 (17), 7941–7953.

Cao, P., Dubé, M.A., Tremblay, A.Y., 2008. High-purity fatty acid methyl ester production from canola, soybean, palm, and yellow grease lipids by means of a membrane reactor. Biomass Bioenergy 32 (11), 1028–1036.

Cargnelutti, D., Tabaldi, L.A., Spanevello, R.M., de Oliveira Jucoski, G., Battisti, V., et al., 2006. Mercury toxicity induces oxidative stress in growing cucumber seedlings. Chemosphere 65, 999–1006.

Carucci, G., Carrasco, F., Trifoni, K., Majone, M., Beccari, M., 2005. Anaerobic digestion of food industry wastes: effect of codigestion on methane yield. J. Environ. Eng. 131 (7), 1037–1045.

Catoire, L., Yahyaoui, M., Osmont, A., et al., 2008. Thermochemistry of compounds formed during fast pyrolysis of lignocellulosic biomass. Energy Fuels 22 (6), 4265–4273.

Chakraborty, S., Dutta, A.R., Sural, S., Gupta, D., Sen, S., 2013. Ailing bones and failing kidneys: a case of chronic cadmium toxicity. Ann. Clin. Biochem. 50 (5), 492–495.

Chalot, M., Blaudez, D., Rogaume, Y., Provent, A.S., Pascual, C., 2012. Fate of trace elements during the combustion of phytoremediation wood. Environ. Sci. Technol. 46, 13361–13369.

Chaturvedi, N., Dhal, N.K., Reddy, P.S.R., 2012. Comparative phytoremediation potential of *Calophyllum inophyllum* L., *Bixa orellana* L. and *Schleichera oleosa* (lour.) Oken on iron ore tailings. Int. J. Min. Reclam. Environ. https://doi.org/10.1080/17480930.2012.655165.

Chaturvedi, N., Dhal, N.K., Patra, H.K., 2015. An assessment of heavy metal remediation potential of *Lawsonia inermis* L. from iron ore tailings. Res. Plant Biol. 5, 1–8.

Chatzikonstantinou, D., Tremouli, A., Papadopoulou, K., Kanellos, G., Lampropoulos, I., Lyberatos, G., 2018. Bioelectricity production from fermentable household waste in a dual-chamber microbial fuel cell. Waste Manag. Res. 36. 0734242X1879693.

Chen, Y., Cheng, J.J., Creamer, K.S., 2008. Inhibition of anaerobic digestion process: a review. Bioresour. Technol. 99, 4044–4064.

Cheraghi, M., Khorasani, N., Yousefi, N., Karami, M., 2009. Findings on the phytoextraction and phytostabilization of soils contaminated with heavy metals. Biol. Trace Elem. Res. 144, 1133–1141.

Chiaramonti, D., Prussi, M., Buffi, M., Rizzo, A.M., Pari, L., 2017. Review and experimental study on pyrolysis and hydrothermal liquefaction of microalgae for biofuel production. Appl. Energy 185, 963–972.

China, S.P., Das, M., Maiti, S., 2014. Phytostabilization of Mosaboni copper mine tailings: a green step towards waste management. Appl. Ecol. Environ. Res. 12, 25–32.

Chisti, Y., 2008. Biodiesel from microalgae beats bioethanol. Trends Biotechnol. 26 (3), 126–131.

Christensen, T., Kjeldsen, P., Bjerg, P., Jensen, D., et al., 2001. Biogeochemistry of landfill leachate plumes. Appl. Geochem. 16, 659–718.

Christopher, L.P., Kumar, H., Zambare, V.P., 2014. Enzymatic biodiesel: challenges and opportunities. Appl. Energy 119, 497–520.

Clare, A., Gou, Y.-Q., Barnes, A., Shackley, S., Smallman, T.L., Wang, W., et al., 2016. Should China subsidize cofiring to meet its 2020 bioenergy target? A spatiotechno-economic analysis. GCB Bioenergy 8, 550–560.

Cornelissen, T., Yperman, J., Reggers, G., Schreurs, S., Carleer, R., 2008. Flash co-pyrolysis of biomass with polylactic acid. Part 1: influence on bio-oil yield and heating value. Fuel 87, 1031–1041.

Cornelissen, T., Jans, M., Stals, M., Kuppens, T., Thewys, T., et al., 2009. Flash co-pyrolysis of biomass: the influence of biopolymers. J. Anal. Appl. Pyrolysis 85, 87–97.

Cui, H., Turn, S.Q., Keffer, V., Evans, D., Tran, T., Foley, M., 2013. Study on the fate of metal elements from biomass in a bench-scale fluidized bed gasifier. Fuel 108, 1–12.

Cundy, A., Bardos, P., Puschenreiter, M., Witters, N., Mench, M., Bert, V., Friesl-Hanl, W., Muller, I., Weyens, N., Vangronsveld, J., 2015. Developing effective decision support for the application of "gentle" remediation options: the GREENLAND project. Remediat. J. 25, 101–114.

DalCorso, G., Manara, A., Furini, A., 2013. An overview of heavy metal challenge in plants: from roots to shoots. Metallomics 5. https://doi.org/10.1039/c3mt00038a.

Danh, L.T., Truong, P., Mammucari, R., Tran, T., Foster, N., 2009. Vétiver grass, *Vetiveria zizanioides*: a choice plant for phytoremediation of heavy metals and organic wastes. Int. J. Phytoremediation 11 (8), 664–691.

Dastyar, W., Raheem, A., Hec, J., Zhao, M., 2019. Biofuel production using thermochemical conversion of heavy metal-contaminated biomass (HMCB) harvested from phytoextraction process. Chem. Eng. J. 358, 759–785.

Davis, A., Shokouhian, M., Shubei, N., 2001. Loading estimates of lead, copper, cadmium, and zinc in urban runoff from specific sources. Chemosphere 44, 997–1009.

de Souza, M.P., Chu, D., Zhao, M., Zayed, A.M., Ruzin, S.E., Schichnes, D., Terry, N., 1999. Rhizosphere bacteria enhance selenium accumulation and volatilization by Indian mustard. Plant Physiol. 119 (2), 565–573.

Delplanque, M., Collet, S., Del, G.F., Schnuriger, B., Gaucher, R., Robinson, B., Bert, V., 2013. Combustion of Salix used for phytoextraction: the fate of metals and viability of the processes. Biomass Bioenergy 49, 160–170.

Demirbaş, A., 2000. Mechanisms of liquefaction and pyrolysis reactions of biomass. Energy Convers. Manag. 41, 633–646.

Demirbas, A., 2007. Recent developments in biodiesel fuels. Int. J. Green Energy 4 (1), 15–26.

Demirbas, A., 2009. Production of biofuels with special emphasis on biodiesel. In: Pandey, A. (Ed.), Handbook of Plant-Based Biofuels. CRC Press, Boca Raton, FL, pp. 45–54.

Dhiman, S.S., Selvaraj, C., Li, J., Singh, R., 2016. Phytoremediation of metal-contaminated soils by the hyperaccumulator canola (*Brassica napus* L.) and the use of its biomass for ethanol production. Fuel 183, 107–114.

Elekes, C.C., Dumitriu, I., Gabriela, B., Nicoleta, S., 2010. The appreciation of mineral element accumulation level in some herbaceous plants species by ICP-AES method. Environ. Sci. Pollut. Res. Int. 17, 1230–1236.

Elliott, D.C., Biller, P., Ross, A.B., Schmidt, A.J., Jones, S.B., 2015. Hydrothermal liquefaction of biomass: developments from batch to continuous process. Bioresour. Technol. 178, 147–156.

Epstein, H.E., Lauenroth, W.K., Burke, I.C., Coffin, D.P., 1996. Ecological responses of dominant grasses along two climatic gradients in the Great Plains of the United States. J. Veg. Sci. 7, 777–788.

Erdei, L., Mezôsi, G., Mécs, I., Vass, I., Fôglein, F., Bulik, L., 2005. Phytoremediation as a program for decontamination of heavy-metal polluted environment. Acta Biol. Szeged. 49, 75–76.

Evangelou, M.W.H., Deram, A., 2014. Phytomanagement: a realistic approach to soil remediating phytotechnologies with new challenges for plant science. Int. J. Plant Biol. Res. 2, 1023.

Fantozzi, F., Buratti, C., 2009. Biogas production from different substrates in an experimental continuously stirred tank reactor anaerobic digester. Bioresour. Technol. 100, 5783–5789.

Farine, D.R., O'Connell, D.A., John, R.R., May, B.M., O'Connor, M.H., Crawford, D.F., et al., 2012. An assessment of biomass for bioelectricity and biofuel, and for greenhouse gas emission reduction in Australia. GCB Bioenergy 4, 148–175.

Ghosh, S., 2010. Wetland macrophytes as toxic metal accumulators. Int. J. Environ. Sci. 1, 523–525.

Ghosh, M., Singh, S.P., 2005. A review on phytoremediation of heavy metals and utilization of it's by products. Appl. Ecol. Environ. Res. 3, 1–18.

Gomes, H.I., 2012. Phytoremediation for bioenergy: challenges and opportunities. Environ. Technol. Rev. 1 (1), 59–66.

Goyal, H.B., Seal, D., Saxena, R.C., 2008. Bio-fuels from thermochemical conversion of renewable resources: a review. Renew. Sustain. Energy Rev. 12, 504–517.

Guo, F., Zhong, Z., 2018. Pollution emission and heavy metal speciation from co-combustion of *Sedum plumbizincicola* and sludge in fluidized bed. J. Clean. Prod. 179, 317–324.

Guo, F., Zhong, Z., Xue, H., Zhong, D., 2017. Migration and distribution of heavy metals during co-combustion of *Sedum plumbizincicola* and coal. Waste Biomass Valorization 9, 2203–2210.

Han, Z., Guo, Z.H., Zhang, Y., Xiao, X.Y., Peng, C., 2018. Potential of pyrolysis for the recovery of heavy metals and bioenergy from contaminated *Broussonetia papyrifera* biomass. Bioresources 13, 2932–2944.

He, Y., Li, X., Xue, X., Switab, M.S., Schmidt, A.Z., Yang, B., 2017. Biological conversion of the aqueous wastes from hydrothermal liquefaction of algae and pine wood by Rhodococci. Bioresour. Technol. 224, 457–464.

He, J., Strezov, V., Zhou, X., Kumar, R., Kan, T., 2020. Pyrolysis of heavy metal contaminated biomass pre-treated with ferric salts: product characterisation and heavy metal deportment. Bioresour. Technol. 313, 123641.

Hossain, N., Mahlia, T.M.I., Saidur, R., 2019. Latest development in microalgae-biofuel production with nano-additives. Biotechnol. Biofuels 12, 125.

Hu, Q., Sommerfeld, M., Jarvis, E., Ghirardi, M., Posewitz, M., Seibert, M., et al., 2008. Microalgal triacylglycerols as feedstocks for biofuel production: perspectives and advances. Plant J. 54, 621–639.

Huber, G.W., Iborra, S., Corma, A., 2006. Synthesis of transportation fuels from biomass: chemistry, catalysts, and engineering. Chem. Rev. 106, 4044–4098.

Jiang, Y., Lei, M., Duan, L., Longhurst, P., 2015. Integrating phytoremediation with biomass valorisation and critical element recovery: a UK contaminated land perspective. Biomass Bioenergy 83, 328–339.

Jiang, Y., Ameh, A., Lei, M., Duan, L., Longhurst, P., 2016. Solid–gaseous phase transformation of elemental contaminants during the gasification of biomass. Sci. Total Environ. 563–564, 724–730.

Jiang, Y., van der Werf, E., van Ierland, E.C., Keesman, K.J., 2017. The potential role of waste biomass in the future urban electricity system. Biomass Bioenergy 107, 182–190.

Karimanal, K.V., Hall, M.J., 1996. Effect of temperature and flow on the volatilization of elemental lead and cadmium. Hazard. Waste Hazard. Mater. 13 (1), 63–71.

Keller, C., Ludwig, C., Davoli, F., Wochele, J., 2005. Thermal treatment of metal-enriched biomass produced from heavy metal phytoextraction. Environ. Sci. Technol. 39, 3359–3367.

Kim, T., Lee, Y.Y., 2005. Pretreatment of corn stover by soaking in aqueous ammonia. Appl. Biochem. Biotechnol. 121–124, 1119–1131.

Ko, C.H., Yu, F.C., Chang, F.C., et al., 2017. Bioethanol production from recovered napier grass with heavy metals. J. Environ. Manag. 203 (3), 1005–1010.

Koppolu, L., Agblevor, F., Clements, L., 2003. Pyrolysis as a technique for separating heavy metals from hyperaccumulators. Part II: lab-scale pyrolysis of synthetic hyperaccumulator biomass. Biomass Bioenergy 25, 651–663.

Kovacs, H., Szemmelveisz, K., 2017. Disposal options for polluted plants grown on heavy metal contaminated brownfield lands—a review. Chemosphere 166, 8–20.

Kumar, K., Patra, D.D., 2012. Alteration in yield and chemical composition of essential oil of *Mentha piperita* L. plant: effect of fly ash amendments and organic wastes. Ecol. Eng. 47. https://doi.org/10.1016/j.ecoleng.2012.06.019.

Kumar, P.B.A.N., Dushenkov, V., Motto, H., Raskin, I., 1995. Phytoextraction: the use of plants to remove heavy metals from soils. Environ. Sci. Technol. 29, 1232–1238.

Kuppens, T., Dael, M.V., Vanreppelen, K., Thewys, T., Yperman, J., Carleer, R., Schreurs, S., Passel, S.V., 2015. Techno-economic assessment of fast pyrolysis for the valorization of short rotation coppice cultivated for phytoextraction. J. Clean. Prod. 88, 336–344.

Lal, K., Minhas, P.S., Chaturvedi, R.K., Yadav, R.K., 2008. Extraction of cadmium and tolerance of three annual cut flowers on Cd-contaminated soils. Bioresour. Technol. 99 (5), 1006–1011.

Lal, K., Yadav, R.K., Kaur, R., Bundela, D.S., Khan, M.I., Chaudhary, M., Meena, R.L., Dar, S.R., Singh, G., 2013. Productivity, essential oil yield, and heavy metal accumulation in lemon grass (*Cymbopogon flexuosus*) under varied wastewater–groundwater irrigation regimes. Ind. Crop. Prod. 45, 270–278.

Lamers, P., Roni, M.S., Tumuluru, J.S., Jacobson, J.J., 2015. Techno-economic analysis of decentralized biomass processing depots. Bioresour. Technol. 194, 205–213.

Leijenhorst, E.J., Wolters, W., van de Beld, L., Prins, W., 2016. Inorganic element transfer from biomass to fast pyrolysis oil: review and experiments. Fuel Process. Technol. 149, 96–111.

Lievens, C., Yperman, J., Vangronsveld, J., Carleer, R., 2008. Study of the potential valorisation of heavy metal contaminated biomass via phytoremediation by fast pyrolysis: part I. Influence of temperature, biomass species and solid heat carrier on the behaviour of heavy metals. Fuel 87, 1894–1905.

Lievens, C., Carleer, R., Cornelissen, T., Yperman, J., 2009. Fast pyrolysis of heavy metal contaminated willow: influence of the plant part. Fuel 88, 1417–1425.

Liu, W.J., Tian, K., Jiang, H., Zhang, X.S., Ding, H.S., Yu, H.Q., 2012. Selectively improving the bio-oil quality by catalytic fast pyrolysis of heavy-metal-polluted biomass: take copper (Cu) as an example. Environ. Sci. Technol. 46, 7849–7856.

Lu, S., Du, Y., Zhong, D., Zhao, B., Li, X., Xu, M., Li, Z., Luo, Y., Yan, J., Wu, L., 2012. Comparison of trace element emissions from thermal treatments of heavy metal hyper accumulators. Environ. Sci. Technol. 46, 5025–5031.

Ma, J., 2011. Techno-Economic Analysis and Engineering Design Consideration of Algal Biofuel in Southern Nevada. https://digitalscholarship.unlv.edu/me_fac_articles/8.

Ma, F., Hanna, M.A., 1999. Biodiesel production: a review. Bioresour. Technol. 70 (1), 1–15.

Machado, M.D., Soares, H.M.V.M., Soares, E.V., 2010. Selective recovery of copper, nickel and zinc from ashes produced from *Saccharomyces cerevisiae* contaminated biomass used in the treatment of real electroplating effluents. J. Hazard. Mater. 184, 357–363.

Maestri, E., Marmiroli, M., Visioli, G., Marmiroli, N., 2010. Metal tolerance and hyperaccumulation: costs and trade-offs between traits and environment. Environ. Exp. Bot. 68 (1), 1–13.

Man, Y., Yang, S., Xiang, D., Li, X., Qian, Y., 2014. Environmental impact and techno-economic analysis of the coal gasification process with/without CO2 capture. J. Clean. Prod. 71, 59–66.

Mani, S., Tabil, L.G., Sokhansanj, S., 2003. An overview of compaction of biomass grinds. Powder Handl. Process 15 (3), 160–168.

Mayer, Z.A., Apfelbacher, A., Hornung, A., 2012. A comparative study on the pyrolysis of metal-and ash enriched wood and the combustion properties of the gained char. J. Anal. Appl. Pyrolysis 96, 196–202.

McKendry, P., 2002. Energy production from biomass (part 1): overview of biomass. Bioresour. Technol. 83 (1), 37–46.

Meers, E., Slycken, S.V., Adriaensen, K., Ruttens, A., et al., 2010. The use of bioenergy crops (*Zea mays*) for 'phytoattenuation' of heavy metals on moderately contaminated soils: a field experiment. Chemosphere 78, 35–41.

Meier, D., Faix, O., 1999. State of the art of applied fast pyrolysis of lignocellulosic materials—a review. Bioresour. Technol. 68, 71–77.

Mench, M., Bes, C., 2009. Assessment of ecotoxicity of topsoils from a wood treatment site. Pedosphere 19 (2), 143–155.

Mishra, J., Singh, R., Arora, N.K., 2017. Alleviation of heavy metal stress in plants and remediation of soil by rhizosphere microorganisms. Front. Microbiol. 8, 1706.

Moqsud, M.A., Omine, K., Yasufuku, N., Hyodo, M., Nakata, Y., 2013. Microbial fuel cell (MFC) for bioelectricity generation from organic wastes. Waste Manag. 33 (11), 2465–2469.

Mudhoo, A., Kumar, S., 2013. Effects of heavy metals as stress factors on anaerobic digestion processes and biogas production from biomass. Int. J. Environ. Sci. Technol. 10, 1383–1398.

Mukherjee, A., Agrawal, S.B., Agrawal, M., 2016. Heavy metal accumulation potential and tolerance in tree and grass species. In: Singh, A., Prasad, S., Singh, R. (Eds.), Plant Responses to Xenobiotics. Springer, Singapore.

Mukherjee, P., Roychowdhury, R., Roy, M., 2017. Phytoremediation potential of rhizobacterial isolates from Kansgrass (*Saccharum spontaneum*) of fly ash ponds. Clean Technol. Environ. Policy. https://doi.org/10.1007/s10098-017-1336-y.

Mullen, C.A., Boateng, A.A., 2013. Accumulation of inorganic impurities on HZSM-5 zeolites during catalytic fast pyrolysis of switchgrass. Ind. Eng. Chem. Res. 52, 17156–17161.

Nzihou, A., Stanmore, B., 2013. The fate of heavy metals during combustion and gasification of contaminated biomass—a brief review. J. Hazard. Mater. 256–257, 56–66A.

Oasmaa, Y., Solantausta, V., Arpiainen, E., Kuoppala, K.S., 2010. Fast pyrolysis bio-oils from wood and agricultural residues. Energy Fuels 24 (2), 1380–1388.

Padmavathiamma, P.K., Li, L.Y., 2007. Phytoremediation technology, hyperaccumulation metals in plants. Water Air Soil Pollut. 184, 105–126.

Pandey, V.C., Bajpai, O., 2019. Phytoremediation: from theory toward practice. In: Pandey, V.C., Bauddh, K. (Eds.), Phytomanagement of Polluted Sites. Elsevier, Netherlands, pp. 1–49.

Pandey, V.C., Bauddh, K., 2018. Phytomanagement of Polluted Sites. Elsevier, ISBN: 9780128139127, https://doi.org/10.1016/C2017-0-00586-4 (Edited Book).

Pandey, V.C., Maiti, D., 2020. Phragmites species—promising perennial grasses for phytoremediation and biofuel production (Authored book with contributors). In: Pandey, V.C., Singh, D.P. (Eds.), Phytoremediation Potential of Perennial Grasses. Elsevier, Amsterdam, pp. 97–114, https://doi.org/10.1016/B978-0-12-817732-7.00005-5.

Pandey, V.C., Praveen, A., 2020. *Vetiveria zizanioides* (L.) Nash—more than a promising crop in phytoremediation (Authored book with contributors). In: Pandey, V.C., Singh, D.P. (Eds.), Phytoremediation Potential of Perennial Grasses. Elsevier, Amsterdam, pp. 31–62, https://doi.org/10.1016/B978-0-12-817732-7.00002-X.

Pandey, V.C., Singh, N., 2015. Aromatic plants versus arsenic hazards in soils. J. Geochem. Explor. 157, 77–80.

Pandey, A.K., Singh, P., 2017. The genus Artemisia: a 2012–2017 literature review on chemical composition, antimicrobial, insecticidal and antioxidant activities of essential oils. Medicines 4 (3), 68.

Pandey, V.C., Souza-Alonso, P., 2019. Market opportunities in sustainable phytoremediation. In: Pandey, V.C., Bauddh, K. (Eds.), Phytomanagement of Polluted Sites. Elsevier, Netherlands, pp. 51–82.

Pandey, V.C., Bajpai, O., Singh, N., 2016. Energy crops in sustainable phytoremediation. Renew. Sust. Energ. Rev. 54, 58–73.

Pandey, J., Verma, R., Singh, S., 2019a. Suitability of aromatic plants for phytoremediation of heavy metal contaminated areas: a review. Int. J. Phytoremediation 21, 1–14.

Pandey, V.C., Rai, A., Korstad, J., 2019b. Aromatic crops in phytoremediation: from contaminated to waste dumpsites. In: Pandey, V.C., Bauddh, K. (Eds.), Phytomanagement of Polluted Sites. Elsevier, Netherlands, pp. 225–275.

Pandey, V.C., Mishra, A., Shukla, S.K., Singh, D.P., 2020a. Reed canary grass (*Phalaris arundinacea* L.)—coupling phytoremediation with biofuel production (Authored book with contributors). In: Pandey, V.C., Singh, D.P. (Eds.), Phytoremediation Potential of Perennial Grasses. Elsevier, Amsterdam, pp. 165–177, https://doi.org/10.1016/B978-0-12-817732-7.00007-9.

Pandey, V.C., Rai, A., Kumari, A., Singh, D.P., 2020b. *Cymbopogon flexuosus*—an essential oil-bearing aromatic grass for phytoremediation (Authored book with contributors). In: Pandey, V.C., Singh, D.P. (Eds.), Phytoremediation Potential of Perennial Grasses. Elsevier, Amsterdam, pp. 195–209, https://doi.org/10.1016/B978-0-12-817732-7.00009-2.

Pandey, V.C., Patel, D., Jasrotia, S., Singh, D.P., 2020c. Potential of napier grass (*Pennisetum purpureum* Schumach.) for phytoremediation and biofuel production (Authored book with contributors). In: Pandey, V.C., Singh, D.P. (Eds.), Phytoremediation Potential of Perennial Grasses. Elsevier, Amsterdam, pp. 283–302, https://doi.org/10.1016/B978-0-12-817732-7.00014-6.

Patel, D., Pandey, V.C., 2020. Switchgrass—an asset for phytoremediation and bioenergy production (Authored book with contributors). In: Pandey, V.C., Singh, D.P. (Eds.), Phytoremediation Potential of Perennial Grasses. Elsevier, Amsterdam, pp. 179–193, https://doi.org/10.1016/B978-0-12-817732-7.00008-0.

Peterson, A.A., Vogel, F., Lachance, R.P., Fröling, M., Michael, J., Antal, J., et al., 2008. Thermochemical biofuel production in hydrothermal media: a review of sub- and super critical water technologies. Energy Environ. Sci. 1, 32–65.

Praveen, A., Pandey, V.C., 2020. Miscanthus—a perennial energy grass in phytoremediation (Authored book with contributors). In: Pandey, V.C., Singh, D.P. (Eds.), Phytoremediation Potential of Perennial Grasses. Elsevier, Amsterdam, pp. 79–95, https://doi.org/10.1016/B978-0-12-817732-7.00004-3.

Pudasainee, D., Paur, H.R., Fleck, S., Seifert, H., 2014. Trace metals emission in syngas from biomass gasification. Fuel Process. Technol. 120, 54–60.

Pulford, I.D., Watson, C., 2003. Phytoremediation of heavy metal-contaminated land by trees—a review. Environ. Int. 29, 529.

Punia, R., Marlar, R.R., Kumar, S., Tyagi, S.K., 2017. Potential of bioelectricity production in India through thermochemical conversion of lignocellulosic biomass. In: Sustainable Biofuels Development in India. Springer International Publishing, Cham, pp. 189–206.

Qian, F., Zhu, X., Liu, Y., Shi, Q., Wu, L., Zhang, S., Chen, J., Ren, Z.J., 2018. Influences of temperature and metal on subcritical hydrothermal liquefaction of hyperaccumulator: implications for the recycling of hazardous hyper accumulators. Environ. Sci. Technol. 52, 2225–2234.

Quintans-Júnior, L.J., Souza, T.T., Leite, B.S., Lessa, N.M.N., Bonjardim, L.R., Santos, M.R.V., Alves, P.B., Blank, A.F., Antoniolli, A.R., 2008. Phytochemical screening and anticonvulsant activity of *Cymbopogon winterianus* Jowitt (Poaceae) leaf essential oil in rodents. Phytomedicine 15 (8), 619–624.

Rascio, N., Navari-Izzo, F., 2011. Heavy metal hyperaccumulating plants: how and why do they do it? And what makes them so interesting? Plant Sci. 180 (2), 169–181.

Reeves, R.D., Baker, A.J.M., Jaffré, T., Erskine, P.D., et al., 2017. A global database for plants that hyperaccumulate metal and metalloid trace elements. New Phytol. 218, 407–411.

Robak, K., Balcerek, M., 2018. Review of second generation bioethanol production from residual biomass. Food Technol. Biotechnol. 56 (2), 174–187.

Roy, M., Dutta, S., Mukherjee, P., Giri, A.K., 2015. Integrated phytobial remediation for sustainable management of arsenic in soil and water. Environ. Int. 75, 180–198.

Said, M., Cassayre, L., Dirion, J.L., Joulia, X., Nzihou, A., 2017. On the relevance of thermodynamics to predict the behaviour of inorganics during CO2 gasification of willow wood. In: Computer Aided Chemical Engineering. Elsevier, pp. 2671–2676.

Said, M., Cassayre, L., Dirion, J., Nzihou, A., Joulia, X., 2018. Influence of nickel on biomass pyro-gasification: coupled thermodynamic and experimental investigations. Ind. Eng. Chem. Res. 57, 9788–9797.

Sas-Nowosielska, A., Kucharski, R., Małkowski, E., Pogrzeba, M., Kuperberg, J.M., Kryński, K., 2004. Phytoextraction crop disposal—an unsolved problem. Environ. Pollut. 128, 373–379.

Scholz, V., Ellerbrock, R., 2002. The growth productivity, and environmental impact of the cultivation of energy crops on sandy soil in Germany. Biomass Bioenergy 23, 81–92.

Schuchardt, U., Sercheli, R., Vargas, R.M., 1998. Transesterification of vegetable oils: a review. J. Braz. Chem. Soc. 9 (1), 199–210.

Shiratori, Y., Yamakawa, T., Sakamoto, M., Yoshida, H., et al., 2017. Biogas production from local biomass feedstock in the Mekong delta and its utilization for a direct internal reforming solid oxide fuel cell. Front. Environ. Sci. https://doi.org/10.3389/fenvs.2017.00025.

Stals, M., Thijssen, E., Vangronsveld, J., Carleer, R., Schreurs, S., Yperman, J., 2010. Flash pyrolysis of heavy metal contaminated biomass from phytoremediation: influence of temperature, entrained flow and wood/leaves blended pyrolysis on the behaviour of heavy metals. J. Anal. Appl. Pyrolysis 87, 1–7.

Summers, A.P., Silver, S., 1978. Microbial transformation of metals. Annu. Rev. Microbiol. 32, 637.

Šyc, M., Pohořelý, M., Jeremiáš, M., Vosecký, M., et al., 2011. Behavior of heavy metals in steam fluidized bed gasification of contaminated biomass. Energy Fuel 25, 2284–2291.

Šyc, M., Pohořelý, M., Kameníková, P., et al., 2012. Willow trees from heavy metals phytoextraction as energy crops. Biomass Bioenergy 37, 106–113.

Tafur-Marinos, J.A., Ginepro, M., Pastero, L., Torazzo, A., Paschetta, E., Fabbri, D., Zelano, V., 2014. Comparison of inorganic constituents in bottom and fly residues from pelletised wood pyro gasification. Fuel 119, 157–162.

Tan, K.T., Lee, K.T., Mohamed, A.R., 2009. Production of FAME by palm oil transesterification via supercritical methanol technology. Biomass Bioenergy 33 (8), 1096–1099.

Tangahu, B.V., Abdullah, S.R.S., Basri, H., Idris, M., Anuar, N., Mukhlisin, M., 2011. A review on heavy metals (As, Pb, and Hg) uptake by plants through phytoremediation. Int. J. Chem. Eng. 2011, 939161.

Tanger, P., Field, J.L., Jahn, C.E., DeFoort, M.W., Leach, J.E., 2013. Biomass for thermochemical conversion: targets and challenges. Front. Plant Sci. 4, 218.

Tao, L., Tan, E.C.D., Aden, A., Elander, R.T., 2014. techno-economic analysis and life-cycle assessment of lignocellulosic biomass to sugars using various pretreatment technologies. In: Biological Conversion of Biomass for Fuels and Chemicals: Explorations from Natural Utilization Systems. The Royal Society of Chemistry, pp. 358–380 (Chapter 19).

Thewys, T., Kuppens, T., 2008. Economics of willow pyrolysis after phytoextraction. Int. J. Phytoremediation 10, 561–583.

Thewys, T., Witters, N., Meers, E., Vangronsveld, J., 2010. Economic viability of phytoremediation of a cadmium contaminated agricultural area using energy maize. Part II: economics of anaerobic digestion of metal contaminated maize in Belgium. Int. J. Phytoremediation 12, 663–679.

Tomović, G.M., Mihailović, N.L., Tumi, A.F., Gajić, B., Mišljenović, T.D., Niketić, M.S., 2013. Trace metals in soils and several Brassicaceae plant species from serpentine sites of Serbia. Arch. Environ. Prot. 39 (4), 29–49.

Unterbrunner, R., Puschenreiter, M., Sommer, P., Wieshammer, G., Tlustošb, P., Zupan, M., Wenzel, W.W., 2007. Heavy metal accumulation in trees growing on contaminated sites in Central Europe. Environ. Pollut. 148 (1), 107–114.

Vander Stelt, M.J.C., Gerhauser, H., Kiel, J.H.A., Ptasinski, K.J., 2011. Biomass upgrading by torrefaction for the production of biofuels: a review. Biomass Bioenergy 35, 3748–3762.

Verma, P., George, K.V., Singh, H.V., Singh, S.K., Juwarkar, A., Singh, R.N., 2006. Modeling rhizofiltration: heavy metal uptake by plant roots. Environ. Model. Assess. 11, 387–394.

Verma, S.K., Singh, K., Gupta, A.K., Pandey, V.C., Trivedi, P., Verma, R.K., et al., 2014. Aromatic grasses for phytomanagement of coal fly ash hazards. Ecol. Eng. 73, 425–428.

Verma, R.S., Pandey, V., Chauhan, A., Tiwari, R., 2015. Essential oil composition of *Mentha longifolia* (L.) L. collected from Garhwal region of western-Himalaya. J. Essent. Oil-Bear. Plants 18 (4), 957–966.

Vervaeke, P., Tack, F.M.G., Navez, F., Martin, J., Verloo, M.G., Lust, N., 2006. Fate of heavy metals during fixed bed downdraft gasification of willow wood harvested from contaminated sites. Biomass Bioenergy 30, 58–65.

Vintila, T., Negrea, A., Barbu, H., Sumalana, R., Kovacs, K., 2016. Metal distribution in the process of lignocellulosic ethanol production from heavy metal contaminated sorghum biomass. J. Chem. Technol. Biotechnol. 91, 1607–1614.

Wang, Y., Lü, J., Feng, D., Guo, S., Jianfa, L., 2020. A biosorption-pyrolysis process for removal of Pb from aqueous solution and subsequent immobilization of Pb in the char. Water 12, 2381.

Werle, S., Bisorca, D., Katelbach-Woźniak, A., et al., 2016. Phytoremediation as an effective method to remove heavy metals from contaminated area–TG/FT-IR analysis results of the gasification of heavy metal contaminated energy crops. J. Energy Inst. 90 (3), 408–417. https://doi.org/10.1016/j.joei.2016.04.002.

White, E.M., Latta, G., Alig, R.J., Skog, K.E., Adams, D.M., 2013. Biomass production from the U.S. forest and agriculture sectors in support of a renewable electricity standard. Energy Policy 58, 64–74.

Wiinikka, H., Carlsson, P., Johansson, A.C., et al., 2015. Fast pyrolysis of stem wood in a pilot-scale cyclone reactor. Energy Fuel 29, 3158–3167.

Wong, M.H., Bradshaw, A.D., 2002. The Restoration and Management of Derelict Land. Modern Approaches. World Scientific Publishing, Singapore, Japan.

Xu, J., Yin, H., Li, X., 2009. Protective effects of proline against cadmium toxicity in micro-propagated hyperaccumulator, *Solanum nigrum* L. Plant Cell Rep. 28, 325–333.

Yang, J., 2010. Heavy metal removal and crude bio-oil upgrading from *Sedum plumbizincicola* harvest using hydrothermal upgrading process. Bioresour. Technol. 101, 7653–7657.

Yu, C., Thy, P., Wang, L., Anderson, S.N., et al., 2014. Influence of leaching pretreatment on fuel properties of biomass. Fuel Process. Technol. 28, 43–53.

Zeng, K., Gauthier, D., Soria, J., Mazza, G., Flamant, G., 2017. Solar pyrolysis of carbonaceous feedstocks: a review. Sol. Energy 156, 73–92.

Zeng, K., Lia, R., Minh, D.P., et al., 2020. Characterization of char generated from solar pyrolysis of heavy metal contaminated biomass. Energy 206, 118128.

Zhang, L., Xu, C., Champagne, P., 2010. Overview of recent advances in thermo-chemical conversion of biomass. Energy Convers. Manag. 51 (5), 969–982.

Zheng, W., Ma, X., Tang, Y., Ke, C., Wu, Z., 2017. Heavy metal control by natural and modified limestone during wood sawdust combustion in a CO2/O2 atmosphere. Energy Fuel 32, 2630–2637.

Zhong, D., Zhong, Z., Wu, L., Ding, K., Luo, Y., Christie, P., 2016. Pyrolysis of *Sedum plumbizincicola*, a zinc and cadmium hyperaccumulator: pyrolysis kinetics, heavy metal behaviour and bio-oil production. Clean Techn. Environ. Policy 18, 2315–2323.

Policy implications and future prospects for adaptive phytoremediation practices

1 Climatic zones and potential shift under future global warming scenario

Climate is the weather conditions that prevail in some area over long period of time. A climatic zone is the world area distinguished from another by climatic characteristics. Climatic classification that remains even today was given by German botanist and climatologist Wladimir Köppen. His first scientific paper was in 1868, and later, he revised his work in 1900, 1918, and 1934. His major categories of climatic zones are based on the average temperature, the amount of precipitation, and the natural vegetation, i.e., each climatic zone corresponds to vegetation type (Köppen, 1936; Beck et al., 2018). However, nowadays classification of the main climatic zones presents modified Köppen-Geiger system (Fig. 1):

Tropical (type A climate)—this is warm, humid climatic zone with temperatures of coolest months 18°C: (a) wet equatorial climate (no dry season)/evergreen rain forest; (b) tropical monsoon (short dry season)/deciduous or monsoon forest; and (c) tropical wet-dry climate (winter dry season)/savanna forest or woodland and tropic grassland.

Dry (type B climate)—this is arid and semiarid climatic zone (subtropical) with temperatures above 20°C or between 10°C and 20°C, with low precipitation and high evaporates rates: (a) hyperarid/tropical and subtropical desert climate/"barren desert"; (b) warm arid/tropic, subtropic climate/steppe/desert savanna, desert short grass; and (c) warm semiarid/steppe, desert/thorn forest, desert savanna.

Temperate (type C climate)—this is warm, humid climatic zone with cool month < 18°C and hot month over 22°C: (a) subtropical climate with moist, hot summers/ warm temperate rain forest and prairie; (b) Mediterranean climate/sclerophyll woodland and scrub; and (c) tropical mountain climates and marine west coastal climate/ subantarctic forest.

Adaptive Phytoremediation Practices. https://doi.org/10.1016/B978-0-12-823831-8.00006-2

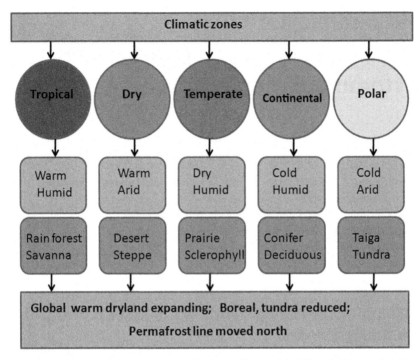

Fig. 1 Climatic zones with vegetation distribution and potential shift under future global warming scenario.

Continental (type D climate)—this is cold climatic zone with month temperatures between 10°C and 22°C, and warm to cool summers and very cold winters (−38°C): (a) humid continental climate/conifer forest/mixed deciduous and conifer forest; (b) continental subarctic climate/cold semiarid/arid desert.

Polar (type E climate)—this is polar climatic zone with temperatures of warmest month between 0°C and 10°C and low precipitation: (a) subpolar climate/taiga forest; (b) polar climate/tundra and snow/ice climate.

Global warming leads to the potential shift in world's climate zones with threat to water, food, and human health (Jylhä et al., 2010; Mahlstein et al., 2013; Li et al., 2021). Researchers noted the following: (a) the tropic climate is extending (Staten et al., 2018); the Sahara becomes larger (10% from 1920) (Thomas and Nigam, 2018); the 100th meridian has moved to east (140 miles from 1980) (Seager et al., 2018); the tornado corridor in the United States has moved to east (500 miles since 1980) (Agee et al., 2016); the plant hardiness zones in the United States have shifted northward, 13 miles per decade since 1990 (USDA, 2012); the permafrost in Canada has shifted to north (80 miles during 50 years) (Bush and Lemmen, 2019); and the wheat production is dropped due to drier than average conditions between 1990 and 2015 (Hochman et al., 2017).

Land area on the globe (5.7%) has expanded; i.e., it is moved to hot and dry climates, whereas tundra regions were reduced in size from 1950 to 2010 (Wang and

Overland, 2004; Chan and Wu, 2015). Jylhä et al. (2010) noted shifts of the Köppen climates from 1950 to 2006 in Europe and imply that from 1950, 12.1% of the land areas has shifted toward warmer and drier climate and it is five times greater than shift toward colder or wetter climate types, except tundra that had the largest percent of changes. Furthermore, 45%–70% of land area will be shifted in climatic zones in Europe: northeast/Alps/Iberia/Black Sea toward hot and dry climates which depends on greenhouse emissions' scenario. It was projected that in the next decades (2040–69 and 2070–99), tundra or boreal climates will shrink whereas temperate and dry climate will expand which will have effects on the ecosystem services and socioeconomic relations (Jylhä et al., 2010) (Fig. 1). According to Mahlstein et al. (2013), shifts in climates raise with temperature; i.e., these shifts will happen in some area with increasing global temperature for 2°C, such as climates at northern latitudes and mountainous regions, whereas polar climate, such as tundra, seems to show no change. Li et al. (2021) used root-layer soil moisture for the projection of potential shift in climates under warming scenario. They showed that by the end of 21st century, global drylands will experience expansion for 10.5%, indicating the high environmental impact and ecosystem vulnerability.

2 Current scenario

2.1 Increase in polluted sites

The global environmental pollution derived from human activities remains the world's greatest issue which humanity is facing. Fossil fuel combustion and the use of fertilizers, pesticides, and chemicals contribute to soil, air, and water pollution (Ukaogo et al., 2020). There are more than 450,000 contaminated sites in America (USEPA, 2017), 342,000 polluted sites in Europe (EEA, 2014), and more than millions of hectares of soil affected by contamination in China (MEP, 2014). Emissions from fossil fuels increased from 6 to 9 billion tonnes per year during 1990–2010, whereas there are more than 100,000 chemicals that are commercially available in Europe and global market (UNEP, 2012a,b). Due to improper agricultural practice, at least 1 million people are unintentionally poisoned every year (FAO, 2013). According to Panagos et al. (2013) in the European Union (EU), 3 billion tonnes of solid waste are dumped among which around 90 million tons are hazardous.

Soil contamination is a result of industrial activities (mining, oil extraction, and thermal power plant), waste disposal (municipal and industrial), storage (oil storage and obsolete chemicals), transport (spill on land), military (explosives), textile industry wastewater, electroplating wastewater, nuclear activities, etc. (Majhi et al., 2021; Mishra et al., 2020; Panagos et al., 2013; Ukaogo et al., 2020). Furthermore, the common contaminants are metal(loid)s (Ag, Al, As, Cd, Cr, Cu, Co, Cs, Hg, Mn, Mo, Ni, Pb, Se, St, U, and Zn); polycyclic aromatic hydrocarbons (PAH); chlorinated hydrocarbons (CHCs); oils, phenols, and aromatic hydrocarbons (benzene, toluene, etc.); and cyanides (McIntyre, 2003; Zhang et al., 2005; Panagos et al., 2013). For 2011–12, it was estimated potential contaminated sites using the

European Environment Information and Observation Network for soils (EIONET-SOIL) and it was registered more than 2,500,000 polluted sites (Panagos et al., 2013). In Europe, soil contamination is derived from the municipal and industrial waste disposal (37.2%), industrial sector (33.3%), storage sector (10.5%), transport spills (7.9%), armed forces (3.4%), and nuclear actions (0.1%) (Panagos et al., 2013). The contributors of soil contamination with percent contamination are as follows: metal industry (13%), chemical, oil industry, energy generation (6%–8%), food industry (6%), glass, ceramics, textile, leather, wood and paper industry (2%–4%), electronic industry (1%); service sector contributes to soil contamination: gasoline stations (14%) and car services (5%); and mining sector (6%) (Panagos et al., 2013). Furthermore, the main contaminant categories that contribute to soil pollution are metal(loids) (34.8%), mineral oil (23.8%), PAH (10.9%), phenols (10.2%), CHCs (8.3%), aromatic hydrocarbons (10.2%), phenols (1.3%), and cyanides (1.1%) (Panagos et al., 2013).

According to Schaider et al. (2007) due to mining processing around the world, 5 billion tonnes of tailings/per year are produced and 2 billion tonnes of ore are in the flotation process using around 4 million tons of chemicals. The beneficiation process of around 205 million tonnes of Cu ore requires 383,000 million flotation chemicals to generate 4.2 million tonnes of materials, and 61,000 tonnes of flotation chemicals are needed to generate 21.5 million tonnes of Fe materials (Plumbo-Roe et al., 2009).

The plastic products (polypropylene, polyethylene, polystyrene, polyamide, and polyesters) used in shopping and food storing items, packed in plastic bottles/boxes, paints, cosmetics, etc. (Auta et al., 2017), and according to Tsiamis et al. (2018), in the United States between 1960 and 2013 the plastic products was 8238%. Microplastics (MP, 0.3–4.7 mm) have been registered as a main threat to marine/coast region (Ukaogo et al., 2020).

The inorganic and organic pollutants have negative effects on human health due to their dispersion, bioavailability, and toxicity (Panagos et al., 2013). They increase cancers (lymphoma, leukemia, liver, breast) and affect the central nervous system, kidneys, heart, lungs, skin, and reproductive systems provoking depression, paralysis, and death (Gossel, 1984; Badawi et al., 2000).

Air pollution from industrial and urban areas derived from road traffic releases gaseous pollutants, such as sulfur oxide (SO_2), nitrogen oxides (NO, NO_2), volatile organic compounds (VOCs), carbon monoxide (CO), PAHs, ozone (O_3), and total suspended particulate matter (TSP) and particulate matter in diameter $< 10 \mu m$ (PM_{10}) and particulate matter in diameter $< 2.5 \mu m$ ($PM_{2.5}$) (Begg et al., 2007; Thurston, 2008; Wang et al., 2015). Air pollutants have different effects and distribution depending on source, types, and weather conditions (Bell and Treshow, 2002). The concentrations of air pollutants in urban areas still exceed public health standards causing pulmonary and cardiac diseases, autism, and death of around 8 million people every year (Saravia et al., 2013; Volk et al., 2013; WHO, 2014). Around 4.2 million deaths annually are related to ambient $PM_{2.5}$ originating from transport, households, industry, energy production, agriculture, forest fires, etc. (Shaddick et al., 2020).

2.2 Status of remediation

Site contamination presents a serious threat to the environment and human health, and the cleanup includes a preliminary assessment of contaminated sites, their inspection, hazard ranking system, placement on the National Priorities List, selection of potential remediation technologies, and long-term site cleanup maintenance (USEPA, 2011). Hazardous waste sites include manufacturing facilities, processing plants, landfills, and mining areas.

United States. In the United States, the law that allows EPA to cooperate with governments to clean up hazardous waste sites is officially called CERCLA (Comprehensive Environmental Response, Compensation, and Liability Act of 1980), and informally called Superfund. This law makes responsible parties pay to perform work at Superfund sites. According to Voltaggio and Adams (2016), around 70% of Superfund cleanup activities are paid by potentially responsible parties ("polluter pays" principle), but 30% of the time the responsible party cannot be found or is unable to pay for cleanup, and instead of them, taxpayers pay for remediation process. Potentially responsible parties (PRP) such as people, companies, and municipalities can be liable for contamination at a Superfund site, and they can be classified as current owner/operator of the site who are responsible for occurrence, transportation, and managing of the disposal of hazardous pollutants at the site (USEPA, 2011). In 2021, there were noted about 1327 hazardous sites for cleanup, 438 had been delisted, and 43 new sites put on the list.

To manage contaminated sites, EPA uses the following: (a) *removal action* (short-term response classified as emergency/time-critical/nontime-critical) is used for the elimination of immediate risk for public safety (chemical releases in environment, such as oil spills); (b) *remedial actions* (long-term response requires several years to study the problem and find the solution and then clean up the sites) are used to reduce risk of contamination in order to remove/treat/neutralize hazardous substances.

When a site is added to the National Priorities List, Remedial Investigation/ Feasibility Study (RI/FS) is required. The RI involves data collecting and analysis of samples of soil, water, and waste from locations throughout the site and includes assessing the risk for the site, whereas the FS involves analysis of various remediation alternatives. After that, site enters into the Remedial Design phase (cleanup method selection and specification) and Remedial Action phase (starting cleanup at a site). EPA provides public comments and develops Community Involvement Plan (CIP) to inform the public of the availability of documents in archives (USEPA, 2011). Furthermore, EPA has the *CERLA Information system* that presents a database of sites and consists of *the Removal/Remedial/Enforcement Inventory*; *The Superfund Innovative Technology Evaluation program*; *The reportable quantity; The source of control action*; *The section 104(e)* that presents a request by the government for information about the contaminated site; *The section 106 order* is the order for performing remedial action when actual release of contaminant in the environment is a threat to public health; *The remedial response* presents a long-term action that can stop/reduce the release of hazardous substances (CERCLA, 2002; CERCLA and EPCRA, 2018).

Making each Superfund cleanup "greener" is the process of minimizing waste production, conserving natural resources, and reducing energy consumption to protect environment/human health, i.e., decrease footprint and maximize the environmental outcome of cleanup (USEPA, 2011). EPA also should provide returning sites to productive use (industrial/commercial) or redevelopment of the site added to ecological/economical/community benefits from restoring the site to productivity. EPA includes *Five-Year Review* of inspecting and monitoring a site and notifies the community about that. Deleting a site from the NPL is finally the cleanup goal that is required for the protection of environment and human health (USEPA, 2011).

The average cleanup costs ("foot the bill", FTB) by type of site in millions of dollars are as follows: (a) mining (170.4), radiological tailings (75.4), chemical manufacturing (41.1), wood preserving (40.6), leaking container (34.4), waste oil (32.3), landfill (29.0), electrical (26.4), drum recycling (19.9), manufacturing (13.5), and metalworking (13.0) (Probst et al., 1995).

Europe. In Europe, cleanup costs, damage of sites, and the potential responsible parties are the same as in the United States, but there are significant differences in how the liability regime works (McGuigan, 2000). The proposed liability regime does not include guidance for remedies, but it has criteria that will need to be developed according to different types of damage. It involves employment of the best available technologies to clean up sites to fit actual/future economic/technical use. In Europe, the number of liability sites that will fund the cleanups depends on the number of cases that plaintiffs bring to the courts, and if the courts do not rule in favor of the public, the responsible parties make no commitment to clean up the sites, as Superfund does. Also, the proposed European liability regime will not be retroactive, as Superfund does (McGuigan, 2000). However, the proposed European liability regimes employ responsibility for damages triggered by dangerous activities. Damage to biodiversity is defined as area that is protected under Natura 2000 network (Flora/Fauna/Habitats Directives). European liability regimes would cover damage to health and property (McGuigan, 2000).

According to Panagos et al. (2013), there are 58,336 remediation sites in Europe (33 European countries). The cost of remediation of contaminated sites is shared by private/public money and requires estimation of annual cost for remediation and according to proposed European directive that can be from 2.4 to 17.3 billion Euros (EC, 2006). The management of polluted sites in 11 European countries can be assessed to around 6.526 million Euros annually. Furthermore, 42% of the total cost is funded by public account whereas 58% comes from private investments. The majority of the budget is spent on the remediation measures (80.6%); 15.1% is used for site examination and 4.3% on site after care and redevelopment (Panagos et al., 2013). It is calculated that the average amount per remediation site annually is around 37.1 thousand Euros and if remediation lasts more than 1 year, 40% of the remediation sites spend 50,000–500,000 Euros (Panagos et al., 2013).

China. Ma et al. (2018) estimated of current remediation status in China should be supported with who, when, what, why, which, where, and how ideology according to comprehensive legislation to handle finance and liability of remediating polluted soils. Risk assessment of polluted sites is based on the guidelines from China's Ministry

Department, EPA, and local governments. The remediation labor includes various technologies to treat contaminated sites: landfill (14%), cement kiln (37%), solubilization/stabilization (18%), and mechanical soil aeration (7%). About 72% of remediated sites were connected to community and business evolvement, and 52% of that was funded by developers, 26% is funded from government and 22% from owners. According to Qu et al. (2016), redevelopment of contaminated sites is in residential areas (abandoned factories); i.e., the most remediation actions are based on changing the use of the land.

2.3 Policy implications

Climate change policy—Greenhouse gas emissions (GHG) derived from human activities cause climatic change with physical and socioeconomic consequences: decreased productivity (lower agricultural yields and lack of water resources); physical damage to human environments from increased storms and floods; and risk to human life (people's death from contaminated water, heat waves, and storms). The Organization for Economic Co-operation and Development (OECD) contributes to the implementation of climatic change policies and indicates that reductions in GHG emissions can be achieved at a low cost with effective and right policy (OECD, 2007). First World Climate Conference was held in 1979, whereas the First Assessment Report of the IPCC (Intergovernmental Panel on Climate Change) was completed in 1990 where it is noted that human activities might affect the climate. In 1992, UNFCC (United Nations Framework Convention on Climate Change) in Rio de Janeiro (Brazil) stated that GHG emissions should be stabilized preventing further damage and nations worldwide should cooperate to improve human adaptation and mitigation of climate changes through financial support and low-emission technologies highlighted the human responsibilities in response measures (Shorgen and Toman, 2000). The Kyoto Protocol of the Framework Convention in 1997 states that industrialized countries (United States, western Europe, and Japan) agreed to legally reduce the GHG emissions by 5% below 1990 levels before 2008 with "flexibility mechanisms" (Shorgen and Toman, 2000). Post-Kyoto Protocol meetings debated about technical, legal, moral foundations of the proposed flexibility mechanisms (Shorgen and Toman, 2000).

Policy instruments can reduce GHG emissions by using carbon and energy taxes, removing environmental dangerous subsides (energy and transport), and using tradable permit schemes (OECD, 2007). Many countries participate in project-based mechanisms, such as Clean Development Mechanisms and Joint Implementation which permit companies in industrialized countries to earn emission credits by investing in projects with the decrease of GHG emissions in other countries. Furthermore, innovative solutions in the reduction of GHG emissions require greater policy attention to accelerate "green technologies" in research and development (R & D) programs. Introduction to climate change concerns in public policy can strengthen the co-benefits of GHG mitigation actions which can involve improved urban air quality and human health (OECD, 2007). Low-carbon economy includes policy options where recycling tax revenues are given back to the sectors or green tax reform may be used for environmental, economic, and public benefits by using the revenues from carbon /energy taxes to decrease taxes on employment (OECD, 2007). Integration of climate change and energy policy presents investments that "lock in" the infrastructure, fuel, and technologies as well as improve vehicle energy efficiency through pricing mechanisms. In addition, policies to reduce environmental harmful agricultural and waste subsides (landfill gas recovery, animal manure, and fertilizer management) can reduce GHG emissions (OECD, 2007). Introduction of climate change adaptation in public policy is a process which involves awareness, and it involves legal and institutional policy changes. Global price on carbon is crucial for reduction in GHG emissions, and global cooperation can be achieved by monitoring emission reductions, reporting and compliance extending international and national initiatives. A key element in national negotiations for reduction of GHG emissions in post-2012 international frameworks will be participation of all major emitters in the coming decades (OECD, 2007).

Pollution policy—In environmental management, pollution prevention includes prevention, recycling, treatment, and disposal/release; i.e., pollution prevention means source reduction or if it cannot be prevented than it should be recycled/treated/disposed in an environmentally safe manner (EPA, 1992). Contamination prevention includes reduction in the amount of hazardous substances/contaminants into environment and reduction of hazard to public health. The main objectives of EPA's Pollution Protection (P2) Program Strategic Plan are as follows: (a) reducing the emissions of GHG to mitigate climate changes; (b) reducing the manufacturing and use of hazardous substances to protect environment and human health; (c) conserving natural resources; (d) creating greener and more sustainable economy; and (e) integrating pollution prevention practice through institution, government, and public services (EPA, 2010). This Plan includes a 5-year strategy in five sectors: Chemical & Manufacturing Industries, Hospitality, Electronics, Building & Construction, and Municipalities & Institutions (EPA, 2010). P2 programs promote "green" chemistry, products, and processes and help the public make green choices easier to understand, involving competence in risk management, source reduction, etc.

In Europe, a number of regulations, instruments, and specific legislations were introduced in order to improve air quality; they especially address emissions from road transport and industrial sectors (power plants, refineries, and manufacturing) that release mostly PM, CO, NO_x, SO_2, O_3, and VOC (EEA, 2010). Despite the Gothenburg Protocol to the UNECE Convention on Long-range Transboundary Air Pollution (UNECE, 1979) and National Emission Ceilings Directive (NEC, 2009) with national limits of key air pollutants in EU, air quality could be further improved. Health benefits in all EEA member countries can be achieved by reduction of PM concentrations from road transport and industrial combustion (the air quality impact indicator, years of life lost, YOLL is 1%–10%) (EEA, 2010). Furthermore, there are three legislations for reducing the emissions of key air pollutants: (a) Euro standards for road vehicles (EC, 2007); (b) EU Large Combustion Plant (LCP) Directive (EC, 2001); and (c) EU Integrated Pollution Prevention and Control (IPPC) Directive (EC, 1996). In road transport sector, diesel vehicles, two-wheelers, and heavy-duty vehicles were targeted. In large combustion plants, a directive was included, the concept of "best available techniques" that are the most effective in protection of environmental and human health (BAT, Remus et al., 2012). Policy response addressing the gaseous air pollutants and PM in order to prevent damages of ecosystems, soils, and cultural heritage and harm to human health can be observed within DPSIR logic (fuel use, emissions, concentrations, and impact on health and ecosystems) (EEA, 2010).

According to Clean Air Scenario, the emissions of SO_2, NOx, PM and NH_3 could be reduced until 2040 with adequate environmental, climate, energy, agricultural, and food policy interventions (Amann et al., 2020). Clean Air Scenario delivers co-benefits in Sustainable Development Goals (SDG):

- Improve human health (SDG 3)
- Mitigate climate change (SDG 13)
- Provide clean energy (SDG 7)
- Enhance nitrogen and phosphorous use efficiency (SDG 2—Zero hunger; SDG 14—Life under water; SDG 15—Life on land; SDG 12—Responsible consumption and production; SDG 13).

Appropriate Cleanaway involves a multisectoral management approach with coordination in energy, climate, agricultural, and food policy domains and includes capital three Cs (Commitment, Capability, and Compatibility) to link people, markets, assets, and finance in the form of formal learning and couching, building organization capabilities, recruitment and selection, and development through challenging assignments (Amann et al., 2020).

3 Climate change effects and adaptive phytomanagement

Climate shift possesses a significant threat to the natural environment and human communities. The increase of GHG emissions predominantly due to combustion of fossil fuels raises the average temperature of atmosphere and concentrations of CO_2, changes the sea level, the amount of rainfall, increases severe heat waves, drought, floods, high-intensity storm, and soil erosion (USEPA, 2011) (Fig. 2). The acceleration of global warming affects food security, depletion of agricultural resources, and production whereas natural ecosystem has a limited capacity to manage all these multiple pressures (Singh, 2017; Dhankher and Foyer, 2018). Water, air, and soil pollution together with climate changes have impact on crop yields, human health, food security, and ecosystem management on the globe (USEPA, 2010; Raza et al., 2019) (Fig. 2).

Climate change and environmental extremes directly affect the agricultural production which cause biochemical, physiological, morphological, and phenotypic plant changes and show a negative impacts on soil fertility, carbon sequestration, and microbial activity having socioeconomic effects on the policy, costs, trade, and food distribution (Dhankher and Foyer, 2018; Raza et al., 2019). In addition, effective emission reduction is achieved by green remediation strategies and engineering evaluation/cost analysis with main goals to minimize/eliminate source of pollution, conserve natural resources, and maximize the environmental outcome of a cleanup (Pavlović et al., 2004; Pandey, 2012, 2013, 2015, 2020, 2021; Pandey et al., 2012, 2016; Gajić et al., 2016, 2018, 2019, 2020a,b; Gajić and Pavlović, 2018; Pandey and Bauddh, 2019).

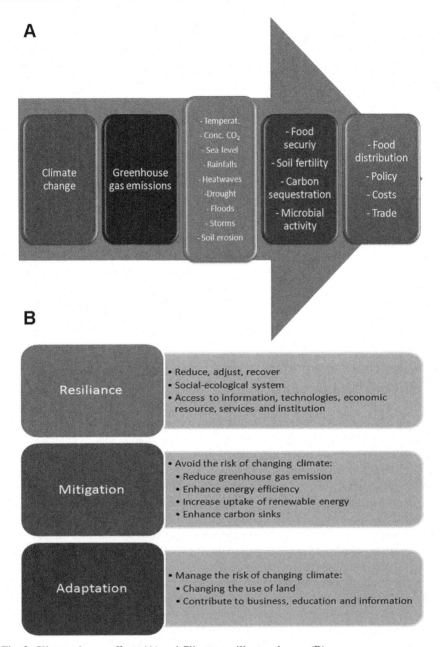

Fig. 2 Climate change effects (A) and Climate-resilient pathways (B).

Climate-Resilient pathways—Risk reduction of climate changes to natural resources (water, land, biodiversity) and human health involves climate-resilient actions, i.e., portfolio of adaptation, mitigation and sustainable development applications, policy, and frameworks (Denton et al., 2014). Vulnerability to climatic change is a combination of exposure to hazards and sensitivity to impacts and effects (Denton et al., 2014). Building *resilience* to climatic change presents the ability of socio-ecological system to reduce, adjust, or recover from the hazards by wide range of actions and ensure that vulnerable communities have reliable access to information, technologies, economic resources, and essential services and institutions (Munasinghe and Swart, 2005; Denton et al., 2014). According to Folke (2006), *resilience* is defined as a process of innovation and development that continuously allows flexibility to adapt to extreme events by social awareness, resource access to scientific and technological expertise and alternatives for problem solving, monitoring of climatic impacts, and contribution of policy, regulatory, and legal frameworks. Climatic change responses include mitigation and adaptation that can be recognized as important for sustainable development (Riahi, 2000).

Climatic change *mitigation* works to avoid the risk of changing climate by reducing climate-related stresses, such as emission of greenhouse gases, enhancing energy efficiency, increased uptake of renewable energy, improvements of the industrial processes, increased uptake of sustainable transport, and enhanced carbon sinks (forest, soil, oceans, atmosphere), and all these depend on levels of development, human, financial and technological capital, education and skills base, and the ability to adopt new technologies (Winkler et al., 2007; Denton et al., 2014).

Climate change *adaptation* works to manage the risks by changing the use of land, upgrading buildings and infrastructure, adjusting human activities and lifestyle, contributing to emergency and business planning, education, and information, and all these depend on adaptive capacity to respond to climatic impact and transformational adaptation policy, and they can promote and support sustainable development (Denton et al., 2014). Some actions may have mitigation and adaptation benefits as win-win and triple-win options by selecting materials that are good for the environment, promoting the conservation of energy and water, reuse, recycling, and decreasing waste production and protecting environment and human well-being (UNFCCC, 2011). However, the best of option on transformation and adaptation requires to be ethical and sustainable (O'Brien, 2012).

Adaptive management frameworks—Adaptive environmental management presents an multidisciplinary approach to improve ecosystem services by learning from outcomes of policy and practices (experience, experimenting, and monitoring) contributing to ecological resilience (Bormann et al., 1999; Bunnell et al., 2007; CEAA, 2009) (Fig. 3). According to Allen et al. (2011), the adaptive management cycle consists of (a) decision making—define the problem, identify objectives, formulate criteria, estimate outcomes, evaluate trade-offs, and decide—and (b) learning to reduce key uncertainties and knowledge gaps—implement, monitor, evaluate, and adjust.

Passive adaptive management includes management strategy, modeling, implementation, monitoring, and evaluation where managers implement the "best" option and monitor to see if they are appropriate and then make adjustments (Walters, 1986)

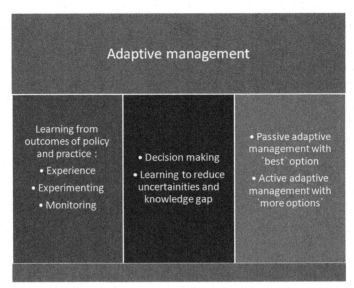

Fig. 3 Adaptive environmental management.

(Fig. 3). Active management implements more alternatives testing different hypotheses where alternatives can be compared more confidently (Linkov et al., 2006; Smith, 2008) (Fig. 3). According to Farr (2000), decision makers and different stakeholders should work together in order to design a management plan and test new options. Adaptive management success model depends on stakeholders' support, progress in resource objectives, monitoring and assessment, and implementation of the model that should be consistent with law and policy to create "win-win" options whenever possible (Williams et al., 2007; Owens, 2009). Adaptive management provides trade-offs through space and time that allow managers to learn about the relationship between system process and different potential suites for ecosystem services that can be important for climate change resilience and available for crop production (Birge et al., 2016).

4 Climate-smart agriculture and phytoremediation

"Designer" plants possess desirable characteristics in order to combat environmental stresses, and they can be developed by different genomics strategies (Basu et al., 2018; Raza et al., 2019). Genomics investigates molecular mechanisms underlying the stress tolerance, and it includes sequencing and phenotyping, transcriptomics, proteomics, metabolomics, functional genomics, DNA fingerprinting, quantitative trait loci mapping, isolation of abiotic and biotic stress-responsive genes, DNA markers (single nucleotide polymorphism, SNPs), and genome editing (Araus and Cairns, 2014; Kole et al., 2015; D'Agostino and Tripodi, 2017). QTL mapping is related to marker-assisted selection (MAS), whereas genome-wide association studies present a significant tool to identify the genetic variants in crops linked with specific traits (Manolio, 2010). Genetic engineering is used to develop plants that are tolerant to environmental stresses (transgenic plants) (Nejat and Mantri, 2017). Crops that are tolerant to abiotic and biotic stresses can be modified through biotechnology decreasing the negative impact of climate change (Singh, 2017).

Therefore, green biotechnology offers climate-resilient crops for global food security and safety, decreases greenhouse gases, uses environmentally friendly energy, and increases soil carbon sequestration (Singh, 2017; Dhankher and Foyer, 2018). Biotech crops have been tolerant to salinity, drought, cold, and heat stress (Dhankher and Foyer, 2018) (Fig. 4). A number of genes for salt tolerance, drought, cold, and heat resistance were identified in *Triticum* sp., *Oryza sativa*, *Hordeum vulgare*, *Glycine max*, *Solanum lycopersicum*, *Solanum tuberosum*, *Arabidopsis*, *Nicotiana tabacum*, etc. (Raza et al., 2019). Plant-specific transcription factors (TFs), such as family of AP2/EREBP, are necessary for plant growth pathways and abiotic and biotic stress responses (Licausi et al., 2010). The subfamilies *DREB* (dehydration-responsive element-binding protein), such as *DREB1* and *DREB2* TFs and *ERF* TF, were overexpressed to develop transgenic plants tolerant to salinity stress, drought, freezing, and high temperature (Liu et al., 1998; Lata and Prasad, 2011). Furthermore, *MYB* TFs family (myeloblastosis oncogene) regulates biochemical and physiological pathways in plants and has function in stress response; i.e., *AtMYB61*, *AtMYB60*, and *AtMYB44* enhance drought tolerance in *Arabidopsis* by controlling the stomata moving (Cominelli et al., 2005) whereas *AtMZB96* gene controls the biosynthesis of cuticle wax under drought conditions (Seo et al., 2011). *WRKY* and *NAC* families of TFs are also involved in plant response to environmental stress (Nuruzzaman et al., 2013; Phukan et al., 2016).

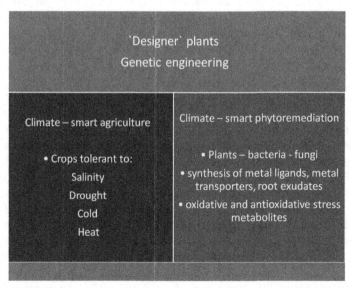

Fig. 4 Designer plants in agriculture and phytoremediation through genetic engineering.

Green biotechnology offers in agriculture and phytoremediation "designer" plants with desirable traits to combat climate change effects (climate smart agriculture/ phytoremediation). Genome editing employs techniques with precise and efficient gene modification manipulating with multiple genes responsible for complex genetic networks of plants' characteristics. Genome editing uses sequence-specific nucleases (SSN) as molecular DNA scissor recognizing specific DNA sequence inducing of double-stranded breaks (DSB) and trigger repair pathways to fix it, such as nonhomologous end joining (NHEJ) and homology-dependent recombination (HR) (Symigton and Gauter, 2011). Nowadays, the most prominent SSNs are zinc-finger nucleases (ZFNs) and transcription activator-like effector nucleases (TALENs). Furthermore, CRISPR/Cas9 system (clustered regularly interspaced palindromic short repeats associated with Cas9 endonuclease) presents a revolutionary editing tool used to modify, regulate, or mark genomic loci of several genes at the same time with high target recognition rate (Shukla et al., 2009; Chen and Gao, 2013; Miglani, 2017). These technologies can be used to create mutations, insertion, deletion, replacement, and chromosome rearrangements in the genome (Doudna and Charpentier, 2014). ZFNs, TALENs, and CRISPR/Cas9 system is used for improving crop growth and yield (tiller number, panicle size, grain number/size/weight, fruit development, etc.), quality/ nutrition (starch content, fragrance, carotenoid, lycopene content, etc.), abiotic stress tolerance (heat, drought, freezing, salt), disease stress resistance, and herbicide tolerance (Shukla et al., 2009; Ainley et al., 2013; Clasen et al., 2016; Jaganathan et al., 2018). However, ZFNs, TALENs, and CRISPR/Cas9-mediated genome modification can be used in phytoremediation designing plants with great potential to clean up polluted sites, i.e., modify the genes responsible for biosynthesis of metal transporters, root exudates, nitrogenase complex, phytohormones, and oxidative stress metabolites (Saxena et al., 2020). The gene editing by ZFNs, TALENs, and CRISPR/Cas9 has been demonstrated in climate-resilient plants, such as *Arabidopsis thaliana*, *O. sativa*, *Triticum aestivum*, *Zea mays*, *G. max*, *Gossypium hirsutum*, *S. lycopersicum*, *S. tuberosum*, *Cucumis sativus*, *Lactuca sativa*, *Citrus*, *Vitis vinifera*, *Linum usitatissimum*, *Manihot esculenta*, *N. tabacum*, *Saccharum officinarum*, *H. vulgare*, *Sorghum bicolor*, *Populus tomentosa*, and *Miscanthus giganteus* (Dhankher et al., 2012; Basharat et al., 2018; Jaganathan et al., 2018).

A wide range of phytoremediation technologies are used for site remediation (Mitrović et al., 2008; Pandey et al., 2015a; Gajić et al., 2016, 2020a; Kostić et al., 2018). Suitable plant phytoremediator should have fast growth, well-developed root system, large biomass, tolerance to metal(loid)s and organic pollutants, and tolerance to abiotic stress (Djurdjević et al., 2006; Pandey, 2012; Kostić et al., 2012; Gajić et al., 2013, 2016; Pandey and Singh, 2014; Pandey et al., 2015b; Grbović et al., 2019, 2020). Therefore, genome engineering can be used for plants involved in phytostabilization/phytoextraction/phytodegradation of inorganic/organic contaminants (Basharat et al., 2018; DalCorso et al., 2019; Patel et al., 2021). The modification of gene expression, pathway, and pollutant homeostasis that include accumulation/hyperaccumulation or degradation has a high potential for genetically engineered phytoremediation (Dhankher et al., 2012; Basharat et al., 2018). Genome editing can be used in gene transcription regulation, i.e., expression of genes responsible for synthesis of metal ligands, production of metal transporters, growth factors, root exudates, or oxidative stress metabolites in plants/bacteria that are applicable in phytoremediation (Basu et al., 2018; Basharat et al., 2018; Saxena et al., 2020; Patel et al., 2021) (Fig. 4). On a wide scale, genome editing strategies enable the improvements in efficiency of phytoremediation under climate change (Basu et al., 2018; DalCorso et al., 2019; Saxena et al., 2020; Patel et al., 2021).

5 Conclusion and future prospects

Köppen-Geiger classification of climate zones combines temperature, precipitation, and vegetation type: (a) tropical, warm, humid climate zone/evergreen rain forest, deciduous forest, grassland; (b) warm, arid climate zone/desert savanna; (c) temperate, warm, humid climate zone/sclerophyll woodland and scrub; (d) continental, cold, humid climate zone/conifer and deciduous forest; and (e) polar, cold climate zone/taiga, tundra, ice. Human activities lead to global warming and shifts of climate zones from colder to warmer and drier climate. Soil and air contamination is derived from fossil fuel combustion, mining, thermal plants, manufacturing, municipal and industrial waste disposal, application of synthetic fertilizers and pesticides, plastics, military, nuclear operations, etc. Huge amounts of SO_2, NOx, CO, VOCs, PM, metal(loid)s, PAH, PCB, CHCs, phenols, cyanides, etc. are released into the environment having negative effects to human health. In the United States, Europe, and China, there are thousands of contaminated areas. Contamination management involves removal and remedial actions, monitoring, hazard ranking system, placement on the National Priorities List, and liability of responsible parties for cleanup sites. Climate change policy contributes to regulations in order to reduce greenhouse gas emissions, and commonly, carbon and energy taxes, tradable permit schemes, and removal of environmental dangerous subsides have been used. Pollution policy includes prevention, recycling, treatment, and disposal of pollutants in an environmentally safe manner. Appropriate policy pollution and cleanaway create a greener and more sustainable economy protecting environment and human health, mitigate climate change, enhance nitrogen and phosphorous use efficiency related to Zero hunger, and link people, market, assets, and finance. Therefore, building resilience to climate change is achieved by mitigation and adaptation that work to avoid/manage the risks by reducing climate-related stresses contributing to ecosystem management, agricultural practice, and phytoremediation of polluted sites, as well as protecting the environment and human well-being. Adaptive management frameworks include management strategy and monitoring, evaluation and assessment of more alternatives, and learning from practice and experiments.

ZFNs, TALENs, and CRISPR/Cas9 are gene editing tools that open opportunities to engineer traits to provide food security in agriculture practice, and they create phytoremediator plants in order to clean up contaminated sites with superior phenotypes. Further efforts are required to optimize genome editing protocols for their marketing. New challenges in plant gene engineering can be genome editing in chloroplasts and mitochondria, or developing plant biosensors in order to integrate/transduce signals in plant synthetic biology. Targeted gene modification can be used to enhance phytoremediation and accelerate bioenergy technology editing genes of interest, their expression, and homeostasis network of pollutants. Despite beneficial aspects of gene editing tools, these "suitable edited" plants need a social and political policy, public acceptance, and government regulation. Ethical concerns are essential for successful implementation of gene editing technologies to engineer food crops and phytoremediators on the globe.

References

Agee, E., Larson, J., Childs, S., Marmo, A., 2016. Spatial redistribution of U.S. tornado activity between 1954 and 2013. J. Appl. Meteorol. Climatol. 55 (8), 1681–1697.

Ainley, W.M., Sastry-Dent, L., Welter, M.E., Murray, M.G., Zeitler, B., Amora, R., Corbin, D.R., Miles, R.R., Arnold, N.L., Strange, T.L., Simpson, M.A., Cao, Z., Caroll, C., Pawelczak, K.S., Blue, R., West, K., Rowland, L.M., Perkins, D., Samuel, P., Dewes, C.M., Shen, L., Sriram, S., Evans, S.L., Rebar, E.J., Zhang, L., Gregory, P.D., Urnov, F.D., Webb, S.R., Petolino, J.E., 2013. Trait stacking via targeted genome editing. Plant Biotechnol. J. 11, 1126–1134.

Allen, C.R., Fontaine, J.J., Pope, K.L., Garmestani, A.S., 2011. Adaptive management for a turbulent future. J. Environ. Manag. 92, 1339–1345.

Amann, M., Kiesewetter, G., Schopp, W., Klimont, Z., Winiwarter, W., Cofala, J., Rafaj, P., Hoglund-Isaksson, L., Gomez-Sabriana, A., Heyes, C., Purohit, P., Borken-Kleefeld, J., Wagner, F., Sander, R., Fagerli, H., Nyiri, A., Cozzi, L., Pavarini, C., 2020. Reducing global air pollution: the scope for further policy interventions. Philos. Trans. R. Soc. A 378, 20190331.

Araus, J.L., Cairns, J.E., 2014. Field high-throughput phenotyping: the new crop breeding frontier. Trends Plant Sci. 19, 52–61.

Auta, H.S., Eminke, C.U., Fauziah, S.H., 2017. Distribution and importance of microplastics in the marine environment: a review of the sources, fate, effects, and potential solutions. Environ. Int. 102, 165–176.

Badawi, A.F., Cavalieri, E.L., Rogan, E.G., 2000. Effect of chlorinated hydrocarbons on expression of cytochrome P450 1A1, 1A2 and 1B1 and 2- and 4-hydroxylation of 17β-estradiol in female Sprague-Dawley rats. Carcinogenesis 21 (8), 1593–1599.

Basharat, Z., Novo, L., Yasmin, A., 2018. Genome editing weds CRISPR: what is in it for phytoremediation? Plan. Theory 7 (3), 51.

Basu, S., Rabarara, R.C., Nwgi, S., Shukla, P., 2018. Engineering PGPMOSs through gene editing and systems biology: a solution for phytoremediation. Trends Biotechnol. 36 (5), 499–510.

Beck, H.E., Zimmermann, N.E., McVicar, T.R., Vergopolan, N., Berg, A., Wood, E.F., 2018. Present and future Köppen-Geiger climate classification maps at 1-km resolution. Sci. Data 5, 180214.

Begg, S., Vos, T., Barker, B., Stevenson, C., Stanley, L., Lopez, A., 2007. The Burden of Disease and Injury in Australia 2003. Cat. No. PHE 82, Australian Institute of Health and Wellbeing (IHW), Canberra.

Bell, J.N.B., Treshow, M., 2002. Air Pollution and Plant Life, second ed. John Wiley & Sons, Ltd., England, p. 465.

Birge, H.E., Allen, C.R., Garmestani, A.S., Pope, K.L., 2016. Adaptive management for ecosystem services. J. Environ. Manag. 183, 343–352.

Bormann, B.T., Martin, J.R., Wagner, F.H., Wood, G., Alegria, J., Cunningham, P.G., Brookes, M.H., Friesema, P., Berg, J., Henshaw, J., 1999. Adaptive management. In: Johnson, N.C., Malk, A.J., Sexton, W., Szaro, R. (Eds.), Ecological Stewasrdship: A Common Reference for Ecosystem Management. Elsevier, Amsterdam, pp. 505–534.

Bunnell, F.L., Dunsworth, B.G., Kremswater, L., Huggard, D., Beese, W.J., Sandford, J.S., 2007. Forestry and Biodiversity—Learning How to Sustain Biodiversity in Managed Forests. UBC Press, Vancouver.

Bush, E., Lemmen, D.S., 2019. Canada's Changing Climate Report. Government of Canada, Ottawa, p. 444.

CEAA, 2009. Operational Policy Statement: Adaptive Management Measures under the Canadian Environmental Assessment Act. Canadian Environmental Assessment Agency, Ottawa, ON.

CERCLA (Comprehensive Environmental Response, Compensation, and Liability Act (Superfund)), 2002. American Bar Association. Section of Natural Resources, Energy, and Environmental Law, Chicago, IL.

CERCLA (Comprehensive Environmental Response, Compensation, and Liability Act (Superfund)), EPCRA (Emergency Planning and Community Right-to-know Act), 2018. Continuous Release Reporting. Environmental Protection Agency, EPA.

Chan, D., Wu, Q.G., 2015. Significant anthropogenic–induced changes of climate classes since 1950. Sci. Rep. 5, 13487.

Chen, K., Gao, C., 2013. TALENs: customizable molecular DNA scissors for genome engineering of plants. J. Genet. Genomics 40, 271–279.

Clasen, B.M., Stoddard, T.J., Luo, S., Demorest, Z.L., Li, J., Cedrone, F., Tibebu, R., Davison, S., Ray, E.E., Daulhac, A., Coffman, A., Yabandith, A., Retterath, A., Haun, W., Baltes, N.J., Mathis, L., Voytas, D.F., Zhang, F., 2016. Improving cold storage and processing traits in potato through targeted gene knockout. Plant Biotechnol. J. 14, 169–176.

Cominelli, E., Galbiati, M., Vavasseur, A., Conti, L., Sala, T., Vuylsteke, M., Leonhardt, N., Dellaporta, S.L., Tonelli, C.A., 2005. A guard-cell-specific MYB transcription factor regulates stomatal movements and plant drought tolerance. Curr. Biol. 15, 1196–1200.

D'Agostino, N., Tripodi, P., 2017. NGS-based genotyping, high-throughput phenotyping and genome-wide association studies laid the foundations for next-generation breeding in horticultural crops. Diversity 9, 38.

DalCorso, G., Fasani, E., Manara, A., Visioli, G., Furini, A., 2019. Heavy metal pollutants: state of the art and innovation in phytoremediation. Int. J. Mol. Sci. 20, 3412.

Denton, F., Wilbanks, T.J., Abeysinghe, A.C., Burton, I., Gao, Q., Lemos, M.C., Masui, T., O'Brien, K.L., Warner, K., 2014. Climate-resilient pathways: adaptation, mitigation, and sustainable development. In: Field, C.B., Barros, V.R., Dokken, D.J., Mach, K.J., Mastrandea, M.D., Bilir, T.E., White, L.L. (Eds.), Climate Change 2014: Impacts, Adaptation, and Vulnerability. Part A: Global and Sectoral Aspects. Contribution of Working Group II to the Fifth Assessment Report of the Intergovermental Panel on Climate Change. Cambridge University Press, Cambridge and New York, NY, pp. 1101–1131.

Dhankher, O.P., Foyer, C.H., 2018. Climate resilient crops for improving global food security and safety. Plant Cell Environ. 41, 877–884.

Dhankher, O.P., Pilon-Smits, E.A.H., Meagher, R.B., Doty, S., 2012. Biotechnological approaches for phytoremediation. In: Altman, A., Hasegawa, P.M. (Eds.), Plant Biotechnology and Agriculture. Academic Press, Elsevier, Oxford, pp. 309–328.

Djurdjević, L., Mitrović, M., Pavlović, P., Gajić, G., Kostić, O., 2006. Phenolic acids as bioindicators of fly ash deposit revegetation. Arch. Environ. Contam. Toxicol. 50, 488–495.

Doudna, J.A., Charpentier, E., 2014. The new frontier of genome engineering with CRISPR-Cas9. Science 346 (6213), 1258096.

EC, 1996. Council Directive 96/61/EC of 24 September 1996 concerning integrating pollution prevention and control. Off. J. L 257, 26–40. 10.10.1996.

EC, 2001. Directive 2001/80/EC of the European Parliament and of the Council of 23 October 2001 on the limitation of emissions of certain pollutants into the air from large combustion plants. Off. J. L 309, 1–21. 27.11.2001.

EC, 2006. Proposal for a Directive of the European Parliament and of the Council establishing a framework for the protection of soil and amending Directive 2004/35/EC. COM, Brussels. 232 final.

EC, 2007. Commission staff working document—Annex to the Proposal for a Regulation of the European Parliament and of the Council on the approximation of the laws of the Member States with respect to emissions from on-road heavy duty vehicles and on access to vehicle repair information—Impact assessment. {COM(2007) 851 final} {SEC(2007) 1720} /* SEC/2007/1718 final */.

EEA, 2010. Impact of selected policy measures on Europe's air quality. European Environmental Agency Report, Publications Office of the European Union, Copenhagen, p. 68. No. 8.

EEA, 2014. Progress in Management of Contaminated Sites (CSI 015/LSI 003). European Environmental Agency, Copenhagen.

EPA (United States Environmental Protection Agency), 1992. EPA Definition of 'Pollution Prevention'. US EPA, Washington, DC, p. 4.

EPA (United States Environmental Protection Agency), 2010. U.S. Environmental Protection Agency 2010–2014 Pollution Prevention (P2) Program Strategic Plan. US EPA, Washington, DC, p. 34.

FAO, 2013. FAOSTAT. Food and Agricultural Organization of the United Nations, Rome.

Farr, D., 2000. Defining adaptive management. In: Document for Adaptive Management Experiment (AME). University of Alberta, Edmonton. http://www.ameteam.ca.

Folke, C., 2006. Resilience: the emergence of a perspective for social-ecological systems analyses. Glob. Environ. Change 16 (3), 253–267.

Gajić, G., Pavlović, P., 2018. The role of vascular plants in the phytoremediation of fly ash deposits. In: Mathichenkov, V. (Ed.), Phytoremediation. Methods, Management and Assessment. Nova Science Publishers, Inc., New York, pp. 151–236.

Gajić, G., Pavlović, P., Kostić, O., Jarić, S., Djurdjević, L., Pavlović, D., Mitrović, M., 2013. Ecophysiological and biochemical traits of three herbaceous plants growing of the disposed coal combustion fly ash of different weathering stage. Arch. Biol. Sci. 65 (1), 1651–1667.

Gajić, G., Djurdjević, L., Kostić, O., Jarić, S., Mitrović, M., Stevanović, B., Pavlović, P., 2016. Assessment of the phytoremediation potential and an adaptive response of *Festuca rubra* L. Sown on fly ash deposits: native grass has a pivotal role in ecorestoration management. Ecol. Eng. 93, 250–261.

Gajić, G., Djurdjević, L., Kostić, O., Jarić, S., Mitrović, M., Pavlović, P., 2018. Ecological potential of plants for phytoremediation and ecorestoration of fly ash deposits and mine wastes. Front. Environ. Sci. 6, 124. https://doi.org/10.3389/fenvs.2018.00124.

Gajić, G., Mitrović, M., Pavlović, P., 2019. Ecorestoration of fly ash deposits by native plant species at thermal power stations in Serbia. In: Pandey, V.C., Bauddh, K. (Eds.), Phytomanagement of Polluted Sites. Elsevier, Amsterdam, pp. 113–177.

Gajić, G., Mitrović, M., Pavlović, P., 2020a. Feasibility of *Festuca rubra* L. native grass in phytoremediation. In: Pandey, V.C., Singh, D.P. (Eds.), Phytoremediation Potential of Perennial Grasses. Elsevier, Amsterdam, pp. 115–164.

Gajić, G., Djurdjević, L., Kostić, O., Jarić, S., Stevanović, B., Mitrović, M., Pavlović, P., 2020b. Phytoremediation potential, photosynthetic and antioxidative response to As-induced stress of *D. glomerata* grown on fly ash deposits. Plants 9 (5), 657.

Gossel, T.A., 1984. Principles of Clinical Toxicology. Raven Press, New York.

Grbović, F., Gajić, G., Branković, S., Simić, Z., Ćirić, A., Rakonjac, L., Pavlović, P., Topuzović, M., 2019. Allelopathic potential of selected woody species growing on fly-ash deposits. Arch. Biol. Sci. 71 (1), 83–94.

Grbović, F., Gajić, G., Branković, S., Simić, Z., Vuković, N., Pavlović, P., Topuzović, M., 2020. Complex effect of *Robinia pseudoacacia* L. and *Ailanthus altissima* (Mill.) Swingle growing on asbestos deposits: allelopathy and biogeochemistry. J. Serb. Chem. Soc. 85 (1), 141–153.

Hochman, Z., Gobbert, D.L., Horan, H., 2017. Climate trends account for stalled wheat yields in Australia since 1990. Glob. Chang. Biol. 23 (5), 2071–2081.

Jaganathan, D., Ramasamy, K., Sellamuthu, G., Jayabalan, S., Venkataraman, G., 2018. CRISPR for crop improvement: an update review. Front. Plant Sci. 9, 985.

Jylhä, K., Tuomenvirta, H., Ruosteenoja, K., Niemi-Hugartes, H., Keisu, K., Karhu, J.A., 2010. Observed and projected future shifts of climate zones in Europe and their use to visualize climate change information. Weather Clim. Soc. 2, 148–167.

Kole, C., Muthamilarasan, M., Henry, R., Edwards, D., Sharma, R., Abberton, M., Batley, J., Bentley, A., Blakney, M., Bryant, J., 2015. Application of gemics-assisted breeding for generation of climate resilient crops. Progress and prospects. Front. Plant Sci. 6, 563.

Köppen, W., 1936. Das geographisca System der Klimate. In: Köppen, W., Geiger, G. (Eds.), Handbuch der Klimatologie. vol. 1. C. Gebr. Borntraeger.

Kostić, O., Mitrović, M., Knežević, M., Jarić, S., Gajić, G., Djurdjević, L., Pavlović, P., 2012. The potential of four woody species for the revegetation of fly ash deposits from the 'Nikola Tesla–A' thermoelectric plant (Obenovac, Serbia). Arch. Biol. Sci. 64 (1), 145–158.

Kostić, O., Jarić, S., Gajić, G., Pavlović, D., Pavlović, M., Mitrović, M., Pavlović, P., 2018. Pedological properties and ecological implications of substrates derived 3 and 11 years after the revegetation of lignite fly ash disposal sites in Serbia. Catena 163, 78–88.

Lata, C., Prasad, M., 2011. Role of DREBs in regulation of abiotic stress responses in plants. J. Exp. Bot. 62, 4731–4748.

Li, M., Wu, P., Sexton, D.M.H., Ma, Z., 2021. Potential shifts in climate zones under a future global warming scenario using soil moisture classification. Clim. Dyn. 56, 2071–2092.

Licausi, F., Giogi, F.M., Zenoni, S., Osti, F., Pezzoti, M., Perata, P., 2010. Genomic and transcriptomic analysis of the AP2/ERF superfamily in *Vitis vinifera*. BMC Genomics 11, 719.

Linkov, I., Satterstrom, F.K., Kiker, G., Batchelor, C., Bridges, T., Ferguson, E., 2006. From comparative risk assessment to multi-criteria decision analysis and adaptive management: recent developments and applications. Environ. Int. 32, 1072–1093.

Liu, Q., Kasuga, M., Sakuma, Y., Abe, H., Miura, S., Yamaguchi-Shinozaki, K., Shinozaki, K., 1998. Two transcription factors, DREB1 and DREB2, with an EREBP/AP2 DNA binding domain separate two cellular signal transduction pathways in drought-and low-temperature-responsive gene expression, respectively, in *Arabidopsis*. Plant Cell 10, 1391–1406.

Ma, Y., Dong, B., Bai, Y., Zhang, M., Xie, Y., Shi, Y., Du, X., 2018. Remediation status and practice for contaminated sites in China: survey-based analysis. Environ. Sci. Pollut. Res. 25, 33216–33224.

Mahlstein, I., Daniel, J.S., Solomon, S., 2013. Pace of shift in climate regions increases with global temperatures. Nat. Clim. Chang. 3, 739–743.

Majhi, P.K., Kothari, R., Arora, N.K., Pandey, V.C., Tyagi, V.V., 2021. Impact of pH on pollutional parameters of textile industry wastewater with use of *Chlorella pyrenoidosa* at lab-scale: a green approach. Bull. Environ. Contam. Toxicol. https://doi.org/10.1007/s00128-021-03208-5.

Manolio, T.A., 2010. Genome wide association studies and assessment of the risk of disease. N. Engl. J. Med. 363, 166–176.

McGuigan, J.S., 2000. The Potential Economic Impact of Environmental Liability: The American and European Context. Economic Analysis Unit, Environment Directorate, European Commission, p. 34.

McIntyre, T., 2003. Phytoremediation of heavy metals from soils. Adv. Biochem. Eng. Biotechnol. 78, 97–123.

MEP, 2014. National Soil Contamination Survey Report. Ministry of Environmental Protection, Beijing.

Miglani, G.S., 2017. Genome editing in crop improvement: present scenario and future prospects. J. Crop Improv. 31 (4), 453–559.

Mishra, T., Pandey, V.C., Praveen, A., Singh, N.B., Singh, N., Singh, D.P., 2020. Phytoremediation ability of naturally growing plant species on the electroplating wastewater-contaminated site. Environ. Geochem. Health 42, 4101–4111. https://doi.org/10.1007/s10653-020-00529-y.

Mitrović, M., Pavlović, P., Lakušić, D., Stevanović, B., Djurdjević, L., Kostić, O., Gajić, G., 2008. The potencial of *Festuca rubra* and *Calamagrostis epigejos* for the revegetation on fly ash deposits. Sci. Total Environ. 72, 1090–1101.

Munasinghe, M., Swart, R., 2005. Primer on climate change and sustainable development. In: Facts, Policy Analysis, and Applications. Cambridge University Press.

NEC, 2009. The National Emission Ceilings Directive Status report. Reporting by the Member States under Directive 2001/81/EC of the European Parliament and of the Council of 23 October 2001 on national emission ceilings for certain atmospheric pollutants.

Nejat, N., Mantri, N., 2017. Plant immune system: crosstalk between responses to biotic and abiotic stresses the missing link in understanding plant defence. Curr. Issues Mol. Biol. 23, 1–16.

Nuruzzaman, M., Sharoni, A.M., Kikuchi, S., 2013. Roles of NAC transcription factors in the regulation of biotic and abiotic stress responses in plants. Front. Microbiol. 4, 248.

O'Brien, K., 2012. Global environmental change II: from adaptation to deliberate transformation. Prog. Hum. Geogr. 36 (5), 667–676.

OECD, 2007. Climate change policies. In: Organisation for Economic Co-operation and Development. Policy Brief. The Public Affairs Divisions Affairs and Communications Directorate, p. 8.

Owens, P.N., 2009. Adaptive management frameworks for natural resource management at the landscape scale: implications and applications for sediment resources. J. Soils Sediments 9, 578–593.

Panagos, P., Van Liedekerke, M., Yigini, Y., Montanarella, L., 2013. Contaminated sites in Europe: review of the current situation based on data collected through a European network. J. Environ. Public Health 2013, 158764.

Pandey, V.C., 2012. Invasive species based efficient green technology for phytoremediation of fly ash deposits. J. Geochem. Explor. 123, 13–18.

Pandey, V.C., 2013. Suitability of *Ricinus communis* L. cultivation for phytoremediation of fly ash disposal sites. Ecol. Eng. 57, 336–341.

Pandey, V.C., 2015. Assisted phytoremediation of fly ash dumps through naturally colonized plants. Ecol. Eng. 82, 1–5.

Pandey, V.C., 2020. Phytomanagement of Fly Ash. Elsevier, Amsterdam, https://doi.org/10.1016/C2018-0-01318-3.

Pandey, V.C., 2021. Direct seeding offers affordable restoration for fly ash deposits. Energy Ecol. Environ. https://doi.org/10.1007/s40974-021-00212-7.

Pandey, V.C., Bauddh, K., 2019. Phytomanagement of Polluted Sites: Market Opportunities in Sustainable Phytoremediation. Elsevier, Amsterdam, https://doi.org/10.1016/C2017-0-00586-4.

Pandey, V.C., Singh, N., 2014. Fast green capping on coal fly ash basins through ecological engineering. Ecol. Eng. 73, 671–675.

Pandey, V.C., Singh, K., Singh, R.P., Singh, B., 2012. Naturally growing *Saccharum munja* L. On the fly ash lagoons: a potential ecological engineer for the revegetation and stabilization. Ecol. Eng. 40, 95–99.

Pandey, V.C., Pandey, D.N., Singh, N., 2015a. Sustainable phytoremediation based on naturally colonizing and economically valuable plants. J. Clean. Prod. 86, 37–39.

Pandey, V.C., Prakash, P., Bajpai, O., Kumar, A., Sing, N., 2015b. Phytodiversity on fly ash deposits: evaluation of naturally colonized species for sustainable phytorestoration. Environ. Sci. Pollut. Res. 22 (4), 2776–2787.

Pandey, V.C., Bajpai, O., Sinhg, N., 2016. Plant regeneration potential in fly ash ecosystem. Urban For. Urban Green. 15, 40–44.

Patel, H., Shakhreliya, S., Maurya, R., Pandey, V.C., Gohil, N., Bhattacharjee, G., Alzahrani, K.J., Singh, V., 2021. CRISPR-assisted strategies for futuristic phytoremediation. In: Pandey, V.C. (Ed.), Assisted Phytoremediation. Elsevier, ISBN: 9780128230831 (Edited Book).

Pavlović, P., Mitrović, M., Djurdjević, L., 2004. An ecophysiological study of plants growing on the fly ash deposits from the "Nikola Tesla–A" thermal power station in Serbia. Environ. Manag. 33, 654–663.

Phukan, U.J., Jeena, G.S., Shukla, R.K., 2016. WRKY transcription factors: molecular regulation and stress responses in plants. Front. Plant Sci. 7, 760.

Plumbo-Roe, B., Klinck, B., Banks, V., Quigley, S., 2009. Prediction of the long-term performance of abandoned lead, zinc mine tailings in a Welsh catchment. J. Geochem. Explor. 100 (2–3), 169–181.

Probst, K.N., Fullerton, D., Litan, D.E., Portney, P.R., 1995. Footing the Bill for Superfund Cleanups: Who Pays and How? The Brookings Institution and Resources for the Future, Washington, DC, p. 176.

Qu, C., Shi, W., Guo, J., Fang, B., Wang, S., Giesy, J.P., Holm, P.E., 2016. China's soil pollution control: choices and challenges. Environ. Sci. Technol. 50, 13181–13183.

Raza, A., Razzaq, A., Mehmood, S.S., Zou, X., Zhang, X., Lv, Y., Xu, J., 2019. Impact of climate change on crops adaptation and strategies to tackle its outcome: a review. Plan. Theory 8, 34.

Remus, R., Aguado Monsonet, M., Roudier, S., Delgado Sancho, L., 2012. Best Available Techniques (BAT) Reference Document for Iron and Steel Production: Industrial Emissions Directive 2010/75/EU: (Integrated Pollution Prevention and Control). EUR 25521 EN, Publications Office of the European Union, Luxembourg. JRC69967.

Riahi, K., 2000. Energy technology strategies for carbon dioxide mitigation and sustainable development. Environ. Econ. Policy Stud. 32 (2), 89–123.

Saravia, J., Lee, G.I., Lomnicki, S., Dellinger, B., Cormeir, S.A., 2013. Particulate matter containing environmentally persistent free radicals and adverse infant respiratory health effects: a review. J. Biochem. Mol. Toxicol. 27, 56–68.

Saxena, P., Singh, N.K., Harish, Singh, A.K., Pandey, S., Thanki, A., Yadav, T.C., 2020. Recent advances in phytoremediation using genome engineering CRISPR-Cas9 technology. In: Pandey, V.C., Singh, V. (Eds.), Bioremediation of Pollutants From Genetic Engineering to Genome Engineering. Elsevier, pp. 125–141.

Schaider, L.A., Senn, D.B., Brabander, D.J., McCartthy, K.D., Shine, J.P., 2007. Characterization of zinc, lead, and cadmium in mine waste: implications for transport, exposure, and bioavailability. Environ. Sci. Technol. 41 (11), 4164–4171.

Seager, R., Lis, N., Feldman, J., Ting, M., Williams, A.P., Nakamura, J., Lin, H., Henderson, N., 2018. Whither the 100th Meridian? The once and future physical and human geography of America's arid–humid divide. Part I: the story so far. Earth Interact. 22 (5), 1–22.

Seo, P.J., Lee, S.B., Suh, M.C., Park, M.J., Go, Y.S., Park, C.M., 2011. The MYB96 transcription factor regulates cuticular wax biosynthesis under drought conditions in *Arabidopsis*. Plant Cell 23, 1138–1152.

Shaddick, G., Thomas, M.L., Mudu, P., Ruggeri, G., Gumy, S., 2020. Half the world's population are exposed to increasing air pollution. Clim. Atmos. Sci. 3, 23.

Shorgen, J., Toman, M., 2000. Climate Change Policy. Resources for the Future. Discussion Paper, p. 22.

Shukla, V.K., Doyon, Y., Miller, J.C., DeKelver, R.C., Moehle, E.A., Worden, S.E., Mitchell, J.C., Arnold, N.L., Gopalan, S., Meng, X., Choi, V.M., Rock, J.M., Wu, Y.-Y., Katibah, G.E., Zhifang, G., McCaskill, D., Simpson, M.A., Blakeslee, B., Greenwalt, S.A., Butler, H.J., Hinkley, S.J., Zhang, L., Rebar, E.J., Gregory, P.D., Urnov, F.D., 2009. Precise genome modification in the crop species *Zea mays* using zinc-finger nucleases. Nature 459, 437–441.

Singh, R.P., 2017. Role of biotechnology for breeding climate resilient varieties of field crops. Seed Times 10 (1), 61–71.

Smith, A., 2008. Defining adaptive management in the BC Ministry of Forests and Range. Link 10, 12–15.

Staten, P.W., Lu, J., Grise, K.M., Davis, S.M., Birner, T., 2018. Re-examining tropical expansion. Nat. Clim. Chang. 8 (9), 768–775.

Symigton, L.S., Gauter, J., 2011. Double-strand break end resection and repair pathway choice. Annu. Rev. Genet. 45, 241–271.

Thomas, N., Nigam, S., 2018. Twentieth-century climate change over Africa: seasonal hydroclimate trends and Sahara desert expansion. J. Clim. 31 (9), 3349–3370.

Thurston, G.D., 2008. Outdoor air pollution: sources, atmospheric transport, and human health effects. In: Kris, H. (Ed.), International Encyclopedia of Public Health. Academic Press, Oxford, pp. 700–712.

Tsiamis, D.A., Torres, M., Castaldi, M.J., 2018. Role of plastics in decoupling municipal solid waste and economic growth in the US. Waste Manag. 77, 147–155.

Ukaogo, P.O., Ewuzie, U., Onwuka, C.V., 2020. Environmental pollution: causes, effects, and the remedies. In: Chaudhary, P., Verma, D., Ray, A., Akhtur, Y. (Eds.), Microorganisms for Sustainable Environment and Health. Elsevier, pp. 419–429.

UNECE, 1979. United Nations Economic Commissions for Europe (UNECE) Convection on Long Range Transboundary Air (LRTAP) of 1979.

UNEP, 2012a. Global Environment Outlook 5—Environment for the Future We Want. United Nations Environment Programme, Nairobi.

UNEP, 2012b. The Global Chemicals Outlook: Towards Sound Management of Chemicals. United Nations Environment Programme, Nairobi.

UNFCCC, 2011. Assessing the Costs and Benefits of Adaptation Options: An Overview of Approaches. United Nations Framework Convention on Climate Change (UFCCC), UNFCCC Secretariat, Bonn, p. 48.

USDA, 2012. Plant Hardiness Zone Map. United States Department of Agriculture.

USEPA, 2011. This Is Superfund. A Community Guide to EPA's Superfund Program. US Environmental Protection Agency, Office of Solid Waste and Emergence Response. EPA-540-R-11-021.

USEPA, 2017. Overview of the Brownfield Program. U.S. Environmental Protection Agency.

USEPA (US Environmental Protection Agency), 2010. Superfund Green Remediation Strategy.

Volk, H.E., Lurman, F., Penfold, B., Herz-Picciotto, I., McConnell, R., 2013. Traffic-related air pollution particulate matter and autism. JAMA Psychiatry 70, 71–76.

Voltaggio, T., Adams, J., 2016. Superfund: A Half Century of Progress. EPA Alumni Association.

Walters, C.J., 1986. Adaptive Management of Renewable Resources. McGraw-Hill, New York.

Wang, M., Overland, J.E., 2004. Detecting Arctic climate change using Köppen climate classification. Climate Change 67, 43–62.

Wang, H., Hui, S., Wang, Y., 2015. Dynamics of the captured quantity of particulate matter by plant leaves under typical weather conditions. Acta Ecol. Sin. 35 (6), 1696–1705.

WHO (World Health Organisation), 2014. Burden of Disease from Ambient Air Pollution for 2012. The World Health Organisation, Geneva.

Williams, B.K., Szaro, R.C., Shapiro, C.D., 2007. Adaptive Management. United States Department of the Interior Technical Guide, Adaptive Management Working Group, Washington, DC.

Winkler, H., Baumert, K., Blanchard, O., Burch, S., Robinson, J., 2007. What factors influence mitigative capacity? Energy Policy 35, 692–703.

Zhang, X.L., Tao, S., Liu, W.X., Yang, Y., Zuo, Q., Liu, S.Z., 2005. Source diagnostics of polycyclic aromatic hydrocarbons based on species ratios: a multimedia approach. Environ. Sci. Technol. 39 (23), 9109–9114.

Index

Note: Page numbers followed by *f* indicate figures and *t* indicate tables.

Printed in the United States
by Baker & Taylor Publisher Services